Lecture Notes in Control and Information Sciences 417

Rolf Johansson and Anders Rantzer (Eds.)

Distributed Decision Making and Control

Springer

Editors

Rolf Johansson
Lund University
Department of Automatic Control
Ole Römers väg 1
22100 Lund
Sweden
Telephone: +46 46 222 87 91
Fax: +46 46 13 81 18
E-mail: Rolf.Johansson@control.lth.se

Prof. Anders Rantzer
Lund University
Department of Automatic Control
Ole Römers väg 1
221 00 Lund
Sweden
Telephone: +46 46 222 87 78
Fax: +46 46 13 81 18
E-mail: anders.rantzer@control.lth.se

ISBN 978-1-4471-2264-7 e-ISBN 978-1-4471-2265-4

DOI 10.1007/978-1-4471-2265-4

Lecture Notes in Control and Information Sciences ISSN 0170-8643

Library of Congress Control Number: 2011937175

Typeset & Cover Design: Scientific Publishing Services Pvt. Ltd., Chennai, India.

Printed on acid-free paper

9 8 7 6 5 4 3 2 1

springer.com

In Memoriam
Ulf T. Jönsson (1963–2011)

performance of multi-agent systems, while closing the local loops with embedded $\mathcal{L}_1$-adaptive controllers. A communication and adaptation co-design scheme is proposed that helps one predict system performance. With this scheme, an agent locally determines its broadcast time instants using distributed event triggering. The embedded $\mathcal{L}_1$-adaptive controller enables each agent to compensate for the local uncertainties and disturbances. Performance bounds are derived on the difference between the signals of the ideal model (in the absence of uncertainties and with perfect communication) and the real system operating with the proposed co-design scheme in the presence of uncertainties and communication constraints. It is shown that these bounds can be arbitrarily reduced by decreasing the thresholds in the local events and increasing the local adaptation gain in each subsystem, subject only to hardware limitations. The proposed co-design scheme can help one predict the performance of multi-agent systems in the presence of uncertainties. The results can be used for design guidelines in safety-critical applications, including air traffic control and collision avoidance in multi-agent systems.

Weight determination by manifold regularization is the topic of Ohlsson and Ljung. A new type of linear kernel smoother is derived and studied. The smoother, referred to as weight determination by manifold regularization, is the solution to a regularized least squares problem. The regularization avoids overfitting and can be used to express prior knowledge of an underlying smooth function. An interesting property of the kernel smoother is that it is well suited to systems governed by the semisupervised smoothness assumption. Several examples are given to illustrate this property, the authors also discuss why these types of techniques can hold a potential interest for the system identification community.

Beck et al. study dynamic coverage and clustering, using a maximum entropy approach. They present a computational framework for solving a large class of dynamic coverage and clustering problems, ranging from those that arise in the deployment of mobile sensor networks to the classification of cellular data for diagnosing cancer stages. This framework provides a way for to identify natural clusters in an underlying dataset, and allows them to address inherent trade-offs such as those between cluster resolution and computational cost. More specifically, they define the problem of minimizing an instantaneous coverage metric as a combinatorial optimization problem in a maximum entropy principle framework, which is formulated specifically for the dynamic setting. Location of cluster centers and their associated velocity fields is cast as a control design problem that ensures the algorithm achieve progressively better coverage with time.

Transverse linearization for underactuated nonholonomic mechanical systems with application to orbital stabilization is studied by Freidovich and Shiriaev, who consider a class of mechanical systems with an arbitrary number of passive (non-actuated) degrees of freedom, which are subject to a set of nonholonomic constraints. They assume that the challenging problem of motion planning is solved, giving rise to a feasible desired periodic trajectory. Their goal is either to analyze orbital stability of this trajectory with a given time-independent feedback control law or to design a controller. They extend their previous work done for mechanical systems without nonholonomic constraint. The main contribution is an analytical method for computing coefficients of linear reduced order control system equations—solutions of which approximate dynamics are transversal to the preplanned trajectory. This linear system is shown to be useful for stability analysis and for design of feedback controllers of orbitally exponentially dynamics.

In Part III, a distributed nonlinear model predictive control (NMPC) scheme without stabilizing terminal constraints is designed by Grüne and Worthmann, who consider a distributed NMPC scheme in which the individual systems are coupled via state constraints. In order to avoid violation of the constraints, subsystems communicate their individual predictions to the other subsystems once in each sampling period. For this setting, Richards and How have proposed a distributed MPC formulation with stabilizing terminal constraints. In this chapter, it is shown how this scheme can be extended to MPC without stabilizing terminal constraints or costs. They show theoretically and by means of numerical simulations that under a suitable controllability condition, stability and feasibility can be ensured even for rather short prediction horizons.

A set theoretic method for verifying feasibility of a fast explicit nonlinear model predictive control is proposed by Raimondo et al.. In this chapter an algorithm for nonlinear explicit model predictive control is presented. A low complexity receding horizon control law is obtained by approximating the optimal control law using multiscale basis function approximation. Simultaneously, feasibility and stability of the approximate control law is ensured through the computation of a capture basin (region of attraction) for the closed-loop system. In previous work, interval methods were used to construct the capture basin (feasible region), yet this approach suffered due to slow computation times and high grid complexity. They suggest an alternative to interval analysis based on zonotopes. The suggested method significantly reduces the complexity of the combined function approximation and verification procedure through the use of DC programming, and recursive splitting. The result is a multiscale function approximation method with improved computational efficiency for fast nonlinear explicit MPC with guaranteed stability and constraint satisfaction. Stability with uniform bounds for online dial-a-ride problems under some reasonable load is considered.

A survey and directions for future research towards parallel implementation of hybrid MPC is presented by Axehill and Hansson. Different methods for achieving parallelism at different levels of the algorithms are surveyed. It is seen that there are many possible ways of obtaining parallelism for hybrid MPC, and it is by no means clear which possibilities should be utilized to achieve the best possible performance. An answer to this question is a challenge for future research.

Hierarchical model predictive control for plug-and-play resource distribution is developed by Bendtsen et al., who deal with hierarchical model predictive control of distributed systems. A three-level hierarchical approach is proposed, consisting of a high level MPC controller, a second level of so-called aggregators, controlled by an on-line MPC-like algorithm, and a lower level of autonomous units. The approach

Acknowledgements

The contributions were presented at the LCCC Workshops held at Lund University during 2010:

- Multi-Agent Coordination and Estimation (Jan. 18–Feb. 19, 2010)
- Distributed Decisions via Games and Price Mechanisms (Feb. 22–Mar. 26, 2010)
- Adaptation and Learning in Autonomous Systems (April 6–30, 2010)
- Distributed Model Predictive Control and Supply Chains (May 3-28, 2010).

This research was supported by the Linnaeus Grant Lund Center for Control of Complex Engineering Systems (LCCC)—Swedish Research Council; Ref. VR 2007-8646. LCCC is a Linnaeus Center at Lund University funded by the Swedish Research Council. The ten principal investigators are from the Department of Automatic Control and the Department of Electrical and Information Technology.

The research vision of LCCC is to make fundamental contributions to a general theory and methodology for control of complex engineering systems. This includes scalable methods and tools for modeling, analysis and control synthesis, as well as reliable implementations using networked embedded systems. Our goal is to maintain a leading role in a worldwide effort involving partners of many kinds.

The editors would like to thank the LCCC participants for rewarding contributions to the workshops and to this book. Finally, thanks are extended to Dr. Eva Westin and Mr. Leif Andersson and the Springer editors for valuable editorial help and advice.

Contents

Francesca Flamini
Department of Economics, University of Glasgow, Adam Smith Building, Glasgow, G12 8RT, UK
e-mail: f.flamini@lbss.gla.ac.uk

Leonid Freidovich
Department of Applied Physics and Electronics, Umeå University, SE-901 87 Umeå, Sweden
e-mail: leonid.freidovich@tfe.umu.se

Lars Grüne
Mathematical Institute, University of Bayreuth, Universitätsstraße 30, 95440 Bayreuth, Germany
e-mail: lars.gruene@uni-bayreuth.de

Anders Hansson
Division of Automatic Control, Linköping University, SE-581 83 Linköping, Sweden
e-mail: hansson@isy.liu.se

Hans Hellendoorn
Delft University of Technology, Delft Center for Systems and Control, Mekelweg 2, 2628 CD Delft, The Netherlands
e-mail: j.hellendoorn@tudelft.nl

Naira Hovakimyan
Department of Mechanical Science and Engineering, University of Illinois at Urbana-Champaign, 1206 West Green Street, Urbana, IL 61801, USA
e-mail: nhovakim@illinois.edu

Rolf Johansson
Lund University, Dept Automatic Control, PO Box 118, SE-221 00 Lund, Sweden
e-mail: Rolf.Johansson@control.lth.se

Colin N. Jones
Automatic Control Laboratory, ETH, Physikstrasse 3, 8092, Zürich, Switzerland
e-mail: cjones@control.ee.ethz.ch

Ulf T. Jönsson
Division of Optimization and Systems Theory, Department of Mathematics, Royal Institute of Technology, 10044 Stockholm, Sweden
Correspondence should be directed to Anders Rantzer
e-mail: Anders.Rantzer@control.lth.se

Tamás Keviczky
Delft Center of Systems and Control, Delft Technical University, Mekelweg 2, 2628 CD Delft, The Netherlands
e-mail: t.keviczky@tudelft.nl

Sven Oliver Krumke
Department of Mathematics University of Kaiserslautern Paul-Ehrlich-Str. 14-434 67663 Kaiserslautern, Germany
e-mail: krumke@mathematik.uni-kl.de

Sanjay Lall
Department of Electrical Engineering and Department of Aeronautics and Astronautics, Stanford University, Stanford, CA 94305, USA
e-mail: lall@stanford.edu

Cédric Langbort
Department of Aerospace Engineering & Coordinated Science Laboratory, University of Illinois at Urbana-Champaign, 306 Talbot Lab 104 S. Wright St., Urbana, IL 61801, USA
e-mail: langbort@illinois.edu

Lennart Ljung
Division of Automatic Control, Department of Electrical Engineering, Linköping University, SE-583 37 Linköping, Sweden
e-mail: ljung@isy.liu.se

John Lygeros
Automatic Control Laboratory, ETH, Physikstrasse 3, 8092 Zürich, Switzerland,
e-mail: lygeros@control.ee.ethz.ch

Manfred Morari
Automatic Control Laboratory, ETH, Physikstrasse 3, 8092 Zürich, Switzerland
e-mail: morari@control.ee.ethz.ch

Henrik Ohlsson
Division of Automatic Control, Department of Electrical Engineering, Linköping University, SE-583 37 Linköping, Sweden
e-mail: ohlsson@isy.liu.se

Daniel P. Palomar
Department of Electronic and Computer Engineering, Hong Kong University of Science and Technology, Clear Water Bay, Kowloon, Hong Kong
e-mail: palomar@ust.hk

Jong-Shi Pang
University of Illinois at Urbana-Champaign, 117 Transportation Building MC-238, 104 S. Mathews Ave., Urbana IL 61801, USA
e-mail: jspang@uiuc.edu

Jörg Rambau
Lehrstuhl für Wirtschaftsmathematik, Universität Bayreuth, D-95440 Bayreuth, Germany
e-mail: joerg.rambau@uni-bayreuth.de

Davide M. Raimondo
Automatic Control Laboratory, ETH, Physikstrasse 3, 8092 Zürich, Switzerland
e-mail: davide.raimondo@control.ee.ethz.ch

Anders Rantzer
Lund University, Dept Automatic Control, PO Box 118, SE-221 00 Lund, Sweden
e-mail: Anders.Rantzer@control.lth.se

Stefano Riverso
Laboratorio di Identificazione e Controllo dei Sistemi Dinamici, Università degli Studi di Pavia, Dipartimento di Informatica e Sistemistica, Via Ferrata, 1, 27100 Pavia, Italy
e-mail: stefano.riverso@unipv.it

Srinivasa Salapaka
Coordinated Science Lab, University of Illinois at Urbana-Champaign & Department of Mechanical Science and Engineering, 362c Mechanical Engineering Building, 1206 West Green Street, Urbana, IL 61801, USA
e-mail: salapaka@illinois.edu

Gesualdo Scutari
Department of Electrical Engineering, State University of New York (SUNY) at Buffalo, Buffalo, NY 14260, USA
e-mail: gesualdo@buffalo.edu

Puneet Sharma
Integrated Data Systems Department, Siemens Corporate Research, 755 College Road East, Princeton, NJ 08540, USA
e-mail: sharma.puneet@gmail.com

Anton S. Shiriaev
Department of Engineering Cybernetics, Norwegian University of Science and Technology (NTNU), NO-7491 Trondheim, Norway, and Department of Applied Physics and Electronics, Umeå University, SE-901 87 Umeå, Sweden
e-mail: anton.shiriaev@itk.ntnu.no

Andrea Simonetto
Delft Center of Systems and Control, Delft Technical University, Mekelweg 2, 2628 CD Delft, The Netherlands
e-mail: a.simonetto@tudelft.nl

Jakob Stoustrup
Department of Electronic Systems, Automation and Control, Aalborg University, Fr. Bajers Vej 7C, 9220 Aalborg, Denmark
e-mail: jakob@es.aau.dk

Sean Summers
Automatic Control Laboratory, ETH, Physikstrasse 3, 8092 Zürich, Switzerland
e-mail: ssummers@control.ee.ethz.ch

John Swigart
Department of Aeronautics and Astronautics, Stanford University, Stanford, CA 94305, USA
e-mail: jswigart@stanford.edu

Alina N. Tarău
Delft University of Technology, Delft Center for Systems and Control, Mekelweg 2, 2628 CD Delft, The Netherlands
e-mail: a.n.tarau@tue.nl

Klaus Trangbaek
Department of Electronic Systems, Automation and Control, Aalborg University, Fr. Bajers Vej 7C, 9220 Aalborg, Denmark
e-mail: ktr@es.aau.dk

Xiaofeng Wang
Department of Mechanical Science and Engineering, University of Illinois at Urbana-Champaign, 1206 West Green Street, Urbana, IL 61801, USA
e-mail: wangx@illinois.edu

Karl Worthmann
Mathematical Institute, University of Bayreuth, 95440 Bayreuth, Germany
e-mail: karl.worthmann@uni-bayreuth.de

Yunwen Xu
Coordinated Science Lab, University of Illinois at Urbana-Champaign & Department of Industrial and Enterprise Systems Engineering, University of Illinois, 1308 W Main Street Urbana, IL 61801-2307, USA
e-mail: xu27@illinois.edu

Chun Zhang
Cisco Systems Inc., Research Triangle Park, NC 27709, USA
e-mail: chunzha@cisco.com

MULTI-AGENT CONTROL AND GAME THEORY

The first part of the book is devoted to a fundamental scientific question with connections to many different branches of science, engineering and society: How can the dynamics of a large number of agents be coordinated to meet desired performance objectives? The agents could, for example, be generators and loads in a power network, cars in a vehicle formation, or computers connected via the Internet. From a science perspective the corresponding question would be: How does nature do such coordination? For example, how do individuals interact in a flock of birds and how do the light emitters of a laser synchronize?

Central issues of decentralized control and coordination are treated in the initial book chapters, both from a mathematical and from an engineering perspective. A mathematical tool of particular interest is game theory. During the last decade game-theoretic methods have found extensive use in networked systems, which are pervasive in automotive, aerospace, biomedical and telecommunication technologies. Simultaneously, important areas of economics, such as mechanism design and social networks, have come to exploit methods from computer science and control theory.

Chapter 1
Primal and Dual Criteria for Robust Stability Applied to Large Scale Systems

Ulf T. Jönsson

Abstract Primal and dual formulations of stability criteria based on multipliers will be discussed. The foundation for multiplier-based stability analysis is the use of a convex cone of multipliers to characterize the uncertainty in a system. The primal and dual stability criteria are formulated as convex feasibility tests involving the nominal dynamics and multipliers from the cone and the polar cone, respectively. The motivation for introducing the dual is that it provides additional insight into the stability criterion and that it is sometimes easier to use than the primal.

The case considered in this chapter is that of uncertainty as it represents the interconnection of a complex network. The multipliers are used to describe characteristic properties of the network such as the spectral location or the structure of the underlying graph.

1.1 Introduction

In this chapter, we derive the dual of a class of primal stability criteria that are defined in terms of multipliers. Multipliers are typically used to characterize complicated or uncertain components in the system. Here, we let the multipliers characterize the network interconnection of a system where single-input single-output (SISO) linear systems are connected over the network. We show that the dual criterion provides insight into the structure of the stability criterion, which sometimes allows us to derive simpler and more explicit criteria.

The traditional point of view in large scale systems analysis characterizes the various subsystems using integral quadratic constraints (IQC) and then combines these into an aggregate IQC that the interconnection operator must satisfy; see [9, 13], where dissipation theory was used and [8] for general IQCs. A bottleneck that

Ulf T. Jönsson (11 May 1963–11 May 2011)
Division of Optimization and Systems Theory, Department of Mathematics, Royal Institute of Technology, 10044 Stockholm, Sweden

R. Johansson and A. Rantzer (Eds.): Distributed Decision Making and Control, LNCIS 417, pp. 3–25.
springerlink.com

sometimes limits applicability is the computational complexity, which does not scale gracefully unless network structure can be explored. Recently, there has been progress in identifying network structures for which the analysis decomposes into equivalent criteria consisting of subproblems that can be solved with low effort; see e.g. [12, 3, 10].

Many applications of recent interest motivate the reverse point of view, i.e., to let the IQC characterize the network structure and then to verify that the subsystems jointly satisfy the complementary IQC. This is the motivation behind the examples in this chapter. To our knowledge there are only a few available works that take this perspective. In [4, 6] multipliers were used to characterize the spectrum of the network, while in [5] bipartite network structures were considered. We limit the discussion to linear systems and IQCs defined pointwise in frequency. This covers many interesting cases and it simplifies the treatment.

The chapter is organized as follows. We first introduce the notation used. In the second section we present our main primal and dual stability criteria. In Sec. 1.3 we discuss a point of view for large scale systems analysis where the focus is on characterizations of the interconnection structure rather than of the individual subsystem dynamics. We illustrate this approach in Sec. 1.4, where primal and dual stability criteria are derived for several frequently appearing interconnection structures.

1.1.1 Notation and Preliminaries

We let $\mathbf{R}$ denote the real numbers, $\mathbf{R}_+ = \{x \in \mathbf{R} : x \geq 0\}$, $\mathbf{C}$ the complex numbers, $\mathbf{C}_+ = \{s \in \mathbf{C} : \mathrm{Re}\, s > 0\}$ and $\mathrm{cl}\,\mathbf{C}_+ = \{s \in \mathbf{C} : \mathrm{Re}\, s \geq 0\} \cup \{\infty\}$. We let[1]

$$\mathcal{A}(\mathbf{C}_+)^{n\times m} = \{H : \mathrm{cl}\,\mathbf{C}_+ \to \mathbf{C}^{n\times m} | H \text{ is analytic in } \mathbf{C}_+ \text{ and continuous on } \mathrm{cl}\,\mathbf{C}_+\}$$

be the algebra of transfer functions that are analytic in the open right half-plane and continuous on $\mathrm{cl}\,\mathbf{C}_+$. This implies continuity on the extended imaginary axis $j\mathbf{R} \cup \{\infty\}$. We equip it with the norm $\|H\| = \max_{\omega \in \mathbf{R} \cup \{\infty\}} (\bar{\sigma}(H(j\omega)))$, where $\bar{\sigma}(\cdot)$ denotes the largest singular value. We will throughout the chapter use the compact notation $\mathcal{A}^{n\times m} \stackrel{\mathrm{def}}{=} \mathcal{A}(\mathbf{C}_+)^{n\times m}$. A transfer function is in this chapter called *stable* if and only if it belongs to $\mathcal{A}^{n\times m}$.

We let $\mathcal{S}_{\mathbf{C}}^{m\times m} = \{X \in \mathbf{C}^{m\times m} : X = X^*\}$ be the Hilbert space of Hermitian matrices equipped with the inner product $\langle X, Y\rangle = \mathrm{tr}(XY)$ and the corresponding norm $\|X\| = \mathrm{tr}(X^2)^{1/2}$ (the Frobenius norm). We use the standard notation $X \succ 0$ ($X \succeq 0$) to denote that the matrix $X \in \mathcal{S}_{\mathbf{C}}^{m\times m}$ is positive definite (positive semidefinite).

Suppose $K \subset \mathcal{S}_{\mathbf{C}}^{m\times m}$ is a convex cone. The negative polar cone is the closed convex cone defined as

$$K^{\ominus} = \left\{Y \in \mathcal{S}_{\mathbf{C}}^{m\times m} : \langle X, Y\rangle \leq 0; \quad \forall X \in K\right\}.$$

[1] This algebra is obtained by Möbius transformation of the frequency variable of the so-called disc algebra of functions that are analytic inside the unit disc.

The sum of two convex cones is defined as $K_1 + K_2 = \{X_1 + X_2 : X_1 \in K_1; X_2 \in K_2\}$. We will use the following property.

Lemma 1.1. *Let $K_1, \ldots, K_N \subset \mathcal{S}_{\mathbf{C}}^{m\times m}$ be nonempty convex cones. Then*

$$(\sum_{j=1}^{N} K_j)^{\ominus} = \cap_{j=1}^{N} K_j^{\ominus}, \qquad (\cap_{j=1}^{N} \mathrm{cl} K_j)^{\ominus} = \mathrm{cl} \sum_{j=1}^{N} K_j^{\ominus}$$

Proof. *The result holds for nonempty convex cones in $\mathbf{R}^n$; see Corollary 16.4.2 in [11]. Since the space $\mathcal{S}_C^{m\times m}$ with inner product $\langle X, Y\rangle = \mathrm{tr}(XY)$ is isometrically isomorphic to $\mathbf{R}^{m^2}$ the result also holds for cones defined on $\mathcal{S}_C^{m\times m}$.* □

Finally, we define the convex hull as

$$\mathrm{co}\{w_1, \ldots, w_n\} := \{\sum_{i=1}^{n} \alpha_i w_i : \alpha_i \geq 0; \sum_{i=1}^{n} \alpha_i = 1\},$$

the convex conic hull as $\mathrm{cone}\{w_1, \ldots, w_n\} := \{\sum_{i=1}^{n} \alpha_i w_i : \alpha_i \geq 0\}$, and the direct sum of matrices $\oplus_{i=1}^{n} M_i = \mathrm{diag}(M_1, \ldots, M_n)$.

1.2 Primal and Dual Stability Criteria

We consider the feedback interconnection defined as

$$\begin{aligned} v &= \Delta w + r_1, \\ w &= Hv + r_2, \end{aligned} \tag{1.1}$$

where $\Delta, H \in \mathcal{A}^{n\times n}$. This interconnection is called stable if and only if

$$[\Delta, H] := \begin{bmatrix} \Delta \\ I \end{bmatrix} (I - H\Delta)^{-1} \begin{bmatrix} H & I \end{bmatrix} \in \mathcal{A}^{2n\times 2n}.$$

The system Δ is, in some of our applications below, a transfer function from $\mathcal{A}^{n\times n}$ but more often a real or complex matrix representing a network interconnection. Since the systems are linear and time-invariant we employ a frequency-wise analysis. Note that the primal and dual stability criteria in the theorems below easily can be generalized to stability criteria for interconnections where Δ is a nonlinear operator. This follows by invoking the theory of integral quadratic constraints.

We further use the operator $M_H : \mathcal{S}_{\mathbf{C}}^{2n\times 2n} \to \mathcal{S}_{\mathbf{C}}^{n\times n}$ and its adjoint $M_H^{\times} : \mathcal{S}_{\mathbf{C}}^{n\times n} \to \mathcal{S}_{\mathbf{C}}^{2n\times 2n}$ defined as

$$M_H \Pi = \begin{bmatrix} I \\ H \end{bmatrix}^* \Pi \begin{bmatrix} I \\ H \end{bmatrix}, \quad M_H^{\times} Z = \begin{bmatrix} I \\ H \end{bmatrix} Z \begin{bmatrix} I \\ H \end{bmatrix}^*. \tag{1.2}$$

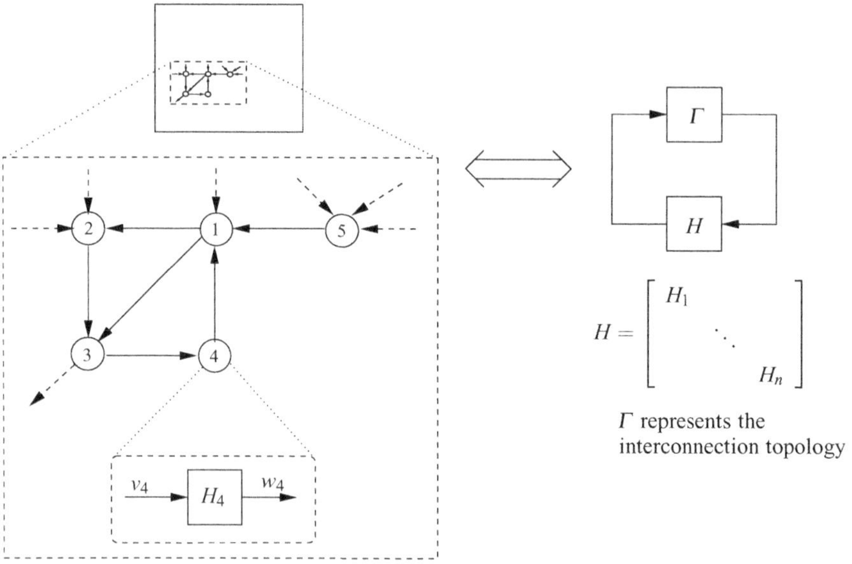

Fig. 1.1 Compact modeling of large scale systems: The figure On the left illustrates a part of a large scale system where the inputs and outputs of the subsystems H_k are interconnected in some complex fashion. We restrict attention to the case when the H_k are SISO linear time-invariant systems. The incoming arcs are summed at the input of H_k and its output is communicated to one or several neighboring subsystems. The figure on the right shows a more compact model of the system where the subsystem dynamics are collected in a diagonal transfer function H and operator Γ represents their interconnection.

We remark that a complete characterization of the network usually is difficult to find and expensive to use. Instead, one has to search for structure in the network that is easy to explore. The approach is well motivated in the analysis of very large scale systems where simple-to-use criteria (e.g., graphical criteria) are valuable even at the expense of possible conservatism.

Some properties of interconnections that recently have been used to derive scalable stability criteria are the spectral characteristic of the interconnection which has been used in the analysis of consensus networks [6, 4], and the bipartite interconnection structure of networks that appear in Internet congestion control [7, 5].

In the examples in the next section we will illustrate how the primal and the dual criteria apply in some structured large scale networks. We discuss, in particular, spectral characterizations of networks, characterization of some bipartite structures and an approach to model the adjacency matrix of sparse weighted graphs with good accuracy. For each case we perform the following three steps:

1. Introduce multipliers to characterize the network interconnection.
2. Formulate the primal criterion. The primal criterion is typically not easy to interpret but it can be tested using convex optimization and it generalizes also to the case when the subsystem dynamics are nonlinear and time-varying.

3. Apply the dual criterion and use it to derive new formulations of the stability criterion. This sometimes provide nice interpretations of the primal multiplier-based criterion.

1.4 Examples

In this section we consider a few examples illustrating the primal and dual stability criteria and the framework suggested in Sec. 1.3. In the first example we discuss the use of identical multipliers for all subsystems. The spectral characterization of the network used in [4] is an interesting case, which can be used when we consider normal interconnection matrices. A similar treatment will then be applied to an aggregate bipartite model that appears in Internet congestion control. The third example presents a simplified version of a result from [5], and it indicates how symmetric bipartite interconnections can be treated in the framework. Finally, we discuss a simple way to find analysis results of relatively low complexity for general networks. We only state the primal and dual stability condition but never any complete theorem statement.

1.4.1 Spectral Characterization of Interconnections

Consider the case when $H = \mathrm{diag}(H_1, \dots, H_n)$, where each $H_k \in \mathcal{A}$ and $\Gamma \in \mathbf{R}^{n\times n}$. The system represents a set of heterogeneous stable linear time-invariant (LTI) single-input single-output (SISO) systems interconnected over a network defined by the interconnection matrix Γ. We will use Theorem 1.2 with $H_0 := h_0 I_n$, where $h_0 \in \mathcal{A}$ and the nominal homogeneous interconnection $[\Gamma, h_0 I_n]$ is assumed stable.

One possibility is to use identical multipliers for the subsystems, i.e.,

$$\Pi_\Gamma = \left\{ \begin{bmatrix} \pi_{11} I_n & \pi_{12} I_n \\ \bar{\pi}_{12} I_n & \pi_{22} I_n \end{bmatrix} : \pi_{22} \leq 0; \Gamma^* \Gamma \pi_{11} + \Gamma^* \pi_{12} + \Gamma \bar{\pi}_{12} + \pi_{22} I_n \preceq 0 \right\}.$$

This can equivalently be written $\Pi_\Gamma = \{\pi \otimes I_n : \pi \in \pi_\Gamma\}$, where π_Γ is defined as

$$\begin{aligned} \pi_\Gamma &= \{\pi \in \mathcal{S}_{\mathbf{C}}^{2\times 2} : \pi_{22} \leq 0;\ \Gamma^* \Gamma \pi_{11} + \Gamma^* \pi_{12} + \pi_{12}^* \Gamma + \pi_{22} I_n \preceq 0\} \\ &= \{\pi \in \mathcal{S}_{\mathbf{C}}^{2\times 2} : \langle \pi, V_0 \rangle \leq 0;\ \langle \pi, V_1(v) \rangle \leq 0,\ v \in \mathbf{C}^n; |v| = 1\}, \end{aligned} \tag{1.7}$$

where

$$V_0 = \begin{bmatrix} 0 & 0 \\ 0 & 1 \end{bmatrix}, \qquad V_1(v) = \begin{bmatrix} v^* \Gamma^* \Gamma v & v^* \Gamma v \\ v^* \Gamma^* v & v^* v \end{bmatrix}.$$

The primal stability condition in Theorem 1.2 reduces to the following condition:

Primal condition: For every $\omega \in \mathbf{R} \cup \{\infty\}$ there exists $\pi \in \pi_\Gamma$ such that

$$\pi_{11} + 2\mathrm{Re}\,\pi_{12} H_k(j\omega) + \pi_{22}|H_k(j\omega)|^2 > 0,$$

for $k = 0, 1, \ldots, n$.

The interpretation is that the subsystems H_k must find one common multiplier $\pi \in \pi_\Gamma$ such that the stability criterion is satisfied. This implies that we only need to search for four parameters in the stability criterion. Another consequence of using identical multipliers for the subsystems is that the dual will be formulated in terms of convex hulls of the subsystem dynamics.

We next derive the dual condition along the lines of [5]. The condition $W \in \Pi_\Gamma^\ominus$ is equivalent to

$$\langle \Pi, W \rangle = \left\langle \begin{bmatrix} \pi_{11} & \pi_{12} \\ \bar{\pi}_{12} & \pi_{22} \end{bmatrix}, \begin{bmatrix} \mathrm{tr}(W_{11}) & \mathrm{tr}(W_{12}) \\ \overline{\mathrm{tr}(W_{12})} & \mathrm{tr}(W_{22}) \end{bmatrix} \right\rangle \leq 0$$

for all $\pi \in \pi_\Gamma$. It can be shown that

$$\pi_\Gamma^\ominus = \mathrm{cone}\{V_0, V_1(v) : v \in \mathbf{C}^n; |v| = 1\}. \tag{1.8}$$

Indeed, it follows from Chapter 14 of [11] that $\pi_\Gamma^\ominus = \mathrm{cl\,cone}\{V_0, V_1(v) : v \in \mathbf{C}^n; |v| = 1\}$, and since $\{V_0, V_1(v) : v \in \mathbf{C}^n; |v| = 1\}$ is a nonempty compact set which does not contain the origin, it follows that the conic hull is closed. It hence follows that the polar cone is

$$\Pi_\Gamma^\ominus = \left\{ \begin{bmatrix} W_{11} & W_{12} \\ W_{12}^* & W_{22} \end{bmatrix} \in \mathcal{S}_{\mathbf{C}}^{2n \times 2n} : \begin{bmatrix} \mathrm{tr}(W_{11}) & \mathrm{tr}(W_{12}) \\ \mathrm{tr}(W_{12}^*) & \mathrm{tr}(W_{22}) \end{bmatrix} \in \pi_\Gamma^\ominus \right\}.$$

and thus the dual condition $M_H^\times Z_1 + M_{H_0}^\times Z_0 \notin \Pi_\Gamma^\ominus$ becomes

$$\begin{bmatrix} \mathrm{tr}(Z_1 + Z_0) & \mathrm{tr}(Z_1 H^* + Z_0 H_0^*) \\ \mathrm{tr}(H Z_1 + H_0 Z_0) & \mathrm{tr}(H Z_1 H^* + H_0 Z_0 H_0^*) \end{bmatrix} \notin \pi_\Gamma^\ominus.$$

Since H is diagonal and H_0 is diagonal and homogeneous, i.e., $H_0 = h_0 I_n$, it is no restriction to let $Z_1 = \mathrm{diag}(z_1, \ldots, z_n)$ and $Z_0 = z_0 I_n / n$, where the elements satisfy $z_k \geq 0$ and $\sum_{k=0}^n z_k = 1$. We have shown that the dual stability condition in Theorem 1.2 reduces to the following criterion:

Dual condition: For every $\omega \in \mathbf{R} \cup \{\infty\}$

$$\mathrm{co} \left\{ \begin{bmatrix} 1 & \overline{H_k(j\omega)} \\ H_k(j\omega) & |H_k(j\omega)|^2 \end{bmatrix} : k = 0, 1, \ldots, n \right\} \cap \pi_\Gamma^\ominus = \emptyset.$$

The primal and dual criteria provide little insight, as they are formulated. They must be verified using computations. However, it turns out that a rather simple criterion can be derived from the dual in special cases. We will here consider the case wherein Γ is a normal matrix, which means that Γ is unitarily diagonalizable. In this case the above multiplier characterization simplifies to

$$\begin{aligned}\pi_\Gamma &= \left\{\pi \in \mathcal{S}_{\mathbf{C}}^{2\times 2} : \pi_{22} \leq 0, |\lambda_k|^2\pi_{11} + 2\mathrm{Re}\,\overline{\lambda}_k\pi_{12} + \pi_{22} \leq 0,\ \forall \lambda_k \in \mathrm{eig}(\Gamma)\right\} \\ &= \left\{\pi \in \mathcal{S}_{\mathbf{C}}^{2\times 2} : \langle \pi, V_k\rangle \leq 0,\ k = 0, 1, \ldots, n\right\}\end{aligned}$$

and

$$\pi_\Gamma^{\ominus} = \mathrm{cone}\{V_k : k = 0, 1, \ldots, n\}.$$

where

$$V_0 = \begin{bmatrix} 0 & 0 \\ 0 & 1 \end{bmatrix}, \qquad V_k = \begin{bmatrix} |\lambda_k|^2 & \lambda_k \\ \overline{\lambda}_k & 1 \end{bmatrix},\ k = 1, \ldots, n.$$

It is now possible to use ideas from [4] to show that the dual can be verified using various three-dimensional Nyquist criteria. Here, we will derive an inverse Nyquist criterion.

The dual condition holds if the following system is violated at each frequency:

$$\begin{aligned}\sum_{k=0}^{n} z_k &= \sum_{k=1} \psi_k|\lambda_k|^2, \\ \sum_{k=0}^{n} z_k H_k(j\omega) &= \sum_{k=1}^{n} \psi_k\bar{\lambda}_k, \\ \sum_{k=0}^{m} z_k|H_k(j\omega)|^2 &= \sum_{k=1}^{n} \psi_k + \psi_0,\end{aligned}$$

for any parameters satisfying $\sum_{k=0}^{n} z_k = 1$, $z_k \geq 0$ and $\psi_k \geq 0$. Our first step is to make the change of variables $\hat{z}_k = z_k|H_k(j\omega)|^2$ for k such that $H_k(j\omega) \neq 0$. Without loss of generality, we may rescale the equation system by multiplying all equations with a positive number. We may thus assume that

$$\sum_{k:H_k(j\omega)\neq 0} \hat{z}_k = 1.$$

This gives the scaled equation system

$$\begin{aligned}\sum_{k:H_k(j\omega)\neq 0} \hat{z}_k\frac{1}{|H_k(j\omega)|^2} &= \sum_{k=1} \psi_k|\lambda_k|^2 - \psi_{n+1} \\ \sum_{k:H_k(j\omega)\neq 0} \hat{z}_k\frac{1}{H_k(j\omega)} &= \sum_{k=1}^{n} \psi_k\bar{\lambda}_k \\ 1 &= \sum_{k=1}^{n} \psi_k + \psi_0,\end{aligned}$$

where $\psi_k \geq 0$ and where $\psi_{n+1} = \sum_{k:H_k(j\omega)=0} z_k / \sum_{k:H_k(j\omega)\neq 0} z_k |H_k(j\omega)|^2$ could be any positive number by proper choice of the z_k. The dual holds when this system is violated for any ω and any choice variables. Hence, the dual stability condition in Theorem 1.2 becomes:

Dual condition: For every $\omega \in \mathbf{R} \cup \{\infty\}$

$$\widehat{\mathcal{N}}[H_0, \ldots, H_n](\omega) \cap \widehat{\Omega} = \emptyset,$$

where the three-dimensional *inverse Nyquist polytope* $\widehat{\mathcal{N}} \subset \mathbf{C} \times \mathbf{R}_+$ is defined as

$$\widehat{\mathcal{N}}[H_0, \ldots, H_n] = \text{co}\left\{\left(\frac{1}{H_k(j\omega)}, \frac{1}{|H_k(j\omega)|^2}\right) : H_k(j\omega) \neq 0\right\}$$

and the *instability region* $\widehat{\Omega} \subset \mathbf{C} \times \mathbf{R}_+$ is defined as

$$\widehat{\Omega} = \{\alpha \cdot \text{co}\{(\lambda_k, |\lambda_k|^2) : \lambda_k \in \text{eig}(\Gamma)\} : \alpha \in [0,1]\} - (0, R_+).$$

Note that the last term in $\widehat{\Omega}$ can be removed if $H_k(j\omega) \neq 0$, for all $k = 1, \ldots, n$. We refer to [4] for a more in-depth discussion about this type of criterion.

1.4.2 Aggregate Bipartite Interconnections

In the next example we consider a model of Internet congestion control where the the routing matrices and the link dynamics have been combined into one block; see e.g., [7] for modeling details. The block diagram is illustrated in Fig. 1.2, where

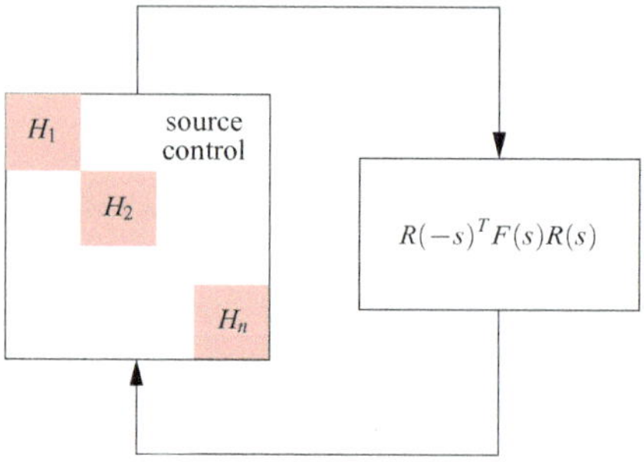

Fig. 1.2 Equilibrium dynamics of Internet congestion control.

$F = \oplus_{l=1}^{L} F_l$ and $\rho(R(j\omega)^* R(j\omega)) \leq 1$. Here $R(s)$ is a routing matrix where the dynamics only are due to delays. We assume for simplicity that $G_k, F_k \in \mathcal{A}$.

Then the system is an interconnection on the form (1.1) with[2]

$$\Gamma(s) = R(-s)^T F(s) R(s).$$

We will use Theorem 1.2 with $H_0 := h_0 I_n$, i.e., the nominal dynamics are homogeneous. Let us use frequency-wise multipliers of the same type as in the previous subsection, i.e., we use identical multipliers for the subsystems H_k.

The primal stability condition in Theorem 1.2 can be formulated as:

Primal condition: For every $\omega \in \mathbf{R} \cup \{\infty\}$ there exists $\pi \in \pi_{\Gamma(j\omega)}$ such that

$$\pi_{11} + 2\mathrm{Re}\,\pi_{12} H_k(j\omega) + \pi_{22}|H_k(j\omega)|^2 > 0$$

for $k = 0, 1, \ldots, n$, where for $\Gamma = \Gamma(j\omega)$ we define π_Γ as in (1.7) in Sec. 1.4.1.

As in Sec. 1.4.1, we obtain the following dual:

Dual condition: For every $\omega \in \mathbf{R} \cup \{\infty\}$

$$\mathrm{co}\left\{ \begin{bmatrix} 1 & H_k(j\omega)^* \\ H_k(j\omega) & |H_k(j\omega)|^2 \end{bmatrix} : k = 0, 1, \ldots, n \right\} \cap \pi^{\ominus}_{\Gamma(j\omega)} = \emptyset,$$

where for $\Gamma = \Gamma(j\omega)$, $\pi^{\ominus}_\Gamma$ is defined as in (1.8).

The primal and dual criteria provide little insight as they are formulated. However, it turns out that a rather simple criterion can be derived from the dual if we accept some additional conservatism.

Our first step is to notice that the change of variables $\hat{z}_k = z_k |H_k(j\omega)|^2$ in

$$\sum_{k=0}^{n} z_k \begin{bmatrix} 1 \\ H_k(j\omega) \end{bmatrix} \begin{bmatrix} 1 \\ H_k(j\omega) \end{bmatrix}^* \notin \pi^{\ominus}_\Gamma \tag{1.9}$$

gives the alternative formulation

$$\sum_{k:H_k(j\omega)\neq 0} \hat{z}_k \begin{bmatrix} H_k(j\omega)^{-1} \\ 1 \end{bmatrix} \begin{bmatrix} H_k(j\omega)^{-1} \\ 1 \end{bmatrix}^*$$

$$\notin \pi^{\ominus}_\Gamma - \left\{ \begin{bmatrix} -\psi_{n+1} & 0 \\ 0 & 0 \end{bmatrix} : \psi_{n+1} \geq 0 \right\}, \tag{1.10}$$

where we may assume $\hat{z}_k \geq 0$ satisfies $\sum_{k=0}^{n} \hat{z}_k = 1$. This follows since it is possible to rescale by multiplying with $1/\sum_{k:H_k(j\omega)\neq 0} z_k |H_k(j\omega)|^2$ on both sides of the equation. It then follows that

---

[2] Note that $\Gamma \notin \mathcal{A}$ but the system can be transformed in such a way that the frequency domain criteria presented in this subsection are valid stability tests. See [5].

$$\psi_{n+1} = \frac{\sum_{k:H_k(j\omega)=0} z_k}{\sum_{k:H_k(j\omega)\neq 0} z_k |H_k(j\omega)|^2},$$

which is an arbitrary positive number by suitable choices of the z_k. This proves (1.10).

There is then no loss of generality to assume that the elements in $\pi_\Gamma^\ominus$ are scaled such that the $(2,2)$ element is one, i.e., we have

$$\psi_0 + \sum_{k=1}^{N} \psi_k |v_k|^2 = 1$$

for some $\psi_k \geq 0, k = 0, 1 \ldots, n$, and $v_k \in \mathbf{C}^n$ with $|v_k| = 1$. Defining $\tilde{v}_k = v_k \sqrt{\sum_{k=1}^{N} \psi_k}$ and $\alpha_k = \psi_k / \sum_{k=1}^{N} \psi_k$, this can be written

$$\sum_{k=1}^{N} \alpha_k |\tilde{v}_k|^2 = 1 - \psi_0.$$

Since $|\tilde{v}_k| \leq 1$, it follows that (1.10) equivalently can be stated as

$$\text{co}\left\{ \left(\frac{1}{H_k(j\omega)}, \frac{1}{|H_k(j\omega)|^2} \right) : H_k(j\omega) \neq 0 \right\} \cap \widehat{\Omega} = \emptyset,$$

where

$$\widehat{\Omega} = \text{co}\,\{(v^*\Gamma v, v^*\Gamma^*\Gamma v) : v \in \mathbf{C}^n; |v| \leq 1\} - (0, \mathbf{R}_+).$$

To simplify this criterion we use an idea which we adopt from [14, 7] to overestimate the set on the right hand side. Let R_l denote the lth row of R. Since $\rho(R^*R) \leq 1$ it follows that $\sum_{l=1}^{L} |R_l v|^2 \leq 1$ for $|v| \leq 1$ and thus

$$\begin{aligned}
(v^*\Gamma v, v^*\Gamma^*\Gamma v) &= (v^*R^*FRv, v^*R^*F^*RR^*FRv) \\
&\in \{(v^*R^*FRv, \alpha v^*R^*F^*FRv) : \alpha \in [0,1]\} \\
&= \left\{ \sum_{l=1}^{L} (F_l, \alpha |F_l|^2) |R_l v|^2 : \alpha \in [0,1] \right\} \\
&= \left\{ \left(\sum_{j=1}^{L} |R_j v|^2\right) \sum_{l=1}^{L} (F_l, \alpha |F_l|^2) \frac{|R_l v|^2}{\sum_{j=1}^{L} |R_j v|^2} : \alpha \in [0,1] \right\} \\
&\subset \left\{ \left(\sum_{j=1}^{L} |R_j v|^2\right) \text{co}\{(F_l, \alpha |F_l|^2) : l = 1, \ldots, L\} : \alpha \in [0,1] \right\} \\
&\subset \{\text{co}\{(0,0), (F_l, \alpha |F_l|^2) : l = 1, \ldots, L\} : \alpha \in [0,1]\} \\
&= \text{co}\{(0,0), (F_l, 0), (F_l, |F_l|^2) : l = 1, \ldots, L\}.
\end{aligned}$$

It follows that

$$\text{co}\,\{(v^*\Gamma v, v^*\Gamma^*\Gamma v) : |v| \leq 1\} \subset \text{co}\{(0,0), (F_l, 0), (F_l, |F_l|^2) : l = 1, \ldots, L\}$$

and hence that

$$\widehat{\Omega} \subset \mathrm{co}\{(0,0),(F_l,0),(F_l,|F_l|^2) : l=1,\ldots,L\} - (0,\mathbf{R}_+).$$

It is easy to see that any element on the right hand side with positive second element also belongs to $\mathrm{co}\{(0,0),(F_l,0),(F_l,|F_l|^2) : l=1,\ldots,L\}$. We arrive at the following sufficent stability condition:

Simple dual condition: For every $\omega \in \mathbf{R}\cup\{\infty\}$, $\widehat{\mathcal{N}}[H_0,H_1,\ldots,H_n]\cap\widehat{\Omega}_e = \emptyset$, where

$$\widehat{\mathcal{N}}[H_0,H_1,\ldots,H_n] = \mathrm{co}\left\{\left(\frac{1}{H_k(j\omega)},\frac{1}{|H_k(j\omega)|^2}\right) : k=1,\ldots,n\right\},$$
$$\widehat{\Omega}_e = \mathrm{co}\{(0,0),(F_l(j\omega),0),(F_l(j\omega),|F_l(j\omega)|^2) : l=1,\ldots,L\}.$$

If we project this to the complex plane we get the following two-dimensional criterion, which typically is sufficient at low frequencies (assuming $H_k(j\omega)\neq 0$)

$$\mathrm{co}\{H_0^{-1},H_1^{-1},\ldots,H_n^{-1}\}(j\omega)\cap\mathrm{co}\{0,F_1,\ldots,F_L\}(j\omega)=\emptyset.$$

The simple dual above is easier to use but more restrictive than our previous results in [5].

1.4.3 Simple Symmetric Bipartite Interconnection

Consider the feedback interconnection of

$$H = \begin{bmatrix} G & 0 \\ 0 & K \end{bmatrix}, \qquad \Gamma = \begin{bmatrix} 0 & 1 \\ -1 & 0 \end{bmatrix},$$

which corresponds to a negative feedback interconnection of a plant $G \in \mathcal{A}$ with a compensator $K \in \mathcal{A}$. We will use Theorem 1.1 with $\mathcal{S}_\Gamma = \{\tau\Gamma : \tau \in [0,1]\}$ and the multipliers

$$\Pi_\Gamma = \left\{\begin{bmatrix} x_1 & & y & \\ & x_2 & & \bar{y} \\ \bar{y} & & -x_2 & \\ & y & & -x_1 \end{bmatrix} : x_1,x_2 \geq 0; y \in \mathbf{C}\right\}. \tag{1.11}$$

In this case the primal condition becomes:

Primal condition: For every $\omega \in \mathbf{R}\cup\{\infty\}$ there exists $x_1,x_2 \geq 0$ and $y \in \mathbf{C}$ such that

$$x_1 - x_2|G(j\omega)|^2 + 2\mathrm{Re}\, yG(j\omega) > 0,$$
$$x_2 - x_1|K(j\omega)|^2 + 2\mathrm{Re}\, \bar{y}K(j\omega) > 0.$$

It is shown in [5] that the dual condition in Theorem 1.1 in this case is reduced to the following simple criterion:
Dual condition: For all $\omega \in \mathbf{R} \cup \{\infty\}$, $G(j\omega)K(j\omega) \notin (-\infty, -1]$.

Note that the dual resembles the classical Nyquist criterion. Hence, by optimizing over the parameters x_1, x_2 and y in the primal we obtain a criterion similar to the classical Nyquist criterion. This is interesting because the the primal criterion is valid also for nonlinear or time-varying operators. The idea in this example can be generalized to systems interconnected over a symmetric bipartite graph. The bipartite graph can be given a characterization analogous to (1.11) with a resulting stability criterion that takes both phase and magnitude of the loop gain into account. We discuss this in further detail in [5].

1.4.4 General Interconnections

Consider the case where the interconnection $\Gamma = [\gamma_{ij}]_{i,j=1}^n$ is defined in terms of the adjacency matrix of a weighted directed graph $\mathcal{G} = (\mathcal{V}, \mathcal{E})$ with vertex set $\mathcal{V} = \{1, 2, \ldots, n\}$ and edge set e. Then γ_{ij} is the weight of the edge from vertex j to vertex i and γ_{ii} is the weight of the loop from vertex i to itself. The dynamics in node i can be represented as

$$w_i = H_i\Big(\sum_{j=1}^{n} \gamma_{ij} w_j + r_{1,i}\Big) + r_{2,i}.$$

If the graph is sparse then this sum contains only a few terms. Next, we discuss an attempt to use the interconnection structure to define multipliers. The results are primarily of interest in the case of sparse graphs. We may consider $\Gamma \in \mathcal{A}^{n \times n}$ by doing frequency-wise analysis.

Let

$$\Gamma = \begin{bmatrix} \gamma_1^* \\ \vdots \\ \gamma_n^* \end{bmatrix},$$

where $\gamma_i^* = \begin{bmatrix} \gamma_{i1} & \cdots & \gamma_{in} \end{bmatrix}$. We may then use the multipliers

$$\Pi_k = \left\{ \begin{bmatrix} \pi_{11} e_k e_k^T & \pi_{12} e_k \gamma_k^* \\ \bar{\pi}_{12} \gamma_k e_k^T & \pi_{22} \gamma_k \gamma_k^* \end{bmatrix} : \pi \in \pi_{\mathrm{set}} \right\},$$

where e_k is the kth unit vector in $\mathbf{R}^n$ and where π_{set} constrains the multipliers appropriately. We may consider the case where

$$\pi_{\text{set}} = \left\{ \pi = \begin{bmatrix} \pi_{11} & \pi_{12} \\ \bar{\pi}_{12} & \pi_{22} \end{bmatrix} : \psi_k^* \pi \psi_k \leq 0,\ k = 1,2; \psi_3^* \pi \psi_3 \geq 0 \right\},$$

where $\psi_1^T = \begin{bmatrix} 1 & 1 \end{bmatrix}$ and either

$$\psi_2^T = \begin{bmatrix} 0 & 1 \end{bmatrix} \text{ and } \psi_3^T = \begin{bmatrix} 0 & 0 \end{bmatrix} \tag{1.12}$$

or

$$\psi_2^T = \begin{bmatrix} 0 & 1 \end{bmatrix} \text{ and } \psi_3^T = \begin{bmatrix} 1 & 0 \end{bmatrix}. \tag{1.13}$$

The first inequality involving ψ_1 ensures that the multiplier characterization is valid, while either of the last two could be included in order to enforce constraints on the multipliers such that the results in the previous sections can be applied. Indeed, in the case (1.12) we have the constraint $\pi_{22} \leq 0$, which allows us to apply Theorem 1.2. In the case (1.13) we have the constraints $\pi_{11} \geq 0$ and $\pi_{22} \leq 0$, which allows us to apply Theorem 1.1 with $\mathcal{S}_\Gamma = \{\tau\Gamma : \tau \in [0,1]\}$.

It is easy to see that the multipliers in Π_k define valid characterizations of the graph, since

$$\Gamma^* \Pi_{11} \Gamma + \Gamma^* \Pi_{12} + \Pi_{12}^* \Gamma + \Pi_{22} = \gamma_k \gamma_k^* \psi_1^* \pi \psi_1 \preceq 0,$$

since $\pi \in \pi_{\text{set}}$.

A special case of multipliers that satisfy (1.13) and that will be used later is

$$\pi_{\text{set}} = \left\{ \pi = \begin{bmatrix} x & iy \\ -iy & -x \end{bmatrix} : x \geq 0;\ y \in \mathbf{R} \right\}. \tag{1.14}$$

Primal Stability Criterion

We restrict attention to the case where the multipliers are defined by π_{set} in (1.14). We can formulate the primal in Theorem 1.1 as

Primal condition: For every $\omega \in \mathbf{R} \cup \{\infty\}$ there exists $\pi_k \in \pi_{\text{set}}$, $k = 1, \ldots, n$, such that

$$\Psi(j\omega) = \sum_{k=1}^{n} M_{H(j\omega)} \Pi_k = \sum_{k=1}^{n} \begin{bmatrix} I \\ H(j\omega) \end{bmatrix}^* \begin{bmatrix} \pi_{k,11} e_k e_k^T & \pi_{k,12} e_k \gamma_k^* \\ \bar{\pi}_{k,12} \gamma_k e_k^T & \pi_{k,22} \gamma_k \gamma_k^* \end{bmatrix} \begin{bmatrix} I \\ H(j\omega) \end{bmatrix} \succ 0$$

This criterion has many more parameters to optimize compared to the previous results, but it nevertheless has some attractive scalability properties: 1) the number of parameters to optimize grows linearly as $4n$ with the dimension of the graph; 2) if one more dynamic enters the network or some system is modified, feasibility of the primal problem can be tested using an inexpensive sufficient test. To see this, let us introduce the notation

$$\check{\Gamma} = \begin{bmatrix} \check{\gamma}_1^* \\ \vdots \\ \check{\gamma}_n^* \\ \check{\gamma}_{n+1}^* \end{bmatrix} = \left[\begin{array}{c|c} \gamma_1^* & \gamma_{1,n+1} \\ \vdots & \vdots \\ \gamma_n^* & \gamma_{n,n+1} \\ \hline \gamma_{n+1}^* & \gamma_{n+1,n+1} \end{array}\right] \tag{1.15}$$

and

$$\check{H} = \oplus_{k=1}^{n+1} H_k = \left[\begin{array}{c|c} H & 0 \\ \hline 0 & H_{n+1} \end{array}\right], \tag{1.16}$$

$$\check{\Pi}_k = \begin{bmatrix} \pi_{k,11}\check{e}_k\check{e}_k^T & \pi_{k,12}\check{e}_k\check{\gamma}_k^* \\ \bar{\pi}_{k,12}\check{\gamma}_k\check{e}_k^T & \pi_{k,22}\check{\gamma}_k\check{\gamma}_k^* \end{bmatrix}, \tag{1.17}$$

where $\check{e}_k$ is the kth unit vector in $\mathbf{R}^{n+1}$ and $\pi_k \in \pi_{\text{set}}$.

The primal condition becomes (we suppress the argument ω)

$$\check{\Psi} = \sum_{k=1}^{n+1} M_{\check{H}}\check{\Pi}_k = \begin{bmatrix} \Psi + \check{\Psi}_{11} & \check{\Psi}_{12} \\ \check{\Psi}_{12}^* & \check{\Psi}_{22} \end{bmatrix} \succ 0,$$

where the blocks are defined in terms of the multipliers as

$$\begin{aligned}
\Psi &= \sum_{k=1}^{n} M_H \Pi_k, \\
\check{\Psi}_{11} &= \pi_{n+1,22} H^* \gamma_{n+1}\gamma_{n+1}^* H, \\
\check{\Psi}_{12} &= \sum_{k=1}^{n} \pi_{k,12} e_k \gamma_{k,n+1} H_{n+1} + \bar{\pi}_{n+1,12} H^* \gamma_{n+1} + \pi_{n+1,22} H^* \gamma_{n+1}\gamma_{n+1,n+1} H_{n+1}, \\
\check{\Psi}_{22} &= M_{H_{n+1}} \Pi_{n+1} + \sum_{k=1}^{n+1} \pi_{k,22} |\gamma_{k,n+1}|^2 |H_{n+1}|^2,
\end{aligned}$$

where

$$\Pi_{n+1} = \begin{bmatrix} \pi_{n+1,11} & \pi_{n+1,12}\gamma_{n+1,n+1} \\ \bar{\pi}_{n+1,12}\bar{\gamma}_{n+1,n+1} & \pi_{n+1,22}|\gamma_{n+1,n+1}|^2 \end{bmatrix}.$$

We will discuss a low complexity and low-dimensional test to verify that the system remains stable when the H_{n+1} enters the network (or is perturbed). We assume that

$$\Psi(j\omega) = \sum_{k=1}^{n} M_{H(j\omega)} \Pi_k \succeq \phi(j\omega) I$$

for some strictly positive function ϕ. Assume further that the dynamics in node $n+1$ only communicate with the dynamics in nodes $\mathcal{N} \subset \{1,\dots,n\}$, i.e., $k \in \mathcal{N}$ if and only if either $\gamma_{k,n+1} \neq 0$ or/and $\gamma_{n+1,k} \neq 0$. Let

$$E = \begin{bmatrix} e_k : & k \in \mathcal{N} \end{bmatrix},$$

Then the system remains stable if $\pi_{n+1} \in \pi_{\text{set}}$ exists such that (all other multipliers are kept fixed)[3]:

$$\begin{bmatrix} \phi I + E^T \check{\Psi}_{11} E & E^T \check{\Psi}_{12} \\ \check{\Psi}_{12}^* E & \check{\Psi}_{22} \end{bmatrix} (j\omega) \succ 0 \tag{1.18}$$

for all $\omega \in \mathbf{R} \cup \{\infty\}$. This "greedy" approach of accommodating new dynamics into the stability test will not always work, but it is simple to use and inexpensive. It works analogously to verify that the system remains stable if some of the dynamics are perturbed.

Dual Stability Criterion

In order to derive the dual stability criterion we first notice that

$$\Pi_k^{\ominus} = \left\{ W = \begin{bmatrix} W_{11} & W_{12} \\ W_{12}^* & W_{22} \end{bmatrix} : \begin{bmatrix} e_k^T W_{11} e_k & e_k^T W_{12} \gamma_k \\ \gamma_k^* W_{12}^* e_k & \gamma_k^* W_{22} \gamma_k \end{bmatrix} \in \pi_{\text{set}}^{\ominus} \right\}.$$

This follows since it is easy to show that for any $\Pi \in \Pi_k$ we have

$$\langle \Pi, W \rangle = \left\langle \begin{bmatrix} \pi_{11} & \pi_{12} \\ \bar{\pi}_{12} & \pi_{22} \end{bmatrix}, \begin{bmatrix} e_k^T W_{11} e_k & e_k^T W_{12} \gamma_k \\ \gamma_k^* W_{12}^* e_k & \gamma_k^* W_{22} \gamma_k \end{bmatrix} \right\rangle,$$

---

[3] This follows since if we let $\check{E} = \begin{bmatrix} \check{E}_1 & \check{E}_2 \end{bmatrix}$, where $\check{E}_1 = \begin{bmatrix} \check{e}_k : & k \notin \mathcal{N} \end{bmatrix}$ and $\check{E}_2 = \begin{bmatrix} \check{e}_k : & k \in \mathcal{N} \end{bmatrix}$, and analogously $\hat{E} = \begin{bmatrix} \hat{E}_1 & E \end{bmatrix}$, where $\hat{E}_1 = \begin{bmatrix} e_k : & k \notin \mathcal{N} \end{bmatrix}$ and $E = \begin{bmatrix} e_k : & k \in \mathcal{N} \end{bmatrix}$, then

$$\begin{aligned} \begin{bmatrix} \check{E} & \check{e}_{n+1} \end{bmatrix}^T \check{\Psi} \begin{bmatrix} \check{E} & \check{e}_{n+1} \end{bmatrix} &= \begin{bmatrix} \hat{E}^T (\Psi + \check{\Psi}_{11}) \hat{E} & \hat{E}^T \check{\Psi}_{12} \\ \check{\Psi}_{12}^* \hat{E} & \check{\Psi}_{22} \end{bmatrix} \\ &\succ \begin{bmatrix} \phi I & 0 & 0 \\ 0 & \phi I + E^T \check{\Psi}_1 E & E^T \check{\Psi}_{12} \\ 0 & \check{\Psi}_{12}^* E & \check{\Psi}_{22} \end{bmatrix}, \end{aligned}$$

where we used that $\hat{E}_1^T \check{\Psi}_{12} = 0$, $\hat{E}_1^T \check{\Psi}_{11} \hat{E} = 0$ by assumption and that $\hat{E}^T \Psi \hat{E} \succ \hat{E}^T \phi \hat{E} = \begin{bmatrix} \phi I_1 & 0 \\ 0 & \phi I_2 \end{bmatrix}$, where the dimensions of the identity matrices I_1 and I are $n - |\mathcal{N}|$ and $|\mathcal{N}|$, respectively. It follows that $\check{\Psi} \succ 0$ provided that (1.18) holds.

which is nonpositive if and only if

$$\begin{bmatrix} e_k^T W_{11} e_k & e_k^T W_{12} \gamma_k \\ \gamma_k^* W_{12}^* e_k & \gamma_k^* W_{22} \gamma_k \end{bmatrix} \in \pi_{\text{set}}^{\ominus}.$$

We focus on the case (1.14), in which

$$\pi_{\text{set}}^{\ominus} = \left\{ W = \begin{bmatrix} w_{11} & w_{12} \\ \bar{w}_{12} & w_{22} \end{bmatrix} : \operatorname{Im} w_{12} = 0;\; w_{11} - w_{22} \leq 0 \right\}.$$

Since the condition $M_H Z \in \Pi_k^{\ominus}$ equivalently can be written as

$$\begin{aligned} \operatorname{Im} e_k^T Z H^* \gamma_k &= 0, \\ e_k^T Z e_k - \gamma_k^* H Z H^* \gamma_k &\leq 0, \end{aligned} \tag{1.19}$$

we get the following dual:

Dual condition: For every $\omega \in \mathbf{R} \cup \{\infty\}$ and $Z \in \mathcal{Z}\rfloor\dashv\updownarrow$ there exists $k \in \{1, \ldots, n\}$ such that the equation system

$$\begin{aligned} \operatorname{Im} e_k^T Z H^* \gamma_k &= 0 \\ e_k^T Z e_k - \gamma_k^* H Z H^* \gamma_k &\leq 0 \end{aligned}$$

is violated.

By using a result on rank one decomposition of Hermitian positive semidefinite matrices in [2] we obtain the following alternative formulation of the dual.

Alternative dual condition: For every $\omega \in \mathbf{R} \cup \{\infty\}$ and $z \in z_{\text{unit}} \stackrel{\text{def}}{=} \{z \in \mathbf{C}^n : |z| = 1\}$ there exists $k \in \{1, \ldots, n\}$ such that

$$\frac{\gamma_k^* H(j\omega) z}{e_k^T z} \notin (-\infty, -1] \cup [1, \infty).$$

Proof. *Suppose that the dual is violated. This implies that for every $Z \in \mathcal{Z}\rfloor\dashv\updownarrow$ the systems (1.19) are satisfied for all $k \in \{1, \ldots, n\}$. Let us consider some particular index $k \in \{1, \ldots, n\}$. The condition (1.19) can equivalently be written*

$$\langle Z, A \rangle = 0 \quad \text{and} \quad \langle Z, B \rangle \leq 0, \tag{1.20}$$

where

$$\begin{aligned} A &= i(H^* \gamma_k e_k^T - e_k \gamma_k^* H), \\ B &= e_k e_k^T - H^* \gamma_k \gamma_k^* H. \end{aligned}$$

Corollary 2.2. in [2] implies that there exists a rank one decomposition $Z = \sum_{l=1}^{r} z_l z_l^*$ *where* $r = \mathrm{rank} Z$, *and* $z_l \in \mathbf{C}^n$ *is such that* $z_l^* A z_l = 0$ *and* $z_l^* B z_l \leq 0$, $l = 1, \dots, r$, *i.e.,*

$$\begin{aligned} Im\,(z_l^* e_k \gamma_k^* H z_l) &= 0, \\ |e_k^T z_l|^2 - |\gamma_k^* H z_l|^2 &\leq 0, \end{aligned} \tag{1.21}$$

for $l = 1, \dots, r$. *Conversely, if* $z_l \in \mathbf{C}^n$ *exists such that (1.21) holds, then (1.20) obviously holds with* $Z = \sum_{l=1}^{r} z_l z_l^*$.

The dual stability criterion holds if and only if for every $Z \in \mathcal{Z}\rfloor\dashv\updownarrow$ *an index* $k \in \{1, \dots, N\}$ *exists such that (1.19) is violated. The above arguments show that this is equivalent to the condition that for every* $z_l \in \mathbf{C}^n$ *there exists* $k \in \{1, \dots, n\}$ *such that (1.21) is violated. Clearly, if the first equation in (1.21) is violated then* $Im\,\gamma_k^* Hz/e_k^T z \neq 0$ *and otherwise we must have* $|\gamma_k^* Hz/e_k^T z| < 1$, *i.e., if (1.21) is violated we must have* $\gamma_k^* Hz/e_k^T z \notin (-\infty, -1] \cup [1, \infty)$.

Note that it is no restriction to normalize z such that it belongs to z_{unit}. □

The alternative dual can easily be interpreted. It implies that the equation

$$(\lambda I - \Gamma H(j\omega))z = 0$$

has no nontrivial solution for any real number with $|\lambda| \geq 1$. Since it holds for any $\omega \in \mathbf{R} \cup \{\infty\}$ it follows that:

1. the closed loop system is well-posed, i.e., $I - \Gamma H(\infty)$ is an invertible matrix;
2. $\phi(j\omega) = \det(I - \tau\Gamma H(j\omega)) \neq 0$ for all $\omega \in \mathbf{R} \cup \{\infty\}$ and any $\tau \in [0, 1]$.

The second condition is a zero exclusion property, which allows us to conclude stability of the network.

Acknowledgements The work was supported by the Swedish Research Council (VR), the ACCESS Linnaeus Centre and the EU FP7 project FeedNetBack.

Appendix

The primal stability condition implies that the system (1.1) is stable. Indeed, by Assumption 1.1 (b) there exists a continuous parametrization Δ_θ, $\theta \in [0, 1]$, such that $\Delta_1 = \Delta$. Let

$$\psi(s, \theta, \alpha) = |\det(I - H(s + \alpha)\Delta_\theta(s + \alpha))|.$$

We will show below that the primal condition together with the inequality in the definition of the cones $\Pi_{k,\Delta}$ in Assumption 1.1 (a) can be shown to imply

$$\psi(j\omega, \theta, 0) = |\det(I - H(j\omega)\Delta_\theta(j\omega))| \geq \varepsilon, \quad \forall \omega \in \mathbf{R} \cup \{\infty\}, \theta \in [0, 1] \tag{1.22}$$

for some $\varepsilon > 0$. From Assumption 1.1 (c) it follows that $(I - H\Delta_0)^{-1} \in \mathcal{A}^{n\times n}$ and hence that $\psi(s,0,0)$ has no zeros in the closed right half-plane. By continuity $\psi(s,\theta,\alpha)$ satisfies $\psi(j\omega,\theta,\alpha) > 0, \forall\omega \in \mathbf{R}\cup\{\infty\}, \theta \in [0,1]$ for small enough $\alpha > 0$. This function is analytic in the closed right half plane and we may apply the zero exclusion principle to conclude that $\psi(s,1,\alpha)$ has no zeros in the closed right half-plane; see e.g. Lemma A.1.18 in [1]. Once again, by continuity it follows that $\psi(s,1,0)$ also has no zeros in closed right half plane and hence $(I - H\Delta)^{-1} \in \mathcal{A}^{n\times n}$ and the claim is proven.

To prove (1.22) we use the following version of Finsler's lemma.

Lemma 1.2. *Let $B \in \mathbf{C}^{n\times m}$, $m < n$ and $B_\perp \in \mathbf{C}^{(n-m)\times n}$ be such that (1) $B_\perp B = 0$, (2) $\operatorname{rank}\begin{bmatrix} B & B_\perp^* \end{bmatrix} = n$, and (3) $B_\perp B_\perp^* \succeq I$. Then the following are equivalent:*

i. $B^*AB \prec 0$,
ii. *there exists $\mu \in (0,\mu^*]$ such that $A \prec \mu B_\perp^* B_\perp$, where*

$$\mu^* = |A| + \frac{|A|^2|B|^2}{\sigma_{\min}(B^*AB)}$$

Remark 1.1. If in (i) we have $B^*AB \preceq -\eta I$ then we may take

$$\mu^* = |A| + \frac{|A|^2|B|^2}{\eta}$$

Proof. *The proof of $(ii) \Rightarrow (i)$ is obvious. For $(i) \Rightarrow (ii)$ suppose there exists $x \in \mathbf{C}^n$ such that $x^*Ax \geq 0$. If $x \in \operatorname{Ker} B_\perp$ it follows by assumption (1) and (2) that $x = Bu$ for some $u \in \mathbf{C}^m$. This would contradict i. Hence, we may assume $B_\perp x \neq 0$. However, then $x^*(A - \mu B_\perp^* B_\perp)x < 0$ whenever $\mu > x^*Ax/|B_\perp x|^2$. In order to derive a bound on μ we notice that $x = Bu + x_\perp$ for some $u \in \mathbf{C}^m$ and some nonzero $x_\perp \in \operatorname{Range} B_\perp^*$. We have*

$$\begin{aligned}\max_{x:B^\perp x\neq 0} \frac{x^*Ax}{|B_\perp x|^2} &= \max_{x_\perp \in \operatorname{Range} B_\perp^*} \frac{x_\perp^*(A - A^*B(B^*AB)^{-1}B^*A)x_\perp}{|B_\perp x_\perp|^2} \\ &\leq |A| + |A|^2|B|^2/\sigma_{\min}(B^*AB), \end{aligned} \tag{1.23}$$

where the equality follows by using (i), which allows us to optimize over $u \in \mathbf{C}^m$ in the decomposition $x = Bu + x_\perp$. In the last inequality we used (3). This concludes the proof. □

By Assumption 1.1 (a) and Theorem 1.1(a) there exists for each $\omega \in \mathbf{R}\cup\{\infty\}$, a multiplier $\Pi_\omega \in \sum_{k=1}^N \Pi_{k,\Delta(j\omega_k)}$ such that for all $\theta \in [0,1]$,

$$\begin{bmatrix} \Delta_\theta(j\omega) \\ I \end{bmatrix}^* \Pi_\omega \begin{bmatrix} \Delta_\theta(j\omega) \\ I \end{bmatrix} \preceq 0$$

$$\text{and} \quad \begin{bmatrix} I \\ H(j\omega) \end{bmatrix}^* \Pi_\omega \begin{bmatrix} I \\ H(j\omega) \end{bmatrix} \succeq (1+\eta)(2+\|H\|^2))I,$$

where $\eta > 0$. We may perturb the multipliers as $\Pi_\omega := \Pi_\omega - (1+\eta)I$ such that

$$\begin{bmatrix} \Delta_\theta(j\omega) \\ I \end{bmatrix}^* \Pi_\omega \begin{bmatrix} \Delta_\theta(j\omega) \\ I \end{bmatrix} \preceq -(1+\eta)I$$

$$\text{and} \quad \begin{bmatrix} I \\ H(j\omega) \end{bmatrix}^* \Pi_\omega \begin{bmatrix} I \\ H(j\omega) \end{bmatrix} \succeq (1+\eta)I.$$

By continuity of H and Δ_θ on $\mathrm{cl}\,\mathbf{C}_+$, there exists a finite grid of frequencies $-\infty = \omega_1 < \omega_2, \cdots \omega_{L-1} < \omega_L = \infty$ such that $\Pi : j\mathbf{R} \cup \{\infty\} \to \mathcal{S}_{\mathbf{C}}^{2n\times 2n}$ defined as

$$\Pi(j\omega) = \begin{cases} \Pi_{\omega_l}, & \omega \in [\omega_l, \omega_{l+1}) \\ \Pi_{\omega_1}, & \omega = \omega_L \end{cases}$$

satisfies

$$\begin{bmatrix} \Delta_\theta(j\omega) \\ I \end{bmatrix}^* \Pi(\omega) \begin{bmatrix} \Delta_\theta(j\omega) \\ I \end{bmatrix} \preceq -I, \begin{bmatrix} I \\ H(j\omega) \end{bmatrix}^* \Pi(\omega) \begin{bmatrix} I \\ H(j\omega) \end{bmatrix} \succeq I$$

for all $\omega \in \mathbf{R} \cup \{\infty\}$.

We may now apply Lemma 1.2 with $A = \Pi(j\omega)$ and

$$B = \begin{bmatrix} \Delta_\theta(j\omega) \\ I \end{bmatrix} \quad \text{and} \quad B_\perp = \begin{bmatrix} I & -\Delta_\theta(j\omega) \end{bmatrix},$$

which satisfies (1)–(3) in Lemma 1.2. This implies that

$$I \prec \begin{bmatrix} I \\ H(j\omega) \end{bmatrix}^* \Pi(j\omega) \begin{bmatrix} I \\ H(j\omega) \end{bmatrix} \prec \mu |I - \Delta_\theta(j\omega)H(j\omega)|^2,$$

where $\mu \le \|\Pi\| + \|\Pi\|^2(1+\|\Delta\|^2)$ and where $\|\Pi\| = \max_{l\in\{1,\dots,N\}} |\Pi(j\omega_l)|$. This choice of μ makes the above inequality valid for all $\omega \in \mathbf{R} \cup \{\infty\}$, which implies that there exists $\varepsilon > 0$ such that

$$|\det(I - H(j\omega)\Delta_\theta(j\omega))| = |\det(I - \Delta_\theta(j\omega)H(j\omega))| \ge \varepsilon \tag{1.24}$$

for all $\omega \in \mathbf{R} \cup \{\infty\}$ and $\theta \in [0,1]$, which proves (1.22).

Finally, to prove that the primal and dual conditions are equivalent, i.e., that (a) and (b) in the theorem statement are equivalent, we first derive a criterion for (a) being violated.

If the criterion in (a) is violated there exists $\omega \in \mathbf{R} \cup \{\infty\}$ such that the sets

$$\begin{aligned} C_1 &= \left\{ M_{H(j\omega)} \Pi : \Pi \in \sum_{k=1}^{N} \Pi_{k,\Delta(j\omega)} \right\} \\ C_2 &= \left\{ P : P \in \mathcal{S}_{\mathbf{C}}^{m\times m},\ P \succ 0 \right\} \end{aligned}$$

are disjoint. By the separating hyperplane theorem [11, Theorem 11.3] a nonzero $Z \succeq 0$ exists such that (the argument ω is suppressed)

$$\begin{aligned} & \langle M_H \Pi, Z \rangle \le 0, \quad \forall \Pi \in \textstyle\sum_{k=1}^{N} \Pi_{k,\Delta} \\ \Leftrightarrow \quad & \mathrm{tr}((M_H \Pi) Z) \le 0, \quad \forall \Pi \in \textstyle\sum_{k=1}^{N} \Pi_{k,\Delta} \\ \Leftrightarrow \quad & \mathrm{tr}\left(\Pi (M_H^{\times} Z) \right) \le 0, \quad \forall \Pi \in \textstyle\sum_{k=1}^{N} \Pi_{k,\Delta}, \end{aligned}$$

which implies

$$M_{H(j\omega)}^{\times} Z \in \Big(\sum_{k=1}^{N} \Pi_{k,\Delta}\Big)^{\ominus} = \cap_{k=1}^{N} \Pi_{k,\Delta(j\omega)}^{\ominus}, \tag{1.25}$$

where we used Lemma 1.1 in the last equality. Note that we may normalize the dual variables such that $\mathrm{tr}(Z) = 1$, i.e., $Z \in \mathcal{Z}\rfloor\dashv\updownarrow$, where

$$\mathcal{Z}\rfloor\dashv\updownarrow = \{ Z \in \mathcal{S}_{\mathbf{C}}^{m\times m} : Z \succeq 0;\ \mathrm{tr}(Z) = 1 \}, \tag{1.26}$$

which is a compact convex set.

In order to prove the theorem we need to establish the reverse direction of the above duality result. Hence, if (1.25) fails, i.e., (1.4) holds, then $\forall \omega \in \mathbf{R} \cup \{\infty\}$ the two convex sets

$$\begin{aligned} C_3 &= \left\{ M_{H(j\omega)}^{\times} Z : Z \in \mathcal{Z}\rfloor\dashv\updownarrow \right\} \\ C_4 &= \cap_{k=1}^{N} \Pi_{k,\Delta(j\omega)}^{\ominus} \end{aligned}$$

are disjoint. From Lemma 1.1, it follows that

$$(\cap_{k=1}^{N} \Pi_{k,\Delta}^{\ominus})^{\ominus} = \mathrm{cl} \sum_{k=1}^{N} \Pi_{k,\Delta},$$

where we used that $\Pi_{k,\Delta}$ is a closed convex cone in the topology defined by the Frobenius norm, and therefore $\Pi_{k,\Delta}^{\ominus\ominus} = \Pi_{k,\Delta}$. Hence, since C_3 is convex and compact and C_4 is closed and convex it follows that a hyperplane exists that separates the two sets strongly [Corollary 11.4.2 in [11]], i.e., a nonzero $\Pi \in \mathrm{cl} \sum_{k=1}^{N} \Pi_{k,\Delta}$ exists such that

$$\begin{aligned} & \left\langle \Pi, M_{H(j\omega)}^{\times} Z \right\rangle > 0, \quad \forall Z \in \mathcal{Z}\rfloor\dashv\updownarrow \\ \Leftrightarrow\ & \left\langle M_{H(j\omega)} \Pi, Z \right\rangle > 0, \quad \forall Z \in \mathcal{Z}\rfloor\dashv\updownarrow \\ \Leftrightarrow\ & M_{H(j\omega)} \Pi \succ 0. \end{aligned}$$

Since the inequality is strict it follows that for each $\omega \in \mathbf{R} \cup \{\infty\}$ there exists $\Pi \in \sum_{k=1}^{N} \Pi_{k,\Delta(j\omega)}$ such that $M_{H(j\omega)} \Pi \succ 0$.

References

1. Curtain, R.F., Zwart, H.: An Introduction to Infinite Dimensional Linear Systems Theory. Springer-Verlag, New York (1995)
2. Huang, Y., Zhang, S.: Complex matrix decomposition and quadratic programming. Mathematics of Operations Research **32**, 758–768 (2007)
3. Jönsson, U.T., Kao, C.Y., Fujioka, H.: A Popov criterion for networked systems. Systems and Control Letters **56**(9–10), 603–610 (2007)
4. Jönsson, U.T., Kao, C.Y.: A scalable robust stability criterion for systems with heterogeneous LTI components. IEEE Trans. Automatic Control **55**(10), 2219–2234 (2010)
5. Jönsson, U.: Primal and dual stability criteria for robust stability and their application to systems interconnected over a bi-partite graph. In: Proc. American Control Conference (ACC2010). Baltimore, MD, USA (2010)
6. Lestas, I., Vinnicombe, G.: Scalable robustness for consensus protocols with heterogenous dynamics. In: Proc. 16th IFAC World Congress. Prague, Czech Republic (2005)
7. Lestas, I., Vinnicombe, G.: Scalable decentralized robust stability certificates for networks of interconnected heterogeneous dynamical systems. IEEE Trans. Automatic Control **51**(10), 1613–1625 (2006)
8. Megretski, A., Rantzer, A.: System analysis via integral quadratic constraints. IEEE Trans. Automatic Control **42**(6), 819–830 (1997)
9. Moylan, P., Hill, D.: Stability criteria for large-scale systems. IEEE Trans. Automatic Control **23**(2), 143–149 (1978)
10. Rantzer, A.: Distributed performance analysis of heterogeneous systems. In: Proc. 49th IEEE Conf. Decision and Control (CDC2010), pp. 2682–2685. Atlanta, GA (2010)
11. Rockafellar, R.: Convex Analysis. Princeton University Press, Princeton, NJ (1970)
12. Sontag, E.: Passivity gains and the "secant condition" for stability. Systems Control Letters **55**(3), 177–183 (2006)
13. Vidyasagar, M.: Input-Output Analysis of Large Scale Interconnecred Systems. Lecture Notes in Control and Information Sciences. Springer-Verlag, Berlin (1981)
14. Vinnicombe, G.: Robust congestion control for the Internet (2002). Unpublished

Chapter 2
Optimal Controller Synthesis for a Decentralized Two-Player Linear-Quadratic Regulator via Spectral Factorization

John Swigart and Sanjay Lall

Abstract We develop controller synthesis algorithms for decentralized control problems. The particular system considered here consists of two interconnected linear subsystems, with communication allowed in only one direction. We develop the concept of spectral factorization, which is the approach used to construct the optimal controllers. Explicit state-space formulae are provided, and we show that each player has to do more than simply estimate the states that they cannot observe. In other words, the simplest separation principle does not hold for this decentralized control problem. Some intuition into the control policies is provided, and the order of the optimal controllers is established.

2.1 Introduction

Decentralized systems, consisting of multiple subsystems interacting over a network with limited communication, are important in many practical problems today. Examples include formation flight, teams of vehicles, or large spatially distributed systems such as the Internet or the power grid.

With the advent of the elegant solution for the centralized control problem, it was generally believed that decentralized problems would have similarly clean solutions. Unfortunately, a simple counterexample disproved this notion, showing linear control policies may be strictly suboptimal, even when the underlying system dynamics are linear and time-invariant [28]. In general, decentralized control problems are currently intractable [2].

This chapter focuses on a specific information structure, consisting of two interconnected systems with dynamics such that player 1's state affects the state of player 2. Our objective is to find a pair of controllers such that player 1 has access

J. Swigart and S. Lall
Department of Aeronautics and Astronautics, Stanford University, Stanford, CA 94305, USA.
e-mail: jswigart@stanford.edu@stanford.edu, lall@stanford.edu

R. Johansson and A. Rantzer (Eds.): Distributed Decision Making and Control, LNCIS 417, pp. 27–53.
springerlink.com

only to the first state, whereas player 2 can measure both states. The controller is chosen to minimize the $\mathcal{H}_2$ norm of the closed-loop transfer function. This system can be visualized by the simple graph in Fig. 2.1. It has been shown, in a number of

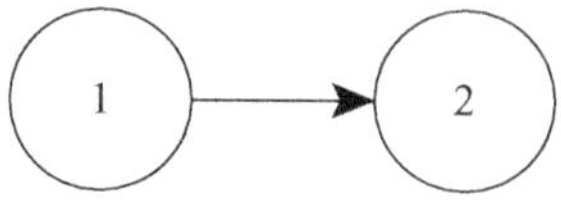

Fig. 2.1 Two-player system.

different ways, that this problem admits linear optimal controllers. Typically, these results reduce the problem to one of convex optimization [26, 6, 17, 18]. Though convex, these formulations remain infinite-dimensional. In particular, the approach in [18] uses a change of variables via the standard Youla parameterization, and optimization over this parameter. Since the parameter itself is a linear stable system, a standard approximation would be via a finite basis for the impulse response function [3]. This is in contrast to the centralized case, for which explicit state-space formulae can be constructed.

The work presented in this chapter is based on recent work in [23, 22, 25]. In [23], explicit formulae for the optimal controllers were constructed for the finite-horizon, time-varying version of this problem. A dynamic programming approach was taken in [22]. Since the approach used here is based on spectral factorization, we begin our results by developing the method for the centralized case. The work presented here focuses on the infinite-horizon $\mathcal{H}_2$ version of the problem.

We show that both controllers separate naturally into a composition of controller and estimator. Moreover, each controller has the same number of states as player 2. Such formulae offer the practical advantages of computational reliability and simplicity. In addition, it provides significant understanding and interpretation of the optimal controller structure. Lastly, it establishes the order of the optimal controller for this system, which is an open problem for general decentralized systems, even in the simplest cases.

The advantage of the spectral factorization methods used here is that they extend naturally to more general networks, and the results in this chapter are a first step towards general state-space solutions.

Previous Work

Some classic results in decentralized control can be found in [27, 7, 11]. As noted above, the general decentralized problem remains intractable. Consequently, most work in this area has been directed at classifying those systems that can be reformulated as convex problems [6, 10, 12, 1]. These results were recently unified and generalized under the concept of quadratic invariance [17]. For distributed systems over networks, conditions for quadratic invariance, and thus tractability, of such systems were provided in [24, 16]. In [20], a poset-based framework was used to obtain similar results.

In terms of controller synthesis, different approaches have been used to find numerical solutions to some of these problems. Some methods were considered, though not implemented, in [26]. A semidefinite programming (SDP) solution for the problem considered here was provided in [19]. Other SDP approaches were presented in [13, 30]. In [5], an approximation scheme for solving decentralized control problems was suggested. For the quadratic case, a vectorization approach can be used to obtain a finite-dimensional problem [18], but this loses the intrinsic structure and results in high-order controllers.

However, in none of these approaches have explicit state-space formulae been derived for this problem. In the work presented here, a spectral factorization approach is taken to construct explicit state-space formulae for the two-player problem. As a result, we can efficiently and analytically compute the optimal controllers for this decentralized problem. Moreover, we gain significant insight into the form of the solution which previous approaches do not provide.

2.2 Problem Formulation

We use the following notation in this chapter. The real and complex numbers are denoted by $\mathbb{R}$ and $\mathbf{C}$, respectively. The complex open unit disc is $\mathbb{D}$; its boundary, the unit circle, is $\mathbb{T}$; and the closed unit disc is $\bar{\mathbb{D}}$. The set $\mathcal{L}_2(\mathbb{T})$ is the Hilbert space of Lebesgue measurable functions on $\mathbb{T}$, which are square integrable, with inner product

$$\langle F, G\rangle = \frac{1}{2\pi}\int_0^{2\pi} \operatorname{tr}(F^*(e^{j\theta})G(e^{j\theta}))\, d\theta$$

As is standard, $\mathcal{H}_2$ denotes the Hardy space

$$H_2 = \left\{ f : \{\infty\} \cup \mathbf{C} \setminus \bar{\mathbb{D}} \to \mathbf{C} \;\middle|\; \exists x \in \ell_2(\mathbb{Z}_+) \text{ s.t. } f(z) = \sum_{k=0}^{\infty} x_k z^{-k} \right\}$$

of functions analytic outside the closed unit disc, and at infinity, with square-summable power series. The set $\mathcal{H}_2^{\perp}$ is the orthogonal complement of $\mathcal{H}_2$ in $\mathcal{L}_2$. The prefix $\mathcal{R}$ indicates the subsets of proper real rational functions. That is, $\mathcal{RL}_2$ is the set of transfer functions with no poles on $\mathbb{T}$, and $\mathcal{RH}_2$ is the set of transfer functions with no poles outside $\mathbb{T}$.

Also, we denote the subspace $\mathcal{L}_\infty(\mathbb{T})$ as the set of Lebesgue measurable functions which are bounded on $\mathbb{T}$. Similarly, $\mathcal{H}_\infty$ is the subspace of $\mathcal{L}_\infty$ with functions analytic outside of $\mathbb{T}$, and $\mathcal{H}_\infty^-$ is the subspace of $\mathcal{L}_\infty$ with functions analytic inside $\mathbb{T}$. Consequently, $\mathcal{RH}_\infty$ is the set of transfer functions with no poles outside of $\mathbb{T}$. Note that, in this case, $\mathcal{RH}_2 = \mathcal{RH}_\infty$; we will use these spaces interchangeably.

Some useful facts about these sets which we will take advantage of in this work are [31]:

- if $G \in \mathcal{L}_\infty$, then $G\mathcal{L}_2 \subset \mathcal{L}_2$,
- if $G \in \mathcal{H}_\infty$, then $G\mathcal{H}_2 \subset \mathcal{H}_2$,
- if $G \in \mathcal{H}_\infty^-$, then $G\mathcal{H}_2^\perp \subset \mathcal{H}_2^\perp$.

For any $F \in \mathcal{RL}_\infty$, we define $F^\sim \in \mathcal{RL}_\infty$ as

$$F^\sim(z) = F^T(z^{-1})$$

It is straightforward to see that the multiplication operator corresponding to $F^\sim$ is the adjoint of the multiplication operator defined by F. For transfer functions $F \in \mathcal{RL}_2$, we use the notation

$$F(z) = \left[\begin{array}{c|c} A & B \\ \hline C & D \end{array}\right] = C(zI - A)^{-1}B + D$$

We are interested in the following discrete time state-space system

$$\begin{bmatrix} x_1(t+1) \\ x_2(t+1) \end{bmatrix} = \begin{bmatrix} A_{11} & 0 \\ A_{21} & A_{22} \end{bmatrix} \begin{bmatrix} x_1(t) \\ x_2(t) \end{bmatrix} + \begin{bmatrix} B_{11} & 0 \\ B_{21} & B_{22} \end{bmatrix} \begin{bmatrix} u_1(t) \\ u_2(t) \end{bmatrix} + \begin{bmatrix} H_1 & 0 \\ 0 & H_2 \end{bmatrix} \begin{bmatrix} w_1(t) \\ w_2(t) \end{bmatrix} \quad (2.1)$$

This corresponds to a two-player system, in which player 1's state can influence player 2's state. We are interested in finding controllers of the form

$$\begin{aligned} q_1(t+1) &= A_{K1}q_1(t) + B_{K1}x_1(t) \\ u_1(t) &= C_{K1}q_1(t) + D_{K1}x_1(t) \end{aligned}$$

and

$$\begin{aligned} q_2(t+1) &= A_{K2}q_2(t) + B_{K2}x(t) \\ u_2(t) &= C_{K2}q_2(t) + D_{K2}x(t) \end{aligned}$$

That is, player 1 makes decision u_1 based only on the history of his own state x_1, while player 2 makes decision u_2 based on the history of both states x_1 and x_2. This controller can be represented by the transfer functions $\mathcal{K}_{11}, \mathcal{K}_{21}, \mathcal{K}_{22} \in \mathcal{RL}_\infty$, such that

$$\begin{bmatrix} u_1 \\ u_2 \end{bmatrix} = \begin{bmatrix} \mathcal{K}_{11} & 0 \\ \mathcal{K}_{21} & \mathcal{K}_{22} \end{bmatrix} \begin{bmatrix} x_1 \\ x_2 \end{bmatrix}$$

For a set S whose elements are partitioned as

$$F = \begin{bmatrix} F_{11} & F_{12} \\ F_{21} & F_{22} \end{bmatrix}$$

for all $F \in S$, we define lower(S) to be the subset of S consisting of elements with lower triangular structure. In other words, $F \in$ lower(S) if and only if $F \in S$ and

$$F = \begin{bmatrix} F_{11} & 0 \\ F_{21} & F_{22} \end{bmatrix}$$

In particular, our desired controllers are in the set $\mathcal{K} \in \text{lower}(\mathcal{RL}_\infty)$.

Note that the space $S = \text{lower}(\mathcal{RH}_2) \subset \mathcal{L}_2$ has an orthogonal complement, such that $G \in S^\perp$ if and only if $G_{11}, G_{21}, G_{22} \in \mathcal{H}_2^\perp$ and $G_{12} \in \mathcal{L}_2$. We will also define $\mathbb{P}_{\mathcal{H}_2} : \mathcal{L}_2 \to \mathcal{H}_2$ as the orthogonal projection onto $\mathcal{H}_2$. Similarly, $\mathbb{P}_S : \mathcal{L}_2 \to S$ is the orthogonal projection onto S.

Our cost is the vector

$$z(t) = \begin{bmatrix} C_1 & C_2 \end{bmatrix} \begin{bmatrix} x_1(t) \\ x_2(t) \end{bmatrix} + \begin{bmatrix} D_1 & D_2 \end{bmatrix} \begin{bmatrix} u_1(t) \\ u_2(t) \end{bmatrix}$$

where, for simplicity, we will assume that $C^T D = 0$ and $D^T D > 0$. Notice that this formulation allows for coupling of the states in the cost. Consequently, our plant can be expressed as the matrix $P \in \mathcal{RL}_\infty$, where

$$\begin{bmatrix} z \\ x \end{bmatrix} = \begin{bmatrix} P_{11} & P_{12} \\ P_{21} & P_{22} \end{bmatrix} \begin{bmatrix} w \\ u \end{bmatrix}$$

and

$$P = \begin{bmatrix} C \\ I \end{bmatrix} (zI - A)^{-1} \begin{bmatrix} H & B \end{bmatrix} + \begin{bmatrix} 0 & D \\ 0 & 0 \end{bmatrix} \tag{2.2}$$

where A and B are lower triangular, and H is block diagonal and invertible, as defined in (2.1). Note that H being invertible simply implies that no component of the state is deterministic. This assumption merely simplifies our presentation while not fundamentally affecting our results.

We define $\mathcal{F}(P, \mathcal{K})$ as the linear fractional transformation

$$\mathcal{F}(P, \mathcal{K}) = P_{11} + P_{12}\mathcal{K}(I - P_{22}\mathcal{K})^{-1} P_{21}$$

Our objective function is the $\mathcal{H}_2$ norm of the closed-loop transfer function from w to z. In other words, we have the following optimization problem:

$$\begin{array}{ll} \text{minimize} & \|\mathcal{F}(P, \mathcal{K})\|_2 \\ \text{subject to} & \mathcal{K} \text{ is stabilizing} \\ & \mathcal{K} \in \text{lower}(\mathcal{RL}_\infty) \end{array} \tag{2.3}$$

2.3 Main Results

Having established our notation and problem formulation, we now present the optimal solution for (2.3). We will develop the proof for this result in the following sections.

Theorem 2.1. *For the system in* (2.2)*, suppose* $C^T D = 0$ *and* $D^T D > 0$*. Suppose* (A_{11}, B_{11}) *and* (A_{22}, B_{22}) *are stabilizable. Also, suppose there exist stabilizing solutions* X *and* Y *to the algebraic Riccati equations*

$$X = C^T C + A^T XA - A^T XB(D^T D + B^T XB)^{-1} B^T XA \tag{2.4}$$

$$Y = C_2^T C_2 + A_{22}^T Y A_{22} - A_{22}^T Y B_{22} (D_2^T D_2 + B_{22}^T Y B_{22})^{-1} B_{22}^T Y A_{22} \tag{2.5}$$

Define

$$K = (D^T D + B^T XB)^{-1} B^T XA$$
$$J = (D_2^T D_2 + B_{22}^T Y B_{22})^{-1} B_{22}^T Y A_{22}$$

and let

$$A^{\mathcal{K}} = A_{22} - B_{21} K_{12} - B_{22} K_{22}$$
$$B^{\mathcal{K}} = A_{21} - B_{21} K_{11} - B_{22} K_{21}$$

Then, there exists a unique optimal $\mathcal{K} \in \text{lower}(\mathcal{RL}_\infty)$ *for* (2.3) *given by:*

- *Controller* 1 *has realization*

$$q_1(t+1) = A^{\mathcal{K}} q_1(t) + B^{\mathcal{K}} x_1(t)$$
$$u_1(t) = -K_{12} q_1(t) - K_{11} x_1(t)$$

- *Controller* 2 *has realization*

$$q_2(t+1) = A^{\mathcal{K}} q_2(t) + B^{\mathcal{K}} x_1(t)$$
$$u_2(t) = \big(J - K_{22}\big) q_2(t) - K_{21} x_1(t) - J x_2(t)$$

Note that there may not always exist stabilizing solutions to the algebraic Riccati equations (2.4–2.5). To simplify our results and avoid confusing the presentation with additional technical assumptions, we will simply assume the existence of stabilizing solutions. For a thorough discussion on algebraic Riccati equations, see [31].

2.4 Analysis

Before trying to find the optimal controllers, it is important we note when stabilization is even possible. The following lemma provides the necessary and sufficient conditions for the existence of any stabilizing controller.

Lemma 2.1. *There exists a controller* $\mathcal{K} \in \text{lower}(\mathcal{RL}_\infty)$ *which stabilizes* P *in* (2.2) *if and only if* (A_{11}, B_{11}) *is stabilizable and* (A_{22}, B_{22}) *is stabilizable.*

Proof. *(⇒) If* (A_{11}, B_{11}) *and* (A_{22}, B_{22}) *are stabilizable, then there exist matrices* F_1 *and* F_2 *such that* $A_{11} + B_{11}F_1$ *and* $A_{22} + B_{22}F_2$ *are stable. Consequently, the controller*

$$\mathcal{K} = \begin{bmatrix} F_1 & 0 \\ 0 & F_2 \end{bmatrix}$$

produces the closed-loop system

$$\begin{bmatrix} x_1(t+1) \\ x_2(t+1) \end{bmatrix} = \begin{bmatrix} A_{11} + B_{11}F_1 & 0 \\ A_{21} + B_{21}F_1 & A_{22} + B_{22}F_2 \end{bmatrix} \begin{bmatrix} x_1(t) \\ x_2(t) \end{bmatrix}$$

which is clearly stable.

(⇐) Suppose that (A_{11}, B_{11}) *is not stabilizable. Then, there exists a transformation* U *such that*

$$U^{-1}A_{11}U = \begin{bmatrix} a_{11} & a_{12} \\ 0 & a_{22} \end{bmatrix}, \qquad U^{-1}B_{11} = \begin{bmatrix} b_1 \\ 0 \end{bmatrix},$$

where a_{22} *has at least one unstable eigenvalue* λ. *Let* v *be the corresponding eigenvector of* a_{22}, *so that* $a_{22}v = \lambda v$. *Then, it can be readily shown that with the initial condition*

$$x_1(0) = U \begin{bmatrix} 0 \\ v \end{bmatrix}$$

the state $x_1(t) \nrightarrow 0$ *as* $t \to \infty$ *for any inputs* u. *A similar argument holds for the case where* (A_{22}, B_{22}) *is not stabilizable.* □

Thus, Lemma 2.1 shows that the two-player system can be stabilized if and only if each individual subsystem can be stabilized.

When the system can be stabilized, by choosing stabilizing matrices F_1 and F_2, we can use the standard Youla parametrization to simplify our optimization problem [29, 4].

Lemma 2.2. *Let* $S = \text{lower}(\mathcal{RH}_2)$. *Suppose* (A_{11}, B_{11}) *and* (A_{22}, B_{22}) *are stabilizable, and let* F_1 *and* F_2 *be matrices, such that* $A_{11} + B_{11}F_1$ *and* $A_{22} + B_{22}F_2$ *have stable eigenvalues. Let*

$$F = \begin{bmatrix} F_1 & 0 \\ 0 & F_2 \end{bmatrix}$$

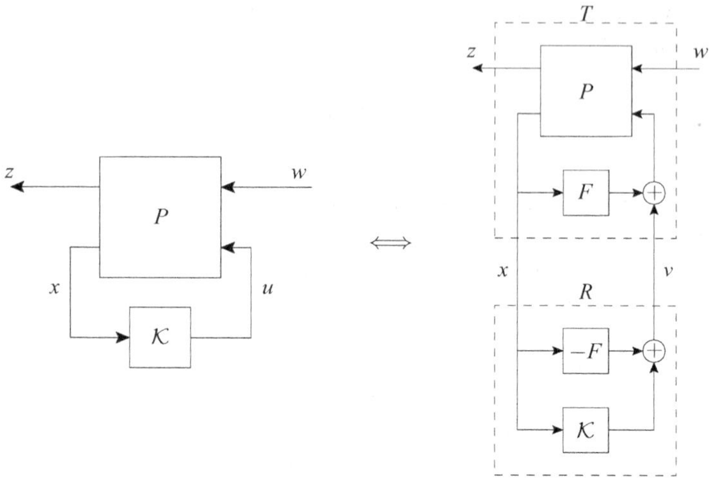

Fig. 2.2 Equivalent feedback systems.

Then, the set of all stabilizing controllers $\mathcal{K} \in \text{lower}(\mathcal{RL}_\infty)$ *is parametrized by*

$$\mathcal{K} = Q(I + MQ)^{-1} + F, \qquad Q \in S,$$

where

$$M = \left[\begin{array}{c|c} A+BF & B \\ \hline I & 0 \end{array}\right]$$

Moreover, the set of stable closed-loop transfer functions satisfies

$$\{\mathcal{F}(P,\mathcal{K}) \mid \mathcal{K} \in \text{lower}(\mathcal{RL}_\infty),\ \mathcal{K}\ \textit{stabilizing}\} = \{N_{11} + N_{12}QN_{21} \mid Q \in S\}$$

where $N_{12} = z^{-1}((C+DF)(zI-(A+BF))^{-1}B+D)$ *and*

$$\begin{bmatrix} N_{11} \\ N_{21} \end{bmatrix} = \left[\begin{array}{c|c} A+BF & H \\ \hline C+DF & 0 \\ A+BF & H \end{array}\right]$$

Proof. *Notice in Fig. 2.2 that we can create an equivalent feedback system by adding and subtracting the gain F. However, in doing so we can create a new feedback system with*

$$\mathcal{F}(T,R) = \mathcal{F}(P,\mathcal{K})$$

where

$$T = \left[\begin{array}{c|cc} A+BF & H & B \\ \hline C+DF & 0 & D \\ I & 0 & 0 \end{array}\right] \qquad R = \mathcal{K} - F$$

Let us now define

$$Q = R(I - T_{22}R)^{-1}$$

Notice that the map $R \mapsto Q$ is bijective. As a result, we now have

$$\mathcal{F}(T,R) = T_{11} + T_{12}QT_{21}$$

Now, suppose $Q \in S$. Since T and Q are stable, then the closed-loop map is stable. In addition, since both T_{22} and Q are lower triangular, then R is lower triangular, where

$$R = Q(I + T_{22}Q)^{-1}$$

Thus,

$$\mathcal{K} = R + F = Q(I + T_{22}Q)^{-1} + F \in \text{lower}(\mathcal{RL}_\infty)$$

and $\mathcal{K}$ stabilizes P. As a result, we have

$$\{T_{11} + T_{12}QT_{21} \mid Q \in S\} \subset \{\mathcal{F}(P,\mathcal{K}) \mid \mathcal{K} \in \text{lower}(\mathcal{RL}_\infty),\ \mathcal{K} \text{ stabilizing}\}$$

Conversely, suppose that $\mathcal{K} \in \text{lower}(\mathcal{RL}_\infty)$ stabilizes P. This implies that

$$\begin{bmatrix} A + BD_K & BC_K \\ B_K & A_K \end{bmatrix}$$

is stable. Thus, letting $R = \mathcal{K} - F \in \text{lower}(\mathcal{RL}_\infty)$, we can show that

$$Q = R(I - T_{22}R)^{-1} = \left[\begin{array}{cc|c} A+BD_K & BC_K & B(D_K - F) \\ B_K & A_K & B_K \\ \hline D_K - F & C_K & D_K - F \end{array}\right]$$

Consequently, we see that Q is stable. Moreover, since T_{22} and R are lower triangular, so is Q. Hence, $Q \in S$. Thus, we've shown that

$$\{T_{11} + T_{12}QT_{21} \mid Q \in S\} = \{\mathcal{F}(P,\mathcal{K}) \mid \mathcal{K} \in \text{lower}(\mathcal{RL}_\infty),\ \mathcal{K} \text{ stabilizing}\}$$

Thus, the result follows by letting

$$\begin{bmatrix} N_{11} & N_{12} \\ N_{21} & M \end{bmatrix} = \begin{bmatrix} T_{11} & z^{-1}T_{12} \\ zT_{21} & T_{22} \end{bmatrix}$$

□

As in the classical case, the Youla parametrization translates the difficult optimization problem (2.3) into an affine optimization problem. Since we have state feedback in our problem, we can simplify the problem even further.

Lemma 2.3. *For the system in* (2.2)*, let N be defined as in Lemma 2.2. Suppose Q is optimal for*

$$\begin{aligned} &\text{minimize} && \|N_{11} + N_{12}Q\|_2 \\ &\text{subject to} && Q \in S \end{aligned} \tag{2.6}$$

Then, there exists $\hat{Q} \in S$ such that $Q = \hat{Q}N_{21}$, and $\hat{Q}$ is optimal for

$$\begin{aligned} &\text{minimize} && \|N_{11} + N_{12}\hat{Q}N_{21}\|_2 \\ &\text{subject to} && \hat{Q} \in S \end{aligned} \tag{2.7}$$

Conversely, if $\hat{Q} \in S$ is optimal for (2.7)*, then $Q = \hat{Q}N_{21}$ is optimal for* (2.6).

Proof. *This follows from the fact that $N_{21}, N_{21}^{-1} \in S$, so that $Q \in S$ if and only if $\hat{Q} \in S$.* □

In order to solve the optimization problem in (2.6), it is convenient to find an equivalent optimality condition, which the following lemma provides.

Lemma 2.4. *Let $S = \text{lower}(\mathcal{RH}_2)$. Suppose $U, G \in \mathcal{RH}_\infty$. Then, $Q \in S$ minimizes*

$$\begin{aligned} &\text{minimize} && \|U + GQ\|_2 \\ &\text{subject to} && Q \in S \end{aligned}$$

if and only if

$$G^*U + G^*GQ \in S^\perp \tag{2.8}$$

Proof. *This result is a version of the classical projection theorem; see for example [8]. We first show that if $Q_0 \in S$ is a minimizer of $\|U + GQ\|_2$, then Q_0 satisfies* (2.8)*. Suppose, to the contrary, that there exists $\Gamma \in S$ such that*

$$\langle \Gamma, G^*U + G^*GQ_0 \rangle = \delta \neq 0$$

We assume, without loss of generality, that $\|\Gamma\| = 1/\|G\|$. Then, letting $Q_1 = Q_0 - \delta\Gamma$, we have

$$\begin{aligned} \|U + GQ_1\|_2^2 &= \|U + GQ_0 - \delta G\Gamma\|_2^2 \\ &= \|U + GQ_0\|_2^2 - \langle U + GQ_0, \delta G\Gamma \rangle - \langle \delta G\Gamma, U + GQ_0 \rangle + |\delta|^2 \|G\Gamma\|_2^2 \\ &= \|U + GQ_0\|_2^2 - 2\delta\,\text{Re}\langle G^*U + G^*GQ_0, \Gamma \rangle + |\delta|^2 \|G\Gamma\|_2^2 \\ &= \|U + GQ_0\|_2^2 - 2|\delta|^2 + |\delta|^2 \|G\Gamma\|_2^2 \\ &\leq \|U + GQ_0\|_2^2 - |\delta|^2 \\ &< \|U + GQ_0\|_2^2 \end{aligned}$$

Thus, if $Q_0 \in S$ does not satisfy (2.8)*, then Q_0 is not a minimizer.*

Conversely, suppose that $Q_0 \in S$ satisfies (2.8)*. Then, for any $Q \in S$, we have*

$$\begin{aligned}\|U+GQ\|_2^2 &= \|U+GQ_0+G(Q-Q_0)\|_2^2\\ &= \|U+GQ_0\|_2^2+\langle U+GQ_0, G(Q-Q_0)\rangle\\ &\quad+\langle G(Q-Q_0), U+GQ_0\rangle+\|G(Q-Q_0)\|_2^2\\ &= \|U+GQ_0\|_2^2+2\,\mathrm{Re}\langle Q-Q_0, G^*U+G^*GQ_0\rangle+\|G(Q-Q_0)\|_2^2\end{aligned}$$

Since S is a subspace, then $Q-Q_0 \in S$. As a result, the above inner product term is zero, so we have

$$\|U+GQ\|_2^2 = \|U+GQ_0\|_2^2+\|G(Q-Q_0)\|_2^2$$

Thus, $Q = Q_0$ is a minimizer. □

While Lemma 2.4 provides necessary and sufficient conditions for optimality of a controller, it does not guarantee existence of such a controller. The existence of an optimal controller will be shown by explicit construction in the following sections.

2.5 Spectral Factorization

The optimality condition (2.8) could be equivalently written as

$$G^*U+G^*GQ = \Lambda \tag{2.9}$$

where $\Lambda \in S^{\perp}$. It is important to note here that Q and Λ are orthogonal, that they possess complementary structures. However, as it is currently written, the G^*G operator is coupling these terms and making a solution difficult to find.

Before we embark on solving our two-player problem, it is necessary to first understand the approach taken in the classical case; that is, when $S = \mathcal{RH}_2$. We assume the classical case, $S = \mathcal{RH}_2$, throughout this section.

2.5.1 Finite Horizon Case

Perhaps the clearest way to understand our approach is to consider the finite horizon version of the problem. In this case, Q is a real block lower triangular matrix and Λ is a strictly block upper triangular matrix. Visually, Eq. (2.9) looks like

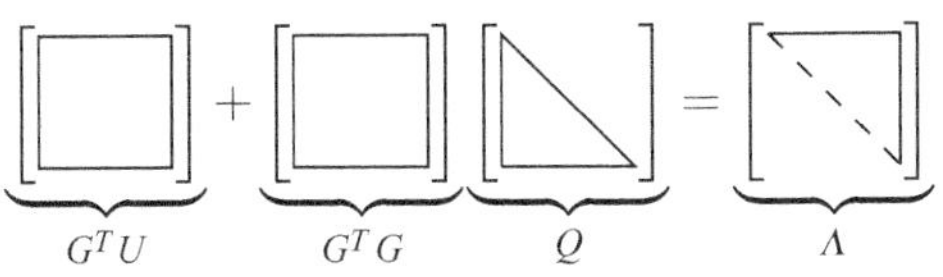

(2.10)

The difficulty again is that $G^T G$ is a full matrix, which ruins the triangular structure of Q. However, since $G^T G > 0$, it is well-known that a Cholesky factorization exists which decomposes $G^T G$ into the product of an invertible upper triangular matrix and an invertible lower triangular matrix. In particular,

$$G^T G = L^T L$$

where L is lower triangular and invertible. With this factorization, the optimality condition is equivalent to

$$L^{-T} G^T U + LQ = L^{-T} \Lambda$$

Since the set of lower triangular matrices is invariant under addition and multiplication (similarly for upper triangular matrices), notice now that the term LQ remains lower triangular, while the $L^{-T}\Lambda$ term is still strictly upper triangular. Thus, our factorization has succeeded in decoupling our variables, so that they may be solved independently. In other words, if we consider only the lower triangular elements of the above optimality condition, we obtain

$$\mathbb{P}_{\text{Lower}}(L^{-T} G^T U) + LQ = 0$$

where $\mathbb{P}_{\text{Lower}}$ is the orthogonal projection onto the set of lower triangular matrices. Solving for Q is now straightforward.

2.5.2 Scalar Transfer Functions

While the infinite horizon/transfer function version of this problem is more complicated, the basic idea remains the same. Since Q is a stable transfer function and Λ is an anti-stable transfer function, our goal is again to factor the $G^* G$ term into the product of a stable transfer function, with stable inverse, and an anti-stable transfer function, with anti-stable inverse. This is known as *spectral factorization*, since the terms stable/anti-stable are references to the spectrum of the operators. The results of this section follow from [14, 15].

To begin our discussion of spectral factorization, we first consider the case of trigonometric polynomials. A *trigonometric polynomial* is a rational function of the form

$$f(z) = \sum_{k=-n}^{n} c_k z^k$$

It is straightforward to show that $f(z)$ is real for all $z \in \mathbb{T}$ if and only if

$$c_k = \overline{c_{-k}}$$

for all k. Consequently, if $f(z)$ is real, then

$$f\left(\frac{1}{\bar{z}}\right) = \overline{f(z)}$$

for all $z \in \mathbf{C}$, so that r is a root of f if and only if $1/\bar{r}$ is a root. Notice that if $|r| < 1$, then $|\frac{1}{\bar{r}}| > 1$, so that f has an equal number of stable and unstable roots. Spectral factorization of f takes advantage of this fact, as seen in the following theorem.

Theorem 2.2. *Suppose f is the trigonometric polynomial*

$$f(z) = \sum_{k=-n}^{n} c_k z^k$$

and $f(z)$ is real for all $z \in \mathbb{T}$. Then,

$$f(z) \geq 0 \qquad \text{for all } z \in \mathbb{T}$$

if and only if there exists a polynomial

$$q(z) = a(z - r_1) \dots (z - r_n)$$

with all $|r_i| \leq 1$ such that

$$f(z) = q^{\sim}(z) q(z)$$

Proof. *Clearly, if there exists such a polynomial q, then for all $z \in \mathbb{T}$,*

$$f(z) = q^{\sim}(z) q(z) = |a|^2 |z - r_1|^2 \dots |z - r_n|^2 \geq 0$$

Conversely, suppose that $f(z)$ is nonnegative for all $z \in \mathbb{T}$. Without loss of generality, we assume that $c_{-n} \neq 0$, and let p be the polynomial

$$p(z) = z^n f(z)$$

Since p is a polynomial of degree $2n$, it has $2n$ nonzero roots, and we can factorize it as

$$p(z) = c \prod_{i=1}^{m} (z - \bar{r}_i^{-1})(z - r_i) \prod_{j=1}^{s} (z - w_j)^2$$

where $|r_i| < 1$ and $|w_j| = 1$. Note that r_i and $\bar{r}_i^{-1}$ are both roots since f is nonnegative on $\mathbb{T}$. Since f is also continuous on $\mathbb{T}$, then each root $w_j \in \mathbb{T}$ must have even multiplicity. Consequently, f may be written as

$$f(z) = d \prod_{i=1}^{m} (z^{-1} - \bar{r}_i)(z - r_i) \prod_{j=1}^{s} (z^{-1} - \bar{w}_j)(z - w_j)$$

where $d > 0$, since each pair of terms above is nonnegative on $\mathbb{T}$. Thus, letting

$$q(z) = \sqrt{d} \prod_{i=1}^{m} (z - r_i) \prod_{j=1}^{s} (z - w_j)$$

we obtain our desired factorization. □

While the above theorem, also known as *Wiener-Hopf factorization*, applies to trigonometric polynomials, the extension to scalar rational transfer functions is straightforward, as the following theorem demonstrates.

Theorem 2.3. *Suppose* $g \in \mathcal{RH}_\infty$, *with no poles or zeros on* $\mathbb{T}$. *Then, there exists* $l \in \mathcal{RH}_\infty$ *such that*

$$g^* g = l^* l$$

and $l^{-1} \in \mathcal{RH}_\infty$.

Proof. *We can write*

$$g(z) = \frac{a(z)}{b(z)}$$

where a and b are polynomials. Consequently, we have

$$g^{\sim}(z)g(z) = \frac{a^{\sim}(z)a(z)}{b^{\sim}(z)b(z)}$$

Using Theorem 2.2, we can find spectral factors for both the numerator and denominator, so that

$$a^{\sim}(z)a(z) = \alpha^{\sim}(z)\alpha(z)$$
$$b^{\sim}(z)b(z) = \beta^{\sim}(z)\beta(z)$$

where α *and* β *are the same order and have all their roots in* $\mathbb{D}$. *Consequently, we let*

$$l(z) = \frac{\alpha(z)}{\beta(z)}$$

Then, $g^* g = l^* l$, *and l has all its poles and zeros in* $\mathbb{D}$, *so that* $l, l^{-1} \in \mathcal{RH}_\infty$. □

With this result, we can extend our previous discussion on solving the optimality condition (2.8) to the case of scalar transfer functions. Specifically, suppose that

$$G(z) = \frac{a(z)}{b(z)}, \qquad U(z) = \frac{c(z)}{d(z)}$$

Then, to solve (2.9), we can find a spectral factorization

$$L^* L = G^* G$$

such that $L, L^{-1} \in \mathcal{RH}_\infty$. Suppose that

$$L(z) = \frac{\alpha(z)}{\beta(z)}$$

Then, since L is invertible, the optimality condition is equivalent to

$$L^{-*}G^{*}U + LQ = L^{-*}\Lambda$$

Once again, since $LQ \in \mathcal{RH}_2$ and $L^{-*}\Lambda \in \mathcal{H}_2^{\perp}$, we can decouple these terms and solve directly for LQ. To this end, we have

$$\mathbb{P}_{\mathcal{H}_2}(L^{-*}G^{*}U) + LQ = 0$$

To determine $\mathbb{P}_{\mathcal{H}_2}(\cdot)$, the projection of the first term onto $\mathcal{H}_2$, we can write this term as

$$L^{-*}G^{*}U = \frac{\beta^{\sim}(z)a^{\sim}(z)c(z)}{\alpha^{\sim}(z)b^{\sim}(z)d(z)}$$

Using a partial fraction decomposition of this term will allow us to write it as the sum of a stable transfer function and an anti-stable transfer function:

$$L^{-*}G^{*}U = \frac{n_s(z)}{d_s(z)} + \frac{n_a(z)}{d_a(z)}, \qquad \frac{n_s(z)}{d_s(z)} \in \mathcal{H}_2, \quad \frac{n_a(z)}{d_a(z)} \in \mathcal{H}_2^{\perp}$$

Thus, the projection onto $\mathcal{H}_2$ is simply the stable term. Lastly, since $L^{-1} \in \mathcal{RH}_\infty$, then the solution for $Q \in \mathcal{RH}_\infty$ is

$$Q = -L^{-1}\mathbb{P}_{\mathcal{H}_2}(L^{-*}G^{*}U) = -\frac{\beta(z)n_s(z)}{\alpha(z)d_s(z)}$$

2.5.3 Matrix Transfer Functions

We are now ready to discuss the general form of (2.8), when U and G are matrix transfer functions. Once again, the approach taken here is very similar to our previous discussions. However, finding the spectral factorization of $G^{*}G$ is considerably more complicated than it is in the scalar case. Unfortunately, a full development of the spectral factorization results for this case would take us too far afield. However, the results can be shown by straightforward algebraic manipulations.

Lemma 2.5. *Suppose $U, G \in \mathcal{RH}_\infty$ have the realizations*

$$U = C(zI - A)^{-1}H$$
$$G = z^{-1}(C(zI - A)^{-1}B + D)$$

Suppose there exists a stabilizing solution X to the algebraic Riccati equation

$$X = C^TC + A^TXA - (A^TXB + C^TD)(D^TD + B^TXB)^{-1}(B^TXA + D^TC)$$

Let $W = D^TD + B^TXB$ and $K = W^{-1}(B^TXA + D^TC)$ and $L \in \mathcal{RH}_\infty$ satisfying

$$L = \left[\begin{array}{c|c} A & B \\ \hline W^{1/2}K & W^{1/2} \end{array}\right]$$

Then, $L^{-1} \in \mathcal{RH}_\infty$, $L^{-*} \in \mathcal{H}_\infty^-$, *and*

$$L^*L = G^*G$$

Moreover,

$$L^{-*}G^*U = W^{-1/2}B^T(z^{-1}I - (A-BK)^T)^{-1}XH + zW^{1/2}K(zI-A)^{-1}H$$

Proof. *This result follows from algebraic manipulations of the Riccati equation. A simple proof follows the approach in [9].* □

With the above spectral factorization, we can now solve the optimality condition (2.9) in the general matrix transfer function case.

Lemma 2.6. *Let* $U, G \in \mathcal{RH}_\infty$ *be defined as in Lemma 2.5. Suppose there exists a stabilizing solution* X *to the algebraic Riccati equation*

$$X = C^TC + A^TXA - (A^TXB + C^TD)(D^TD + B^TXB)^{-1}(B^TXA + D^TC)$$

Let K *and* L *be defined as in Lemma 2.5. Then, the unique* $Q \in \mathcal{RH}_\infty$ *satisfying*

$$G^*U + G^*GQ \in \mathcal{H}_2^\perp$$

is given by

$$Q = -zK(zI - (A-BK))^{-1}H$$

Proof. *From Lemma 2.5, we have the spectral factorization* $G^*G = L^*L$. *Since we showed that* $L^{-*} \in \mathcal{H}_\infty^-$, *then* $L^{-*}\mathcal{H}_2^\perp \subset \mathcal{H}_2^\perp$. *Hence, the optimality condition is equivalent to*

$$L^{-*}G^*U + LQ \in \mathcal{H}_2^\perp$$

Since $LQ \in \mathcal{RH}_2$, *we can project the optimality condition onto* $\mathcal{H}_2$ *to obtain*

$$\mathbb{P}_{\mathcal{H}_2}(L^{-*}G^*U) + LQ = 0$$

From Lemma 2.5, we have

$$\mathbb{P}_{\mathcal{H}_2}(L^{-*}G^*U) = zW^{1/2}K(zI-A)^{-1}H$$

Consequently, we have

$$\begin{aligned} Q &= -L^{-1}\mathbb{P}_{\mathcal{H}_2}(L^{-*}G^*U) \\ &= -zK(zI - (A-BK))^{-1}H \end{aligned}$$

□

This completes our discussion of the classical case; that is, when $S = \mathcal{RH}_2$. As we will show in the next section, the solution of the two-player problem relies on the spectral factorization approach that was developed for the classical case.

2.6 Two-Player Solution

We turn back now to the two-player system in (2.3). Our goal is now to find a solution $Q \in \text{lower}(\mathcal{RH}_2)$ that satisfies the optimality condition (2.8). To this end, we have the following result.

Lemma 2.7. *Let $S = \text{lower}(\mathcal{RH}_2)$, and suppose $U, G \in \mathcal{RH}_\infty$. Then, $Q \in S$ satisfies*

$$G^*U + G^*GQ \in S^\perp$$

if and only if both following two conditions hold:

i) $(G^*U)_{22} + (G^*G)_{22}Q_{22} \in \mathcal{H}_2^\perp$;

ii) $\begin{bmatrix} (G^*U)_{11} \\ (G^*U)_{21} \end{bmatrix} + G^*G \begin{bmatrix} Q_{11} \\ Q_{21} \end{bmatrix} \in \mathcal{H}_2^\perp.$

Proof. *Let $G^*U + G^*GQ = \Lambda$ where $\Lambda \in S^\perp$. Note that Λ is partitioned as*

$$\Lambda = \begin{bmatrix} \Lambda_{11} & \Lambda_{12} \\ \Lambda_{21} & \Lambda_{22} \end{bmatrix}$$

where $\Lambda_{11}, \Lambda_{21}, \Lambda_{22} \in \mathcal{H}_2^\perp$. Consequently, (i) comes from the fact that $\Lambda_{22} \in \mathcal{H}_2^\perp$, and (ii) because

$$\begin{bmatrix} \Lambda_{11} \\ \Lambda_{21} \end{bmatrix} \in \mathcal{H}_2^\perp$$

□

The important aspect of Lemma 2.7 is that it decomposes our optimality condition (2.8) over $S^\perp$ into two separate conditions over $\mathcal{H}_2^\perp$. Each of these conditions can be solved via the spectral factorization approach of Lemmas 2.5 and 2.6.

If we now want to apply this spectral factorization approach to our problem, our Riccati equations would be in terms of the precompensator F. However, this difficulty can be avoided if we apply the following result.

Lemma 2.8. *Let $A_F = A + BF$ and $C_F = C + DF$. Then, X is a stabilizing solution of the algebraic Riccati equation*

$$X = C^TC + A^TXA - (A^TXB + C^TD)(D^TD + B^TXB)^{-1}(B^TXA + D^TC)$$

if and only if X is a stabilizing solution of the algebraic Riccati equation

$$X = C_F^T C_F + A_F^T X A_F - (A_F^T X B + C_F^T D)(D^T D + B^T X B)^{-1}(B^T X A_F + D^T C_F)$$

for any matrix F.

Proof. *By substitution of A_F and C_F, it can be readily shown that the two Riccati equations are equivalent.* □

Thus, our ability to solve the optimality condition (2.8) is independent of our choice of the precompensator F.

For convenience, we define

$$E_1 = \begin{bmatrix} I \\ 0 \end{bmatrix}, \qquad E_2 = \begin{bmatrix} 0 \\ I \end{bmatrix},$$

where the dimensions are defined by the context. We can now solve for the $Q \in S$ satisfying the optimality condition (2.8)

Lemma 2.9. *For the system in (2.2), suppose $C^T D = 0$ and $D^T D > 0$. Suppose (A_{11}, B_{11}) and (A_{22}, B_{22}) are stabilizable, and let F_1, F_2 be matrices such that $A_{11} + B_{11}F_1$ and $A_{22} + B_{22}F_2$ have stable eigenvalues. Furthermore, suppose there exist stabilizing solutions X and Y to the algebraic Riccati equations*

$$X = C^T C + A^T X A - A^T X B (D^T D + B^T X B)^{-1} B^T X A \tag{2.11}$$

$$Y = C_2^T C_2 + A_{22}^T Y A_{22} - A_{22}^T Y B_{22} (D_2^T D_2 + B_{22}^T Y B_{22})^{-1} B_{22}^T Y A_{22} \tag{2.12}$$

Define

$$K = (D^T D + B^T X B)^{-1} B^T X A \tag{2.13}$$

$$J = (D_2^T D_2 + B_{22}^T Y B_{22})^{-1} B_{22}^T Y A_{22} \tag{2.14}$$

and let

$$A_F = A + BF$$
$$A_K = A - BK$$
$$A_J = A_{22} - B_{22} J$$

Finally, let N_{11} and N_{12} be defined as in Lemma 2.2. Then, the unique optimal $Q \in S$ for (2.6) is given by

$$Q_{22} = \left[\begin{array}{c|c} A_J & A_J H_2 \\ \hline -J - F_2 & -(J+F_2)H_2 \end{array}\right] \tag{2.15}$$

$$\begin{bmatrix} Q_{11} \\ Q_{21} \end{bmatrix} = \left[\begin{array}{c|c} A_K & A_K E_1 H_1 \\ \hline -K - F & -(K+F)E_1 H_1 \end{array}\right] \tag{2.16}$$

Proof. *From Lemma 2.4, we know that the optimal $Q \in S$ for (2.6) satisfies the optimality condition (2.8). Using Lemma 2.7, this can be solved as two separate problems. Condition (i) of the lemma can be solved via Lemma 2.6, where*

$$U = (C_2 + D_2 F_2)(zI - (A_F)_{22})^{-1} H_2$$
$$G = z^{-1}\left((C_2 + D_2 F_2)(zI - (A_F)_{22})^{-1} B_{22} + D_2\right)$$

to obtain the optimal Q_{22} in (2.15). Note that (2.12) and Lemma 2.8 imply the existence of the required algebraic Riccati equation needed in Lemma 2.6, for whatever stabilizing F is chosen. A similar argument is used to solve for Q_{11} and Q_{21} in condition (ii) of Lemma 2.7, via Lemma 2.6, where we let $U = N_{11}E_1$ and $G = N_{12}$. □

With the optimal $Q \in S$ obtained in Lemma 2.9, the optimal controller for our decentralized problem in (2.3) can be found with the following result.

Theorem 2.4. *For the system in (2.2), suppose the conditions of Lemma 2.9 hold, with X, Y, K, J defined by the Riccati equations (2.11–2.14). Let $A_K = A - BK$. Then, the unique optimal $\mathcal{K} \in \mathrm{lower}(\mathcal{RL}_\infty)$ for (2.3) is*

$$\mathcal{K} = \begin{bmatrix} -K_{11} - K_{12}\Phi & 0 \\ -K_{21} - (K_{22} - J)\Phi & -J \end{bmatrix} \tag{2.17}$$

where

$$\Phi = (zI - (A_K)_{22})^{-1}(A_K)_{21}$$

Proof. *From Lemma 2.9, the unique optimal $Q \in S$ for (2.6) is given by (2.15) and (2.16). Lemma 2.3 then implies that $\hat{Q} = QN_{21}^{-1}$ is optimal for (2.7), where N_{21} is defined in Lemma 2.2. Using Lemma 2.2, the unique optimal $\mathcal{K}$ for (2.3) is given by*

$$\mathcal{K} = \hat{Q}(I + M\hat{Q})^{-1} + F$$

with M defined in the lemma, and the result follows. □

Proof of Theorem 2.1. The result follows directly from Theorem 2.4, where we let

$$A^{\mathcal{K}} = (A_K)_{22} = A_{22} - B_{21}K_{12} - B_{22}K_{22}$$
$$B^{\mathcal{K}} = (A_K)_{21} = A_{21} - B_{21}K_{11} - B_{22}K_{21}$$

2.7 Estimation Structure

Notice that the solution in (2.17) is given in terms of the transfer function Φ. While this answer is sufficient to run the controller in practice, we can obtain some additional insight to the optimal policy by further analysis. Accordingly, we let

$$\eta = \Phi x_1$$

This represents the following state-space system:

$$\eta(t+1) = (A_K)_{22}\eta(t) + (A_K)_{21}x_1(t)$$

with the initial condition $\eta(0) = 0$. Consequently, the optimal policy can be written as

$$\begin{bmatrix} u_1(t) \\ u_2(t) \end{bmatrix} = -\begin{bmatrix} K_{11} & 0 & K_{12} \\ K_{21} & J & K_{22} - J \end{bmatrix} \begin{bmatrix} x_1(t) \\ x_2(t) \\ \eta(t) \end{bmatrix}$$

Combining this with the dynamics in (2.1), it is straightforward to show that the closed-loop dynamics of the overall system become

$$\begin{bmatrix} x_1(t+1) \\ \eta(t+1) \\ x_2(t+1) \end{bmatrix} = \begin{bmatrix} (A_K)_{11} & (A_K)_{12} & 0 \\ (A_K)_{21} & (A_K)_{22} & 0 \\ (A_K)_{21} & (A_K)_{22} - A_J & A_J \end{bmatrix} \begin{bmatrix} x_1(t) \\ \eta(t) \\ x_2(t) \end{bmatrix} + \begin{bmatrix} H_1 & 0 \\ 0 & 0 \\ 0 & H_2 \end{bmatrix} \begin{bmatrix} w_1(t) \\ w_2(t) \end{bmatrix} \tag{2.18}$$

With this system in mind, we now attempt to construct the minimum-mean square error estimator of $x_2(t)$ based on measurements of $x_1(0),\dots,x_1(t)$ and $\eta(0),\dots,\eta(t)$. This is given by the conditional mean

$$\mathrm{E}\big(x_2(t) \mid x_1(0),\dots,x_1(t),\eta(0),\dots,\eta(t)\big)$$

To this end, we have the following lemma.

Lemma 2.10. *Suppose x_1, x_2 represent the state of the following autonomous system driven by noise*

$$\begin{bmatrix} x_1(t+1) \\ x_2(t+1) \end{bmatrix} = \begin{bmatrix} A_{11} & 0 \\ A_{21} & A_{22} \end{bmatrix} \begin{bmatrix} x_1(t) \\ x_2(t) \end{bmatrix} + \begin{bmatrix} w_1(t) \\ w_2(t) \end{bmatrix} \tag{2.19}$$

where $x_1(0), x_2(0), w_1(t), w_2(t)$ are independent, zero-mean random variables for all $t \geq 0$. Define μ_t such that

$$\mu_t(z_0,\dots,z_t) = \mathrm{E}\big(x_2(t) \mid x_1(0) = z_0,\dots,x_1(t) = z_t\big)$$

Then, $\mu_0(z_0) = 0$, and for each $t \geq 0$,

$$\mu_{t+1}(z_0,\ldots,z_{t+1}) = A_{22}\mu_t + A_{21}z_t$$

Proof. *For each t, let q_t be the conditional probability density function*

$$q_t(y) = p^{x_2(t)|x_1(0)\cdots x_1(t)}(y)$$

so that μ_t is the mean of this distribution. Since $x_1(0)$ and $x_2(0)$ are independent, then clearly $\mu_0(z_0) = 0$.

Now, using our definition for q_{t+1} and Bayes' law, we can show that

$$q_{t+1}(y) = \int_v g^{x_2(t+1)|x_1(t)x_2(t)}(y)\, q_t(v)\, dv$$

where $g^{x_2(t+1)|x_1(t)x_2(t)}$ is the transition pdf of x_2 defined by (2.19). *Consequently, we can recursively compute the mean μ_{t+1} as*

$$\mu_{t+1}(z_0,\ldots,z_{t+1}) = A_{22}\mu_t(z_0,\ldots,z_t) + A_{21}z_t$$

The result follows by induction. □

With Lemma 2.10, a very simple representation for the optimal controller can be obtained.

Theorem 2.5. *Suppose x_1, x_2, η are the states of the autonomous system in* (2.18). *Then,*

$$\eta(t) = \mathrm{E}\big(x_2(t) \mid x_1(0),\ldots,x_1(t)\big)$$

Proof. *From* (2.18)*, we see that the state transition matrix is lower triangular. Thus, we can use the results of Lemma 2.10 to get*

$$\begin{aligned}
\mu_{t+1} &= \mathrm{E}\big(x_2(t+1) \mid x_1(0),\ldots,x_1(t+1),\eta(0),\ldots,\eta(t+1)\big) \\
&= A_J\mu_t + \begin{bmatrix}(A_K)_{21} & (A_K)_{22} - A_J\end{bmatrix}\begin{bmatrix}x_1(t) \\ \eta(t)\end{bmatrix} \\
&= \eta(t+1) + A_J\big(\mu_t - \eta(t)\big)
\end{aligned}$$

where we have used the definition of $\eta(t+1)$ in the last expression. Now, since $\mu_0 = \eta(0) = 0$, we inductively see that $\mu(t) = \eta(t)$ for all $t \geq 0$. Lastly, since $\eta(t)$ can be deterministically computed given $x_1(0),\ldots,x_1(t)$, we have

$$\begin{aligned}
\eta(t) &= \mathrm{E}\big(x_2(t) \mid x_1(0),\ldots,x_1(t),\eta(0),\ldots,\eta(t)\big) \\
&= \mathrm{E}\big(x_2(t) \mid x_1(0),\ldots,x_1(t)\big)
\end{aligned}$$

as desired. □

Having established the form of the optimal controller, a number of remarks are in order.

We have shown that the optimal controllers are in terms of the minimum-mean square error estimate of x_2 given the history of x_1. In other words, letting $\eta(t) = \mathrm{E}\big(x_2(t) \mid x_1(0),\dots,x_1(t)\big)$, the optimal control policy can be written as

$$\begin{aligned} u_1(t) &= -K_{11}x_1(t) - K_{12}\eta(t) \\ u_2(t) &= -K_{21}x_1(t) - K_{22}\eta(t) + J\big(\eta(t) - x_2(t)\big) \end{aligned} \tag{2.20}$$

Thus, the optimal policy is, in fact, attempting to perform the optimal centralized policy, though using η instead of x_2. However, there is an additional term in u_2 which represents the error between x_2 and its estimate η. We also see that in the case where x_2 is deterministic, so that $\eta = x_2$, then the optimal distributed controller reduces to the optimal centralized solution, as it should.

In addition, with the inclusion of η, the optimal controller is not a static gain, despite the fact that we have state feedback in each subsystem and player 2 has complete state information. Contrast this result with the classical LQR controller in which the optimal centralized controller would be the static gain K. In fact, both controllers have dynamics, and each has the same number of states as system 2.

2.8 Examples

We conclude our discussion of this two-player problem with a couple of examples. In the first example, we compare the optimal decentralized policy with a standard heuristic. The second example compares the centralized and decentralized policies for a particular system.

2.8.1 A Standard Heuristic

It is worth comparing the optimal decentralized solution that was obtained in (2.20) with a standard heuristic solution to this problem. To motivate the heuristic, consider the classic centralized problem. In the state feedback case, it is known that the optimal policy is the static gain

$$u(t) = -Kx(t)$$

where K is the same gain, given by (2.13) and the associated Riccati equation (2.11). In this case, when the state is not known directly, but is instead measured via some noisy output, then the optimal policy becomes

$$u(t) = -Kx^{\mathrm{est}}(t)$$

where x^{est} is the minimum mean square error estimate of x. In other words, the state x is replaced by its estimate in the optimal policy. This is the certainty equivalence principle for the classic LQG problem [31].

Following this logic, an intuitively reasonable heuristic policy for the two-player problem here would be

$$u_1(t) = -K_{11}x_1(t) - K_{12}x_2^{est}(t)$$
$$u_2(t) = -K_{21}x_1(t) - K_{22}x_2(t)$$

For this heuristic policy, since system 1 cannot directly measure state 2, it uses an estimate of x_2 based on the measurements of its own state; this matches the optimal decentralized solution. Additionally, system 2 has complete state information, so it should not need to estimate anything and uses the centralized LQR policy.

However, it turns out that this heuristic policy can perform arbitrarily poorly. To see this, consider the simple two-player system with the following system matrices:

$$A = \begin{bmatrix} 0 & 0 \\ 0 & 2 \end{bmatrix}, \quad B = \begin{bmatrix} 0.1 & 0 \\ 1 & 1 \end{bmatrix}, \quad C = \begin{bmatrix} I \\ 0 \end{bmatrix}, \quad D = \begin{bmatrix} 0 \\ I \end{bmatrix}$$

As a result, the centralized gain is found to be

$$K \approx \begin{bmatrix} 0 & -0.8 \\ 0 & -0.8 \end{bmatrix}$$

For the heuristic approach, it is straightforward to show that the closed-loop dynamics for player 2 evolve as

$$x_2(t+1) = (2-0.8)x_2(t)$$

which is clearly unstable. Thus, the heuristic approach can destabilize a system.

The optimal decentralized solution given here provides the correct structure for the optimal policy. While the motivation for the heuristic is reasonable and intuitive, this intuition is misleading. A more accurate rationale for the optimal policy is that player 2 must correct for errors that player 1 makes in estimation.

2.8.2 Decentralized Policy

For the second example, consider the following system of two decoupled masses.

$$A = \begin{bmatrix} 1 & .05 & 0 & 0 \\ 0 & 1 & 0 & 0 \\ 0 & 0 & 1 & .05 \\ 0 & 0 & 0 & 1 \end{bmatrix} \qquad B = \frac{1}{100} \cdot \begin{bmatrix} 0.1 & 0 \\ 5 & 0 \\ 0 & 0.1 \\ 0 & 5 \end{bmatrix} \qquad H = 0.01 \cdot I$$

The cost objective has the form

$$100 \cdot (x_1 - x_2)^2 + 0.001 \cdot (x_1^2 + x_2^2 + \dot{x}_1^2 + \dot{x}_2^2) + \mu(u_1^2 + u_2^2)$$

Here, μ is a parameter that we will vary. Note that this cost penalizes the difference between the two masses' positions. We will use μ to trade-off the cost of error with the control effort required. The second term above is simply a regularization term, which drives the system back to zero.

Figure 2.3 shows the state trajectories in the case where $\mu = 0.01$. Here, we apply three impulses to the system: an impulse to system 1 at $t = 50$, an impulse to system 2 at $t = 200$, and an impulse to both systems at $t = 300$. Notice the difference

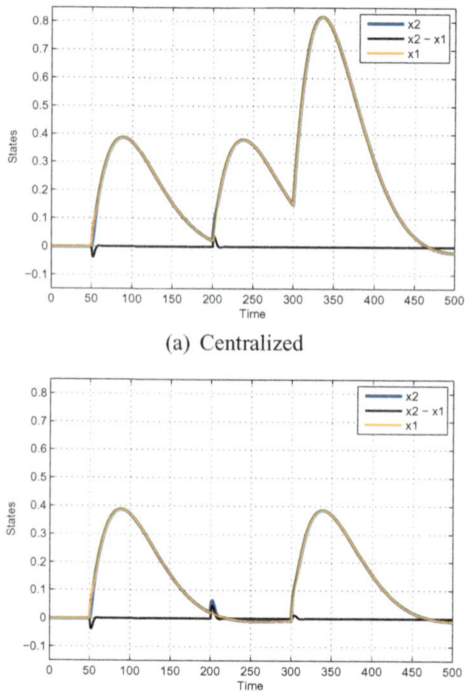

(a) Centralized

(b) Decentralized.

Fig. 2.3 State trajectories.

between the two plots. For the centralized problem in Fig. 2.3(a), the trajectories resulting from the first two impulses are very similar since both players can observe the other's change in position. At the last impulse again, they can observe both impulses and move together.

Contrast this with the decentralized case in Fig. 2.3(b). For the first impulse, since both systems can observe the change in system 1's position, there is not much difference between the centralized and decentralized cases. However, at the second impulse, only player 2 can observe the position error, so it must quickly return to zero in order to avoid errors. At the third impulse, system 1 behaves as it did at the first impulse; recognizing this, system 2 behaves to follow system 1. Consequently, the first and third impulses result in very similar trajectories in the decentralized case.

An alternate way to view the optimal decentralized policy is by looking at the trade-off between position error and control authority as we vary μ. This is shown in Fig. 2.4(a), and demonstrates that the optimal decentralized policy is not significantly different from the centralized policy.

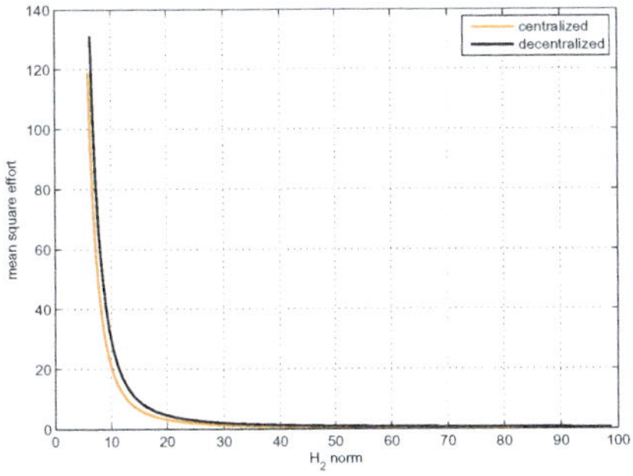

(a) Cost trade-off

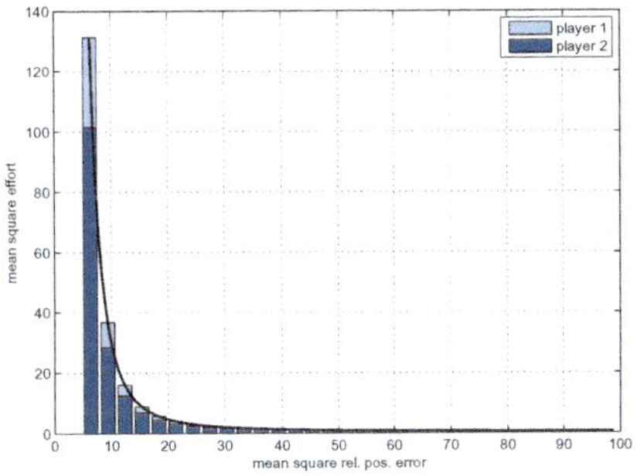

(b) Input effort.

Fig. 2.4 Controller effort.

Figure 2.4(b) shows the relative control effort for both players in the decentralized policy. Notice that player 2 accounts for the vast majority of control authority used, since it is responsible for more information.

2.9 Conclusion

In this chapter, we solved a decentralized, two-player control problem with a particular information structure. The development of this solution required a number of steps. To begin, it was shown that stabilization of the whole system was possible if and only if each of the individual subsystems could be stabilized. Using a particular Youla parametrization for the optimization problem, we converted the problem into a convex one. In contrast to other results for this problem which involve numerical approximations or SDP methods, a spectral factorization approach was developed. This approach admits an explicit state-space solution and provides significant insight into the optimal policies, which was previously unknown. In particular, it was shown that both players implement estimators, and the dimension of the optimal controllers is equal to the dimension of player 2's state.

Perhaps the most important aspect of this result is that it can be extended to more general networks. A detailed discussion of some recent results can be found in [21]. Future work will consider systems with output feedback and time delays. It will also be interesting to see if the spectral factorization results here can be extended to other system norms.

References

1. Bamieh, B., Voulgaris, P.G.: Optimal distributed control with distributed delayed measurements. Proc. IFAC World Congress (2002)
2. Blondel, V.D., Tsitsiklis, J.N.: A survey of computational complexity results in systems and control. Automatica **36**(9), 1249–1274 (2000)
3. Boyd, S., Barratt, C.: Linear Controller Design: Limits of Performance. Prentice Hall, Englewood Cliffs, NJ (1991)
4. Desoer, C., Liu, R.W., Murray, J., Saeks, R.: Feedback system design: The fractional representation approach to analysis and synthesis. IEEE Trans. Automatic Control **25**(3), 399–412 (1980)
5. Gupta, V.: Distributed estimation and control in networked systems. Ph.D. thesis, California Institute of Technology, Pasadena, CA (2006)
6. Ho, Y.C., Chu, K.C.: Team decision theory and information structures in optimal control problems – Part I. IEEE Trans. Automatic Control **17**(1), 15–22 (1972)
7. Ikeda, M., Šiljak, D.D., White, D.E.: Decentralized control with overlapping information sets. J. Optimization Theory and Applications **34**(2), 279–310 (1981)
8. Luenberger, D.: Optimization by Vector Space Methods. John Wiley & Sons, Inc., New York (1969)
9. Mee, D.H.: Factorisation result for optimal discrete-time systems. Electronics Letters **6**(8), 233–234 (1970)
10. N. R. Sandell, J., Athans, M.: Solution of some nonclassical LQG stochastic decision problems. IEEE Trans. Automatic Control **19**(2), 108–116 (1974)
11. N. R. Sandell, J., Varaiya, P., Athans, M., Safonov, M.G.: Survey of decentralized control methods for large scale systems. IEEE Trans. Automatic Control **AC-23**(2), 108–128 (1978)
12. Qi, X., Salapaka, M., Voulgaris, P., Khammash, M.: Structured optimal and robust control with multiple criteria: A convex solution. IEEE Trans. Automatic Control **49**(10), 1623–1640 (2004)
13. Rantzer, A.: Linear quadratic team theory revisited. In: Proc. American Control Conference, pp. 1637–1641. Minneapolis, MN (2006)
14. Rosenblatt, M.: A multi-dimensional prediction problem. Arkiv för Matematik **3**, 407–424 (1958)
15. Rosenblum, M., Rovnyak, J.: The factorization problem for nonnegative operator valued functions. Bull. American Mathematical Society **77**, 287–318 (1971)
16. Rotkowitz, M., Cogill, R., Lall, S.: A simple condition for the convexity of optimal control over networks with delays. In: Proc. IEEE Conference on Decision and Control, pp. 6686–6691. Seville, Spain (2005)
17. Rotkowitz, M., Lall, S.: A characterization of convex problems in decentralized control. IEEE Trans. Automatic Control **51**(2), 274–286 (2002)
18. Rotkowitz, M.: Information structures preserved under nonlinear time-varying feedback. In: Proc. American Control Conference, pp. 4207–4212. Minneapolis, MN (2006)
19. Scherer, C.W.: Structured finite-dimensional controller design by convex optimization. Linear Algebra and its Applications **351**(352), 639–669 (2002)
20. Shah, P., Parrilo, P.: A partial order approach to decentralized control. In: Proc. IEEE Conference on Decision and Control, pp. 4351–4356. Cancun, Mexico (2008)

21. Swigart, J.: Optimal controller synthesis for decentralized systems. Ph.D. thesis, Stanford University, Stanford, CA (2010)
22. Swigart, J., Lall, S.: An explicit dynamic programming solution for a decentralized two-player optimal linear-quadratic regulator. In: Proc. Mathematical Theory of Networks and Systems. Budapest, Hungary (2010)
23. Swigart, J., Lall, S.: An explicit state-space solution for a decentralized two-player optimal linear-quadratic regulator. In: Proc. American Control Conference. Baltimore, MD (2010)
24. Swigart, J., Lall, S.: A graph-theoretic approach to distributed control over networks. In: Proc. IEEE Conference on Decision and Control. Shanghai, China (2009)
25. Swigart, J., Lall, S.: Optimal synthesis and explicit state-space solution for a decentralized two-player linear-quadratic regulator. In: Proc. IEEE Conference on Decision and Control. Atlanta, GA (2010)
26. Voulgaris, P.: Control of nested systems. In: Proc. American Control Conference, vol. 6, pp. 4442–4445. Chicago, IL (2000)
27. Wang, S.H., Davison, E.J.: On the stabilization of decentralized control systems. IEEE Trans. Automatic Control **AC-18**(5), 473–478 (1973)
28. Witsenhausen, H.S.: A counterexample in stochastic optimum control. SIAM J. Control **6**(1), 131–147 (1968)
29. Youla, D., Jabr, H., Bongiorno Jr., J.: Modern Wiener–Hopf design of optimal controllers–Part II: The multivariable case. IEEE Trans. Automatic Control **21**(3), 319–338 (1976)
30. Zelazo, D., Mesbahi, M.: $\mathcal{H}_2$ analysis and synthesis of networked dynamic systems. In: Proc. American Control Conference, pp. 2966–2971. St. Louis, MO (2009)
31. Zhou, K., Doyle, J., Glover, K.: Robust and Optimal Control. Prentice Hall, Upper Saddle River, NJ (1995)

Chapter 3
Decentralized Control with Communication Bandwidth Constraints

Chun Zhang and Geir E. Dullerud

Abstract In this chapter, we investigate the decentralized control problem in the setting of limited bandwidth sensing channels. Specifically, we consider the decentralized stabilization problem of a linear time-invariant (LTI) plant by multiple control stations that receive sensing information through rate-limited channels, and these stations are not capable of communicating with each other directly. The main result of the this chapter is a sufficient condition on the data rate of respective channels to guarantee system stabilizability. We provide an explicit way to construct the associated stabilizing encoder, decoder, and controller. We also present a robustness analysis showing that this control algorithm is structurally robust against model mismatch.

3.1 Introduction

When one builds a geographically-distributed feedback control system that utilizes a communication network, issues of bandwidth limitation, latency, and packet loss become inevitable challenges, adding to the challenge already presented by structural constraints imposed by the communication graph. It is particularly important that these information exchange issues be systematically addressed when aggressive network control schemes are to be deployed. In this chapter, we investigate a stabilization problem in which a multistation control system operating over rate-limited data links must stabilize a plant through the collaboration of its subsys-

Chun Zhang
Cisco Systems Inc., Research Triangle Park, NC 27709, USA
e-mail: chunzha@cisco.com

Geir E. Dullerud
Department of Mechanical Science and Engineering, University of Illinois at Urbana-Champaign, 340 Mechanical Engineering Building, 1206 West Green Street, Urbana, IL 61801, USA
e-mail: dullerud@illinois.edu

R. Johansson and A. Rantzer (Eds.): Distributed Decision Making and Control, LNCIS 417, pp. 55–81.
springerlink.com

tems, but where these control stations do not have the capability to communicate directly with each other. Namely, in the formulation studied the decentralized controller structure poses a *topological* constraint on information exchange; meanwhile, the limited bandwidth of the communication channels over which system measurements are transmitted to controllers (without latency or packet loss) give rise to a *nontopological* limitation. Our goal is to develop a stabilizing algorithm that can simultaneously accommodate these two different types of constraints.

The problem of stabilizing an LTI system with a centralized sensor and a centralized controller connected by a bandwidth-limited channel has been extensively studied; see [3], [7], [11], [12], [16] and references therein. The important information lower bound $R = \Sigma_{\lambda(A)} \max(0, \log|\lambda(A)|)$, where $\lambda(A)$ denotes the eigenvalue of the system matrix, has been derived by several researchers. It is also proved to be tight in [12]. In a slightly different setting, a simple quantizer is constructed to obtain a looser upper bound, [7]. This simple quantizer plays an important role in deriving our results here. In [8] and [13], the problem of stabilizing an LTI system with distributed sensors and a centralized controller is solved via the Slepian–Wolf coding approach, where the centralized decoder is the key point in the derivation.

In all above works, a centralized control station (decoder and controller) is assumed. However, when one is dealing with large-scale systems, centralized control is prohibitive due to quickly increasing measurement costs and system complexity. Decentralized or distributed control is the natural alternative. The problem of decentralized stabilization subject to finite data rate feedbacks has been considered in [10], where necessary and sufficient conditions on stabilizing data rates are derived for systems with a *diagonalizable* system matrix. The problem of decentralized control over capacity constrained networks is studied in [9], where an existence condition on stabilizability is given. A similar problem as ours is investigated via an information theoretical approach in [15].

In this chapter, we study the system depicted in Fig. 3.1. A generic LTI plant G with no special structural assumption is controlled by v decentralized control stations. The state-space representation of the system is given in Sec. 3.2.

In the established research literature on decentralized control, one assumes the measurement y_i of the LTI plant G is available to the ith control station precisely and instantaneously. It is shown that the system can be stabilized by a set of LTI controllers if and only if all *decentralized fixed modes*[1] *(DFMs)* are stable; see [2] and [14]. Meanwhile, if time-varying controllers are allowed, a larger class of systems can be stabilized under the decentralized information structure; see [1], [4], and [5]. Information exchange between stations is the key in these approaches.

In this chapter, instead of the perfect measurement, we consider the decentralized stabilization problem in the setting of limited bandwidth sensing channels. Each local measurement y_i is quantized, encoded and then sent to its respective control station over a noiseless digital memoryless channel with capacity R_i bits per step, without latency or packet loss. We derive coupled *upper bounds* on the data rates required on each channel to guarantee global asymptotic stabilizability.

[1] Eigenvalues of the system matrix that cannot be moved by any controller with the decentralized information structure.

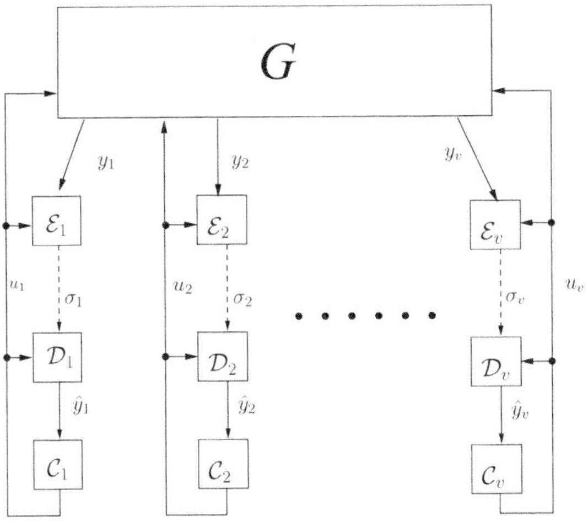

Fig. 3.1 Decentralized system over communication channels with v control stations

We also provide an explicit way to construct the associated stabilizing encoder, decoder, communication code and controller. This unique feature of our work makes the developed algorithm easy to implement. Furthermore, we also show that our stabilizing algorithm is robust against model mismatch. This is very important in implementation, since model mismatch between plant and controllers is generally unavoidable, and our result shows that the stabilization property provided by the presented controller is a well-posed one.

The rest of the chapter is organized as follows. Section 3.2 sets up the problem formally and introduces the notation used. Section 3.3 presents the main result of the chapter, where the decentralized stabilization problem is solved in a two-station setting. The stabilizing algorithm is developed along with the design of stabilizing encoder, decoder, and controller. Robustness analysis of the control algorithm against model mismatch is presented in Sec. 3.4. Then, Sec. 3.5 discusses extending the algorithm to multistation systems. The main result is illustrated with an example in Sec. 3.6. Some concluding remarks are given in the last section.

3.2 Problem Set-Up

The set of integers, non-negative integers, natural numbers, real numbers and non-negative real numbers is denoted by $\mathbb{Z}$, $\mathbb{N}_0$, $\mathcal{N}$, $\mathbb{R}$ and $\mathbb{R}^+$, respectively.

The infinity norm of a vector $x \in \mathbb{R}^n$ is defined as $\|x\|_\infty := \max_{1\leq i\leq n} |x_i|$. The infinity norm of a matrix $A \in \mathbb{R}^{n\times m}$ is defined as $\|A\|_\infty := \max_{1\leq i\leq n} \sum_{j=1}^{m} |A_{ij}|$. We denote $B_\infty^{p_i}(a,b)$ as a p_i-dimensional hypercube, which is centered at point a with edge length $2b$.

The decentralized control system with v control stations as shown in Fig. 3.1 has the following state-space representation:

$$x(t+1) = Ax(t) + \sum_{i=1}^{v} B_i u_i(t), \quad \|x(0)\|_\infty \le E_0$$
$$y_i(t) = C_i x(t), \qquad i = 1,2,\ldots,v \tag{3.1}$$

where state $x \in \mathbb{R}^n$, controls $u_i \in \mathbb{R}^{m_i}$, measurements $y_i \in \mathbb{R}^{p_i}$; and A, B_i, C_i are compatible real matrices.

We assume the system is unstable to make the problem non-trivial; that is, $1 \le \Lambda := \|A\|_\infty$. We also assume centralized controllability and observability, that is, the matrix pair $(A, [B_1, \cdots, B_v])$ is controllable, and the pair $([C_1^T, \cdots, C_v^T]^T, A)$ is observable. Otherwise, there is no way to stabilize the system with decentralized controller. To avoid trivial cases, we assume no single station enjoys joint controllability and observability. In other words, no (C_i, A, B_i) triple for all $i = 1, 2, \cdots, v$ is both controllable and observable. Collaboration among stations is necessary for stabilization purposes.

We use $\mathcal{K}_i = [C_i|A]$ to denote the unobservable subspace of the pair (C_i, A), and use $\mathcal{R}_i = \langle A|B_i\rangle$ to denote the controllable subspace of (A, B_i). If $\mathcal{R}_i \not\subseteq \mathcal{K}_j$, then we say Station i is connected to Station j, in the sense that control action from station i can be observed by Station j, [5]. The system is *strongly connected* if all stations are connected to each other, possibly through intermediate stations, [2]. Strong connectivity together with joint controllability and observability ensures the decentralized stabilization problem is solvable with time varying control laws; see reference [1], [4] and [5].

Remark 3.1. A non-strongly connected system can be partitioned into a unique set of strongly connected subsystems, and if treating each subsystem as a single unit, the resulting system is called a *quotient decentralized system* (QDS). Assuming joint controllability and observability, if either the system is strongly connected or the QDS has no unstable DFM, the decentralized stabilization problem can be solved with time varying control laws; see [4] and [6].

Define $s_i := \inf\{m \text{ such that } \dim(\cap_{i=0}^{m} \ker CA^i) = n - \dim(\mathcal{K}_i^\perp)\}$, which is the *generalized observability index* of (C_i, A); in other words, the least number of steps needed to observe the state in $\mathcal{K}_i^\perp$ from measurements y_i.

Define

$$W_i := \begin{bmatrix} C_i^T & (C_iA)^T & \cdots & (C_iA^{s_i-1})^T \end{bmatrix}^T$$

where $i = 1, \ldots, v$ and define W_i^+ as its generalized inverse.

We use $x_i(\cdot) = P_i x(\cdot)$ to denote the *projection* of the state vector in $\mathcal{K}_i^\perp$, where P_i is the projection matrix on $\mathcal{K}_i^\perp$ along $\mathcal{K}_i$. Notice that P_i here is not the natural projection since $x_i(\cdot)$ and $x(\cdot)$ are of the same size. We use $\tilde{x}_i(\cdot)$ and $\hat{x}_i(\cdot)$ to denote Station i's estimate of the state $x(\cdot)$ and its projection $x_i(\cdot)$ in $\mathcal{K}_i^\perp$, respectively.

The ith local encoder at time t is a map $\mathcal{E}_i(t) : \mathbb{R}^{p_i} \times \Sigma_i^{[0,t-1]} \times \mathbb{R}^{m_i t}$ to Σ_i, taking values $(y_i(t), \sigma_i[0,t-1], u_i[0,t-1]) \mapsto \sigma_i(t)$, the new codeword. Notation Σ_i is the

ith codeword space and $\Sigma_i^{[0,t-1]}$ denotes the union of all past codeword spaces up to time t. Notation $\sigma_i[0,t-1]$ is used to denote past codewords and $u_i[0,t-1]$ is used to denote all past local controls. The local encoder knows the local decoding policy $\mathcal{D}_i$ but not the local control policy $\mathcal{C}_i$.

The ith local decoder at time t is a map $\mathcal{D}_i(t) : \Sigma_i^{[0,t]} \times \mathbb{R}^{m_i t}$ to $\mathbb{R}^{p_i}$, taking values $(\sigma_i[0,t], u_i[0,t-1]) \mapsto \hat{y}_i(t)$, an estimate of the respective measurement y_i. The decoder knows the local encoding policy $\mathcal{E}_i$ but not the local control policy $\mathcal{C}_i$.

The ith local controller at time t is a map $\mathcal{C}_i(t) : \mathbb{R}^{p_i(t+1)}$ to $\mathbb{R}^{m_i}$, taking values $\hat{y}_i[0,t] \mapsto u_i(t)$, the local control, which depends causally on the local decoder's outputs.

The encoder and the decoder are assumed to have unlimited memory, therefore they can store and use all the past information. A digital noiseless memoryless channel connects each local encoder and decoder pair.

The major difference in our set-up from reference [12] is that the local encoder, decoder controller triple does not have information such as measurements, codewords, or controls of other stations. Thus, extra communication cost incurs when they exchange these local information with their peers. This is not an issue in the centralized single-station system.

3.3 Stabilizing Algorithm

In this section, we assume all local stations have the exact model of the plant. The robustness issue of this algorithm against model mismatch is investigated in the next section. In order to keep the notation clean and describe the algorithm more efficiently, we present the main idea of this chapter in the two-control-station setting, that is, system (3.1) with $\nu = 2$. Direct extension of the algorithm to the multistation case is briefly discussed in Sec. 3.5.

The control algorithm adopted here is motivated by [5]. It is divided into three phases: observation, communication, and control. First, both stations listen to the system with no controls applied in order to compute their own estimates of the initial state. Then these stations exchange their estimates by coding them into control signals. Finally, both controllers use their own noisy estimates of the initial state to design controls, and try to drive the state of the system back to zero.

The logic behind this algorithm is that if the data rate on each channel is high enough, then measurements y_i get sufficiently finely quantized so that each station will have quite accurate information about the initial state. Then the control action taken will at least bring the state to a smaller uncertainty set than the initial one and then eventually lead it to zero.

Without loss of generality, we assume $s_1 \geq s_2$.

3.3.1 Observation

For the observation stage, we have the following lemma.

Lemma 3.1. *Within at most s_i steps, Station i can estimate $x_i(0) = P_i x(0)$ with error bounded by*

$$\frac{\varepsilon_i}{\sqrt[p_i]{N_i}} E_0$$

where $p_i = \dim(y_i)$, $\varepsilon_i = \|W_i^+\|_\infty \|C_i\|_\infty \Lambda^{s_i-1}$ and $N_i = 2^{R_i}$ represents the number of quantization levels on channel i.

Remark 3.2. From now on, we assume that $\sqrt[p_i]{N_i}$ is an integer; otherwise, we simply replace N_i with the smallest integer $\tilde{N}_i \geq N_i$ such that $\sqrt[p_i]{\tilde{N}_i}$ is an integer.

Proof. *Set the control $u_1(t) = u_2(t) \equiv 0$ for all $0 \leq t < s_1$ ($s_1 \geq s_2$ is assumed). Since $y_i(0) = C_i x(0)$, thus $\|y_i(0)\|_\infty \leq \|C_i\|_\infty \|x(0)\|_\infty$. Then it is clear that $y_i(0) \in B_\infty^{p_i}(0, \|C_i\|_\infty E_0)$.*

Therefore, for all $t = 0, 1, \ldots, s_1 - 1$, we have

$$y_i(t) \in B_\infty^{p_i}(0, \|C_i\|_\infty \Lambda^t E_0)$$

Divide the p_i-cube at time t into N_i equal smaller p_i-cubes, let $\sigma_i(t, y_i(t))$ be the codeword that represents the small cube containing the true value of $y_i(t)$, and let $\hat{y}_i(t)$ denote the reconstruction of the codeword by the decoder. We have

$$\|y_i(t) - \hat{y}_i(t)\|_\infty \leq \frac{\|C_i\|_\infty \Lambda^t E_0}{\sqrt[p_i]{N_i}}$$

For convenience, let us define the following shorthand notation

$$\mathcal{Y}_i = \begin{bmatrix} y_i(0) \\ y_i(1) \\ \vdots \\ y_i(s_i - 1) \end{bmatrix}; \quad \hat{\mathcal{Y}}_i = \begin{bmatrix} \hat{y}_i(0) \\ \hat{y}_i(1) \\ \vdots \\ \hat{y}_i(s_i - 1) \end{bmatrix} \tag{3.2}$$

If perfect measurements are available, then from linear system theory, we know that $x_i(0)$ can be computed as follows:

$$x_i(0) = W_i^+ \mathcal{Y}_i$$

However, since Station i only has quantized measurements $\hat{y}_i(\cdot)$, it can only estimate $x_i(0)$ via the following reconstruction with noisy measurements:

$$\hat{x}_i(0) = W_i^+ \hat{\mathcal{Y}}_i \tag{3.3}$$

where the error is bounded by

$$\begin{aligned}
\|x_i(0)-\hat{x}_i(0)\|_\infty &\leq \|W_i^+\|_\infty \max_{0\leq t\leq s_i-1} \|\hat{y}_i(t)-y_i(t)\|_\infty \\
&\leq \frac{\|W_i^+\|_\infty \|C_i\|_\infty \Lambda^{s_i-1}E_0}{\sqrt[p_i]{N_i}} && (3.4)\\
&=: \frac{\varepsilon_i}{\sqrt[p_i]{N_i}}E_0 && (3.5)
\end{aligned}$$

Of course, we want $\varepsilon_i/\sqrt[p_i]{N_i} < 1$*; otherwise,* $\mathbf{0}$ *is a better estimate. Notice that* ε_i *is entirely determined by the open-loop system parameters. Also notice that Eq. (3.4) is valid only when* $\Lambda \geq 1$ *as we assumed.* □

Remark 3.3. Stations 1 and 2 may use different numbers of steps to reconstruct their own estimates of the initial state. If $s_1 > s_2$, Station 2 simply waits after $s_2 - 1$.

Let us consider the error in estimating $x_i(s_i)$ at time $t = s_i - 1$. This result will be used in later derivation. Let $\hat{x}_i(s_i)$ be the s_i-step ahead estimate, then

$$\|x_i(s_i)-\hat{x}_i(s_i)\|_\infty \leq \Lambda^{s_i}\|x_i(0)-\hat{x}_i(0)\|_\infty \leq \frac{\Lambda^{s_i}\varepsilon_i E_0}{\sqrt[p_i]{N_i}} \quad (3.6)$$

3.3.2 Communication

At time $t = s_1$, each station has an estimate $\hat{x}_i(0)$ of the initial state. This information needs to be exchanged between them for stabilizing purposes. From Station i to Station j, we only need to send $(I-P_j)\hat{x}_i(0)$, where $i, j = 1, 2$, and $i \neq j$. The remaining part is directly available to Station j. Since direct communication between stations is not available, one station has to encode its estimate into control signals and the counter party has to decode it based on its local measurements. This is effectively to explore the *perfect* channel through the plant.

Recall that Station i's control action can be observed by Station j only if $\mathcal{R}_i \not\subseteq \mathcal{K}_j$. Assume this is true, then there exists a positive integer t_i[2] such that

$$\alpha_i := \mathtt{rank}\left\{C_j \sum_{\ell=0}^{t_i-1} A^{t_i-1-\ell}B_iB_i^T(A^T)^{t_i-1-\ell}\right\} \neq 0 \quad (3.7)$$

Let β_i be the smallest positive integer such that $\alpha_i\beta_i \geq n$, where $n = \dim(x)$. Define the following *encoding matrix* from Station i to Station j

$$E_{ji} = \begin{bmatrix} C_jM_i(t_i,t_i) & \cdots & 0 \\ \vdots & \ddots & \vdots \\ C_jM_i(\beta_i t_i, t_i) & \cdots & C_jM_i(t_i,t_i) \end{bmatrix} \quad (3.8)$$

[2] As we will see later, the smallest t_i should be used to minimize error propagation.

where $M_i(p,q) = \sum_{\ell=0}^{t_i-1} A^{p-1-\ell} B_i B_i^T (A^{q-1-\ell})^T$. It is clear that $\text{rank}\{E_{ji}\} \geq n$. Thus, there exist matrices S_i and T_i such that $S_i E_{ji} T_i = I_n$. All these matrices can be pre-computed and stored in both local control stations.

We have the following lemma on information exchange between multiple stations when there are *no* nontopological constraints; in other words, all channels are perfect and are able to transmit real numbers precisely and instantaneously.

Lemma 3.2. *Given system (3.1) with no communication constraints, and assume* $\mathcal{R}_i \not\subseteq \mathcal{K}_j$ *for* $i,j = 1,2,\cdots,v$, *then any vector* $\phi \in \mathbb{R}^n$ *can be encoded into a control sequence* $u_i(\ell)$, $0 \leq \ell \leq \beta_i t_i - 1$ *by Station i and decoded exactly by Station j from its measurements* $y_j(kt_i)$, *where* $k = 1,2,\cdots,\beta_i$.

Proof. *Define* $z^p := T_i\phi = \begin{bmatrix} z_1^T & \cdots & z_{\beta_i}^T \end{bmatrix}^T$, *where each* $z_i \in \mathbb{R}^n$. *Then we have* $S_i E_{ji} z^p = \phi$. *We can treat* z^p *as the* communication codeword.

As a discrete-time counterpart to [5], design controls as follows:

$$u_i(\ell) = \begin{cases} B_i^T (A^T)^{t_i-1-\ell} z_1 & \text{for } 0 \leq \ell \leq t_i - 1 \\ B_i^T (A^T)^{2t_i-1-\ell} z_2 & \text{for } t_i \leq \ell \leq 2t_i - 1 \\ \cdots & \\ B_i^T (A^T)^{\beta_i t_i-1-\ell} z_{\beta_i} & \text{for } (\beta_i - 1)t_i \leq \ell \leq \beta_i t_i - 1 \end{cases} \tag{3.9}$$
$$u_j(\ell) = 0 \qquad \text{for } 0 \leq \ell \leq \beta_i t_i - 1$$

Station j can reconstruct ϕ *from its measurements* $y_j(kt_i)$, $k = 1,2,\ldots,\beta_i$ *and the system initial state* $x(0)$ *as follows*

$$\phi = S_i \begin{bmatrix} y_j(t_i) - C_j A^{t_i} x(0) \\ \vdots \\ y_j(\beta_i t_i) - C_j A^{\beta_i t_i} x(0) \end{bmatrix} \tag{3.10}$$

It can be easily verified that we can replace $x(0)$ *with* $x_j(0)$ *in Eq. (3.10) since the part of* $x(0)$ *in* $\mathcal{K}_j$ *is filtered by* C_j.

Figuratively, the above scheme encodes the information into β_i *pieces, while each one of them requires* t_i *steps of transmission time.* □

Now let us consider the information exchange in the two station system ($v = 2$) with the bandwidth constraint described in the previous sections. We have,

Lemma 3.3. *Given* $\mathcal{R}_1 \not\subseteq \mathcal{K}_2$, *Station 2 can reconstruct the initial state* $x(0)$ *at time* $t = s_1 + \beta_1 t_1$ *with error bounded by*[3]

$$\left(\frac{\eta_1}{\sqrt[p_2]{N_2}} + \frac{\varepsilon_1}{\sqrt[p_1]{N_1}} + \frac{\varepsilon_2}{\sqrt[p_2]{N_2}} \right) E_0$$

where ε_1 *and* ε_2 *are given in Lemma 3.1, and*

[3] Notice that all constants are completely determined by the open-loop system parameters.

$$\eta_1 = \|S_1\|_\infty \|C_2\|_\infty \left(\Lambda^{\beta_1 t_1} \left(\frac{\Lambda^{s_1}\varepsilon_2}{\sqrt[p_2]{N_2}} \right) (1 + \sqrt[p_2]{N_2}) + \frac{1-\Lambda^{\beta_1 t_1}}{1-\Lambda} \|B_1\|_\infty \rho_1 \right) \quad (3.11)$$

$$\rho_1 = \|B_1^T\|_\infty \Lambda^{t_1 - 1} \|T_1\|_\infty \left(\frac{\varepsilon_1}{\sqrt[p_1]{N_1}} + 1 \right) \quad (3.12)$$

Proof. *Encode* $(I - P_2)\hat{x}_1(0)$ *into controls* $u_1(\ell)$ *as in Eq. (3.9) and set* $u_2(\ell) \equiv 0$ *for all* $s_1 \le \ell \le s_1 + \beta_1 t_1 - 1$. *It is easy to verify that* $\|u_1(\ell)\|_\infty \le \rho_1 E_0$.

According to Lemma 3.2, if Station 2 knows exact measurements and initial state, it can then reconstruct $(I - P_2)\hat{x}_1(0)$ *exactly. However, due to limited bandwidth, it only has the decoder's estimate* $\hat{y}_2(\cdot)$ *and the noisy estimate* $\hat{x}_2(0)$ *computed at time* $s_2 - 1$. *Thus, it can only compute a noisy estimate* $(I - P_2)\bar{x}_1(0)$ *using the available data, such that*

$$(I - P_2)\bar{x}_1(0) := S_1 \begin{bmatrix} \hat{y}_2(s_1 + t_1) - C_2 A^{s_1 + t_1} \hat{x}_2(0) \\ \vdots \\ \hat{y}_2(s_1 + \beta_1 t_1) - C_2 A^{s_1 + \beta_1 t_1} \hat{x}_2(0) \end{bmatrix}$$

The error between $(I - P_2)\hat{x}_1(0)$ *and* $(I - P_2)\bar{x}_1(0)$ *can be bounded as follows:*

$$\begin{aligned} \varepsilon_\infty &= \|(I - P_2)\bar{x}_1(0) - (I - P_2)\hat{x}_1(0)\|_\infty \\ &\le \|S_1\|_\infty \left\| \begin{bmatrix} \hat{y}_2(s_1 + t_1) - C_2 A^{s_1 + t_1} \hat{x}_2(0) \\ \vdots \\ \hat{y}_2(s_1 + \beta_1 t_1) - C_2 A^{s_1 + \beta_1 t_1} \hat{x}_2(0) \end{bmatrix} \right. \\ &\quad \left. - \begin{bmatrix} y_2(s_1 + t_1) - C_2 A^{s_1 + t_1} x(0) \\ \vdots \\ y_2(s_1 + \beta_1 t_1) - C_2 A^{s_1 + \beta_1 t_1} x(0) \end{bmatrix} \right\|_\infty \\ &\le \|S_1\|_\infty \max_{1 \le k \le \beta_1} \{ \|y_2(s_1 + kt_1) - \hat{y}_2(s_1 + kt_1)\|_\infty \\ &\quad + \|C_2 A^{s_1 + kt_1} x(0) - C_2 A^{s_1 + kt_1} \hat{x}_2(0)\|_\infty \} \end{aligned} \quad (3.13)$$

In order to compute $\|y_2(s_1 + kt_1) - \hat{y}_2(s_1 + kt_1)\|_\infty$, *notice that both the encoder* $\mathcal{E}_2$ *and the decoder* $\mathcal{D}_2$ *can compute the following* $\check{y}_2(\cdot)$, *which is an estimate of the autonomous part of* $y_2(\cdot)$,

$$\check{y}_2(s_1 + kt_1) = C_2 A^{s_1 + kt_1} \hat{x}_2(0), \quad k = 1, \cdots, \beta_1 \quad (3.14)$$

On the other hand, the actual output $y_2(s_1 + kt_1)$ *evolves as follows*

$$y_2(s_1 + kt_1) = C_2 A^{s_1 + kt_1} x(0) + C_2 \sum_{\ell = s_1}^{s_1 + kt_1 - 1} A^{s_1 + kt_1 - 1 - \ell} B_1 u_1(\ell)$$

It is clear that $y_2(s_1+kt_1) \in B_\infty^{p_2}(\check{y}_2(s_1+kt_1), r_1+r_2)$, *where*

$$r_1 = \|C_2\|_\infty \Lambda^{s_1+kt_1}\|x_2(0)-\hat{x}_2(0)\|_\infty \geq \|C_2A^{s_1+kt_1}x(0) - C_2A^{s_1+kt_1}\hat{x}_2(0)\|_\infty$$

$$r_2 = \|C_2 \sum_{\ell=s_1}^{s_1+kt_1-1} A^{s_1+kt_1-1-\ell}B_1u_1(\ell)\|_\infty$$

Divide this p_2*-cube into* N_2 *smaller cubes and pick the one containing the true value of output* $y_2(s_1+kt_1)$*, the error of this coding scheme is then bounded by*

$$\|y_2(s_1+kt_1)-\hat{y}_2(s_1+kt_1)\|_\infty \leq \frac{1}{\sqrt[p_2]{N_2}}(r_1+r_2)$$

$$\leq \|C_2\|_\infty \frac{1}{\sqrt[p_2]{N_2}}\left[\Lambda^{s_1+kt_1}\frac{\varepsilon_2}{\sqrt[p_2]{N_2}} + \frac{1-\Lambda^{kt_1}}{1-\Lambda}\|B_1\|_\infty \rho_1\right] E_0$$

Notice that r_1 *is also the upper bound for the second part of Eq. (3.13). Since we assumed* $\Lambda \geq 1$*, Eq. (3.13) gives the following error bound*

$$\|(I-P_2)\bar{x}_1(0)-(I-P_2)\hat{x}_1(0)\|_\infty \leq \frac{\eta_1}{\sqrt[p_2]{N_2}}E_0$$

where η_1 *is defined in Eq. (3.11).*

Then the error between $(I-P_2)\bar{x}_1(0)$ *and* $(I-P_2)x_1(0)$ *can be bounded by simply applying the triangle inequality*

$$\begin{aligned}\xi_\infty &= \|(I-P_2)\bar{x}_1(0)-(I-P_2)x_1(0)\|_\infty \\ &\leq \|(I-P_2)\bar{x}_1(0)-(I-P_2)\hat{x}_1(0)\|_\infty + \|(I-P_2)\hat{x}_1(0)-(I-P_2)x_1(0)\|_\infty \\ &\leq \frac{\eta_1}{\sqrt[p_2]{N_2}}E_0 + \frac{\varepsilon_1}{\sqrt[p_1]{N_1}}E_0\end{aligned}$$

At this time, Station 2 has decoded enough information from its measurements about the part of the initial state in $\mathcal{K}_2$*. Together with* $\hat{x}_2(0)$ *computed at time* s_2-1*, it can reconstruct the initial state*

$$\tilde{x}_2(0) = P_1^+(I-P_2)\bar{x}_1(0) + P_2^+\hat{x}_2(0)$$

where P_i^+ *denotes the insertion from* $\mathcal{K}_i^\perp$ *to* $\mathbb{R}^n$*.*

Thus, the ultimate error bound between the estimate computed by Station 2 and the real initial state is as follows:

$$\|\tilde{x}_2(0)-x(0)\|_\infty \leq \left(\frac{\eta_1}{\sqrt[p_2]{N_2}} + \frac{\varepsilon_1}{\sqrt[p_1]{N_1}} + \frac{\varepsilon_2}{\sqrt[p_2]{N_2}}\right)E_0$$

□

During time $s_1+\beta_1t_1 \leq t \leq s_1+(\beta_1+1)t_1-1$, Station 1 applies the following controls to the system to offset the previously accumulated control effects

$$u_1(t) = -B_1^T (A^T)^{s_1+(\beta_1+1)t_1-1-t} z$$

$$z := M_1^+(t_1,t_1) \sum_{\ell=s_1}^{s_1+\beta_1 t_1-1} A^{s_1+\beta_1 t_1-1-\ell} B_1 u_1(\ell)$$

where $M_i^+(\cdot,\cdot)$ is the generalized inverse of $M_i(\cdot,\cdot)$ defined in Eq. (3.8).

Meanwhile, set $u_2(t) = 0$. Then at time $t = s_1 + (\beta_1 + 1)t_1$, the system state $x(t)$ is driven to

$$x(s_1 + (\beta_1 + 1)t_1) = A^{s_1+(\beta_1+1)t_1} x(0)$$

If $(h-1)s_1 < s_1 + (\beta_1+1)t_1 \leq hs_1 - 1$ for some $h \in \mathbb{N}_0$, then let $u_1(t) = u_2(t) = 0$ for time $s_1 + (\beta_1+1)t_1 \leq t \leq hs_1 - 1$ in order to take advantage of the repetitive estimation described below.

Both the encoder $\mathcal{E}_1$ and the decoder $\mathcal{D}_1$ know exactly the controls applied up to time hs_1, so they can run a simulating observation process and compute a tighter state estimate. The basic idea is to observe $\hat{y}_1(s_1), \cdots, \hat{y}_1(2s_1 - 1)$, and then estimate $x_1(s_1)$ using the same method as in the proof of Lemma 3.1. By repeating the process, we have the following error bound for s_1-*step ahead* estimate similar to that computed in Eq. (3.6)

$$\begin{aligned} \|x_1(ms_1) - \hat{x}_1(ms_1)\|_\infty &\leq \frac{\|W_1^+\|_\infty^m \|C_1\|_\infty^m \Lambda^{m(2s_1-1)} E_0}{(\sqrt[p_1]{N_1})^m} \\ &= \frac{\Lambda^{ms_1} \varepsilon_1^m E_0}{(\sqrt[p_1]{N_1})^m}, \quad 1 \leq m \leq h \end{aligned}$$

Then we start the information transmission from Station 2 to Station 1 at time $t = hs_1$. We have the following result:

Lemma 3.4. *Given $\mathcal{R}_2 \not\subseteq \mathcal{K}_1$, Station 1 can reconstruct the initial state $x(0)$ at time $t = hs_1 + \beta_2 t_2$ with error bounded by*

$$\left(\frac{\eta_2}{\sqrt[p_1]{N_1}} + \frac{\varepsilon_1}{\sqrt[p_1]{N_1}} + \frac{\varepsilon_2}{\sqrt[p_2]{N_2}} \right) E_0$$

where ε_1 and ε_2 are given in Lemma 3.1, and

$$\eta_2 = \|S_2\|_\infty \|C_1\|_\infty \left(\Lambda^{\beta_2 t_2} \left(\frac{\Lambda^{s_1} \varepsilon_1}{\sqrt[p_1]{N_1}} \right)^h (1 + \sqrt[p_1]{N_1}) + \frac{1 - \Lambda^{\beta_2 t_2}}{1 - \Lambda} \|B_2\|_\infty \rho_2 \right)$$

$$\rho_2 = \|B_2^T\|_\infty \Lambda^{t_2 - 1} \|T_2\|_\infty \left(\frac{\varepsilon_2}{\sqrt[p_2]{N_2}} + 1 \right)$$

Proof. *The proof is similar to Lemma 3.3. The only difference here is that Station 1 uses the estimate $\hat{x}_1(hs_1)$ to replace $\hat{x}_1(0)$ wherever it is applicable.* □

Finally, we can also design controls $u_2(t)$, for $hs_1+\beta_2 t_2 \le t \le hs_1+(\beta_2+1)t_2-1$ to drive the state of the system back to $x(t_3)=A^{hs_1+(\beta_2+1)t_2}x(0)$ at time $t_3 = hs_1+(\beta_2+1)t_2$.

3.3.3 Control

At time t_3, both stations have their estimates about the initial state, namely $\tilde{x}_1(0)$ and $\tilde{x}_2(0)$. They can then compute their own estimates of the state $x(t_3)$ and design the following controls indepedently to bring the state back to zero

$$u_i(t) = -B_i^T (A^T)^{t_3+n-1-t} z_i, \quad t_3 \le t \le t_3+n-1$$

where

$$z_i := \theta_i M_i^+(n,n) A^{t_3+n} \tilde{x}_i(0)$$

with $M_i^+(\cdot,\cdot)$ defined as the generalized inverse of $M_i(\cdot,\cdot)$ in Eq. (3.8) and $\theta_i = \|M_i\|_\infty (\Sigma_{i=1}^2 \|M_i\|_\infty)^{-1}$. The norm of $x(t_3+n)$ can be easily bounded as follows.

For simplicity, let us omit the parameter n in $M_i(n,n)$; then we have

$$\begin{aligned} x(t_3+n) &= A^{t_3+n}x(0) - \sum_{i=1}^{2} \theta_i M_i M_i^+ A^{t_3+n}\tilde{x}_i(0) \\ &= \sum_{i=1}^{2} \theta_i (I - M_i M_i^+) A^{t_3+n}(x(0)-\tilde{x}_i(0)) \end{aligned}$$

Since M_i is positive semi-definite, from singular value decomposition it is clear that $\|I-M_iM_i^+\|_\infty \le 1$; therefore,

$$\begin{aligned} \|x(t_3+n)\|_\infty &\le \sum_{i=1}^{2} \theta_i \|I-M_iM_i^+\|_\infty \Lambda^{t_3+n} \max_{i=1,2}\|x(0)-\tilde{x}_i(0)\|_\infty \\ &\le \Lambda^{t_3+n} \max_{i=1,2}\|x(0)-\tilde{x}_i(0)\|_\infty \end{aligned}$$

Now, we can state the main result of this chapter.

Theorem 3.1. *If we follow the above control algorithm, then the two-station decentralized system (3.1) with bandwidth-limited sensing channels can be asymptotically stabilized if the following inequality is satisfied for some* $0 \le \delta < 1$*:*

$$\max_{\substack{i,j=1,2 \\ i\ne j}} \left(\frac{\eta_j}{\sqrt[p_i]{N_i}} + \frac{\varepsilon_i}{\sqrt[p_i]{N_i}} + \frac{\varepsilon_j}{\sqrt[p_j]{N_j}} \right) < \delta \Lambda^{-(t_3+n)} \tag{3.15}$$

The stabilizing channel data rate R_i *is then given by* $\log_2 N_i$.

Proof. *To ensure the closed-loop asymptotic stability, we want* $\|x(t)\|_\infty \to 0$ *as* $t \to \infty$. *A sufficient condition is to impose* $\|x(t_3+n)\|_\infty < \delta E_0$ *for some* $0 \le \delta < 1$, *which is guaranteed by imposing* $\max_{i=1,2}(\|x_i(0) - \tilde{x}_i(0)\|_\infty) < \Lambda^{-(t_3+n)}\delta E_0$. *In other words, we want the uncertainty set of state contracts after each cycle. Since the infinity norm of all states between time* 0 *and* t_3+n *is parameterized by* E_0, *and if we can bring the uncertainty set at time* $k(t_3+n)$ *to zero as* $k \to \infty$, *we can thus drive the state to the equilibrium point eventually.*

Since all the terms on the left hand side of inequality (3.15) is parameterized by N_1 *and* N_2, *we can then compute the sufficient data rate on each channel to guarantee the asymptotic stabilizability of the system.* □

We can of course formulate some optimization problems on the required channel rate, for example, $\min(w_1R_1 + w_2R_2)$ subject to Eq. (3.15) where w_i are different weights put on the data rate of channel i.

Remark 3.4. Notice that the bandwidth requirement computed here is the peak value. The channels are idle during part of the cycle; thus, by using better coding strategies or by sharing the channel with other systems, the average bandwidth required can be reduced, perhaps significantly.

There is an immediate enhancement for the two-station case. In the algorithm above, when Station 1 transmits information to Station 2, the latter does nothing but listen. However, since both local loops know their own control values, Station 2 can actually transmit information to Station 1 simultaneously. By doing so, the whole cycle is reduced to $s_1 + \max\{(\beta_1+1)t_1, (\beta_2+1)t_2\} + n$ steps. This certainly tightens the bounds in Eq. (3.15), and consequently lowers the data rates required for stabilization. The only modification to the above derivation is to replace Eq. (3.14) with

$$\check{y}_2(s_1+kt_1) = C_2A^{s_1+kt_1}\hat{x}_2(0) + C_2\sum_{\ell=s_1}^{s_1+kt_1-1} A^{s_1+kt_1-1-\ell}B_2u_2(\ell)$$

3.4 Robustness Analysis against Model Mismatch

In this section, we consider the robustness issue of the stabilizing algorithm developed in the previous section against model mismatch. More specifically, we are interested in stabilizing the decentralized system (3.1) with channel bandwidth constraints and possible mismatched plant model programmed in control stations. In practice, accuracy of system modeling is limited due to factors such as the inevitability of measurement errors, system nonlinearities, change of system characteristics, etc. Thus the model that is being programmed into controllers may have slight errors from the real plant. Control algorithms must be robust against these mismatches for successful deployment. We show that there exists a neighborhood around the nominal system, within which our algorithm is stabilizing for all systems.

Instead of having the exact nominal system as in Eq. (3.1), we assume Station j, where $j \in \{1, \cdots, v\}$ has the following model of the system:

$$x(t+1) = (A + \Delta A^j)x(t) + \sum_{i=1}^{v} (B_i + \Delta B_i^j)u_i(t), \quad \|x(0)\|_\infty \le E_0$$
$$y_i(t) = (C_i + \Delta C_i^j)x(t), \qquad i = 1, 2, \ldots, v \tag{3.16}$$

where A, B_i, C_i are defined as in Sec. 3.2. Errors $\Delta A^j \in \Delta \mathcal{A}^j$, $\Delta B_i^j \in \Delta \mathcal{B}_i^j$, and $\Delta C_i^j \in \Delta \mathcal{C}_i^j$ are the *additive perturbations* to the nominal system, representing the model mismatch programmed in the jth controller.

In order to clearly illustrate the idea, we use the two-station system in the derivation as in the previous section. For simplicity, we also assume $\Delta B_i^j = \Delta C_i^j = 0$ for all $i, j = 1, 2$. Similar analysis can be done when they are nonzero and the result still holds true. From previous discussion in the section, it is reasonable to assume that all control stations have the same mismatched system model since they are most likely programmed with the same parameters in practice. [4]

To summarize, in this section we investigate the stabilization problem of system (3.1) with finite bandwidth limitation, while Station $j = 1, 2$ has the following model of the system:

$$x(t+1) = (A + \Delta A)x(t) + \sum_{i=1}^{2} B_i u_i(t), \quad \|x(0)\|_\infty \le E_0$$
$$y_i(t) = C_i x(t), \qquad i = 1, 2. \tag{3.17}$$

Again, we assume $s_1 \ge s_2$ as in the previous section. In addition, we assume $\Delta A \in \Delta \mathcal{A}$, which is the admissible perturbation set around the nominal system. One task here is to identify its impact on the required stabilizing data rates.

Similar to W_i, we define the *perturbed* observation matrix used by Station i as

$$\hat{W}_i := \begin{bmatrix} C_i^T & (C_i(A+\Delta A))^T & \ldots & (C_i(A+\Delta A)^{s_i - 1})^T \end{bmatrix}^T,$$

where $i = 1, 2$. Its generalized inverse is given by $\hat{W}_i^+ := (\hat{W}_i^* \hat{W}_i)^{-1} \hat{W}_i^* =: W_i^+ + \Delta W_i^+$. Since both the inverse and conjugate transpose are continuous operations, the size of the error, $\|\Delta W_i^+\|_\infty$ is positively related to $\|\Delta A\|_\infty$, in the sense that $\|\Delta W_i^+\|_\infty \to 0$ as $\|\Delta A\|_\infty \to 0$.

In order to simplify the notation, we define the following functions for convenience. They will be used as shorthand for some complex expressions in this section, where only properties listed below are relevant.

Let $\mu_{ik}(\Delta A)$, where $i = 1, 2$ and $k \in \mathcal{N}$, be a nonnegative function of ΔA such that $\mu_{ik}(\Delta A) \to 0$ as $\|\Delta A\|_\infty \to 0$. Functions $\xi_k(\Delta A)$ are similarly defined.

[4] The algorithm we proposed is still robust even if stations have different models of the system. Interested readers can refer to [17] for details.

Define $h^t(\Delta A) := (A+\Delta A)^t - A^t$. From the binomial expansion, we know it is a function of ΔA and $\|h^t(\Delta A)\|_\infty \to 0$ as $\|\Delta A\|_\infty \to 0$.

Let $e_k(\Delta A, x)$, where $k \in \mathcal{N}$, be a function of ΔA and x such that $e_k(\Delta A, x)$ goes to the origin as either ΔA or x does so.

Similar to the previous section, we describe the algorithm in three steps.

3.4.1 Observation

Parallel to Lemma 3.1, we have the following version of the observation lemma when the system model is not exact.

Lemma 3.5. *Within at most s_i steps, Station i can estimate $x_i(0) = P_i x(0)$ with error bounded by*

$$\left(\frac{\varepsilon_i}{\sqrt[p_i]{N_i}} + \mu_{i1}(\Delta A)\right) E_0$$

where p_i, ε_i, and N_i are defined as in Lemma 3.1. The function $\mu_{i1}(\Delta A)$ is as defined above.

Proof. *During the observation stage, we set the controls $u_1(t) = u_2(t) \equiv 0$ for $0 \le t < s_1$. Therefore, we have*

$$\|y_i(t) - \hat{y}_i(t)\|_\infty \le \frac{\|C_i\|_\infty \Lambda^t E_0}{\sqrt[p_i]{N_i}}$$

This error is only determined by the encoder/decoder resolution and range. It is not affected by the model mismatch. However, Station i cannot use Eq. (3.3) to estimate the initial state now. Instead, it has to use the following estimate due to model mismatch,

$$\hat{x}_i(0) = \hat{W}_i^+ \hat{\mathcal{Y}}_i$$

The associated estimation error is

$$\begin{aligned} x_i(0) - \hat{x}_i(0) &= W_i^+ \mathcal{Y}_i - (W_i^+ + \Delta W_i^+)\hat{\mathcal{Y}}_i \\ &= W_i^+(\mathcal{Y}_i - \hat{\mathcal{Y}}_i) - \Delta W_i^+ \hat{\mathcal{Y}}_i \end{aligned}$$

where $\mathcal{Y}_i$ and $\hat{\mathcal{Y}}_i$ are defined in Eq. (3.2).

The bound of the estimation error can be computed as

$$\begin{aligned}
\|x_i(0)-\hat{x}_i(0)\|_\infty &\le \frac{\varepsilon_i E_0}{\sqrt[p_i]{N_i}} + \|\Delta W_i^+\|_\infty \|\hat{\mathcal{Y}}_i\|_\infty \\
&\le \frac{\varepsilon_i E_0}{\sqrt[p_i]{N_i}} + \|\Delta W_i^+\|_\infty (\|\hat{\mathcal{Y}}_i - \mathcal{Y}_i\|_\infty + \|\mathcal{Y}_i\|_\infty) \\
&\le \frac{\varepsilon_i E_0}{\sqrt[p_i]{N_i}} + \|\Delta W_i^+\|_\infty \|C_i\|_\infty \Lambda^{s_i-1}\left[1+\frac{1}{\sqrt[p_i]{N_i}}\right] E_0 \\
&=: \left(\frac{\varepsilon_i}{\sqrt[p_i]{N_i}} + \mu_{i1}(\Delta A)\right) E_0
\end{aligned}$$

□

3.4.2 Communication

Now let us consider the communication stage, during which stations exchange their estimates of the initial state with possibly mismatched models of the system.

Define the following notation:

1. $\hat{\mathcal{R}}_i = \langle A+\Delta A | B_i\rangle, i = 1,2$, as the *perturbed* ith controllable subspace.
2. $\hat{\mathcal{K}}_i = [C_i | A+\Delta A], i = 1,2$, as the *perturbed* ith unobservable subspace.

Recall that $\mathcal{R}_i = \langle A|B_i\rangle$, and $\mathcal{K}_i = [C_i|A], i = 1,2$. Since we assumed that $\mathcal{R}_i \not\subseteq \mathcal{K}_j$, therefore,

$$\begin{bmatrix} C_j \\ C_jA \\ \vdots \\ C_jA^{n-1} \end{bmatrix} \begin{bmatrix} B_i & AB_i & \cdots & A^{n-1}B_i \end{bmatrix} \neq \mathbf{0} \tag{3.18}$$

If the admissible perturbation set $\Delta\mathcal{A}$ is sufficiently small, then Eq. (3.18) continues to hold with A being replaced by any value $A+\Delta A \in A+\Delta\mathcal{A}$, since the perturbation ΔA is continuous. In other words, there exists a $\Delta\mathcal{A}$ such that for all $\Delta A \in \Delta\mathcal{A}$, we have $\hat{\mathcal{R}}_i \not\subseteq \hat{\mathcal{K}}_j$, for $i,j = 1,2$.

This means that even though there are model mismatches, the channel through the plant still exists. Thus, control actions from Station i can still be observed by Station j. We now define

$$\hat{\alpha}_i := \text{rank}\left\{ C_j \sum_{\ell=0}^{t_i-1} (A+\Delta A)^{t_i-1-\ell} B_i B_i^T ((A+\Delta A)^T)^{t_i-1-\ell} \right\} \neq 0 \tag{3.19}$$

Since `rank` is an upper-continuous operation, therefore $\hat{\alpha}_i \geq \alpha_i$, where α_i and t_i are defined in Eq. (3.7).

Remark 3.5. It is possible that there exists a number $\tilde{t}_i \in \mathcal{N}$ such that $\tilde{t}_i < t_i$, and Eq. (3.19) still holds if we replace t_i with $\tilde{t}_i$. Without loss of generality, we choose to use t_i here to simplify the derivation.

Similarly there exists a smallest $\hat{\beta}_i \in \mathcal{N}$ such that $\hat{\alpha}_i\hat{\beta}_i \geq n$. We define the following *coding matrix*, which is used by Station i

$$\hat{E}_{ji}(\Delta A) = \begin{bmatrix} C_j\hat{M}_i(t_i,t_i) & \cdots & 0 \\ \vdots & \ddots & \vdots \\ C_j\hat{M}_i(\hat{\beta}_1 t_i,t_i) & \cdots & C_j\hat{M}_i(t_i,t_i) \end{bmatrix} \tag{3.20}$$

where $\hat{M}_i(p,q) = \sum_{\ell=0}^{t_i-1}(A+\Delta A)^{p-1-\ell}B_iB_i^T((A+\Delta A)^T)^{q-1-\ell}$.

Clearly, when $\Delta A \to 0$, all parameters $\hat{\alpha}_i$, $\hat{\beta}_i$, $\hat{M}_i$ and $\hat{E}_{ji}$ approach their respective nominal values due to the continuity of the perturbation. For example,

$$\begin{aligned}\hat{M}_i(p,q) &= \sum_{\ell=0}^{t_i-1}(A+\Delta A)^{p-1-\ell}B_iB_i^T((A+\Delta A)^T)^{q-1-\ell} \\ &= M_i(p,q) + \Delta M_i(p,q)\end{aligned}$$

where the nominal value $M_i(p,q)$ is given in Eq. (3.8) and

$$\begin{aligned}\Delta M_i(p,q) &= \sum_{\ell=0}^{t_i-1} h^{p-1-\ell}(\Delta A)B_iB_i^T(A^T)^{q-1-\ell} + A^{p-1-\ell}B_iB_i^T(h^{q-1-\ell}(\Delta A))^T \\ &\quad + h^{p-1-\ell}(\Delta A)B_iB_i^T(h^{q-1-\ell}(\Delta A))^T \\ &\to \mathbf{0} \text{ as } \Delta A \to \mathbf{0}\end{aligned}$$

Again, since rank $\hat{E}_{ji}(\Delta A) \geq n$, there exist matrices $\hat{S}_i$ and $\hat{T}_i$ such that $\hat{S}_i\hat{E}_{ji}\hat{T}_i = I_n$.

We show the transmission from Station 1 to Station 2 to illustrate that even with model mismatch, the initial state can still be reconstructed with a predetermined error bound using the algorithm we introduced in the previous section.

Lemma 3.6. *Given $\mathcal{R}_1 \not\subseteq \mathcal{K}_2$, there exists a neighborhood around the nominal system such that for all mismatched models within this neighborhood, Station 2 can reconstruct the initial state $x(0)$ at time $s_1 + \hat{\beta}_1 t_1$ with error bounded by*

$$\left(\frac{\eta_1}{\sqrt[p_2]{N_2}} + \frac{\varepsilon_1}{\sqrt[p_1]{N_1}} + \frac{\varepsilon_2}{\sqrt[p_2]{N_2}} + \xi_1(\Delta A)\right) E_0$$

where η_1 is defined as in Eq. (3.11) with S_1 replaced by $\hat{S}_1$, and ε_1 and ε_2 are given in Lemma 3.1. The function $\xi_1(\cdot)$ is a function defined as in the beginning of this section, which represents the extra estimation error introduced by model mismatch.

Proof. *Again, we only need to encode $(I - P_2)\hat{x}_1(0)$ and send it over the channel through the plant. We design the following codewords:*

$$z^p = \hat{T}_1(I-P_2)\hat{x}_1(0) = \begin{bmatrix} z_1^T & \cdots & z_{\hat{\beta}_1}^T \end{bmatrix}^T$$

Define controls as in Eq. (3.9) with A being replaced with $A+\Delta A$. We have

$$u_1(\ell) = \begin{cases} B_1^T((A+\Delta A)^T)^{s_1+t_1-1-\ell} z_1 & \text{for } s_1 \le \ell \le s_1+t_1-1 \\ B_1^T((A+\Delta A)^T)^{s_1+2t_1-1-\ell} z_2 & \text{for } s_1+t_1 \le \ell \le s_1+2t_1-1 \\ \cdots & \\ B_1^T((A+\Delta A)^T)^{s_1+\hat{\beta}_1 t_1-1-\ell} z_{\hat{\beta}_1} & \text{for } s_1+(\hat{\beta}_1-1)t_1 \le \ell \le s_1+\hat{\beta}_1 t_1-1 \end{cases}$$

$$u_2(\ell) = 0 \qquad \text{for } s_1 \le \ell \le s_1+\hat{\beta}_1 t_1-1$$

It can be easily computed that

$$\begin{aligned} &\max_{s_1\le\ell\le s_1+\hat{\beta}_1 t_1-1} \|u_1(\ell)\|_\infty \\ &\quad \le \|B_1^T\|_\infty \|A+\Delta A\|_\infty^{t_1-1} \|\hat{T}_1\|_\infty \left(1+\frac{\varepsilon_1}{\sqrt[p_1]{N_1}}+\mu_{11}(\Delta A)\right) E_0 \\ &\quad =: (\rho_1+\mu_{12}(\Delta A))E_0 \end{aligned}$$

where ρ_1 is given in Lemma 3.3.

Due to the fact that $\hat{S}_1\hat{E}_{21}\hat{T}_1 = I_n$, we know

$$(I-P_2)\hat{x}_1(0) = \hat{S}_1 \begin{bmatrix} \tilde{y}_2(s_1+t_1) - C_2(A+\Delta A)^{s_1+t_1}x(0) \\ \vdots \\ \tilde{y}_2(s_1+\hat{\beta}_1 t_1) - C_2(A+\Delta A)^{s_1+\hat{\beta}_1 t_1}x(0) \end{bmatrix}$$

where for $0 < k \le \hat{\beta}_1$, the fictitious *output $\tilde{y}_2(\cdot)$ on channel 2 if the system matrix is $A+\Delta A$ is given by*

$$\tilde{y}_2(s_1+kt_1) = C_2(A+\Delta A)^{s_1+kt_1}x(0) + \sum_{\ell=s_1}^{s_1+kt_1-1} C_2(A+\Delta A)^{s_1+kt_1-1-\ell}B_1u_1(\ell)$$

However, the actual measurement $y_2(\cdot)$ during this period before encoding is

$$y_2(s_1+kt_1) = C_2A^{s_1+kt_1}x(0) + \sum_{\ell=s_1}^{s_1+kt_1-1} C_2A^{s_1+kt_1-1-\ell}B_1u_1(\ell)$$

Denote $\hat{y}_2(\cdot)$ as the decoder's estimate of $y_2(\cdot)$ as in Sec. 3.3.2; then Station 2 can decode the message sent by Station 1 in the following way

$$(I-P_2)\bar{x}_1(0) := \hat{S}_1 \begin{bmatrix} \hat{y}_2(s_1+t_1) - C_2(A+\Delta A)^{s_1+t_1}\hat{x}_2(0) \\ \vdots \\ \hat{y}_2(s_1+\hat{\beta}_1 t_1) - C_2(A+\Delta A)^{s_1+\hat{\beta}_1 t_1}\hat{x}_2(0) \end{bmatrix}$$

Now, let us compute the error between the decoded estimate and the data sent by Station 1:

$$\begin{aligned}
\xi_\infty &= \|(I-P_2)\bar{x}_1(0)-(I-P_2)\hat{x}_1(0)\|_\infty \\
&\le \|\hat{S}_1\|_\infty \left\| \begin{bmatrix} \hat{y}_2(s_1+t_1)-C_2(A+\Delta A)^{s_1+t_1}\hat{x}_2(0) \\ \vdots \\ \hat{y}_2(s_1+\hat{\beta}_1 t_1)-C_2(A+\Delta A)^{s_1+\hat{\beta}_1 t_1}\hat{x}_2(0) \end{bmatrix} \right. \\
&\quad \left. - \begin{bmatrix} \tilde{y}_2(s_1+t_1)-C_2(A+\Delta A)^{s_1+t_1}x(0) \\ \vdots \\ \tilde{y}_2(s_1+\hat{\beta}_1 t_1)+C_2(A+\Delta A)^{s_1+\hat{\beta}_1 t_1}x(0) \end{bmatrix} \right\|_\infty \\
&\le \|\hat{S}_1\|_\infty \max_{1\le k\le \hat{\beta}_1} \Big(\|\tilde{y}_2(s_1+kt_1)-\hat{y}_2(s_1+kt_1)\|_\infty \\
&\quad +\|C_2(A+\Delta A)^{s_1+kt_1}\|_\infty \|x_2(0)-\hat{x}_2(0)\|_\infty\Big)
\end{aligned} \tag{3.21}$$

Clearly, by the triangle inequality, we have

$$\begin{aligned}
\|\tilde{y}_2(s_1+kt_1)-\hat{y}_2(s_1+kt_1)\|_\infty &\le \|y_2(s_1+kt_1)-\tilde{y}_2(s_1+kt_1)\|_\infty \\
&\quad + \|\hat{y}_2(s_1+kt_1)-y_2(s_1+kt_1)\|_\infty
\end{aligned} \tag{3.22}$$

where the first term signifies the extra error introduced by model mismatch and the second term is the coding error as in Sec. 3.3. We can compute the first term as follows:

$$\begin{aligned}
&\|y_2(s_1+kt_1)-\tilde{y}_2(s_1+kt_1))\|_\infty \\
&\quad = \|C_2 h^{s_1+kt_1}(\Delta A)x(0)+\sum_{\ell=s_1}^{s_1+kt_1-1} C_2 h^{s_1+kt_1-1-\ell}(\Delta A)B_1 u_1(\ell)\|_\infty \\
&\quad \le \|C_2\|_\infty \Bigg(\|h^{s_1+kt_1}(\Delta A)\|_\infty E_0 \\
&\qquad + \sum_{\ell=s_1}^{s_1+kt_1-1} \|h^{s_1+kt_1-1-\ell}(\Delta A)\|_\infty \|B_1\|_\infty \|u_1(\ell)\|_\infty \Bigg) \\
&\quad =: \mu_{23}(\Delta A)E_0 \to 0 \text{ as } \|\Delta A\|_\infty \to 0
\end{aligned}$$

Now consider the quantization error $\|y_2(s_1+kt_1)-\hat{y}_2(s_1+kt_1)\|_\infty$ in Eq. (3.22). Again, the encoder $\mathcal{E}_2$ and the decoder $\mathcal{D}_2$ can compute $\check{y}_2(s_1+kt_1)=C_2(A+\Delta A)^{s_1+kt_1}\hat{x}_2(0)$, which is their estimate of the autonomous part of the local measurements, in an effort to reduce the quantization range and ultimately to reduce the quantization error. Therefore, the encoding region can be computed as

$$y_2(s_1+kt_1)-\check{y}_2(s_1+kt_1) = C_2A^{s_1+kt_1}(x(0)-\hat{x}_2(0))+C_2h^{s_1+kt_1}(\Delta A)\hat{x}_2(0) + \sum_{\ell=s_1}^{s_1+kt_1-1} C_2A^{s_1+kt_1-1-\ell}B_1u_1(\ell)$$

whose size can be bounded as follows:

$$\begin{aligned}
&\|y_2(s_1+kt_1)-\check{y}_2(s_1+kt_1)\|_\infty \\
&\le \|C_2\|_\infty \Bigg[\Lambda^{s_1+kt_1}\left(\frac{1}{\sqrt[p_2]{N_2}}+\mu_{21}(\Delta A)\right)E_0 \\
&\quad + \|h^{s_1+kt_1}(\Delta A)\|_\infty\left(1+\frac{1}{\sqrt[p_2]{N_2}}+\mu_{22}(\Delta A)\right)E_0 \\
&\quad + \frac{1-\Lambda^{s_1+kt_1}}{1-\Lambda}\|B_1\|_\infty(\rho_1+\mu_{12}(\Delta A))E_0\Bigg] \\
&=: \hat{\eta}_1E_0+\mu_{24}(\Delta A)E_0
\end{aligned}$$

where $\hat{\eta}_1 = \|C_2\|_\infty\left(\Lambda^{s_1+kt_1}\frac{1}{\sqrt[p_2]{N_2}}+\frac{1-\Lambda^{s_1+kt_1}}{1-\Lambda}\|B_1\|_\infty\rho_1\right)$ *and* $\mu_{24}(\Delta A)$ *denotes the remaining portion of the above inequality.*

With the same encoding and decoding strategy as in Sec. 3.3.2, the coding error is given by

$$\|y_2(s_1+kt_1)-\hat{y}_2(s_1+kt_1)\|_\infty \le \frac{1}{\sqrt[p_2]{N_2}}(\hat{\eta}_1+\mu_{24}(\Delta A))E_0$$

It is clear that

$$\begin{aligned}
&\|C_2(A+\Delta A)^{s_1+kt_1}\|_\infty\|x(0)-\hat{x}_2(0)\|_\infty \\
&\le \|C_2\|_\infty(\Lambda^{s_1+kt_1}+\|h^{s_1+kt_1}(\Delta A)\|_\infty)\left(\frac{1}{\sqrt[p_2]{N_2}}+\mu_{21}(\Delta A)\right)E_0 \\
&=: \left(\|C_2\|_\infty\Lambda^{s_1+kt_1}\frac{1}{\sqrt[p_2]{N_2}}+\mu_{25}(\Delta A)\right)E_0
\end{aligned}$$

Combining all pieces we computed above together, the bound on the estimation error between Station 2's estimate and the data sent from Station 1 is given by

$$\begin{aligned}
\xi_\infty &= \|(I-P_2)\hat{x}_1(0)-(I-P_2)\bar{x}_1(0)\|_\infty \\
&\le \|\hat{S}_1\|_\infty\left(\mu_{23}(\Delta A)+\frac{\hat{\eta}_1+\mu_{24}(\Delta A)}{\sqrt[p_2]{N_2}}+\frac{\|C_2\|_\infty\Lambda^{s_1+kt_1}}{\sqrt[p_2]{N_2}}+\mu_{25}(\Delta A)\right)E_0 \\
&= \|\hat{S}_1\|_\infty\left(\frac{1}{\sqrt[p_2]{N_2}}\left(\hat{\eta}_1+\|C_2\|_\infty\Lambda^{s_1+kt_1}\right)+\mu_{26}(\Delta A)\right)E_0 \\
&:= \frac{\eta_1}{\sqrt[p_2]{N_2}}E_0+\mu_{27}(\Delta A)E_0
\end{aligned}$$

where η_1 *is defined as in Lemma 3.3 with* S_1 *replaced by* $\hat{S}_1$.

Thus Station 2 now can compute the full estimate of the initial state as in Sec. 3.3.2, The estimation error is bounded as follows:

$$\|\tilde{x}_2(0) - x(0)\|_\infty \leq \left[\frac{\eta_1}{\sqrt[p_2]{N_2}} + \frac{\varepsilon_1}{\sqrt[p_1]{N_1}} + \frac{\varepsilon_2}{\sqrt[p_2]{N_2}} \right] E_0 + \xi_1(\Delta A) E_0$$

where ε_i, $i = 1,2$, are defined in Lemma 3.3, η_1 is defined as above and

$$\xi_1(\Delta A) = \mu_{11}(\Delta A) + \mu_{21}(\Delta A) + \mu_{27}(\Delta A).$$

□

The sheer effect of the model mismatch is captured in this predetermined additional error term $\xi_1(\Delta A)E_0$, whose size positively depends on the size of the admissible perturbation.

We now design the following controls for $s_1 + \hat{\beta}_1 t_1 \leq t < s_1 + (\hat{\beta}_1 + 1)t_1$ to drive the system back to its autonomous state in order to offset the accumulated control effects.

$$u_1(t) = -B_1^T((A + \Delta A)^T)^{s_1 + (\hat{\beta}_1 + 1)t_1 - 1 - t} z$$

where

$$z := \hat{M}_1^+(t_1, t_1) \sum_{\ell = s_1}^{s_1 + \hat{\beta}_1 t_1 - 1} (A + \Delta A)^{s_1 + \hat{\beta}_1 t_1 - 1 - \ell} B_1 u_1(\ell)$$

Set $u_2(\ell) = 0$. It can be easily verified that the state at time $s_1 + (\hat{\beta}_1 + 1)t_1$ is

$$x(s_1 + (\hat{\beta}_1 + 1)t_1) = A^{s_1 + (\hat{\beta}_1 + 1)t_1} x(0) + e_1(\Delta A, x(0))$$

where $e_1(\cdot,\cdot)$ is a function defined as in the beginning of this section, which goes to the origin as either ΔA or $x(0)$ does so.

In contrast to the nominal system, we have an extra term $e_1(\Delta A, x(0))$ in the previous equation, which means we cannot drive the state back to the autonomous trajectory exactly. There are going to be some errors, which disappear as the model mismatch goes to none.

Now Station 2 starts to transmit its estimate $\hat{x}_2(0)$ to Station 1 in the same way. Then, at time $t_3 = hs_1 + (\hat{\beta}_2 + 1)t_2$, where h is defined as in the previous section, both stations have full estimate about the initial state, namely $\tilde{x}_1(0)$ and $\tilde{x}_2(0)$. And the state at time t_3 is located at

$$x(t_3) = A^{t_3} x(0) + e_2(\Delta A, x(0))$$

3.4.3 Control

After exchanging estimates of the initial state, we can design the following controls, similar to those in the previous section, trying to drive the state back to zero in order to shrink the uncertainty set.

$$u_i(t) = -B_i^T((A+\Delta A)^T)^{t_3+n-1-t}z_i, \quad t_3 \le t \le t_3+n-1$$

where

$$z_i := \hat{\theta}_i \hat{M}_i^+(n,n)(A+\Delta A)^{t_3+n}\tilde{x}_i(0)$$

with $\hat{M}_i^+(\cdot,\cdot)$ defined as the generalized inverse of $\hat{M}_i(\cdot,\cdot)$ in Eq. (3.20) and $\theta_i = \|\hat{M}_i\|_\infty(\Sigma_{i=1}^2\|\hat{M}_i\|_\infty)^{-1}$.

The size of $x(t_3+n)$ can be similarly bounded as

$$\|x(t_3+n)\|_\infty \le \Lambda^{t_3+n}\max_{i=1,2}\|x(0)-\tilde{x}_i(0)\|_\infty + \xi_3(\Delta A)E_0$$

where $\xi_3(\cdot)$ is defined as in the beginning of this section and it captures the effects of model mismatch.

Now, we can state the main result of this section.

Theorem 3.2. *There exists a neighborhood around the nominal system such that for all mismatched models within this neighborhood, by employing the above control algorithm, the two-station decentralized system (3.17) with bandwidth limited sensing channels can still be asymptotically stabilized if the following inequality is satisfied for some number* $0 \le \delta < 1$,

$$\max_{\substack{i,j=1,2\\ i\ne j}}\left(\frac{\eta_j}{\sqrt[p_i]{N_i}} + \frac{\varepsilon_i}{\sqrt[p_i]{N_i}} + \frac{\varepsilon_j}{\sqrt[p_j]{N_j}} + \xi_i\right) < \Lambda^{-(t_3+n)}(\delta-\xi_3) \tag{3.23}$$

where ξ_1 *and* ξ_2 *capture the estimation errors, and* ξ_3 *represents the control errors due to model mismatch. The stabilizing channel data rate* R_i *is then given by* $\log_2 N_i$.

Proof. *Again, to ensure asymptotic stability, we want* $\|x(t)\|_\infty \to 0$ *as* $t\to\infty$. *A sufficient condition to impose is* $\|x(t_3+n)\|_\infty < \delta E_0$, *which is guaranteed by*

$$\max_{i=1,2}(\|x_i(0)-\tilde{x}_i(0)\|_\infty) < \Lambda^{-(t_3+n)}(\delta-\xi_3)E_0$$

The rest is obvious. Notice that $\xi_i \to 0$ *as* $\|\Delta A\|_\infty \to 0$, *so there exists* $\xi_3 \le \delta$. □

Remark 3.6. Again, we have computed the peak value of the bandwidth requirement here and for this two-station case, we can still improve the result by communicating simultaneously. Also, it is worth noting that the larger the model mismatch is, the higher the stabilizing data rates are demanded.

3.5 Multistation Case

For the multistation decentralized control systems, the observation and the control phases stay the same. But the communication scheme becomes more complicated as the number of stations grows.

If the system is strongly connected or the quotient decentralized system contains no unstable decentralized fixed mode—in other words, if it is stabilizable under the decentralized information structure—then our algorithm proceeds as follows: After the observation stage, each station computes its own estimate of the initial state; then it starts to transmit its estimate to others in a round robin fashion; i.e., Station 1 starts to *talk* first, and all the other stations just *listen*; then Station 2 starts to transmit, and so on. After one round, the unobservable subspace of every station gets reduced or, in the worst case, remains the same. We need at most $\nu - 1$ rounds for a ν-station system to achieve full knowledge of the initial state by all stations. Then we can start to design controls to bring the state back to the equilibrium point. The robustness analysis presented in Sec. 3.4 can be directly extended to the multistation case as well.

The drawback of this algorithm is the error propagation. As the number of stations grows, we need to wait longer and longer for the information to be exchanged. Meanwhile, the error accumulates. This can be reduced, maybe significantly, by allowing simultaneous information exchange, as we show at the end of Sec. 3.3. Our enhanced algorithm can handle this naturally in the two-station case. But for multistation systems, the simple encoding/decoding strategy must be refined to allow a station to differentiate measurements triggered by control actions from different stations. This can be achieved, for example, by adding a different predetermined base signal into the control signal sent from each station, or by allowing stations that have disjoint controllable subspaces to transmit simultaneously.

3.6 Example

In this section, we use the algorithm developed in Sec. 3.3 to stabilize the following system and derive the corresponding bandwidth requirements:

$$\begin{aligned} x(t+1) &= \begin{bmatrix} 1 & 0 \\ 0 & 2 \end{bmatrix} x(t) + \begin{bmatrix} 1 \\ 0 \end{bmatrix} u_1(t) + \begin{bmatrix} 0 \\ 1 \end{bmatrix} u_2(t) \\ y_1(t) &= \begin{bmatrix} 0 & 1 \end{bmatrix} x(t) \\ y_2(t) &= \begin{bmatrix} 1 & 0 \end{bmatrix} x(t) \qquad \|x(0)\|_\infty \le E_0 \end{aligned} \tag{3.24}$$

This is a simplest nontrivial example. The system is jointly controllable and observable and it is strongly connected. Therefore, it is stabilizable under the decentralized information structure. However, since the controllable subspace of any single station coincides with its own unobservable subspace, no station can stabilize any mode by itself. Therefore, collaboration is mandatory for stabilization. It is interesting to note that although A is already in the diagonal form, the result in [10] does not apply to this example.

The information lower bound for this system is $R \ge 1$ bit/step for stabilization purposes if *centralized* control is allowed.

Let us compute the upper bound on bandwidth requirement under the decentralized information structure using our algorithms: first the original one, then the enhanced version with simultaneous information exchange.

It is clear that $s_1 = s_2 = 1$ in this example. Thus, outputs $y_i(0), i = 1,2$, are sufficient for stations to reconstruct their estimates of the initial state. If there is no quantization, then $y_i(0) = x(0)_j$, $i,j = 1,2$,and $i \ne j$, where $x(0)_j$ denotes the jth component of the initial state $x(0)$. However, the output y_i is quantized with an N_i-level quantizer on the ith channel. Then, from Lemma 3.1, Station 1 and 2 have

$$\begin{bmatrix} 0 & \hat{x}_{20} \end{bmatrix}^T \text{ and } \begin{bmatrix} \hat{x}_{10} & 0 \end{bmatrix}^T$$

respectively, as their estimate of the initial state, where $|\hat{x}_{i0} - x(0)_i| \le E_0/N_i$ for $i = 1,2$.

Now we can exchange information about the initial state. Instead of constructing β_i, t_i and E_{ij} as in Sec. 3.3, we simply choose $u_1(0) = \hat{x}_{20}$ and $u_2(0) = 0$ since the only difference would just be a constant coefficient. Following Lemma 3.3, we have

$$|x(0)_2 - \bar{x}_{20}| \le \frac{1}{N_2}(2 + \frac{1}{N_2} + \frac{1}{N_1})E_0 + \frac{1}{N_1}E_0$$

which is also the upper bound of $\|x(0) - \tilde{x}_2(0)\|_\infty$ due the structure of $\mathcal{K}_2$.

Now Station 1 applies a control $u_1(1) = -\hat{x}_{20}$ to drive the state $x(2)$ to $A^2x(0)$.

During the same time, $\mathcal{E}_1$ and $\mathcal{D}_1$ run the simulation process and estimate $x(2)_2$ as $\hat{x}_{22}$ with error bounded by $|\hat{x}_{22} - x(2)_2| \le \frac{4}{N_1^3}E_0$.

Now, let $u_1(2) = 0$ and $u_2(2) = \hat{x}_{10}$. We can compute the error between Station 1's estimate $\bar{x}_{10}$ and $x(0)_1$, which is bounded by

$$|\bar{x}_{10} - x(0)_1| \le \frac{8}{N_1^4}E_0 + \frac{1}{N_1}(1 + \frac{1}{N_2})E_0 + \frac{8}{N_1^3}E_0 + \frac{1}{N_2}E_0$$

This is also the upper bound of $\|x(0)-\tilde{x}_1(0)\|_\infty$.

Station 2 now applies a control $u_2(3) = -2\hat{x}_{10}$ to drive the state back to $x(4) = A^4x(0)$.

Now, both stations have a full estimate of the initial state, so we can design controls as follows:

$$\begin{bmatrix} u_1(4) \\ u_2(4) \end{bmatrix} = -\begin{bmatrix} 1 & 0 \\ 0 & 2^5 \end{bmatrix}\begin{bmatrix} \bar{x}_{10} \\ \bar{x}_{20} \end{bmatrix}$$

The condition of $\|x(5)\|_\infty < E_0$ is equivalent to

$$\max\left\{\frac{8}{N_1^4}+\frac{1}{N_1}(1+\frac{1}{N_2})+\frac{8}{N_1^3}+\frac{1}{N_2},\quad 2^5\left(\frac{1}{N_2}(2+\frac{1}{N_1}+\frac{1}{N_2})+\frac{1}{N_1}\right)\right\}<1$$

The feasible rate region is given in Fig. 2(a).

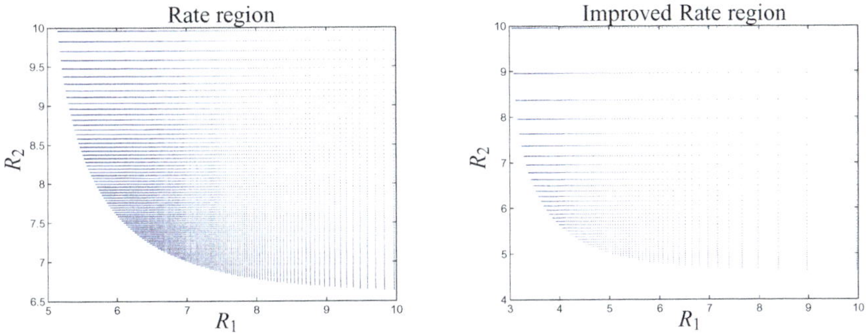

(a) Rate region (b) Improved rate region

Fig. 3.2 Admissible rate regions for system (3.24).

There is a significant increase in the rate requirements compared to the centralized result. The upper bound of total required stabilizing data rate is $R_1 + R_2 \approx 15.0736$ bits/step.

We now proceed to use the enhanced algorithm. The first step of observation does not change. However, we can design control $u_1(0) = \hat{x}_{20}$ and $u_2(0) = \hat{x}_{10}$.

Both stations can compute their estimates $\bar{x}_{10}$ and $\bar{x}_{20}$ from their measurements and local controls. The errors are bounded by

$$\begin{aligned} |\hat{x}_{10}-\bar{x}_{10}| &\le |y_1(1)-\hat{y}_1(1)|+2|x(0)_2-\hat{x}_{20}| \\ |\hat{x}_{20}-\bar{x}_{20}| &\le |y_2(1)-\hat{y}_2(1)|+|x(0)_1-\hat{x}_{10}| \end{aligned}$$

This time the control action will be completed at time $t = 3$. The inequality to satisfy is

$$\max\left\{\frac{1}{N_1}(3+\frac{1}{N_2}+2\frac{1}{N_1})+\frac{1}{N_2},\quad 2^3\left(\frac{1}{N_2}(2+\frac{1}{N_2}+\frac{1}{N_1})+\frac{1}{N_1}\right)\right\}<1$$

The improved rate region is plotted in Fig. 2(b). The upper bound of total required stabilizing data rate is dropped by around 2 bits/step to 13.0008 bits/step.

3.7 Summary

In this chapter, we considered the decentralized control problem subject to finite bandwidth constraints on sensing channels. We developed an open-loop stabilizing control algorithm taking into account both the topological and nontopological constraints of the information exchange. We also provided an explicit way to construct the associated stabilizing encoder, decoder and controller. Sufficient condition on stabilizing data rate required on each channel was derived, while the lower bound is currently given by the centralized result. In addition, we showed that our algorithm is structurally robust against model mismatch; although potential measurement inaccuracy has not been explicitly studied here, we can expect a similar degree of algorithm tolerance to such noise.

References

1. Anderson, B.D.O., Moore, J.: Time-varying feedback laws for decentralized control. IEEE Trans. Automatic Control **26**(5), 1133–1139 (1981)
2. Corfmat, J., Morse, A.: Decentralized control of linear multivariable systems. Automatica **12**, 479–495 (2006)
3. Fu, M., Xie, L.: Finite-level quantized feedback control for linear systems. In: Proc. 45th IEEE Conf. Decision and Control (CDC2006), pp. 1117–1122. San Diego, CA (2006)
4. Gong, Z., Aldeen, M.: Stabilization of decentralized control systems. J. Mathematical Systems, Estimation, and Control **7**(1), 1–16 (1997)
5. Kobayashi, H., Hanafusa, H., Yoshikawa, T.: Controllability under decentralized information structure. IEEE Trans. Automatic Control **23**(2), 182–188 (1978)
6. Kobayashi, H., Yoshikawa, T.: Graphic-theorectic approach to controllability and localizability of decentralized control systems. IEEE Trans. Automatic Control **27**(5), 1096–1108 (1982)
7. Liberzon, D.: On stabilization of linear systems with limited information. IEEE Trans. Automatic Control **48**(2), 304–307 (2003)
8. Matveev, A., Savkin, A.: Stabilization of multisensor networked control systems with communication constraints. In: Proc. 5th Asian Control Conf., pp. 1905–1913. Melbourne, Australia (2004)
9. Matveev, A., Savkin, A.: Decentralized stabilization of linear systems via limited capacity communication networks. In: Proc. 44th IEEE Conf. Control and Decision (CDC-ECC2005), pp. 1155–1161. Seville, Spain (2005)
10. Nair, G., Evans, R., Caines, P.: Stabilising decentralized linear systems under data rate constraints. In: Proc. 43rd IEEE Conf. Control and Decision (CDC2004), pp. 3992–3997. Atlantis, Bahamas (2004)
11. Nair, G., Evans, R.: Exponential stabilisability of finite-dimensional linear systems with limited data rates. Automatica **39**, 585–593 (2003)

12. Tatikonda, S., Mitter, S.: Control under communication constraints. IEEE Trans. Automatic Control **49**(7), 1056–1068 (2004)
13. Tatikonda, S.: Some scaling properties of large distributed control systems. In: Proc. 42nd IEEE Conf. Control and Decision (CDC2003), pp. 3142–3147. Maui, Hawaii (2003)
14. Wang, S., Davison, E.: On the stabilization of decentralized control systems. IEEE Trans. Automatic Control **18**(5), 473–478 (1973)
15. Yuksel, S., Basar, T.: Optimal signaling policies for decentralized multicontroller stabilizability over communication channels. IEEE Trans. Automatic Control **52**(10), 1969–1974 (2007)
16. Zhang, C., Dullerud, G.E.: Uniform stabilization of Markovian jump linear systems with logarithmic quantization—A convex approach. In: Proc. IEEE Conf. Decision and Control (CDC2009). Atlanta, GA (2009)
17. Zhang, C.: Centralized and decentralized control with limited information. Ph.D. thesis, University of Illinois at Urbana-Champaign, Urbana-Champaign, IL (2008)

Chapter 4
Monotone Games for Cognitive Radio Systems

Gesualdo Scutari, Daniel P. Palomar, Francisco Facchinei and Jong-Shi Pang

Abstract Noncooperative game theory is a branch of game theory for the resolution of conflicts among interacting decision makers (called players), each behaving selfishly to optimize his own well-being. In this chapter, we present a mathematical treatment of (generalized) Nash equilibrium problems based on the variational inequality and complementarity approach, covering the topics of existence and uniqueness of an equilibrium, and the design of distributed algorithms using best-response iterations along with their convergence properties. We then apply the developed machinery to the distributed design of cognitive radio systems. The proposed equilibrium models and resulting algorithms differ in performance of the secondary users, level of protection of the primary users, computational effort and signaling among primary and secondary users, convergence analysis, and convergence speed; which makes them suitable for many different CR systems.

Gesualdo Scutari
Department of Electrical Engineering, State University of New York (SUNY) at Buffalo, Buffalo, NY 14260, USA
e-mail: gesualdo@buffalo.edu

Daniel P. Palomar
Department of Electronic and Computer Engineering, Hong Kong University of Science and Technology, Clear Water Bay, Kowloon, Hong Kong
e-mail: palomar@ust.hk

Francisco Facchinei
Dipartimento di Informatica e Sistemistica, University of Rome La Sapienza, Rome Via Buonarroti 12, 00185 Rome, Italy
e-mail: facchinei@dis.uniroma1.it

Jong-Shi Pang
University of Illinois at Urbana-Champaign, 117 Transportation Building MC-238, 104 S. Mathews Ave., Urbana IL 61801, USA
e-mail: jspang@uiuc.edu

R. Johansson and A. Rantzer (Eds.): Distributed Decision Making and Control, LNCIS 417, pp. 83–112.
springerlink.com

4.1 Introduction

In recent years, there has been a growing interest in the use of noncooperative games to model many communications and networking problems, where the interaction among several agents is by no means negligible and centralized approaches are not suitable. Examples are power control and resource sharing in wireless/wired and peer-to-peer networks (e.g., [40, 20, 34, 6, 32, 28, 33, 36]), cognitive radio systems (e.g., [31, 35, 26]) and distributed routing, flow and congestion control in communication networks (e.g., [1] and references therein). Two recent special issues on the subject are [17, 18]. A more general framework suitable for investigating and solving various optimization problems and equilibrium models, even when classical game theory may fail, is known to be the variational inequality (VI) problem, which constitutes a very general class of problems in nonlinear analysis [11].

Building on the VI framework, in this chapter, we present a brief treatment of two classes of Nash problems and their application to the design of cognitive radio (CR) systems [22]. The first class of Nash problems is Nash equilibrium problems (NEP), where the interactions among players take place at the level of objective functions only. The second is the class of generalized Nash equilibrium problems (GNEP), where in addition we have that the choices available to each player also depend on the actions taken by his rivals. We focus on the existence and global uniqueness of equilibria and on distributed algorithms based on the best-response mapping. The results discussed in this chapter are based on [13, 35, 38], to which we refer the interested reader for a more comprehensive treatment of the subject.

The chapter is organized as follows. The first part—Sec. 4.2 for NEPs and Sec. 4.3 for GNEPs—is devoted to the development of general results. The second part of the chapter—Sec. 4.4—applies those results to the design of CR systems.

4.2 Nash Equilibrium Problems (NEPs)

In a general noncooperative game, there are Q players, each of whom has a cost function and a strategy set that may depend on the other players' actions. In a NEP, player i's strategy set $\mathcal{Q}_i \subseteq \mathbb{R}^{n_i}$ is independent of the other players' strategies; player i's cost function $f_i(\mathbf{x}_i, \mathbf{x}_{-i})$ depends on all players' strategies, which are described by a vector $\mathbf{x} \triangleq (\mathbf{x}_1, \ldots, \mathbf{x}_Q)$, where $\mathbf{x}_i$ is the action of the player i and $\mathbf{x}_{-i} \triangleq (\mathbf{x}_1, \ldots, \mathbf{x}_{i-1}, \mathbf{x}_{i+1}, \ldots, \mathbf{x}_Q)$ denotes the vector of all players' strategies variables except that of player i. The joint strategy set of the NEP is given by $\mathcal{Q} = \prod_{i=1}^{Q} \mathcal{Q}_i$, whereas $\mathcal{Q}_{-i} \triangleq \prod_{j \neq i} \mathcal{Q}_j$. The NEP is formally defined by the tuple $\mathcal{G} = \langle \mathcal{Q}, \mathbf{f} \rangle$, with $\mathbf{f} \triangleq (f_i)_{i=1}^{Q}$: The aim of player i, given the other players' strategies $\mathbf{x}_{-i}$, is to choose an $\mathbf{x}_i \in \mathcal{Q}_i$ that minimizes his cost function $f_i(\mathbf{x}_i, \mathbf{x}_{-i})$, i.e.,

$$\begin{array}{ll} \underset{\mathbf{x}_i}{\text{minimize}} & f_i(\mathbf{x}_i, \mathbf{x}_{-i}) \\ \text{subject to} & \mathbf{x}_i \in \mathcal{Q}_i. \end{array} \tag{4.1}$$

Roughly speaking, a NEP is a set of *coupled* optimization problems.

Definition 4.1. A pure strategy Nash equilibrium (NE), or simply a solution of the NEP, is a feasible point $\mathbf{x}^\star$ such that

$$f_i(\mathbf{x}_i^\star, \mathbf{x}_{-i}^\star) \leq f_i(\mathbf{x}_i, \mathbf{x}_{-i}^\star), \qquad \forall \mathbf{x}_i \in \mathcal{Q}_i \tag{4.2}$$

holds for each player $i = 1, \ldots, Q$. □

In words, a NE is a feasible strategy profile $\mathbf{x}^\star$ with the property that no *single* player can benefit from a *unilateral* deviation from $\mathbf{x}_i^\star$. A useful way to see a NE is as a fixed-point of the best-response mapping for each player. Let $\mathcal{B}_i(\mathbf{x}_{-i})$ be the (possibly empty) set of optimal solutions of the ith optimization problem (4.1) and set $\mathcal{B}(\mathbf{x}) \triangleq \mathcal{B}_1(\mathbf{x}_{-1}) \times \mathcal{B}_2(\mathbf{x}_{-2}) \times \cdots \times \mathcal{B}_Q(\mathbf{x}_{-Q})$. It is clear that a point $\mathbf{x}^\star$ is a NE if and only if it is a fixed point of $\mathcal{B}(\mathbf{x})$, i.e., if and only if $\mathbf{x}^\star \in \mathcal{B}(\mathbf{x}^\star)$. This observation is the key to the standard approach to the study of NEPs: the so called fixed-point approach, which is based on the use of the well-developed machinery of fixed-point theory. This approach is adopted in the analysis of several games proposed in the signal processing and communication literature (see, e.g., [40, 8, 34, 6, 28, 33, 32, 36] and [31, 35]). However, the applicability of the fixed-point based analysis as used in the aforementioned papers requires that one be able to compute the best-response mapping $\mathcal{B}(\mathbf{x})$ in closed form, a fact that strongly limits the applicability of this methodology.

In this chapter, we overcome this limitation by studying NEPs through their reduction to a VI problem. This approach is pursued also in [25, 27, 13, 38, 39] and, resting on the well developed theory of VIs, has the advantage of permitting an easy derivation of many results about existence, uniqueness, and stability of the solutions. But its main benefit is probably that it leads quite naturally to the derivation of implementable solution algorithms along with their convergence properties [38].

4.2.1 Connection to Variational Inequalities

The basis of the VI approach to NEPs is the easy equivalence between a NEP and a suitably defined partitioned VI. The (partitioned) VI problem is defined next. Given $\mathcal{Q}$ defined as in Sec. 4.2, let $\mathbf{F} : \Omega \supset \mathcal{Q} \to \mathbb{R}^n$ be a continuous function on

$$\Omega \triangleq \prod_{i=1}^{Q} \Omega_i$$

with each Ω_i being an open subset of $\mathbb{R}^{n_i}$ containing $\mathcal{Q}_i$, and

$$n \triangleq \sum_{i=1}^{Q} n_i$$

We write $\mathbf{F}(\mathbf{x}) = (\mathbf{F}_i(\mathbf{x}))_{i=1}^Q$ where $\mathbf{F}_i : \Omega \to \mathbb{R}^{n_i}$ is the ith component block function of $\mathbf{F}$.

Definition 4.2. The variational inequality, denoted by $\mathrm{VI}(\mathcal{Q},\mathbf{F})$, is to find a vector $\mathbf{x}^\star \in \mathcal{Q}$ such that [11, Def. 1.1.1]

$$(\mathbf{x}-\mathbf{x}^\star)^T \mathbf{F}(\mathbf{x}^\star) \geq 0, \qquad \forall \mathbf{x} \in \mathcal{Q}. \tag{4.3}$$

The set of solutions to this problem is denoted $\mathrm{SOL}(\mathcal{Q},\mathbf{F})$.

Several standard problems in nonlinear programming, game theory and nonlinear analysis can be formulated naturally as VI problems; many examples can be found in [11, Ch. 1], [19, 37]. In particular, the equivalence between NEPs and VIs is given in the following proposition, whose proof follows readily from the minimum principle for convex problems and the Cartesian structure of the joint strategy set $\mathcal{Q}$ [11, Prop. 1.4.2].

Proposition 4.1. *Given the NEP $\mathcal{G} = \langle \mathcal{Q}, \mathbf{f} \rangle$, suppose that for each player i the following hold:*

i) The (nonempty) strategy set $\mathcal{Q}_i$ is closed and convex;
ii) For every fixed $\mathbf{x}_{-i} \in \mathcal{Q}_{-i}$, the payoff function $f_i(\mathbf{x}_i, \mathbf{x}_{-i})$ is convex and continuously differentiable in $\mathbf{x}_i \in \Omega_i \supset \mathcal{Q}_i$.

Then, the game $\mathcal{G}$ is equivalent to the $\mathrm{VI}(\mathcal{Q},\mathbf{F})$*, where* $\mathbf{F}(\mathbf{x}) \triangleq (\nabla_{\mathbf{x}_i} f_i(\mathbf{x}))_{i=1}^Q$.

Building on the VI reformulation above and the well-developed framework of VIs, we focus on the main properties of the NEP, namely: the existence and uniqueness of the solution and the design of distributed algorithms along with their convergence properties. To this end, throughout the chapter we make the following convexity/smoothness assumption.

Assumption 4.1. *For each $i = 1,\ldots,Q$, the set $\mathcal{Q}_i \subset \Omega_i$ is a nonempty, closed and convex subset of $\mathbb{R}^{n_i}$ and the function $f_i(\mathbf{x}_i, \mathbf{x}_{-i})$ is convex in $\mathbf{x}_i \in \mathcal{Q}_i$ for every fixed $\mathbf{x}_{-i} \in \mathcal{Q}_{-i}$ and twice continuously differentiable in $\mathbf{x} \in \Omega \supset \mathcal{Q} = \prod_i \mathcal{Q}_i$ with bounded second derivatives on $\mathcal{Q}$.*

Remark 4.1 (On Assumption 4.1). For the purpose of this chapter, it is enough to focus only on games that satisfy Assumption 4.1 (all the games we study in the second part of the chapter satisfy this condition indeed). However, the assumption can be relaxed, following the techniques developed in [13].

4.2.2 Solution Analysis of the NEP

The solution analysis of a NEP can be carried out in several ways. Here, we address this issue by using the equivalence between the NEP $\mathcal{G}$ and the $\mathrm{VI}(\mathcal{Q},\mathbf{F})$ illustrated in Proposition 4.1. Some of the results in this section (especially those related to existence) could be obtained under weaker assumptions; however, our derivation based on the VI reformulation is interesting in its own right and furthermore prepares the ground work for the algorithmic developments of the next sections. We begin our analysis by introducing some basic definitions.

Definition 4.3. A mapping $\mathbf{F} = (\mathbf{F}_i(\mathbf{x}))_{i=1}^{Q} : \Omega \supset \mathcal{Q} \ni \mathbf{x} \to \mathbb{R}^n$ is

(i) *monotone* on $\mathcal{Q}$ if for all $\mathbf{x}$ and $\mathbf{y}$ in $\mathcal{Q}$,

$$(\mathbf{x}-\mathbf{y})^T(\mathbf{F}(\mathbf{x})-\mathbf{F}(\mathbf{y})) \geq 0; \tag{4.4}$$

(ii) *strictly monotone* on $\mathcal{Q}$ if for all $\mathbf{x} \neq \mathbf{y}$ in $\mathcal{Q}$ the inequality in (4.4) is strict;

(iii) *strongly monotone* on $\mathcal{Q}$ if a constant $c_{\text{sm}} > 0$ exists such that for all $\mathbf{x}$ and $\mathbf{y}$ in $\mathcal{Q}$,

$$(\mathbf{x}-\mathbf{y})^T(\mathbf{F}(\mathbf{x})-\mathbf{F}(\mathbf{y})) \geq c_{\text{sm}} \|\mathbf{x}-\mathbf{y}\|^2. \tag{4.5}$$

The constant c_{sm} is called the strong monotonicity constant of $\mathbf{F}$;

(iv) a *uniform P-function* on $\mathcal{Q} = \prod_i \mathcal{Q}_i$ if a constant $c_{\text{uP}} > 0$ exists such that for all $\mathbf{x} = (\mathbf{x}_i)_{i=1}^{Q}$ and $\mathbf{y} = (\mathbf{y}_i)_{i=1}^{Q}$ in $\mathcal{Q}$,

$$\max_{1\leq i\leq Q} (\mathbf{x}_i-\mathbf{y}_i)^T(\mathbf{F}_i(\mathbf{x})-\mathbf{F}_i(\mathbf{y})) \geq c_{\text{uP}} \|\mathbf{x}-\mathbf{y}\|^2. \tag{4.6}$$

Among the above monotonicity properties, the following relations hold:

$$\boxed{\text{strongly monotone} \Rightarrow \text{uniform P} \Rightarrow \text{strictly monotone} \Rightarrow \text{monotone.}} \tag{4.7}$$

Monotonicity properties in the VI realm play the same role that convex functions play in optimization. In fact, we recall that a differentiable function f is convex (strictly convex, strongly convex) on a convex set $\mathcal{Q}$ if and only if its gradient is monotone (strictly monotone, strongly monotone) on $\mathcal{Q}$. The next theorem collects some rather standard results on solution properties of a VI [11]; thanks to Proposition 4.1, these results readily extend to NEPs.

Theorem 4.1. *Given the NEP $\mathcal{G} = <\mathcal{Q}, \mathbf{f}>$, suppose that $\mathcal{G}$ satisfies Assumption 4.1 and let $\mathbf{F}(\mathbf{x}) \triangleq (\nabla_{\mathbf{x}_i} f_i(\mathbf{x}))_{i=1}^{Q}$. Then the following statements hold:*

(a) The $\text{VI}(\mathcal{Q},\mathbf{F})$ *(the NEP $\mathcal{G}$) has a (possibly empty) closed solution set. If all strategy sets $\mathcal{Q}_i$ are bounded, the solution set is nonempty and thus compact;*

(b) If $\mathbf{F}(\mathbf{x})$ is monotone on $\mathcal{Q}$, then the $\text{VI}(\mathcal{Q},\mathbf{F})$ *(the NEP $\mathcal{G}$) has a convex solution set (possibly empty);*

(c) If $\mathbf{F}(\mathbf{x})$ is strictly monotone on $\mathcal{Q}$, then the $\text{VI}(\mathcal{Q},\mathbf{F})$ *(the NEP $\mathcal{G}$) has at most one solution;*

(d) If $\mathbf{F}(\mathbf{x})$ is a uniformly-P function (or strongly monotone) on $\mathcal{Q}$, then the $\text{VI}(\mathcal{Q},\mathbf{F})$ *(the NEP $\mathcal{G}$) has a unique solution.*

Note that the uniqueness results stated in parts (c) and (d) do not require that the set $\mathcal{Q}$ be bounded. Some sufficient conditions for $\mathbf{F}(\mathbf{x})$ being a (strictly, strongly) monotone or a uniformly-P function are given in the next section (see [38] for more details).

4.2.3 Monotonicity Conditions for the Vector Function F

Assuming that $\mathbf{F}$ is continuously differentiable on Ω, let $\mathbf{JF}(\mathbf{x}) = \left(\mathrm{J}_{\mathbf{x}_j}\mathbf{F}_i(\mathbf{x})\right)_{i,j=1}^{Q}$ be the Jacobian of $\mathbf{F}$, where $\mathrm{J}_{\mathbf{x}_j}\mathbf{F}_i(\mathbf{x})$ is the partial Jacobian matrix of $\mathbf{F}_i$ with respect to the $\mathbf{x}_j$ vector. Note that when $\mathbf{F}(\mathbf{x}) = (\nabla_{\mathbf{x}_i} f_i(\mathbf{x}))_{i=1}^{Q}$, with each $f_i : \Omega_i \mapsto \mathbb{R}$ being a continuously differentiable function on Ω_i, we have $\mathrm{J}_{\mathbf{x}_j}\mathbf{F}_i(\mathbf{x}) = \nabla^2_{\mathbf{x}_i\mathbf{x}_j} f_i(\mathbf{x})$, for $i, j = 1, \ldots, Q$, where $\mathrm{J}_{\mathbf{x}_i}\mathbf{F}_i(\mathbf{x}) = \nabla^2_{\mathbf{x}_i\mathbf{x}_i} f_i(\mathbf{x})$ is the Hessian matrix of f_i.

It is well known and easy to show [24] that the following relations exist among the monotonicity properties of $\mathbf{F}$ and the definiteness properties of the Jacobian matrix $\mathbf{JF}$.

i)	$\mathbf{F}(\mathbf{x})$ is monotone on $\mathcal{Q}$	$\Leftrightarrow$	$\mathbf{JF}(\mathbf{x}) \succeq \mathbf{0},\ \forall \mathbf{x} \in \mathcal{Q};$
ii)	$\mathbf{F}(\mathbf{x})$ is strictly monotone on $\mathcal{Q}$	$\Leftarrow$	$\mathbf{JF}(\mathbf{x}) \succ \mathbf{0},\ \forall \mathbf{x} \in \mathcal{Q};$
iii)	$\mathbf{F}(\mathbf{x})$ is strongly monotone on $\mathcal{Q}$	$\Leftrightarrow$	$\mathbf{JF} - c_{\mathrm{sm}}\mathbf{I} \succeq \mathbf{0},\ \forall \mathbf{x} \in \mathcal{Q}$

(4.8)

where $\mathbf{A} \succeq \mathbf{B}$ ($\mathbf{A} \succ \mathbf{B}$) means that $\mathbf{A} - \mathbf{B}$ is a positive semidefinite (definite) matrix. In some applications we consider later (cf. Sec. 4.4), it is useful to have at hand some further sufficient conditions that guarantee monotonicity properties of $\mathbf{F}$. Below we give two such conditions. Let define the matrix $\mathbf{JF}_{\mathrm{low}}$ having the same dimension of $\mathbf{JF}(\mathbf{x})$ as

$$[\mathbf{JF}_{\mathrm{low}}]_{rs} \triangleq \begin{cases} \inf\limits_{\mathbf{x}\in\mathcal{Q}} [\mathbf{JF}(\mathbf{x})]_{rr}, & \text{if } r = s, \\ -\sup\limits_{\mathbf{x}\in\mathcal{Q}} \left|[\mathbf{JF}(\mathbf{x})]_{rs}\right|, & \text{otherwise}, \end{cases} \tag{4.9}$$

and let introduce the "condensed" $Q \times Q$ real matrices $\Upsilon_{\mathbf{F}}$ and Γ_F, given by

$$[\Upsilon_{\mathbf{F}}]_{ij} \triangleq \begin{cases} \alpha_i^{\min}, & \text{if } i = j, \\ -\beta_{ij}^{\max}, & \text{otherwise}, \end{cases} \tag{4.10}$$

and

$$[\Gamma_{\mathbf{F}}]_{ij} \triangleq \begin{cases} \dfrac{1}{1+\alpha_i^{\min}}, & \text{if } i = j, \\ \dfrac{\beta_{ij}^{\max}}{1+\alpha_i^{\min}}, & \text{otherwise}, \end{cases} \tag{4.11}$$

where

$$\alpha_i^{\min} \triangleq \inf_{\mathbf{z}\in\mathcal{Q}} \lambda_{\mathrm{least}}\left(\mathrm{J}_{\mathbf{x}_i}\mathbf{F}_i(\mathbf{z})\right) \quad \text{and} \quad \beta_{ij}^{\max} \triangleq \sup_{\mathbf{z}\in\mathcal{Q}} \left\|\mathrm{J}_{\mathbf{x}_j}\mathbf{F}_i(\mathbf{z})\right\|, \tag{4.12}$$

with $\lambda_{\text{least}}(\mathbf{A})$ denoting the least eigenvalue of $\mathbf{A}$.[1] In order to explore the relationship between the two matrices $\Upsilon_{\mathbf{F}}$ and Γ_F, we need the following definition.

Definition 4.4. A matrix $\mathbf{M} \in \mathbb{R}^{n\times n}$ is called *P*-matrix if every principal minor of $\mathbf{M}$ is positive.

Many equivalent characterizations for a P-matrix can be given. The interested reader is referred to [7, 4] for more details. Here we note that any positive definite matrix is a P-matrix, but the reverse does not hold (unless the matrix is symmetric).

Building on the properties of the P-matrices [7, Lemma 13.14], one can show that $\Upsilon_{\mathbf{F}}$ is a P-matrix if and only if $\rho(\Gamma_{\mathbf{F}}) < 1$, where $\rho(\mathbf{A})$ denotes the spectral radius of $\mathbf{A}$. The P-property of matrix $\Upsilon_{\mathbf{F}}$ will be used to prove the uniqueness of the equilibrium of Nash games as well as the convergence of some of the proposed distributed algorithms. The positive definiteness property of $\Upsilon_{\mathbf{F}}$ and the strong monotonicity constant c_{sm} of $\mathbf{F}$ will be exploited to study the GNEPs in Sec. 4.3. Matrices $\mathbf{JF}_{\text{low}}$ and $\Upsilon_{\mathbf{F}}$ are also instrumental towards obtaining sufficient conditions for the monotonicity of the mapping $\mathbf{F}$ (more results and milder conditions can be found in [38]), as given next.

Proposition 4.2. *Let* $\mathbf{F} : \Omega \supset \mathcal{Q} \to \mathbb{R}^n$ *be continuously differentiable with bounded derivatives on* $\mathcal{Q}$. *The following statements hold:*

(a) *If either* $\mathbf{JF}_{low}$ *or* $\Upsilon_{\mathbf{F}}$ *are positive semidefinite, then* $\mathbf{F}$ *is* monotone *on* $\mathcal{Q}$;
(b) *If either* $\mathbf{JF}_{low}$ *or* $\Upsilon_{\mathbf{F}}$ *are positive definite, then* $\mathbf{F}$ *is* strongly monotone *on* $\mathcal{Q}$, *with strong monotonicity constant given by either* $c_{\text{sm}} = \lambda_{least}(\mathbf{JF}_{low})$ or $c_{\text{sm}} = \lambda_{least}(\Upsilon_F)$;
(c) *Let* $\mathbf{F} = (\mathbf{F}_i(\mathbf{x}))_{i=1}^{Q} : \Omega = \prod_i \Omega_i \supset \mathcal{Q} = \prod_i \mathcal{Q}_i \ni \mathbf{x} \to \mathbb{R}^n$. *If* $\Upsilon_{\mathbf{F}}$ *is a P-matrix, then* $\mathbf{F}$ *is a uniformly P-function on* $\mathcal{Q}$, *for some positive* c_{uP}.

A lower bound of c_{uP} is given in [38].

Remark 4.2 (On the uniqueness of the NE (cont'd)). Invoking Theorem 4.1 and Proposition 4.2, one can readily obtain sufficient conditions for the uniqueness of the NE of the game $\mathcal{G}$. According to Theorem 4.1(d) and Proposition 4.2(c) indeed, the NE of $\mathcal{G}$ is unique if the matrix $\Upsilon_{\mathbf{F}}$ in (4.10) with $\mathbf{F}(\mathbf{x}) = (\nabla_{\mathbf{x}_i} f_i(\mathbf{x}))_{i=1}^{Q}$ is a *P*-matrix. It turns out that this condition is sufficient also for global convergence of best-response asynchronous distributed algorithms described in Sec. 4.2.4.1. To give additional insight into the uniqueness of the NE of $\mathcal{G}$ we provide the following diagonal dominance-type conditions for the matrix $\Upsilon_{\mathbf{F}}$ to be a *P*-matrix (positive definite).

Proposition 4.3. *The matrix* $\Upsilon_{\mathbf{F}}$ *in* (4.10) *is a P-matrix (positive definite) if one of (both) the following two sets of conditions are satisfied: for some* $\mathbf{w} = (w_i)_{i=1}^{Q} > \mathbf{0}$,

$$\frac{1}{w_i}\sum_{j\neq i} w_j \frac{\beta_{ij}^{\max}}{\alpha_i^{\min}} < 1, \ \forall i = 1,\cdots,Q, \quad \frac{1}{w_j}\sum_{i\neq j} w_i \frac{\beta_{ij}^{\max}}{\alpha_i^{\min}} < 1, \ \forall j = 1,\cdots,Q. \qquad (4.13)$$

[1] The least eigenvalue of a real (not necessarily symmetric) matrix $\mathbf{A}$ is the smallest eigenvalue of the symmetric part of $\mathbf{A}$.

Note that if $\Upsilon_{\mathbf{F}}$ is a P-matrix, it must be

$$\alpha_i^{\min} = \inf_{\mathbf{z}\in\mathcal{Q}} \left[\lambda_{\min}(\nabla^2_{\mathbf{x}_i} f_i(\mathbf{z}))\right] > 0$$

for all i, where $\lambda_{\min}(\nabla^2_{\mathbf{x}_i} f_i(\mathbf{z}))$ denotes the minimum eigenvalue of $\nabla^2_{\mathbf{x}_i} f_i(\mathbf{z})$. Thus an implicit consequence of the P assumption of the matrix $\Upsilon_{\mathbf{F}}$ is the uniform positive definiteness of the matrices $\nabla^2_{\mathbf{x}_i} f_i$ on $\mathcal{Q}$, which implies the uniformly strong convexity of $f_i(\cdot, \mathbf{x}_{-i})$ for all $\mathbf{x}_{-i} \in \mathcal{Q}_{-i}$.

4.2.4 *Distributed Algorithms for Nash Equilibria*

In this section, we discuss some iterative algorithms for computing a NE of NEP (a solution of the VI). For the purposes of this chapter we restrict our attention to distributed algorithms, with special emphasis to best-response iterative algorithms.

4.2.4.1 Best Response Decomposition Algorithms

We focus on asynchronous-iterative algorithms, since they are particularly suitable for CR applications. More specifically, we consider *totally asynchronous* schemes (in the sense specified in [5]), where some players may update their strategies more frequently than others and they may even use an outdated information about the strategy profile used by the others. To provide a formal description of the algorithm, we need to introduce some preliminary definitions. Let $\mathcal{T}_i \subseteq \mathcal{T} \subseteq \{0,1,2,\ldots\}$ be the set of times at which player i updates his own strategy $\mathbf{x}_i$, denoted by $\mathbf{x}_i^{(n)}$ (thus, implying that, at time $n \notin \mathcal{T}_i$, $\mathbf{x}_i^{(n)}$ is left unchanged). Let $t_j^i(n)$ denote the most recent time at which the strategy profile of player j is perceived by player i at the nth iteration (observe that $t_j^i(n)$ satisfies $0 \le t_j^i(n) \le n$). Hence, if player i updates his strategy at the nth iteration, then he minimizes his cost function using the following (possibly) outdated strategy profile of the other players:

$$\mathbf{x}_{-i}^{(\mathbf{t}^i(n))} \triangleq \left(\mathbf{x}_1^{(t_1^i(n))}, \ldots, \mathbf{x}_{i-1}^{(t_{i-1}^i(n))}, \mathbf{x}_{i+1}^{(t_{i+1}^i(n))}, \ldots, \mathbf{x}_Q^{(t_Q^i(n))}\right). \tag{4.14}$$

Some standard conditions in asynchronous convergence theory, which are fulfilled in any practical implementation, need to be satisfied by the schedule $\mathcal{T}_i$ and $t_j^i(n)$; we refer to [5, 32] for the details. Throughout the chapter we assume that these conditions are satisfied and call feasible such an updating schedule. Using the above definitions, the totally asynchronous algorithm based on the best-responses of the players is described in Algorithm 4.1. The convergence properties of the algorithm are given in Theorem 4.2.

Theorem 4.2 ([38]). *Let $\mathcal{G} = <\mathcal{Q}, \mathbf{f}>$ satisfy Assumption 4.1 and let $\mathbf{F} = (\nabla_{\mathbf{x}_i} f_i)_{i=1}^{Q}$. If $\Upsilon_{\mathbf{F}}$ defined in (4.10) is a P-matrix, any sequence $\{\mathbf{x}^{(n)}\}_{n=0}^{\infty}$ generated by the asynchronous best-response algorithm described in Algorithm 4.1 converges to the unique NE of $\mathcal{G}$, for any given updating feasible schedule of the players.*

Algorithm 4.1: Asynchronous Best-Response Algorithm

Data : Choose any feasible starting point $\mathbf{x}^{(0)} = (\mathbf{x}_i^{(0)})_{i=1}^{Q}$; set $n = 0$.
(S.1) : If $\mathbf{x}^{(n)}$ satisfies a suitable termination criterion: STOP
(S.2) : For $i = 1, \ldots, Q$, compute $\mathbf{x}_i^{(n+1)}$ as

$$\mathbf{x}_i^{(n+1)} = \begin{cases} \mathbf{x}_i^{\star} \in \underset{\mathbf{x}_i \in \mathcal{Q}_i}{\operatorname{argmin}} f_i\left(\mathbf{x}_i, \mathbf{x}_{-i}^{(\mathbf{t}^i(n))}\right), & \text{if } n \in \mathcal{T}_i \\ \mathbf{x}_i^{(n)}, & \text{otherwise} \end{cases} \tag{4.15}$$

end
(S.3) : $n \leftarrow n + 1$; go to (S.1).

Remark 4.3 (Flexibility of the algorithm). Algorithm 4.1 contains as special cases a plethora of algorithms, each one obtained by a possible choice of the scheduling of the users in the updating procedure (i.e., the parameters $\{t_r^q(n)\}$ and $\{\mathcal{T}_q\}$). Examples are the *sequential* (Gauss–Seidel scheme) and the *simultaneous* (Jacobi scheme) updates, where the players update their own strategies *sequentially* and *simultaneously*, respectively. Interestingly, Theorem 4.2 states that all these algorithms are robust against missing or outdated updates of the players and are guaranteed to converge to the unique NE of the game under the same set of convergence conditions, since the matrix $\Upsilon_{\mathbf{F}}$ (or $\Gamma_{\mathbf{F}}$) does not depend on the particular choice of $\{t_r^q(n)\}$ and $\{\mathcal{T}_q\}$. This feature strongly relaxes the constraints on the synchronization of the players' updates, which makes this class of algorithms appealing in many practical distributed systems.

Remark 4.4 (On the convergence conditions of best-response algorithms). We have pointed out that the P property of $\Upsilon_{\mathbf{F}}$ (or equivalently $\rho(\Gamma_{\mathbf{F}}) < 1$) cannot be satisfied even if there is just one point where one player has a payoff function with singular Hessian. In fact, if this is the case, we have, $\alpha_i^{\min} = 0$ for some i, let us say $i = 1$ without loss of generality, which implies that the matrix $\Gamma_{\mathbf{F}}$ has a 1 in the left upper corner. Since the matrix $\Gamma_{\mathbf{F}}$ is nonnegative, we have that this implies $\rho(\Gamma_{\mathbf{F}}) \geq 1$ [3, Th. 1.7.4]. Assuming that the element 1 is contained in an irreducible principal matrix, we will actually have $\rho(\Gamma_{\mathbf{F}}) > 1$. Note that the irreducibility assumption is extremely weak and trivially satisfied if the matrix $\Gamma_{\mathbf{F}}$ is positive, which is true in most of our applications. We thus focus in the next subsection on alternative distributed algorithms that are guaranteed to converge under milder assumptions that do not require the strict or strong convexity of the payoff functions. These milder conditions required on $\mathbf{F}$ to have convergence are traded for a slightly increasing computational/signaling complexity.

4.2.4.2 Proximal Decomposition Algorithms for Monotone VIs

We focus now on distributed algorithms whose convergence is guaranteed under the monotonicity assumption on the mapping $\mathbf{F} = (\nabla_{\mathbf{x}_i} f_i)_{i=1}^Q$. Let us introduce first the following assumption.

Assumption 4.2. *The mapping* $\mathbf{F} = (\nabla_{\mathbf{x}_i} f_i)_{i=1}^Q$ *is monotone on* $\mathcal{Q} = \mathcal{Q}_1 \times \ldots \times \mathcal{Q}_Q$.

According to Proposition 4.1, solving the NEP $\mathcal{G} =< \mathcal{Q}, \mathbf{f} >$ is equivalent to solving the VI$(\mathcal{Q}, \mathbf{F})$ that, under Assumption 4.2, is monotone. This can be done in a host of ways (see, e.g., [11, Vol. II]) but not directly by decomposition methods. Here our interest is on devising distributed solution methods. To pursue this goal, we propose the following approach. We consider a regularization of the VI$(\mathcal{Q}, \mathbf{F})$, given by VI$(\mathcal{Q}, \mathbf{F} + \tau(\mathbf{I} - \mathbf{y}))$, where $\mathbf{I}$ is the identity map (i.e., $\mathbf{I} : \mathbf{x} \to \mathbf{x}$), $\mathbf{y}$ is a fixed vector in $\mathbb{R}^n$, and τ is a positive constant. Under Assumption 4.2, this regularized problem is strongly monotone and thus has a unique solution (cf. Theorem 4.1); we denote by $\mathbf{S}_\tau(\mathbf{y}) \triangleq \text{SOL}(\mathcal{Q}, \mathbf{F} + \tau(\mathbf{I} - \mathbf{y}))$ such a unique solution. The relationship between the original game $\mathcal{G} =< \mathcal{Q}, \mathbf{f} >$ and the regularized VI is given in the following.

Lemma 4.1. *Given the game* $\mathcal{G} =< \mathcal{Q}, \mathbf{f} >$, *suppose that Assumptions 4.1 and 4.2 hold. A tuple* $\mathbf{x}^\star \in \mathcal{Q}$ *is a NE of the game if and only if it is a fixed point of the vector function* $\mathbf{S}_\tau(\mathbf{y})$, *i.e.,* $\mathbf{x}^\star = \mathbf{S}_\tau(\mathbf{x}^\star)$.

Under the monotonicity of $\mathbf{F}$, the mapping $\mathbf{S}_\tau(\mathbf{y})$ can be shown to be nonexpansive, meaning that, starting at a given iterate $\mathbf{y}^{(0)} \in \mathcal{Q}$, the sequence generated by a proper averaging of $\mathbf{S}_\tau(\mathbf{y}^{(n)})$ and $\mathbf{y}^{(n)}$ converges to a solution of the VI$(\mathcal{Q}, \mathbf{F})$. This idea is formalized in Algorithm 4.2 below, whose convergence properties are given in Theorem 4.3. Note that the convergence of the algorithm requires only the monotonicity of $\mathbf{F}$. Moreover, one can also replace the exact computation of the solution $\mathbf{S}_\tau(\mathbf{x}^{(n)})$ (see step 2) of the regularized VI$(\mathcal{Q}, \mathbf{F} + \tau(\mathbf{I} - \mathbf{x}^{(n)}))$ with an inexact solution, without affecting the convergence of Algorithm 4.2 (provided that the error bound goes to zero as $n \to \infty$).

Algorithm 4.2: Proximal Decomposition Algorithm (PDA)

Data : Let $\{\varepsilon_n\}_{n=0}^\infty$, $\{\rho_n\}_{n=0}^\infty$, and $\tau > 0$ be given, and choose any feasible starting point $\mathbf{x}^{(0)}$; set $n = 0$.
(S.1) : If $\mathbf{x}^{(n)}$ satisfies a suitable termination criterion: STOP.
(S.2) : Find a point $\mathbf{z}^{(n)}$ such that $\|\mathbf{z}^{(n)} - \mathbf{S}_\tau(\mathbf{x}^{(n)})\| \leq \varepsilon_n$.
(S.3) : Set $\mathbf{x}^{(n+1)} \triangleq (1 - \rho_n)\mathbf{x}^{(n)} + \rho_n \mathbf{z}^{(n)}$.
(S.4) : $n \leftarrow n + 1$; go to (S.1).

Theorem 4.3 ([38]). *Let* $\mathcal{G} =< \mathcal{Q}, \mathbf{f} >$ *satisfy Assumptions 4.1 and 4.2 and let* $\mathbf{F} \triangleq (\nabla_{\mathbf{x}_i} f_i)_{i=1}^Q$. *Let* $\{\varepsilon_n\} \subset [0, \infty)$ *be a sequence such that* $\sum_{n=1}^\infty \varepsilon_n < \infty$, *and let* ρ_n *be such that* $\{\rho_n\} \subset [R_m, R_M]$ *with* $0 < R_m \leq R_M < 2$. *Then, the sequence* $\{\mathbf{x}^{(n)}\}_{n=0}^\infty$ *generated by the proximal decomposition algorithm described in Algorithm 4.2 converges to a solution of the game* $\mathcal{G}$.

A key point now becomes how to compute, for any given $\mathbf{x}^{(n)}$, a(n approximated) solution $\mathbf{S}_\tau(\mathbf{x}^{(n)})$ of the regularized VI$(\mathcal{Q}, \mathbf{F} + \tau(\mathbf{I} - \mathbf{x}^{(n)}))$ in a distributed way. The interesting point is that, going backwards by applying Proposition 4.1 to the VI$(\mathcal{Q}, \mathbf{F} + \tau(\mathbf{I} - \mathbf{x}^{(n)}))$, one can see that $\mathbf{S}_\tau(\mathbf{x}^{(n)})$ coincides with the unique (under Assumption 4.2) NE of the following regularized game:

$$\begin{array}{ll} \underset{\mathbf{x}_i}{\text{minimize}} & f_i(\mathbf{x}_i, \mathbf{x}_{-i}) + \frac{\tau}{2}\|\mathbf{x}_i - \mathbf{x}_i^{(n)}\|^2 \\ \text{subject to} & \mathbf{x}_i \in \mathcal{Q}_i \end{array} \qquad \forall i = 1, \ldots, Q. \tag{4.16}$$

It turns out that $\mathbf{S}_\tau(\mathbf{x}^{(n)})$ can be computed in a distributed way using any iterative algorithm falling in class of asynchronous algorithms described in Algorithm 4.1 and applied to the regularized game in (4.16); Theorem 4.2 states that such a class of algorithms globally converges if the matrix

$$\Upsilon_{\mathbf{F},\tau} \triangleq \Upsilon_{\mathbf{F}} + \tau\mathbf{I}, \tag{4.17}$$

with $\Upsilon_{\mathbf{F}}$ defined in (4.10), is a P-matrix, which is guaranteed for any τ sufficiently large. Stated in mathematical terms, we have the following.

Corollary 4.1 ([38]). *In the setting of Theorem 4.3, if τ is chosen sufficiently large so that $\Upsilon_{\mathbf{F},\tau}$ is a P-matrix, then any asynchronous best-response algorithm (see Algorithm 4.1) applied to the game in (4.16) converges to $\mathbf{S}_\tau(\mathbf{x}^{(n)})$.*

The only thing left to discuss at this point is how to check whether the condition $\|\mathbf{z}^{(n)} - \mathbf{S}_\tau(\mathbf{x}^{(n)})\| \leq \varepsilon_n$ in step 2 is satisfied. This certainly can be accomplished, but it is a rather technical issue and we refer to [38] for practical implementations of this check, under a different level of signaling and supervision.

Remark 4.5 (On the structure of Algorithm 4.2). Algorithm 4.2 is *only conceptually* a double loop method. It is indeed very close to the iterative best-response algorithm applied to game (4.16) (see Algorithm 4.1); the only difference being that "from time to time" (more precisely when the inner termination test $\|\mathbf{z}^{(n)} - \mathbf{S}_\tau(\mathbf{x}^{(n)})\| \leq \varepsilon_n$ is satisfied), the objective functions of the players are changed by changing the regularizing term from $\frac{\tau}{2}\|\mathbf{x}_i - \mathbf{x}_i^{(n)}\|^2$ to $\frac{\tau}{2}\|\mathbf{x}_i - \mathbf{x}_i^{(n+1)}\|^2$.

Note that Algorithm 4.2 does not suffer of the main drawback of best-response based schemes (cf. Remark 4), since the convergence conditions as given in Theorem 4.3 do not require the strong convexity of the payoff functions $f_i(\cdot, \mathbf{x}_{-i})$. The only (sufficient) condition we need is the monotonicity of $\mathbf{F} = (\nabla_{\mathbf{x}_i} f_i)_{i=1}^Q$ on $\mathcal{Q}$ (cf. Proposition 4.2).

4.3 Generalized Nash Equilibrium Problems (GNEP)

In all previous developments we have assumed that the feasible set of each player is independent of the rival players' choices, but this is not always the case. There are many applications of interest where the feasible sets naturally depend on the variables of the player's rivals (see, e.g., Sec. 4.4). The GNEP extends the classical NEP setting described so far precisely by assuming that each player's strategy set can depend on the rival players' strategies $\mathbf{x}_{-i}$. In order to describe a GNEP we denote by $\mathcal{Q}_i(\mathbf{x}_{-i}) \subseteq \mathbb{R}^{n_i}$ the feasible set of player i when the other players choose $\mathbf{x}_{-i}$. Analogously to the NEP case, the aim of each player i, given $\mathbf{x}_{-i}$, is to choose a strategy $\mathbf{x}_i \in \mathcal{Q}_i(\mathbf{x}_{-i})$ that solves the problem

$$\begin{array}{ll} \underset{\mathbf{x}_i}{\text{minimize}} & f_i(\mathbf{x}_i, \mathbf{x}_{-i}) \\ \text{subject to} & \mathbf{x}_i \in \mathcal{Q}_i(\mathbf{x}_{-i}). \end{array} \tag{4.18}$$

Definition 4.5. A generalized Nash equilibrium (GNE), or simply a solution of the GNEP, is a feasible point $\mathbf{x}^\star$ such that

$$f_i(\mathbf{x}_i^\star, \mathbf{x}_{-i}^\star) \leq f_i(\mathbf{x}_i, \mathbf{x}_{-i}^\star), \quad \forall \mathbf{x}_i \in \mathcal{Q}_i(\mathbf{x}_{-i}^\star) \tag{4.19}$$

holds for each player $i = 1, \ldots, Q$.

Due to the variability of the feasible sets, the GNEP is a much harder problem than an ordinary NEP. Indeed, in its full generality, the GNEP problem is almost intractable and also the VI approach is of no great help. We then restrict our attention to particular classes of (more tractable) equilibrium problems: the so-called GNEPs with *jointly convex shared constraints* (see [10] for a survey on GNEPs).

Definition 4.6. A GNEP is termed as *GNEP with jointly convex shared constraints* if the feasible sets are defined as

$$\mathcal{Q}_i(\mathbf{x}_{-i}) \triangleq \left\{\mathbf{x}_i \in \overline{\mathcal{Q}}_i \,:\, \mathbf{g}(\mathbf{x}_i, \mathbf{x}_{-i}) \leq \mathbf{0}\right\}, \tag{4.20}$$

where $\overline{\mathcal{Q}}_i \subseteq \mathbb{R}^{n_i}$ is the closed and convex set of individual constraints of player i and $\mathbf{g}(\mathbf{x}_i, \mathbf{x}_{-i}) \leq \mathbf{0}$ represents the set of shared coupling constraints (equal for all the players), with $\mathbf{g} \triangleq (g_j)_{j=1}^m : \mathbb{R}^n \to \mathbb{R}^m$ (jointly) convex in $\mathbf{x}$.

Note that if there are no shared constraints the problem reduces to a standard NEP. We can give a geometric interpretation to (4.20), as shown next. For a GNEP with shared constraints, let us define a set $\mathcal{Q}$ in the product space of all players:

$$\mathcal{Q} \triangleq \left\{\mathbf{x} \in \mathbb{R}^n \,:\, \mathbf{g}(\mathbf{x}) \leq \mathbf{0} \text{ and } \mathbf{x}_i \in \overline{\mathcal{Q}}_i, \, \forall i = 1, \ldots, Q\right\}. \tag{4.21}$$

It is easy to check that the set $\mathcal{Q}$ is closed and convex and that each feasible set is just a "slice" of the "big" set $\mathcal{Q}$:

$$\mathcal{Q}_i(\mathbf{x}_{-i}) = \left\{\mathbf{x}_i \in \overline{\mathcal{Q}}_i \,:\, \mathbf{g}(\mathbf{x}_i, \mathbf{x}_{-i}) \leq \mathbf{0}\right\} = \left\{\mathbf{x}_i \in \mathbb{R}^n \,:\, (\mathbf{x}_i, \mathbf{x}_{-i}) \in \mathcal{Q}\right\}. \tag{4.22}$$

Similarly to the NEP, throughout the chapter we make the following convexity/smoothness assumption for the GNEP with shared constraints.

Assumption 4.3. *For each $i = 1, \ldots, Q$, the set $\overline{\mathcal{Q}}_i$ is a nonempty, closed and convex subset of $\mathbb{R}^{n_i}$ and the function $f_i(\mathbf{x}_i, \mathbf{x}_{-i})$ is twice continuously differentiable in $\mathbf{x}$ and* *convex in* $\mathbf{x}_i$ *for every fixed* $\mathbf{x}_{-i}$*; the functions* $\mathbf{g}(\mathbf{x}) = (g_j(\mathbf{x}))_{j=1}^m$ *are continuously differentiable and jointly convex in* $\mathbf{x}$.

4.3.1 Connection to VIs: The Variational Solutions

GNEPs with shared constraints are still very difficult, but some types of solutions can be studied and calculated relatively easily by using a VI approach. More specifically, invoking the minimum principle, one can readily obtain the following connection between GNEPs and VIs (see, e.g., [9, 2]).

Lemma 4.2. *Let* $\mathcal{G} =< \mathcal{Q}, \mathbf{f} >$ *be a GNEP with shared constraints. Suppose that* $\mathcal{G}$ *satisfies Assumption 4.3 and let VI*$(\mathcal{Q},\mathbf{F})$ *be the VI with* $\mathcal{Q}$ *defined in (4.21) and* $\mathbf{F} \triangleq (\nabla_{\mathbf{x}_i} f_i)_{i=1}^Q$. *Then, every solution of the VI*$(\mathcal{Q},\mathbf{F})$ *is a solution of the GNEP* $\mathcal{G}$.

Note that in passing from the GNEP to the associated VI not all the GNEP solutions are preserved: Lemma 4.2 in fact does not state that any solution of the GNEP is also a solution of the VI (see [9, 2, 12] for further details and examples). Solutions of the GNEP that are also solutions of the VI$(\mathcal{Q},\mathbf{F})$ are termed as *variational solutions* [10] or *normalized solutions* [29]. Variational solutions enjoy some remarkable properties that make them particularly appealing in many applications. To discuss this issue further, assume that we have a GNEP with jointly convex shared constraints satisfying Assumption 4.3 and that some constraint qualification (CQ) (see, e.g., [11, Sec. 3.2]) holds at all elements in every set $\mathcal{Q}_i(\mathbf{x}_{-i})$ defined in (4.22). Under these assumptions the GNEP is equivalent to its KKT system that is obtained by concatenating the Karush-Kuhn-Tucker (KKT) conditions of the convex optimization problem in (4.18) of each player: $\mathbf{x}^\star$ is a solution of the GNEP if and only if there exist multipliers $\boldsymbol{\lambda}^\star = (\boldsymbol{\lambda}^{(i)\star})_{i=1}^Q \in \mathbb{R}_+^{Qm}$ such that

$$\begin{array}{ll} \mathbf{0} \in \nabla_{\mathbf{x}_i}\mathcal{L}_i(\mathbf{x}_i^\star, \mathbf{x}_{-i}^\star, \boldsymbol{\lambda}^{(i)\star}) + \mathcal{N}(\mathbf{x}_i^\star, \overline{\mathcal{Q}}_i) & \\ \mathbf{0} \leq \boldsymbol{\lambda}^{\star(i)} \perp \mathbf{g}(\mathbf{x}^\star) \leq \mathbf{0} & \end{array} \quad \forall i = 1,\ldots,Q, \tag{4.23}$$

where

$$\mathcal{L}_i(\mathbf{x}_i, \mathbf{x}_{-i}, \boldsymbol{\lambda}^{(i)}) \triangleq f_i(\mathbf{x}) + \boldsymbol{\lambda}^{(i)T}\mathbf{g}(\mathbf{x}) \tag{4.24}$$

is the Lagrangian function of player i's optimization problem (4.18), with $\boldsymbol{\lambda}^{(i)} = (\lambda_k^{(i)})_{k=1}^m$ denoting the multipliers of the shared constraints $\mathbf{g}(\mathbf{x}) \leq \mathbf{0}$, and $\mathcal{N}(\mathbf{x}_i^\star, \overline{\mathcal{Q}}_i) \triangleq \{\mathbf{d} \in \mathbb{R}^{n_i} : \mathbf{d}^T(\mathbf{y} - \mathbf{x}_i^\star) \leq \mathbf{0}, \quad \forall \mathbf{y} \in \overline{\mathcal{Q}}_i\}$ is the normal cone to $\overline{\mathcal{Q}}_i$ at $\mathbf{x}_i^\star$. Similarly, assume some suitable CQ at all elements in the set $\mathcal{Q}$ defined in (4.21). Then, the VI$(\mathcal{Q},\mathbf{F})$, with $\mathbf{F} \triangleq (\nabla_{\mathbf{x}_i} f_i)_{i=1}^Q$, is equivalent to its KKT system [11]: $\bar{\mathbf{x}}$ is a solution of the VI$(\mathcal{Q},\mathbf{F})$ if and only if there exist multipliers $\bar{\boldsymbol{\lambda}} \in \mathbb{R}_+^m$ such that

$$\begin{array}{ll} \mathbf{0} \in \nabla_{\mathbf{x}_i}\mathcal{L}_i\left(\bar{\mathbf{x}}_i, \bar{\mathbf{x}}_{-i}, \bar{\boldsymbol{\lambda}}\right) + \mathcal{N}(\bar{\mathbf{x}}_i, \overline{\mathcal{Q}}_i), & \forall i = 1,\ldots,Q \\ \mathbf{0} \leq \bar{\boldsymbol{\lambda}} \perp \mathbf{g}(\bar{\mathbf{x}}) \leq \mathbf{0} & \end{array} \tag{4.25}$$

Comparing (4.23) with (4.25) it is not difficult to see that the KKT system (4.25) is a special case of (4.23) with $\lambda^{\star(1)} = \cdots = \lambda^{\star(Q)} = \overline{\lambda}$, meaning that a solution $\overline{\mathbf{x}}$ of the GNEP is a variational solution if and only if the shared constraints have the same multipliers for all the players. More formally, we have the following.

Lemma 4.3. *Let $\mathcal{G} =< \mathcal{Q}, \mathbf{f} >$ be a GNEP with shared constraints satisfying Assumption 4.3, and let VI($\mathcal{Q}, \mathbf{F}$) be the VI with $\mathcal{Q}$ defined in (4.21) and $\mathbf{F} \triangleq (\nabla_{\mathbf{x}_i} f_i)_{i=1}^{Q}$. Then, the following hold:*
(i) Suppose that $\overline{\mathbf{x}}$ is a solution of the VI($\mathcal{Q}, \mathbf{F}$) at which the KKT (4.25) holds with multipliers $\overline{\lambda}$. Then $\overline{\mathbf{x}}$ is a solution of the GNEP at which the KKT (4.23) holds with $\lambda^{\star(1)} = \cdots = \lambda^{\star(Q)} = \overline{\lambda}$;
(ii) Conversely, suppose that $\mathbf{x}^\star$ is a solution of the GNEP at which the KKT (4.23) holds with $\lambda^{\star(1)} = \cdots = \lambda^{\star(Q)}$. Then $\mathbf{x}^\star$ os a solution of the VI($\mathcal{Q}, \mathbf{F}$) and the pair $(\mathbf{x}^\star, \lambda^{\star(1)})$ satisfies the KKT (4.25).

The importance of Lemmas 4.2 and 4.3 is twofold. First, we can use VI results as given in Sec. 4.2 (see also [11, Sec. 2]) to obtain practical conditions ensuring the existence and the uniqueness of variational solutions for a GNEP with shared constraints; we leave to the reader the easy task of duplicating these results. Second, Lemma 4.3 gives rise to an interesting game theoretical pricing-based interpretation of the variational solutions, useful for design of distributed algorithms, as detailed in the next section.

4.3.2 Distributed Algorithms for Variational Solutions

In this section, we develop distributed algorithms to compute the variational solutions of a GNEP with jointly convex shared constraints. The presence of the coupling constraints $\mathbf{g}(\mathbf{x}) \leq \mathbf{0}$ prevents a direct application of the decomposition methods presented in Sec. 4.2.4 to the VI($\mathcal{Q}, \mathbf{F}$), because $\mathcal{Q}$ does not have a Cartesian structure. To overcome this issue, we rewrite the VI($\mathcal{Q}, \mathbf{F}$) in a more convenient form, as detailed next. Consider the problem of finding a couple $(\bar{\mathbf{x}}, \bar{\lambda})$, with $\bar{\lambda} = (\bar{\lambda}_k)_{k=1}^m$ such that $\bar{\mathbf{x}}$ solves the NEP

$$\mathcal{G}_{\bar{\lambda}} : \quad \begin{array}{ll} \underset{\mathbf{x}_i}{\text{minimize}} & f_i(\mathbf{x}_i, \mathbf{x}_{-i}) + \bar{\lambda}^T \mathbf{g}(\mathbf{x}) \\ \text{subject to} & \mathbf{x}_i \in \overline{\mathcal{Q}}_i, \end{array} \qquad \forall i = 1, \ldots, Q, \tag{4.26}$$

and furthermore

$$\mathbf{0} \leq \bar{\lambda} \perp \mathbf{g}(\bar{\mathbf{x}}) \leq \mathbf{0}. \tag{4.27}$$

We can interpret the $\bar{\lambda}$ as prices paid by the players for using the common "resource" represented by the shared constraints. Condition (4.27) says that the players actually have to pay only when the resource becomes scarce. In the following we refer to the NEP in (4.26) with fixed vector λ as $\mathcal{G}_\lambda$. A direct comparison of the KKT conditions of NEP (4.26) augmented with condition (4.27) and the KKT conditions (4.25) leads to the following interesting result.

Lemma 4.4. *Suppose that the GNEP with shared constraints $\mathcal{G} = <\mathcal{Q}, \mathbf{f}>$ satisfies assumptions in Lemma 4.2 and some CQ holds at all the elements in the set $\mathcal{Q}$ defined in (4.21). Then, $(\bar{\mathbf{x}}, \bar{\lambda})$ is a solution of the problem (4.26)–(4.27) if and only if $\bar{\mathbf{x}}$ is a variational solution of the GNEP—the VI$(\mathcal{Q}, \mathbf{F})$, with $\mathbf{F} = (\nabla_{\mathbf{x}_i} f_i)_{i=1}^{Q}$—and $\bar{\lambda}$ is the corresponding multiplier of the shared constraints.*

Based on Lemma 4.4 we are now able to compute the variational solutions of the GNEP as solutions of (4.26)–(4.27), to which we can apply in principle the theory developed in Sec. 4.2. Similarly to the NEP case, we consider both cases of strongly monotone (uniformly P) and monotone games.

4.3.2.1 Algorithms for Strongly Monotone Pricing Games

To describe the proposed algorithms we need the following preliminary definitions and results. Under the convexity of the shared constraints $\mathbf{g}(\mathbf{x})$ and the positive definiteness of matrix $\Upsilon_{\mathbf{F}}$ defined in (4.10), which implies the strongly monotonicity of $\mathbf{F} = (\nabla_{\mathbf{x}_i} f_i)_{i=1}^{Q}$ (see Proposition 4.2), the game $\mathcal{G}_\lambda$ in (4.26)—the strongly monotone VI$(\overline{\mathcal{Q}}, \mathbf{F} + \nabla_{\mathbf{x}}\mathbf{g}\lambda)$, with $\overline{\mathcal{Q}} \triangleq \prod_{i=1}^{Q} \overline{\mathcal{Q}}_i$ and $\nabla_{\mathbf{x}}\mathbf{g}$ denoting the matrix whose ith column is equal to $\nabla_{\mathbf{x}} g_i$—has a unique NE $\mathbf{x}^\star(\lambda)$ for any $\lambda \geq \mathbf{0}$. Under this condition, let define the map

$$\Phi(\lambda) : \mathbb{R}_+^m \ni \lambda \rightarrow -\mathbf{g}(\mathbf{x}^\star(\lambda)) \tag{4.28}$$

which measures the (negative) violation of the shared constraints at $\mathbf{x}^\star(\lambda)$.

Based on (4.26)–(4.27), the key result to devise distributed algorithms is given in Theorem 4.4 below, where $\mathbf{F} = (\nabla_{\mathbf{x}_i} f_i)_{i=1}^{Q}$, $\mathbf{x}^\star(\lambda)$ denotes the unique NE of $\mathcal{G}_\lambda$ (under the positive definiteness of matrix $\Upsilon_{\mathbf{F}}$), and $\Upsilon_{\mathbf{F}}$, $\mathcal{Q}$, and Φ are defined in (4.10), (4.21) and (4.28), respectively.

Theorem 4.4 ([38]). *Given the problem (4.26)–(4.27), suppose that Assumption 4.3 and some CQ at all elements of the set $\mathcal{Q}$ hold true. If $\Upsilon_{\mathbf{F}} \succ \mathbf{0}$, then the following hold:*
(a) The problem (4.26)–(4.27) is equivalent to the nonlinear complementarity problem (NCP) in the price tuple λ

$$\text{NCP}(\Phi): \qquad 0 \leq \lambda \perp \Phi(\lambda) \geq 0 \tag{4.29}$$

The equivalence is in the following sense: the NCP(Φ) must have a solution, and for any such a solution λ^{NCP} the pair $(\mathbf{x}^\star(\lambda^{\text{NCP}}), \lambda^{\text{NCP}})$ is a solution of (4.26)–(4.27), with $\mathbf{x}^\star(\lambda^{\text{NCP}}) = \mathbf{x}^{\text{VI}}$, where $\mathbf{x}^{\text{VI}}$ is the unique solution of the VI$(\mathcal{Q}, \mathbf{F})$*; conversely, if $(\mathbf{x}^{\text{NE}}, \lambda^{\text{NE}})$ is a solution of (4.26)-(4.27) $[\mathbf{x}^{\text{NE}} = SOL(\mathcal{Q}, \mathbf{F})]$, then λ^{NE} is a solution of the* NCP(Φ) *with $\mathbf{x}^\star(\lambda^{\text{NE}}) = \mathbf{x}^{\text{NE}}$; (b) The problem (4.26)–(4.27) has a unique least-norm price tuple, denoted by $\lambda^{\text{NE,ln}}$, such that $\|\lambda^{\text{NE,ln}}\|_2 \leq \|\lambda^{\text{NE}}\|_2$ for any price solution λ^{NE} of (4.26)-(4.27).*

Remark 4.6 (On the uniqueness of the solution). Observe that, under $\Upsilon_{\mathbf{F}} \succ \mathbf{0}$, Theorem 4.4(a) implies only the uniqueness of the variables $\mathbf{x}$ of the problem (4.26)–(4.27) [the primal variables of the VI$(\mathcal{Q}, \mathbf{F})$], but not of the price tuple $\boldsymbol{\lambda}$. The interesting result is that, in such a case, all these prices $\boldsymbol{\lambda}^{\text{NCP}}$—the solutions of the NCP$(\boldsymbol{\Phi})$—yield solution pairs $(\mathbf{x}^{\star}(\boldsymbol{\lambda}^{\text{NCP}}), \boldsymbol{\lambda}^{\text{NCP}})$ of (4.26)–(4.27) having the same optimal $\mathbf{x}^{\text{NE}}$, i.e., $\mathbf{x}^{\star}(\boldsymbol{\lambda}^{\text{NCP}}) = \mathbf{x}^{\text{NE}}$. Part (b) of the theorem identifies a unique special price tuple $\boldsymbol{\lambda}^{\text{NE,lm}}$.

The NCP reformulation of the problem (4.26)–(4.27) as stated by Theorem 4.4 offers the possibility of devising iterative algorithms that can be implemented in a distributed fashion among all players (because the feasible set of the NCP has a Cartesian structure) and whose convergence can be studied using known results from the theory of VIs (cf. [11, Chapter 12]). An example is the *projection algorithm with variable steps* [11, Alg. 12.1.4] applied to the NCP$(\boldsymbol{\Phi})$ in (4.29) and formally described in Algorithm 4.3 (under the assumption $\Upsilon_{\mathbf{F}} \succ \mathbf{0}$), where $\overline{\mathcal{Q}} = \prod_i \overline{\mathcal{Q}}_i$ and $\mathbf{F} = (\nabla_{\mathbf{x}_i} f_i)_{i=1}^{Q}$. The convergence properties of the algorithm are given in Theorem 4.5. For other algorithms we refer to [26, 38, 39].

Algorithm 4.3: Projection Algorithm with Variable Steps (PAVS)

Data : Choose any $\boldsymbol{\lambda}^{(0)} \geq \mathbf{0}$; set $n = 0$.

(S.1) : If $\boldsymbol{\lambda}^{(n)}$ satisfies a suitable termination criterion: STOP.

(S.2) : Given $\boldsymbol{\lambda}^{(n)}$, compute $\mathbf{x}^{\star}\left(\boldsymbol{\lambda}^{(n)}\right)$ as the unique NE of $\mathcal{G}_{\boldsymbol{\lambda}^{(n)}}$:

$$\mathbf{x}^{\star}(\boldsymbol{\lambda}^{(n)}) = \text{SOL}(\overline{\mathcal{Q}}, \mathbf{F} + \nabla_{\mathbf{x}}\mathbf{g}\boldsymbol{\lambda}^{(n)}). \tag{4.30}$$

(S.3) : Choose $\tau_n > 0$ and update the price vectors $\boldsymbol{\lambda}$ according to

$$\boldsymbol{\lambda}^{(n+1)} = \left[\boldsymbol{\lambda}^{(n)} - \tau_n \boldsymbol{\Phi}\left(\boldsymbol{\lambda}^{(n)}\right)\right]^{+}. \tag{4.31}$$

(S.4) : Set $n \leftarrow n+1$; go to (S.1).

Theorem 4.5 ([38]). *Suppose* $\Upsilon_{\mathbf{F}} \succ \mathbf{0}$. *If the scalars* τ_n *are chosen so that* $0 < \inf_n \tau_n \leq \sup_n \tau_n < 2c_{\text{sm}}/c_{\text{Lip}}^2$, *with* $c_{\text{Lip}} \triangleq \max_{\mathbf{x}\in\overline{\mathcal{Q}}} \left\|\nabla_{\mathbf{x}}\mathbf{g}(\mathbf{x})^T\right\|_2$ *and* c_{sm} *defined in Proposition 4.2(b), then the sequence* $\{\boldsymbol{\lambda}^{(n)}\}_{n=0}^{\infty}$ *generated by Algorithm 4.3 converges to a solution of the* NCP$(\boldsymbol{\Phi})$.

Remark 4.7 (On the convergence of the inner loop via distributed algorithms). The implementation of Algorithm 4.3 requires the computation of the solution of the $\mathcal{G}_{\boldsymbol{\lambda}^{(n)}}$ (4.30) for a given set of prices, possibly in a distributed way. Given $\boldsymbol{\lambda} \geq \mathbf{0}$, the game $\mathcal{G}_{\boldsymbol{\lambda}}$ is a NEP and thus one can solve it by using any of the algorithms proposed in Sec. 4.2.4 for NEPs. We refer to [38] for a detailed study of the convergence of asynchronous algorithms applied to the VI$(\overline{\mathcal{Q}}, \mathbf{F} + \nabla_{\mathbf{x}}\mathbf{g}\boldsymbol{\lambda}^{(n)})$. The interesting result is that, when the vector function $\mathbf{g}(\mathbf{x})$ is separable, i.e., $\mathbf{g}(\mathbf{x}) = \sum_{i=1}^{Q} \mathbf{g}_i(\mathbf{x}_i)$, conditions in Theorem 4.5 are sufficient for the convergence of both loops in Algorithm 4.3.

4.3.2.2 Algorithms for Monotone Pricing Games

We focus now on the case in which the problem (4.26)–(4.27) (the associated VI) is monotone. In such a case, Algorithm 4.3 is not longer guaranteed to converge; instead, the outer loop has to be complicated. To avoid this complication, here we consider a different approach, based on an equivalent reformulation of (4.26)–(4.27). To this end, observe first that the price complementarity condition in (4.27) is equivalent to

$$\boldsymbol{\lambda} \in \underset{\boldsymbol{\lambda}' \geq \mathbf{0}}{\operatorname{argmin}} \left\{ -\boldsymbol{\lambda}'^T g(\mathbf{x}) \right\}$$

Then, consider the NEP with $Q+1$ players in which the "new" $(Q+1)$-th player controls the price variables $\boldsymbol{\lambda}$:

$$\begin{array}{ll} \underset{\mathbf{x}_i}{\text{minimize}} & f_i(\mathbf{x}_i, \mathbf{x}_{-i}) + \boldsymbol{\lambda}^T \mathbf{g}(\mathbf{x}) \\ \text{subject to} & \mathbf{x}_i \in \overline{\mathcal{Q}}_i \end{array} \qquad \forall i = 1, \ldots, Q, \qquad (4.32)$$

$$\underset{\boldsymbol{\lambda} \geq 0}{\text{minimize}} \quad -\boldsymbol{\lambda}^T \mathbf{g}(\mathbf{x}).$$

The difference between the problem (4.26)–(4.27) and the NEP (4.32) is that in the latter game there are no side constraints, but the complementarity condition is treated as an additional player of the game (at the same level of the other Q players), who solves a nonnegatively constrained linear program in the variable $\boldsymbol{\lambda}$ parametrized by $\mathbf{x}$. It is not difficult to see that this new extended game has the same solution set of the problem (4.26)–(4.27) (note that $\mathbf{g}(\mathbf{x}) \leq \mathbf{0}$ in (4.32) is implicitly satisfied at the equilibrium). Therefore, one can compute a variational solution of a GNEP with jointly convex shared constraints by finding a solution of the NEP in (4.32). This game is equivalent to the *partitioned* VI($\widetilde{\mathcal{Q}}, \widetilde{\mathbf{F}}$), where $\widetilde{\mathcal{Q}} \triangleq \prod_{i=1}^{Q} \overline{\mathcal{Q}}_i \times \mathbb{R}_+^m$ and $\widetilde{\mathbf{F}}(\mathbf{x}, \boldsymbol{\lambda})$ is defined as

$$\widetilde{\mathbf{F}}(\mathbf{x}, \boldsymbol{\lambda}) \triangleq \begin{pmatrix} (\nabla_{\mathbf{x}_i} f_i(\mathbf{x}) + \nabla_{\mathbf{x}_i} \mathbf{g}(\mathbf{x}) \boldsymbol{\lambda})_{i=1}^{Q} \\ -\mathbf{g}(\mathbf{x}) \end{pmatrix} = \begin{pmatrix} \mathbf{F}(\mathbf{x}) + \nabla_{\mathbf{x}} \mathbf{g}(\mathbf{x}) \boldsymbol{\lambda} \\ -\mathbf{g}(\mathbf{x}) \end{pmatrix}. \qquad (4.33)$$

Proposition 4.4. *If* $\mathbf{F}(\mathbf{x}) = (\nabla_{x_i} f_i(\mathbf{x}))_{i=1}^{Q}$ *is monotone on* $\prod_{i=1}^{Q} \overline{\mathcal{Q}}_i$, *then also* $\widetilde{\mathbf{F}}(\mathbf{x}, \boldsymbol{\lambda})$ *is monotone on* $\widetilde{\mathcal{Q}}$.

It follows from Proposition 4.4 that, under Assumptions 4.2 and 4.3, one can compute a variational equilibrium of the GNEP by applying Algorithm 4.2 described in Sec. 4.2.4.2 to the VI($\widetilde{\mathcal{Q}}, \widetilde{\mathbf{F}}$), where

$$\widetilde{\mathbf{S}}_{\tau}(\mathbf{y}^{(n)}) \triangleq \text{SOL} \left(\widetilde{\mathcal{Q}}, \widetilde{\mathbf{F}} + \tau (\mathbf{I} - \mathbf{y}^{(n)}) \right)$$

in step 2 of the algorithm is the unique solution of the regularized

$$\mathrm{VI}\big(\widetilde{\mathcal{Q}}, \widetilde{\mathbf{F}} + \tau(\mathbf{I} - \mathbf{y}^{(n)})\big),$$

given $\mathbf{y}^{(n)} \triangleq (\mathbf{x}^{(n)}, \boldsymbol{\lambda}^{(n)})$. This is formalized in the following.

Theorem 4.6 ([38]). *Suppose that the GNEP with shared constraints* $\mathcal{G} = <\mathcal{Q}, \mathbf{f}>$ *satisfies Assumptions 4.2 and 4.3, and some CQ holds at all the elements in the set* $\mathcal{Q}$ *defined in (4.21). Let* $\{\varepsilon_n\} \subset [0, \infty)$ *be a sequence such that* $\sum_{n=1}^{\infty} \varepsilon_n < \infty$*, and let* ρ_n *be such that* $\{\rho_n\} \subset [R_m, R_M]$*, with* $0 < R_m \leq R_M < 2$*. Then, the sequence* $\{(\mathbf{x}^{(n)}, \boldsymbol{\lambda}^{(n)})\}_{n=0}^{\infty}$ *generated by the PDA described in Algorithm 4.2 and applied to the VI(*$\widetilde{\mathcal{Q}}, \widetilde{\mathbf{F}}$*) converges to a variational solution of the GNEP.*

The last thing left to discuss is how to compute, for any given $\mathbf{y}^{(n)} \triangleq (\mathbf{x}^{(n)}, \boldsymbol{\lambda}^{(n)})$, a(n approximated) solution $\widetilde{\mathbf{S}}_\tau(\mathbf{y}^{(n)})$ of the regularized $\mathrm{VI}\big(\widetilde{\mathcal{Q}}, \widetilde{\mathbf{F}} + \tau(\mathbf{I} - \mathbf{y}^{(n)})\big)$. Under Assumptions 4.2 and 4.3, the $\mathrm{VI}\big(\widetilde{\mathcal{Q}}, \widetilde{\mathbf{F}} + \tau(\mathbf{I} - \mathbf{y}^{(n)})\big)$ is equivalent to the following regularized NEP:

$$\begin{array}{ll} \underset{\mathbf{x}_i}{\text{minimize}} & f_i(\mathbf{x}_i, \mathbf{x}_{-i}) + \boldsymbol{\lambda}^T \mathbf{g}(\mathbf{x}) + \frac{\tau}{2} \left\| \mathbf{x}_i - \mathbf{x}_i^{(n)} \right\|^2 \\ \text{subject to} & \mathbf{x}_i \in \overline{\mathcal{Q}}_i, \end{array} \qquad \forall i = 1, \ldots, Q \tag{4.34}$$
$$\underset{\boldsymbol{\lambda} \geq 0}{\text{minimize}} \quad -\boldsymbol{\lambda}^T \mathbf{g}(\mathbf{x}) + \frac{\tau}{2} \left\| \boldsymbol{\lambda} - \boldsymbol{\lambda}^{(n)} \right\|^2$$

whose Nash equilibria can be distributively computed by using the asynchronous best-response algorithms described in Algorithm 4.1. The global convergence of such a class of algorithms is guaranteed for sufficiently large $\tau > 0$, as stated in the following corollary, restricted to the case of separable $\mathbf{g}(\mathbf{x}) = \sum_{i=1}^{Q} \mathbf{g}_i(\mathbf{x}_i)$ (the more general case of nonseparable $\mathbf{g}$ is addressed in [38]), where $\mathbf{F} = (\nabla_{\mathbf{x}_i} f_i)_{i=1}^{Q}$, the matrix $\bar{\Upsilon}_{\mathbf{F},\tau}$ is defined as

$$\bar{\Upsilon}_{\mathbf{F},\tau} \triangleq \left[\begin{array}{c|c} \Upsilon_{\mathbf{F}} + \tau \mathbf{I} & -\gamma \\ \hline -\gamma^T & \tau \end{array} \right] \tag{4.35}$$

with $\Upsilon_{\mathbf{F}}$ defined in (4.10), $\gamma \triangleq (\gamma_i)_{i=1}^{Q}$ and $\gamma_i \triangleq \sup_{\mathbf{z}_i \in \overline{\mathcal{Q}}_i} \|\nabla_{\mathbf{x}_i} \mathbf{g}_i(\mathbf{z}_i)\|_2$.

Corollary 4.2 ([38]). *In the setting of Theorem 4.6, if* $\mathbf{g}(\mathbf{x}) = \sum_{i=1}^{Q} \mathbf{g}_i(\mathbf{x}_i)$ *and* τ *is chosen sufficiently large so that* $\bar{\Upsilon}_{\mathbf{F},\tau}$ *is a P-matrix, then any asynchronous best-response algorithm (see Algorithm 4.1) applied to the game in (4.34) converges to* $\widetilde{\mathbf{S}}_\tau(\mathbf{y}^{(n)})$.

Remark 4.8 (Partial regularization schemes). In some situations, the optimization problems of some players might be "convex enough" to require no regularization. The study of the convergence properties of distributed schemes based on "partial regularizations" is addressed in [38].

4.4 Design of Cognitive Radio Systems Based on Game Theory

In the last decade, CR has received considerable attention as a way to improve the efficiency of radio networks [21, 15]. CR networks adopt a hierarchical access structure where the primary users (PUs) are the legacy spectrum holders while the secondary users (SUs) are the unlicensed users who sense the electromagnetic environment and adapt their transceivers' parameters as well as the resource allocation decisions in order to dynamically access temporally unoccupied spectrum regions.

We consider a hierarchical CR network composed of P PUs and Q SUs, each formed by a transmitter-receiver pair, coexisting in the same area and sharing the same band. We focus on (block) transmissions over single-input single-output (SISO) frequency-selective channels; more general results valid for multi-input multi-output (MIMO) channels can be found in [31, 35]. Because of the lack of coordination among the CR users and the competitive nature of the system, the set of SUs can be naturally modeled as a frequency-selective N-parallel Gaussian interference channel, where N is the number of available subcarriers. The transmission strategy of each SU i is then the power allocation vector $\mathbf{p}_i = (p_i(k))_{k=1}^N$ over the N subcarriers, subject to the transmit power constraints $\sum_{k=1}^N p_i(k) \leq P_i$. Under mild conditions (see, e.g., [26]), the maximum information rate on link i for a specific power allocation profile $\mathbf{p}_1, \ldots, \mathbf{p}_Q$ is

$$r_i(\mathbf{p}_q, \mathbf{p}_{-q}) = \sum_{k=1}^{N} \log\left(1 + \frac{|H_{ii}(k)|^2 p_i(k)}{\sigma_i^2(k) + \sum_{j\neq i} |H_{ij}(k)|^2 p_j(k)}\right), \tag{4.36}$$

where $\sigma_i^2(k)$ is the thermal noise power over carrier k, $H_{ij}(k)$ is the channel transfer function between the secondary transmitter j and the receiver i, and $\mathbf{p}_{-i} \triangleq (\mathbf{p}_j)_{j\neq i}$ is the set of all the users power allocation vectors, except the ith one.

Temperature-interference constraints: In a manner different from traditional static or centralized spectrum assignment, opportunistic communications in CR systems enable SUs to transmit with overlapping spectra with PUs, provided that the degradation induced on the PU's performance is null or tolerable [15]. In this chapter, we envisage the use of two classes of interference constraints termed *individual conservative* and *global flexible constraints*. For each $i = 1, \ldots, Q$,

Individual per-carrier interference constraints,

$$p_i(k) \leq p_i^{\max}(k) \triangleq \min_{p=1,\ldots,P} \frac{I_{p,i}^{\text{peak}}(k)}{|H_{pi}^{(P,S)}(k)|^2}, \quad \forall k = 1, \cdots, N; \tag{4.37}$$

Individual overall bandwidth interference constraints,

$$\sum_{k=1}^{N} |H_{pi}^{(P,S)}(k)|^2 p_i(k) \leq I_{p,i}^{\text{tot}} \quad \forall p = 1, \ldots, P; \tag{4.38}$$

Global per-carrier interference constraints,

$$\sum_{i=1}^{Q} |H_{pi}^{(P,S)}(k)|^2\, p_i(k) \leq I_p^{\text{peak}}(k), \quad \forall k = 1,\cdots,N,\ \forall p = 1,\ldots,P; \tag{4.39}$$

Global overall bandwidth interference constraints,

$$\sum_{i=1}^{Q}\sum_{k=1}^{N} |H_{pi}^{(P,S)}(k)|^2\, p_i(k) \leq I_p^{\text{tot}} \quad p = 1,\ldots,P \tag{4.40}$$

where $H_{pi}^{(P,S)}(k)$ is the channel transfer function between the secondary transmitter i and the primary receiver p over carrier k; $I_{p,i}^{\text{peak}}(k)$ ($I_p^{\text{tot}}(k)$) and $I_{p,i}^{\text{tot}}$ (I_p^{tot}) are the maximum interferences allowed to be generated by the SU i (all the SUs) that is tolerable at the primary receiver p over carrier k and over the whole spectrum (licensed to the PU p), respectively. The values of the $I_{p,i}^{\text{peak}}(k)$ and $I_{p,i}^{\text{tot}}$ can be obtained at the SUs' transmitters if the type of PUs are known. Methods to obtain the interference limits when the SUs do not have this knowledge are discussed in [15]. To avoid a trivial solution, we assume that, for all $k = 1,\ldots,N$ and $i = 1,\ldots,Q$, and some $p = 1,\ldots,P$,

$$\begin{aligned} &\text{i)} \quad p_i^{\max}(k) < \min\left\{ \frac{I_{p,i}^{\text{peak}}(k)}{|H_{pi}^{(P,S)}(k)|^2}, P_i \right\} \\ &\text{ii)} \quad \textstyle\sum_{k=1}^{N} p_i^{\max}(k) > P_i, \quad \text{or} \quad \sum_{k=1}^{N} |H_{pi}^{(P,S)}(k)|^2 p_i^{\max}(k) > I_{p,i}^{\text{tot}}. \end{aligned} \tag{4.41}$$

Individual interference constraints are motivated by all CR scenarios where primary terminals are oblivious to the presence of SUs (also called common model [14, 15]). These constraints lead to totally distributed algorithms with no signaling among the SUs, as we will show later on. However, sometimes they may be too restrictive and thus marginalize the potential gains offered by the dynamic resource assignment mechanism. The global interference constraints limit instead the *aggregate* interference, which is indeed the interference experimented by the PUs. These constraints will be shown to lead to better performance of SUs than those achievable by imposing individual constraints. However, this gain comes at a price: The resulting algorithms require some signaling between the PUs and SUs. Thus, they can be employed in CR networks where an interaction between the PUs and the SUs is allowed, as, e.g., in the so-called property-right CR model (or spectrum leasing) [14].

Game Theoretical formulation under individual interference constraints. Let us define the strategy set of each SU i as

$$\mathcal{P}_i \triangleq \left\{ \mathbf{p}_i \in \mathbb{R}_+^N : \mathbf{1}^T \mathbf{p}_i \leq P_i,\ \sum_{k=1}^{N} |H_{pi}^{(P,S)}(k)|^2 p_i(k) \leq I_{p,i}^{\text{tot}}\ \forall p, \quad \mathbf{p}_i \leq \mathbf{p}_i^{\max} \right\} \tag{4.42}$$

and consider the NEP $\mathcal{G}_{\text{ind}} =< \prod_i \mathcal{P}_i, (r_i)_{i=1}^{Q} >$, where each SU i, given the strategy profile $\mathbf{p}_{-i}$ of the other users, aims at maximizing his information rate $r_i(\mathbf{p})$ in (4.36) under local power and interference constraints in $\mathcal{P}_i$: for all $i = 1,\ldots,Q$,

$$\begin{array}{ll} \underset{\mathbf{p}_i}{\text{maximize}} & r_i(\mathbf{p}_i,\mathbf{p}_{-i}) \\ \text{subject to} & \mathbf{p}_i \in \mathcal{P}_i. \end{array} \tag{4.43}$$

First of all, observe that for any fixed $\mathbf{p}_{-i} \geq \mathbf{0}$, the single-user optimization problem in (4.43) admits a unique solution (indeed, the feasible set is convex and compact and $r_i(\mathbf{p}_i,\mathbf{p}_{-i})$ is strictly concave in $\mathbf{p}_i \in \mathcal{P}_i$]), given by the (multilevel) waterfilling expression [38]:

$$p_i^\star(k) = \left[\frac{1}{\mu_i + \sum_{p=1}^{P} \lambda_{ip} |H_{pi}^{(P,S)}(k)|^2} - \frac{\sigma_i^2(k) + \sum_{j\neq i} |H_{ij}(k)|^2 p_j(k)}{|H_{ii}(k)|^2}\right]_0^{p_i^{\max}(k)} \tag{4.44}$$

with $k = 1,\ldots,N$, where $[x]_a^b \triangleq \min(\max(a,x),b)$ for $a \leq b$ and the water levels μ_i and $\{\lambda_{ip}\}$ are chosen to satisfy the following complementarity constraints: $0 \leq \mu_i \perp P_i - \sum_{k=1}^{N} p_i^\star(k) \geq 0$ and $0 \leq \lambda_{ip} \perp I_{pi}^{\text{tot}} - \sum_{k=1}^{N} |H_{pi}^{(P,S)}(k)|^2 p_i^\star(k) \geq 0$, for all $p = 1,\ldots,P$; see [38] for efficient algorithms to compute these water levels. The Nash equilibria $\mathbf{p}^\star$ of the NEP $\mathcal{G}_{\text{ind}}$ are thus the fixed-points of the waterfilling mapping (4.44). The study of such a game can be carried out using the VI framework developed in Sec. 4.2. Before stating the main results, we introduce the following definitions. For the NEP $\mathcal{G}_{\text{ind}}$, matrix $\Upsilon_{\mathbf{F}}$ in (4.10) becomes

$$[\Upsilon_{\text{ind}}]_{ij} \triangleq \begin{cases} \displaystyle\min_{k=1,\ldots,N} \left(\frac{|H_{ii}(k)|^2}{\sigma_i^2(k) + \sum_{j=1}^{N} |H_{ij}(k)|^2 p_j^{\max}(k)}\right)^2, & \text{if } i = j, \\ \displaystyle -\max_{k=1,\ldots,N} \frac{|H_{ij}(k)|^2}{\sigma_i^2(k)} \frac{|H_{ii}(k)|^2}{\sigma_i^2(k)}, & \text{otherwise.} \end{cases} \tag{4.45}$$

We also introduce the per-tone $\Upsilon_{\text{ind}}(k)$, defined for each $k = 1,\ldots,N$, as

$$[\Upsilon_{\text{ind}}(k)]_{ij} \triangleq \begin{cases} \displaystyle\left(\frac{|H_{ii}(k)|^2}{\sigma_i^2(k) + \sum_{j=1}^{N} |H_{ij}(k)|^2 p_j^{\max}(k)}\right)^2, & \text{if } i = j, \\ \displaystyle -\frac{|H_{ij}(k)|^2}{\sigma_i^2(k)} \frac{|H_{ii}(k)|^2}{\sigma_i^2(k)}, & \text{otherwise,} \end{cases} \tag{4.46}$$

Note that $\Upsilon_{\text{ind}} \leq \Upsilon_{\text{ind}}(k)$ for all k. Building on Theorem 4.1 and Theorem 4.2 we obtain the following results for the game $\mathcal{G}_{\text{ind}}$.

Theorem 4.7. *Given the NEP $\mathcal{G}_{ind} = <\prod_i \mathcal{P}_i,\ (r_i)_{i=1}^{Q}>$, suppose without loss of generality that condition i) and (the first of) ii) in (4.41) are satisfied. Then the following hold:*

(a) The NEP has a nonempty and bounded solution set;
(b) Suppose that Υ_{ind} in (4.45) is a P-matrix. Then the NEP has a unique NE and any sequence $\{\mathbf{p}^{(n)}\}_{n=0}^{\infty}$ generated by the asynchronous algorithm described in Algorithm 4.1 and based on the waterfilling best-response (4.44) converges to this equilibrium for any given feasible updating schedule of the players.

Remark 4.9 (On the uniqueness/convergence conditions). Theorem 4.7 provides a physical interpretation of the conditions guaranteeing the uniqueness of the NE (and convergence of the asynchronous iterative waterfilling algorithm (IWFA)): the uniqueness of the NE is ensured if the interference among the SUs is sufficiently small. Sufficient conditions for Υ_{ind} being a P-matrix easier to be checked can be readily obtained applying Corollary 4.3 to Υ_{ind}; we leave this task to the reader.

If the channels and the power/interference constraints in the game $\mathcal{G}_{\text{ind}}$ are such that matrix Υ_{ind} is not a P-matrix, we can still guarantee convergence of distributed algorithms to a NE of the game at the cost of some additional computational complexity. In fact, instead of the waterfilling best-response algorithm described in Algorithm 4.1, we can use the proximal algorithm described in Algorithm 4.2. According to Theorem 4.3, the monotonicity of $\mathbf{F}(\mathbf{p}) = (-\nabla_{\mathbf{p}_i} r_i(\mathbf{p}_i, \mathbf{p}_{-i}))_{i=1}^{Q}$ on $\prod_i \mathcal{P}_i$ is enough to guarantee the global convergence of this algorithm.

Theorem 4.8. *Given the NEP $\mathcal{G}_{ind} =< \prod_i \mathcal{P}_i, (r_i)_{i=1}^{Q} >$, suppose that the mapping $\mathbf{F}(\mathbf{p}) = (-\nabla_{\mathbf{p}_i} r_i(\mathbf{p}_i, \mathbf{p}_{-i}))_{i=1}^{Q}$ is monotone on $\prod_i \mathcal{P}_i$. Let $\{\varepsilon_n\} \subset [0,\infty)$ be a sequence such that $\sum_{n=1}^{\infty} \varepsilon_n < \infty$, let ρ_n be such that $\{\rho_n\} \subset [R_m, R_M]$ with $0 < R_m \leq R_M < 2$, let τ be sufficiently large so that $\Upsilon_{ind} + \tau\mathbf{I}$ is a P matrix, and let $\mathbf{S}_\tau(\mathbf{p}^{(n)})$ in step 2 be computed using the asynchronous best-response algorithm described in Algorithm 4.1. Then, the sequence $\{\mathbf{p}^{(n)}\}_{n=0}^{\infty}$ generated by Algorithm 4.2 applied to $\mathcal{G}_{ind}$ converges to a NE of $\mathcal{G}_{ind}$.*

A sufficient condition for $\mathbf{F}(\mathbf{p})$ being a monotone mapping on $\prod_i \mathcal{P}_i$ is that $\Upsilon_{ind}(k) \succ \mathbf{0}$ for all $k = 1, \ldots, N$ (cf. Proposition 4.2).

Numerical results. In Fig. 4.1, we compare the convergence properties of the following algorithms: i) The simultaneous (multilevel) IWFA (best-response algorithm) described in Algorithm 4.2 and based on the mapping in (4.44); ii) the simultaneous proximal-response algorithm applied to game $\mathcal{G}_{\text{ind}}$, according to which the players solve their regularized optimization problems simultaneously, while changing the "center" of their regularization at each iteration; iii) the simultaneous version of the proximal decomposition algorithm described in Algorithm 4.2 and applied to game $\mathcal{G}_{\text{ind}}$; and iv) the iterative Tikhonov algorithm applied to the VI$(\prod_i \mathcal{P}_i, (-\nabla_{\mathbf{p}_i} r_i)_{i=1}^{Q})$ associated to $\mathcal{G}_{\text{ind}}$ and proposed in [39] (see also [19, Ch. 15.2]). We refer to these algorithms as simultaneous iterative waterfilling algorithm (SIWFA), Jacobi proximal-response algorithm (JPRA), Jacobi proximal decomposition algorithm (JPDA), and iterative distributed Tikhonov algorithm (IDTA), respectively. Note that JPRA and JPDA differ only in the rule used to update the center of the regularization: in the former, the players change the center at each iteration, whereas in the latter the center is kept fixed for a certain number of iterations (until the condition in step 2 of the algorithm is satisfied).

We consider a hierarchical CR network where there are two PUs (the base stations of two cells) and ten SUs, randomly distributed in the cell. The (cross)channels among the secondary links and between the primary and the secondary links are simulated as FIR filter of order $L = 10$, where each tap has variance equal to $1/L^2$; the available bandwidth is divided in $N = 64$ subchannels. We focus on two scenarios, namely low/medium interference scenario and high interference scenario.

Low/medium interference means that secondary links are far enough from each other so that matrix Υ_{ind} defined in (4.45) is a P-matrix; whereas in the high interference scenario, matrices $\Upsilon_{\text{ind}}(k)$ are positive definite, but Υ_{ind} is not a P-matrix. In Fig. 4.1(a) we plotted the average rate evolution of the SUs as a function of the iteration index in the low/medium interference case, corresponding to $\text{snr}_i \triangleq P_i/(\sigma_i^2 d_{ii}^2) = 5$ dB and $\text{inr}_{ij} \triangleq P_j/(d_{ij}^2 \sigma_i^2) = 0$ dB for all i and $j \neq i$; whereas in Fig. 4.1(b) we compared the SIWFA with the JPDA in an high interference scenario, corresponding to $\text{snr}_i \triangleq P_i/(\sigma_i^2 d_{ii}^2) = 0\text{dB}$ and $\text{inr}_{ij} \triangleq P_j/(d_{ij}^2 \sigma_i^2) = 5\text{dB}$ for all i and $j \neq i$. To make the picture not excessively overcrowded, we plot only the curves of two out of the ten links. We examined the performance of the above algorithms under the following set-up (guaranteeing the fastest convergence speed of each algorithm for the given scenario). In JPRA, we chose $\tau = 1$; in JPDA we set $\tau = 0.2$ and the error sequence defining the inner termination criterion has been chosen as $\varepsilon_n = \varepsilon_0{}^{-n}$ with $\varepsilon_0 = 1e-2$, where n is the outer iteration index; in IDTA we chose the variable step-size sequences $\gamma_n = n^{-0.4}$ and $\delta_n = n^{-0.49}$ so that (sufficient) conditions for the convergence of IDTA given in [19, Prop. 15.1] are satisfied (we use the same notation as in [39]; see therein for the details).

From Fig. 4.1(a) we infer that in the low/medium interference regime, the SIWFA, the JPRA and the JPDA converge quite fast (less than 10 iterations) and exhibit almost the same convergence speed. The IDTA instead requires many more iterations (about 1500 iterations) to converge, which makes it impractical. The same convergence properties as in Fig. 4.1(a) has been experienced for all the channel realizations we simulated. Figure 4.1(b) shows an example where the SIWFA does not converge because of the high interference among the SUs, whereas the proposed JPDA still converges in a few iterations. Interestingly, we experienced convergence of our JPDA even when the matrices $\Upsilon_{\text{ind}}(k)$ are not positive semidefinite, provided that the error sequence $\{\varepsilon_n\}$ are properly chosen.

Game theoretical formulation under individual/global interference constraints We focus now on the power control problem among the SUs in the presence of both individual and global interference constraints. Because of the global constraints, the game theoretical formulation of this problem leads to a GNEP with shared constraints (cf. Sec. 4.3), whose shared constrained set is given by

$$\hat{\mathcal{P}} \triangleq \mathcal{P} \cap \left\{ \begin{array}{ll} \mathbf{p}: & \displaystyle\sum_{i=1}^{Q}\sum_{k=1}^{N} |H_{pi}^{(P,S)}(k)|^2\, p_i(k) \le I_p^{\text{tot}} \quad \forall p = 1,\ldots,P \\ & \displaystyle\sum_{i=1}^{Q} |H_{pi}^{(P,S)}(k)|^2\, p_i(k) \le I_p^{\text{peak}}(k) \quad \forall p = 1,\ldots,P,\ k = 1,\ldots,N \end{array} \right\} \tag{4.47}$$

where $\mathcal{P} \triangleq \prod_i \mathcal{P}_i$ and $\mathcal{P}_i$ is defined in (4.42).

Aiming at finding distributed algorithms, we focus our interest on the variational solutions of the GNEP, which, according to Lemma 4.4, correspond to the solutions of the following NEP with pricing:

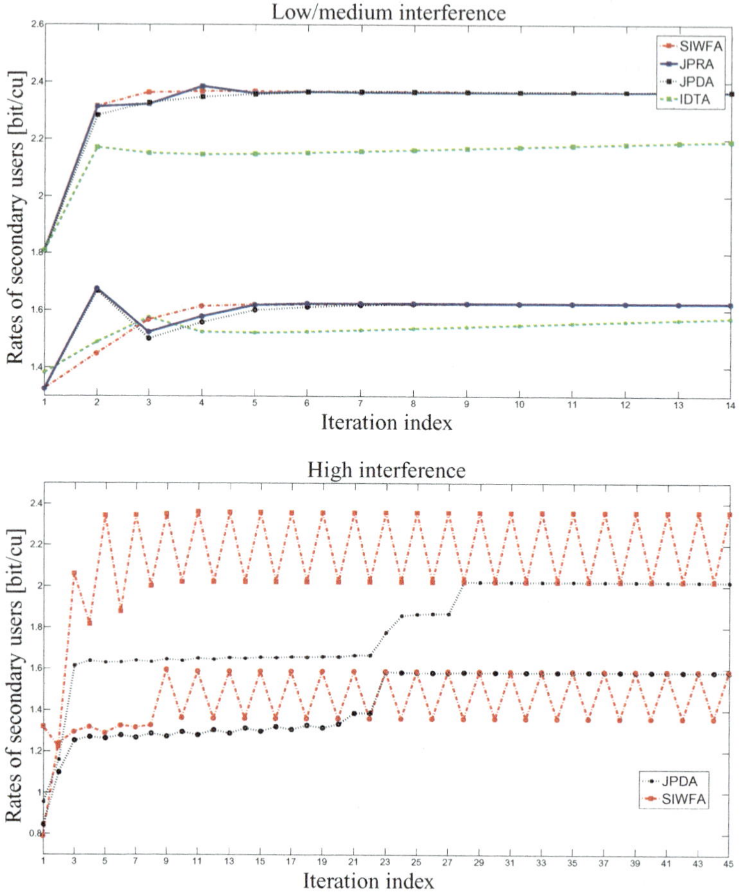

Fig. 4.1 Comparison of distributed algorithms solving the game $\mathcal{G}_{\text{ind}}$ in (4.43): Rates of the SUs versus the iteration index of two out ten users, achieved by the Simultaneous Iterative Waterfilling Algorithm (SIWFA), the Jacobi Proximal-response Algorithm (JPRA), the Jacobi Proximal Decomposition Algorithm (JPDA), and the Iterative Distributed Tikhonov Algorithm (IDTA) in the low/medium interference regime [subplot (a)] and high interference regime [subplot (b)].

$$\begin{aligned} \underset{\mathbf{p}_i}{\text{maximize}} \quad & r_i(\mathbf{p}_i, \mathbf{p}_{-i}) - \sum_{p=1}^{P}\sum_{k=1}^{N} \lambda_{p,k}^{\text{peak}} \, |H_{pi}^{(P,S)}(k)|^2 \, p_i(k) \\ & \qquad - \sum_{p=1}^{P} \lambda_{p,\text{tot}} \sum_{k=1}^{N} |H_{pi}^{(P,S)}(k)|^2 \, p_i(k) \\ \text{subject to} \quad & \mathbf{p}_i \in \mathcal{P}_i \end{aligned} \tag{4.48}$$

for all $i = 1, \cdots, Q$, where $r_i(\mathbf{p}_i, \mathbf{p}_{-i})$ is defined in (4.36), and the prices

$$\boldsymbol{\lambda} \triangleq (\boldsymbol{\lambda}_{\text{tot}}; \boldsymbol{\lambda}^{\text{peak}}), \text{ with } \boldsymbol{\lambda}_{\text{tot}} \triangleq (\lambda_{p,\text{tot}})_{p=1}^{P} \text{ and } \boldsymbol{\lambda}^{\text{peak}} \triangleq ((\lambda_{p,k}^{\text{peak}})_{p=1}^{P})_{k=1}^{N}$$

are chosen such that the following complementary conditions are satisfied:

$$\mathbf{0} \leq \boldsymbol{\lambda} \perp \boldsymbol{\Psi}(\mathbf{p}) \geq \mathbf{0} \quad \Leftrightarrow \quad \underset{\boldsymbol{\lambda} \geq 0}{\text{minimize}}\, \boldsymbol{\lambda}^T \boldsymbol{\Psi}(\mathbf{p}) \tag{4.49}$$

where

$$\boldsymbol{\Psi}(\mathbf{p}) \triangleq \begin{pmatrix} \left(I_p^{\text{tot}} - \sum_{i=1}^{Q} \sum_{k=1}^{N} |H_{pi}^{(P,S)}(k)|^2\, p_i(k) \right)_{p=1}^{P} \\ \left(\left(I_p^{\text{peak}}(k) - \sum_{i=1}^{Q} |H_{pi}^{(P,S)}(k)|^2\, p_i(k) \right)_{k=1}^{N} \right)_{p=1}^{P} \end{pmatrix} \tag{4.50}$$

With a slight abuse of terminology, we will refer in the following to the NEP (4.48) with the complementarity constraints (4.49) as game $\mathcal{G}_{\text{glob}}$.

To study the above game we need the following intermediate definitions, based on results in Sec. 4.3.2. Given the column vector $\mathbf{h} \triangleq (||\sum_{p=1}^{P} \mathbf{H}_{pi}^{(P,S)}||_2)_{i=1}^{Q}$ with $\mathbf{H}_{pi}^{(P,S)} \triangleq (|H_{pi}^{(P,S)}(k)|^2)_{k=1}^{N}$, the matrix Υ_τ in (4.35) associated to the problem (4.48)-(4.49) becomes

$$\Upsilon_{\text{ind},\tau} \triangleq \left[\begin{array}{c|c} \Upsilon_{\text{ind}} + \tau \mathbf{I} & -\mathbf{h} \\ \hline -\mathbf{h}^T & \tau \end{array} \right] \tag{4.51}$$

where Υ_{ind} is defined in (4.45). Among the several distributed schemes proposed in the first part of the chapter to compute the solutions of $\mathcal{G}_{\text{glob}}$, here we focus on the following, leaving to the reader the task of specializing the other proposed algorithms to game $\mathcal{G}_{\text{glob}}$: i) projection algorithm with variable steps (PAVS), described in Algorithm 4.3; and ii) proximal decomposition algorithm (PDA) described in Algorithm 4.2. Step 2 of Algorithm 4.3—the computation of the NE of the game in (4.48) for $\boldsymbol{\lambda} = \boldsymbol{\lambda}^{(n)}$—and the Step 2 of Algorithm 4.2—the computation of the NE $\mathbf{S}_\tau((\mathbf{p}^{(n)}, \boldsymbol{\lambda}^{(n)}))$ of the game obtained by the proximal regularization of (4.48)–(4.49) (see (4.34))—can be efficiently computed using the asynchronous best-response algorithm described in Algorithm 4.1, whose best-response for both games has a closed form (multilevel waterfilling) expression [38]. Building on Theorems 4.4, 4.5, and 4.6, we obtain the following results for the problem $\mathcal{G}_{\text{glob}}$.

Theorem 4.9. *Given the problem $\mathcal{G}_{glob}$, suppose w.l.o.g. that conditions (4.41) are satisfied; let $\mathbf{F}(\mathbf{p}) \triangleq (-\nabla_{\mathbf{p}_i} r_i(\mathbf{p}_i, \mathbf{p}_{-i}))_{i=1}^{Q}$. Then, the following statements hold.*

(a) *$\mathcal{G}_{glob}$ has a nonempty and bounded solution set;*

(b) *Suppose that $\Upsilon_{ind} \succ \mathbf{0}$. Then: i) the power vector $\mathbf{p}^\star$ at every solution of $\mathcal{G}_{glob}$ is unique; and ii) any sequence $\{\boldsymbol{\lambda}^{(n)}, \mathbf{p}^\star(\boldsymbol{\lambda}^{(n)})\}_{n=0}^{\infty}$ generated by the PAVS converges to a solution of $\mathcal{G}_{glob}$, provided that the scalars τ_n are chosen so that $0 < \inf_n \tau_n \leq \sup_n \tau_n < 2\,\lambda_{least}(\Upsilon_{ind})/\,||\mathbf{h}||_2^2$;*

(c) *Suppose that the mapping $\mathbf{F}(\mathbf{p})$ is monotone on $\hat{\mathcal{P}}$. Let $\{\varepsilon_n\} \subset [0, \infty)$ be a sequence such that $\sum_{n=1}^{\infty} \varepsilon_n < \infty$, let ρ_n be such that $\{\rho_n\} \subset [R_m, R_M]$ with $0 <$*

$R_m \leq R_M < 2$, and let τ be sufficiently large so that $\Upsilon_{\mathrm{ind},\tau}$ defined in (4.51) is a P-matrix. Then, the sequence $\{\mathbf{p}^{(n)}, \lambda^{(n)}\}_{n=0}^{\infty}$ generated by the PDA converges to a solution of $\mathcal{G}_{glob}$.

A sufficient condition for $\mathbf{F}(\mathbf{p})$ being a monotone mapping on $\hat{\mathcal{P}}$ is that $\Upsilon_{ind}(k) \succ \mathbf{0}$ for all $k = 1, \ldots, N$ (cf. Proposition 4.2).

Remark 4.10 (Implementation of the algorithms). In the PAVS and PDA, there are two levels of updates: 1) the computation of the optimal power allocations of the SUs, given the prices λ; and 2) the updates of the price vector, given the interference generated by the SUs over the N subcarriers. The former can be performed directly by the SUs via the asynchronous best-response algorithm described in Algorithm 4.1. Note that, once $\sum_{p=1}^{P}(\lambda_{p,k}^{\mathrm{peak}} + \lambda_{p,\mathrm{tot}})|H_{pi}^{(P,S)}(k)|^2$ are given, these algorithms are totally distributed, since the SUs only need to measure the received multiuser interference (MUI) over the N subcarriers [26]. The update of the price vector can be performed either by the PUs or by the SUs themselves, depending on the debate position assumed for the CR network (the property-right model or the common model [14]). More specifically, in a property-right model, an interaction between the PUs and the SUs is allowed. It is thus natural that the update of the prices is performed by the PUs. Note that the signaling from the SUs to the PUs is implicit, since the PUs to update the prices only need to locally measure the global received interference (the function $\Psi(\mathbf{p})$). The signaling from the PUs to the SUs users, however, is explicit: the PUs have to broadcast the prices and the SUs receive and estimate their values. In a CR network based on the common model, the PUs are oblivious to the presence of the SUs and thus the update of prices needs to be performed by the SUs. Building on consensus algorithms [23, 30], at the price of additional signaling among the SUs and computational complexity, the proposed algorithms still can be implemented in a distributed fashion by the SUs. Details can be found in [26].

Numerical results. In Fig. 4.2 we compare the convergence properties of the following algorithms applied to the game $\mathcal{G}_{\mathrm{glob}}$: i) PAVS, ii) PDA, and iii) the algorithm proposed in [16]; we refer to these algorithms as *gradient projection IWFA* (GP-IWFA), *full proximal IWFA* (FP-IWFA) and *price-based IWFA* (PB-IWFA), respectively. The setup is the same as the one considered in Fig. 4.1 for the low/medium interference regime. In addition to the individual constraints, now the PUs impose global per-carrier interference constraints, assumed for the sake of simplicity equal over all the subcarriers. For the scenario considered in the picture, one can see that the convergence of both GP-IWFA and FP-IWFA is reasonably fast (less than 30 iterations), whereas the PB-IWFA requires many more iterations (more than 500).

Thanks to less stringent constraints on the transmission powers of the SUs, the performance of the SUs achievable under global interference constraints (game $\mathcal{G}_{\mathrm{glob}}$) are expected to be much better than those achievable under the more conservative individual interference constraints (game $\mathcal{G}_{\mathrm{ind}}$). Figure 4.3 confirms this intuition, where we plot the sum-rate of the SUs (averaged over 500 random i.i.d. Gaussian channel realizations) achievable in the two aforementioned cases as a function of the maximum tolerable interference at the PUs, within the same set-up of Fig. 4.1.

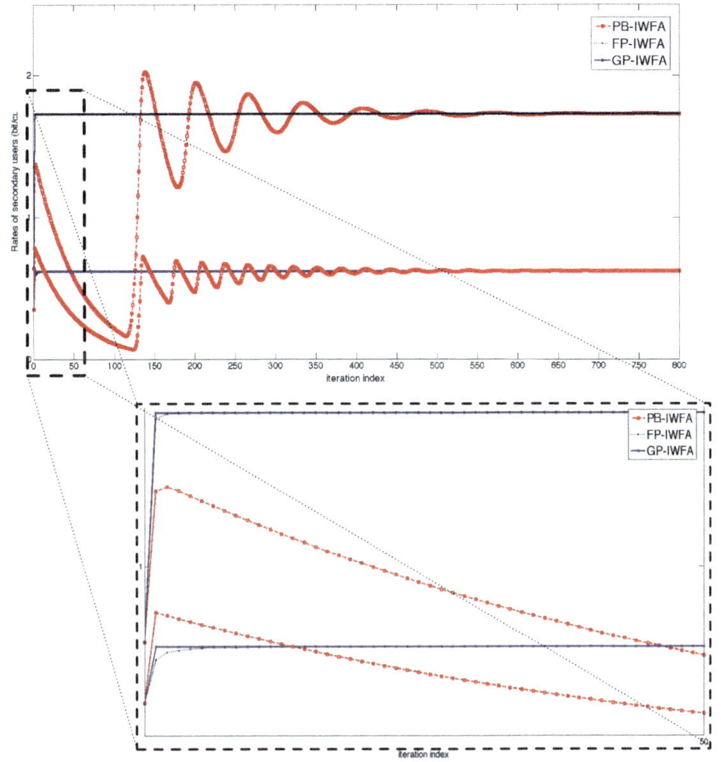

Fig. 4.2 Comparison of distributed algorithms solving the problem $\mathcal{G}_{\text{glob}}$ in (4.48)–(4.49): Rates of the SUs versus the iteration index of two out ten users, achieved by the price-based IWFA (PB-IWFA) [16], the gradient projection IWFA (GP-IWFA) and the full proximal IWFA (FP-IWFA) in the low/medium interference regime.

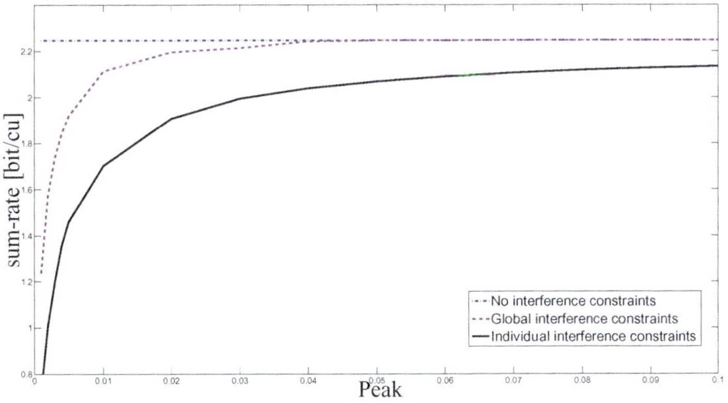

Fig. 4.3 Average sum-rate versus the (peak) interference constraint achievable under no interference, local and global interference constraints.

Concluding Remarks

The first part of this chapter was devoted to providing the (basic) theoretical tools and methods based on VIs to analyze some fundamental issues of NEPs and GNEPs with jointly shared convex constraints, such as the existence and uniqueness of a solution, and the design of iterative distributed algorithms along with their convergence properties. The second part of the chapter made these theoretical results practical by showing how the proposed framework can be successfully applied to solving some challenging equilibrium problems in CR networks.

Acknowledgements The work of Scutari and Pang is based on research supported by the US National Science Foundation grants CMMI 0802022 and 0969600 and by the Air Force Office of Sponsored Research, award No. FA9555-09-10329. The work of Palomar was supported by the Hong Kong RGC 618709 research grant. The work of Facchinei was partially supported by MIUR, Research program PRIN 2007-9PLLN7, Nonlinear Optimization Variational Inequalities and Equilibrium Problems, Italy.

References

1. Altman, E., Boulogne, T., Azouzi, R.E., Jimenez, T., Wynter, L.: A survey on networking games. Computers and Operations Research **33**, 286–311 (2006)
2. Aubin, J.P.: Mathematical Method for Game and Economic Theory. Dover Publications (2007)
3. Bapat, R.B., Raghavan, T.E.S.: Nonnegative Matrices and Applications. Cambridge University Press (1997)
4. Berman, A., Plemmons, R.J.: Nonnegative Matrices in the Mathematical Sciences. Society for Industrial Mathematics (SIAM) (1987)
5. Bertsekas, D.P., Tsitsiklis, J.N.: Parallel and Distributed Computation: Numerical Methods, 2nd edn. Athena Scientific Press, Belmont, MA (1989)
6. Cendrillon, R., Huang, J., Chiang, M., Moonen, M.: Autonomous spectrum balancing for digital subscriber lines. IEEE Trans. Signal Processing **55**(8), 4241–4257 (2007)
7. Cottle, R.W., Pang, J.S., Stone, R.E.: The Linear Complementarity Problem. Academic Press, San Diego, CA (1992)
8. Etkin, R., Parekh, A., Tse, D.: Spectrum sharing for unlicensed bands. IEEE J. Selected Areas in Communications **25**(3), 517–528 (2007)
9. Facchinei, F., Fischer, A., Piccialli, V.: On generalized Nash games and variational inequalities. Operations Research Letters **35**, 159 –164 (2007)
10. Facchinei, F., Kanzow, C.: Generalized Nash equilibrium problems. A Quarterly Journal of Operations Research (4OR) **5**(3), 173–210 (2007)
11. Facchinei, F., Pang, J.S.: Finite-Dimensional Variational Inequalities and Complementarity Problem. Springer-Verlag, New York (2003)
12. Facchinei, F., Sagratella, S.: On the computation of all solutions of jointly convex generalized Nash equilibrium problems. Optimization Letters **4**(3) (2010)
13. Facchinei, F., Pang, J.S.: Nash equilibria: The variational approach. In: D.P. Palomar, Y.C. Eldar (eds.) Convex Optimization in Signal Processing and Communications, chap. 12, pp. 443–493. Cambridge University Press, London (2009)
14. Goldsmith, A., Jafar, S.A., Maric, I., Srinivasa, S.: Breaking spectrum gridlock with cognitive radios: An information theoretic perspective. Proc. IEEE **97**(5), 894–914 (2009)
15. Haykin, S.: Cognitive radio: Brain-empowered wireless communications. IEEE J. Selected Areas in Communications **23**(2), 201–220 (2005)

16. Hong, M., Garcia, A.: Dynamic pricing of interference in cognitive radio networks. IEEE Trans. Signal Processing (submitted) (2010)
17. Huang, J., Palomar, D.P., Mandayam, N., Walrand, J., Wicker, S.B., Basar, T.: (Special Issue on Game Theory in Communication Systems). IEEE J. Selected Areas in Communications **26**(7) (2008)
18. Jorswieck, E.A., Larsson, E.G., Luise, M., Poor, H.V.: (Special Issue on Game Theory in Signal Proc. and Comm.). IEEE Signal Processing Magazine **26**(5) (2009)
19. Konnov, I.V.: Equilibrium Models and Variational Inequalities. Elsevier B.V., Amsterdam (2007)
20. Luo, Z.Q., Pang, J.S.: Analysis of iterative waterfilling algorithm for multiuser power control in digital subscriber lines. EURASIP J. Applied Signal Processing **2006**, 1–10 (2006)
21. Mitola, J.: Cognitive radio for flexible mobile multimedia communication. In: Proc. IEEE 1999 Int. Workshop on Mobile Multimedia Communications (MoMuC 1999), pp. 3–10. San Diego, CA (1999)
22. Nash, J.: Equilibrium points in n-person game. In: National Academy of Science, vol. 36, pp. 48–49 (1950)
23. Olfati-Saber, R., Fax, J.A., Murray, R.M.: Consensus and cooperation in networked multi-agent systems. Proc. IEEE **95**(1), 215–223 (2007)
24. Ortega, J.M., Rheinboldt, W.C.: Iterative Solution of Nonlinear Equations in Several Variables. Society for Industrial Mathematics (SIAM), Philadelphia, PA (1987)
25. Pang, J.S., Chan, D.: Iterative methods for variational and complementarity problems. Mathematical Programming **24**(1), 284–313 (1982)
26. Pang, J.S., Scutari, G., Palomar, D.P., Facchinei, F.: Design of cognitive radio systems under temperature-interference constraints: A variational inequality approach. IEEE Trans. Signal Processing **58**(6), 3251–3271 (2010)
27. Pang, J.S.: Asymmetric variational inequality problems over product sets: Applications and iterative methods. Mathematical Programming **31**(2), 206–219 (1985)
28. Pang, J.S., Scutari, G., Facchinei, F., Wang, C.: Distributed power allocation with rate constraints in Gaussian parallel interference channels. IEEE Trans. Information Theory **54**(8), 3471–3489 (2008)
29. Rosen, J.: Existence and uniqueness of equilibrium points for concave n-person games. Econometrica **33**(3), 520–534 (1965)
30. Scutari, G., Barbarossa, S., Pescosolido, L.: Distributed decision through self-synchronizing sensor networks in the presence of propagation delays and asymmetric channels. IEEE Trans. Signal Processing **56**(4), 1667–1684 (2008)
31. Scutari, G., Palomar, D.P.: MIMO cognitive radio: A game theoretical approach. IEEE Trans. Signal Processing **58**(2), 761–780 (2010)
32. Scutari, G., Palomar, D.P., Barbarossa, S.: Asynchronous iterative water-filling for Gaussian frequency-selective interference channels. IEEE Trans. Information Theory **54**(7), 2868–2878 (2008)
33. Scutari, G., Palomar, D.P., Barbarossa, S.: Competitive design of multiuser MIMO systems based on game theory: A unified view. IEEE J. Selected Areas in Communications **26**(7), 1089–1103 (2008)
34. Scutari, G., Palomar, D.P., Barbarossa, S.: Optimal linear precoding strategies for wideband noncooperative systems based on game theory—Part I & II: Nash equilibria & Algorithms. IEEE Trans. Signal Processing **56**(3), 1230–1249 & 1250–1267 (2008)
35. Scutari, G., Palomar, D.P., Barbarossa, S.: Competitive optimization of cognitive radio MIMO systems via game theory. In: D.P. Palomar, Y.C. Eldar (eds.) Convex Optimization in Signal Processing and Communications, chap. 11, pp. 387–442. Cambridge University Press, London (2009)
36. Scutari, G., Palomar, D.P., Barbarossa, S.: The MIMO iterative waterfilling algorithm. IEEE Trans. Signal Processing **57**(5), 1917–1935 (2009)
37. Scutari, G., Palomar, D.P., Facchinei, F., Pang, J.S.: Convex optimization, game theory, and variational inequality theory in multiuser communication systems. IEEE Signal Processing Magazine **27**(4), 35–49 (2010)

38. Scutari, G., Facchinei, F., Pang, J.S., Palomar, D.P.: Monotone communication games: Theory, algorithms, and models. IEEE Trans. Information Theory (submitted) (2010)
39. Yin, H., Shanbhag, U.V., Mehta, P.G.: Nash equilibrium problems with congestion costs and shared constraints. IEEE Trans. Automatic Control **56**(7), 1702–1708 (2010)
40. Yu, W., Ginis, G., Cioffi, J.M.: Distributed multiuser power control for digital subscriber lines. IEEE J. Selected Areas in Communications **20**(5), 1105–1115 (2002)

Chapter 5
A Mechanism Design Approach to Dynamic Price-Based Control of Multi-Agent Systems

Cédric Langbort

Abstract We show how ideas and tools from the field of mechanism design in economics can be brought to bear on the problem of price-based control of dynamical systems. Specifically, we take inspiration from the Vickrey–Clarkes–Groves mechanism to design *strategy-proof* dynamic price-functions, which can induce subsystems to apply socially efficient control inputs even though they are self-interested and possibly strategically misreport their cost and dynamics' models to the control designer.

5.1 Introduction

Price-based control uses incentives, instead of instructions or direct actions, to induce the self-interested parties involved in a system to make socially desirable decisions. When these decisions affect a dynamical system (as is the case, e.g., when each party is a provider controlling a power plant) or when the environment is time-varying (e.g., when agents can join and leave at any time) and prices are updated dynamically as the system evolves, this control process is usually referred to as *real-time pricing* or *dynamic price-based control*. Examples of infrastructures already partially controlled through incentives include power grids, where prices are used to regulate both production and demand, and communication networks, in which prices are used, e.g., to induce collaboration in peer-to-peer networks or guard against congestion (see [14, 17] and references therein). More recently the use of prices has also been suggested to help incorporate airline preferences in weather-induced rescheduling operations in air traffic flow control [19] and for the coordination of windmills in large-scale wind farms [15], although such schemes have not been implemented yet.

Department of Aerospace Engineering & Coordinated Science Laboratory
University of Illinois at Urbana-Champaign, Urbana, IL 61801, USA
e-mail: langbort@illinois.edu

R. Johansson and A. Rantzer (Eds.): Distributed Decision Making and Control, LNCIS 417, pp. 113–129.
springerlink.com

One of the main obstacles precluding the generalized deployment of dynamic price-based control to these infrastructures in general, and future generations of power networks in particular, is the relative lack of theoretical foundations to guarantee these systems' performance when used in a realistic and uncertain environment. In particular, most existing schemes can only be proved to achieve their regulation goal under the very restrictive assumptions of compliance and truthfulness of the self-interested parties. To reap the expected benefits of real-time pricing in practice, it is thus essential to construct pricing schemes that are strategy-proof, i.e., induce the socially desired behavior, even when agents purposefully misreport information to the designer so as to induce individually advantageous prices.

In this chapter, we show how the ideas and tools of mechanism design theory in economics can be leveraged to construct such provably strategy-proof dynamic price-based control methods. We start with the example of integrated power networks as a way to motivate the need for price-based control and strategy-proofness in Sec. 5.2. We then present an existing technique and demonstrate its lack of strategy-proofness in Sec. 5.3.1, before reviewing the Vickrey–Clarke–Groves (VCG) scheme in Sec. 5.3.2.1. Finally, in Sec. 5.4, we exploit the insight provided by VCG to construct satisfactory dynamic price functions for the power network example. We end with some considerations about the remaining shortcomings of this approach and directions for future work.

5.2 Motivation—Price-Based Control in Integrated Networks

Integrated power networks, such as the ones currently in operation in New York State, New England and New Zealand, are typically regulated through the resource allocation process pictured in Fig. 5.1 [20, 5]. Each provider has a cost function parametrized by the characteristics of and operational constraints on its generators. It transmits these characteristics to the utility or electric control center, which computes the desired power to be generated accounting for global constraints (such as Kirchhoff's laws, cross-area frequency coupling, transmission line overload, etc.) and forecast demand. The utility publishes prices, which serve as indirect control signals and to which providers respond by generating the power level that minimizes their *net* individual cost ("cost-plus-payment"). This is in contrast with so-called "unbundled power markets," in which prices are determined by a market with no central authority [20].

In most cases, this resource allocation process is carried out over horizons of different duration, incorporating different information for each time scale, and repeated up to several times an hour. *Day-ahead pricing*, which computes a single energy price for the next 24 hours based on the demand forecasts and providers' cost function over the same period, is present in most systems. In addition, some utilities (e.g., in Florida [3]) also use *spot pricing* schemes, in which prices are computed for much shorter horizons and in which the global constraints and providers' character-

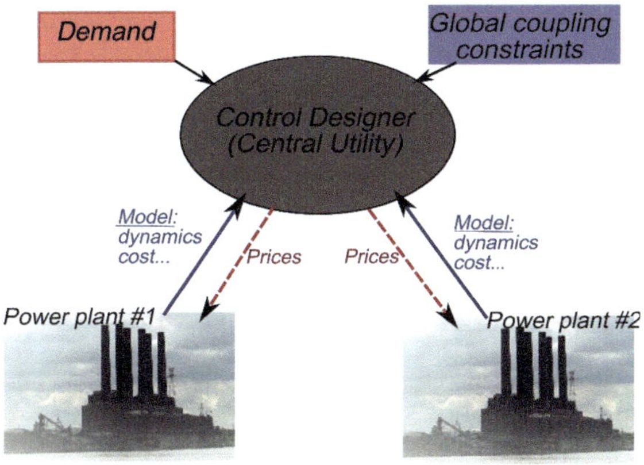

Fig. 5.1 Price-based control in an integrated power network. Note that power plants have *a priori* no reason to report their model information truthfully to the designer.

istics are updated before computations for a new horizon start. In this computation, it is typically assumed that the generators are in quasi-steady state.

In contrast, in real-time pricing (also known as dynamic price-based control [3, 16, 11]), prices are computed and published at the fastest time scale present in the system to be controlled, without assuming that the system is in equilibrium. A specific regulation problem, where dynamic price-based control has been applied [3, 7] is multi-area load frequency control, as discussed next.

5.2.1 Multi-Area Load Frequency Control

Consider two spatial areas, indexed by $i = 1,2$, both equipped with a privately owned generator (several generators per area can also be considered but require more cumbersome notation). In the neighborhood of a fixed set point, each generator's dynamics can be captured by the following linear time-invariant system:

$$x_i(k+1) = A_i x_i(k) + B_i u_i(k) \tag{5.1}$$

$$M_i(k) = C_i x_i(k),\ x_i(0) \text{ given}, \tag{5.2}$$

where x_i is the state of the boiler-turbine generator, u_i is the (scalar) governor control valve position and M_i is the mechanical power to the turbine-generator shaft of the generator in area i. The deviation of area i's current frequency from its set point value of 50 or 60 Hz (depending on which part of the world the regulation problem is set up), denoted by f_i, satisfies

$$f_i(k+1) = f_i(k) + \frac{\Delta k M_i(k) - E_i(k)}{H_i}, \tag{5.3}$$

where $E_i(k)$ is the total energy produced in area i during the time step Δk and H_i is the inertial constant of area i's generator. Areas are coupled through a tie line and energy E_i can be expressed as

$$E_i(k) = D_i(k) + \frac{\delta_i(k) - \delta_{-i}(k)}{X}, \tag{5.4}$$

where the power demand profile in area i over the time window T of interest, $\{D_i(k)\}_{k=0}^{T}$, is assumed to be known to the generator and central planning utility, and X is the inductance of the tie line. The term $(\delta_i - \delta_{-i})$ in (5.4) is the difference between voltage phase angle of area i and the voltage phase angle of the other area. In particular, we have introduced the notation $-i$ to represent $\{1,2\}\backslash\{i\}$. Finally, if we make the typical assumption that the mechanical angle of the generator's rotor is equal to the phase angle of the voltage in the corresponding area, we can write

$$\delta_i(k+1) = \delta_i(k) + \Delta k f_i(k). \tag{5.5}$$

Equations (5.1–5.5) constitute the so called (discrete time) average system frequency model, which is often used for load frequency control problems in the power systems literature (see, e.g., [16]). We note that, although generator i's dynamics (5.1) are typically assumed to be linear time-invariant in such models, most of the results presented next can be derived for nonlinear dynamics as well. This set of equations is captured by the "physical block" subsystem in the block diagram of Fig. 5.2.

5.2.2 Dynamic Price-Based Control

Over a horizon of T time steps, power plant i has a production cost

$$J_i\left(\{u_i(k)\}_{k=0}^{T-1}\right) := \sum_{k=0}^{T-1} \ell_i(x_i(k), u_i(k)) + \Phi_i(x_i(T)), \tag{5.6}$$

where state x_i and input u_i are related through dynamics (5.1). *To simplify further developments, we assume that J_i is a convex function but is otherwise arbitrary.* In load frequency control, the goal of the central utility is to keep the frequency deviations f_1 and f_2 in both areas small, and the voltage phase difference $|\delta_1 - \delta_2|$, which determines the power flow between areas, set to a desired value, while minimizing the global production cost. Introducing functions ℓ_f and ℓ_δ to gauge the amplitude of frequency and phase difference deviations, respectively, amounts to minimizing

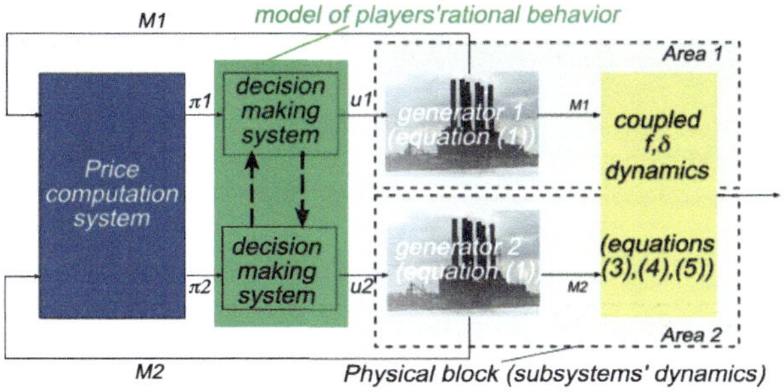

Fig. 5.2 Power network with dynamic price-based control. The "decision making" systems represent the generators' model of rational behavior, i.e., the way in which they react to prices.

$$
\begin{aligned}
J\left(\{u_1(k)\}_{k=0}^{T-1},\{u_2(k)\}_{k=0}^{T-1}\right) &:= J_1\left(\{u_1(k)\}_{k=0}^{T-1}\right)+J_2\left(\{u_2(k)\}_{k=0}^{T-1}\right)\\
&\underbrace{+\sum_{k=0}^{T-1}\Big(\ell_f(f_1(k),f_2(k))+\ell_\delta(\delta_1(k)-\delta_2(k))\Big) + \Phi_f(f_1(T),f_2(T))+\Phi_\delta(\delta_1(T)-\delta_2(T))\Big)}_{G\left(\{u_1(k)\}_{k=0}^{T-1},\{u_2(k)\}_{k=0}^{T-1}\right)}
\end{aligned}
\tag{5.7}
$$

subject to dynamics (5.1-5.5).

In an integrated network, the minimization of cost (5.7) cannot be performed directly by the utility because (i) the optimization does not have access to individual costs J_1 and J_2 and, (ii) even if it could compute the optimal inputs $u_1^\star$ and $u_2^\star$ minimizing (5.7), the algorithm would not have the authority to set inputs u_1 and u_2 to these values, since the governor valve position is controlled by the generator's operator. Instead, the utility can offer the generators an incentive to try to induce them to choose $u_1^\star$ and $u_2^\star$ as their inputs. This, in essence, is the central conceptual idea underlying price-based control. An incentive for generator i is a price function π_i, which, to every decision $\{u_1(k)\}_{k=0}^{T-1},\{u_2(k)\}_{k=0}^{T-1}$ of the generators over the window T of interest associates a monetary payment. We assume that this payment can take on both positive and negative values, with positive values corresponding to payments made from the generator to the utility (a tax or fee), and negative values corresponding to payments from the utility to the generators (a reward). When offered such a price, generator i's *net* production cost becomes

$$
C_i\left(\{u_i(k)\}_{k=0}^{T-1}\right)=J_i\left(\{u_i(k)\}_{k=0}^{T-1}\right)+\pi_i\left(\{u_i(k)\}_{k=0}^{T-1},\{u_{-i}(k)\}_{k=0}^{T-1}\right) \tag{5.8}
$$

Following [3], we assume that providers respond to a sequence of prices by minimizing their net cost C_i over a compact and convex set of acceptable control inputs $\mathcal{U}_i$. This is sometimes referred to as the players being 'price-takers' [10]. With this model of players' rational behavior, the main question of dynamic price-based control is whether the utility can find incremental prices $\{\pi_i^k\}_{k=0}^{T-1}$ such that conditions (5.9) below are satisfied:

$$\{u_i^\star(k)\}_{k=0}^{T-1} = \operatorname*{arg\,min}_{\{u_i(k)\}_{k=0}^{T-1}\in\mathcal{U}_i} \quad \mathcal{C}_i, \tag{5.9}$$

$$\pi_i = \sum_{i=0}^{T-1} \pi_i^k, \text{ for all } k,\ \pi_i^k \text{ is a function of } \{u_i(t)\}_{t=0}^{k} \tag{5.10}$$

Note that there is a slight abuse of notation in (5.8) and (5.9), since cost $\mathcal{C}_i$ can depend on $\{u_{-i}(k)\}_{k=0}^{T-1}$ in addition to player i's decisions. In this case, the argmin is to be understood as the minimization of function $\mathcal{C}_i(.,\{\tilde{u}_{-i}(k)\}_{k=0}^{T-1})$ for a specific choice $\{\tilde{u}_{-i}(k)\}_{k=0}^{T-1}$ of player $-i$'s decisions, to be defined shortly.

When (5.9) holds, we say that price functions $\{\pi_i\}_{i=1}^{2}$ are *efficient*. Equation (5.10) imposes that a price be paid by the utility at every time step, instead of a lump sum at the end of the time window. It is in this sense that the pricing scheme is dynamic. Condition (5.10) ensures that prices are causal functions of the generators' control inputs, which is necessary to guarantee that the utility can compute the price at each step. One can thus think of a dynamic price-based control scheme as a causal system mapping generators' state and input signals into price signals, such that the closed-loop system pictured in Fig. 5.2 minimizes cost function J. In the block diagram of Fig. 5.2, such a dynamic price-based control scheme is captured by the 'price computation' system. The decision making systems capture the generators' *model of rational behavior*, i.e., the specification of how they react to prices.

We propose several examples of efficient price functions in the next sections. The first one, presented in Sec. 5.3 (and similar to the original proposal of Berger and Schweppe [2, 16]) depends only on player i's decisions, but requires the utility to have access to the plant's models and cost functions. In contrast, the price functions explored in Sec. 5.4 are efficient for all cost functions and models, but require the notion of Nash efficiency (which corresponds to a particular choice of $\{u_{-i}(k)\}_{k=0}^{T-1}$ in (5.8)–(5.9)).

5.3 A Formal Model

Motivated by the example of integrated networks, we start by considering the class of static optimization problems:

$$\begin{aligned} &\min && J_1(\xi_1)+J_2(\xi_2)+G(\zeta) && && (5.11a)\\ &\text{subject to} && \zeta = \underbrace{H_1(\xi_1)+H_2(\xi_2)}_{H(\xi_1,\xi_2)}, && \xi_i \in \mathcal{U}_i, i=1,2. && (5.11b)\end{aligned}$$

The load frequency control problem of Sec. 5.2.1 can be captured in the form of (5.11) with appropriate linear functions H_1 and H_2 if we represent the sequence of inputs $\{u_i(k)\}_{k=0}^{T-1}$ by the T-dimensional vector ξ_i, concatenate all variables $\{\delta_1(k)\}_{k=0}^{T-1}$, $\{\delta_2(k)\}_{k=0}^{T-1}$, $\{f_1(k)\}_{k=0}^{T-1}$ and $\{f_2(k)\}_{k=0}^{T-1}$ into vector ζ, and rewrite the cost functions in terms of these variables. For convenience we also henceforth assume that the set $\mathcal{U}_i$ is defined as

$$\mathcal{U}_i = \{\xi \mid g_{ij}(\xi) \leq 0,\ j = 1,\ldots,m_i\},$$

for some convex functions $g_{i1},\ldots,g_{im_i}$. From this analogy, it is natural to think of quantities (ξ_i, J_i) as pertaining to a stand-alone subsystem labeled i (akin to the power plant considered earlier), while (ζ, G) pertains to a leader subsystem similar to the power utility of Sec. 5.2. We can also think of functions H_1 and H_2 as being known by the leader since they are the counterpart of coupling dynamics (5.4, 5.3, 5.5).

As before, we say that a price function π_i of variable ξ_i is efficient if

$$\arg\min_{\xi \in \mathcal{U}_i} J_i(\xi_i) + \pi_i(\xi_i) \tag{5.12}$$

is a solution to problem (5.11). This price function is to be computed by the leader while decision ξ_i is ultimately made by subsystem i according to (5.7). For now, we will not concern ourselves with the requirement that the prices be decomposable into causal increments. We will revisit this issue in Sec. 5.4.

5.3.1 KKT Price-Based Control and Its Inadequacy

We are now in a position to review Berger and Schweppe's original price-based control strategy [2].

Proposition 5.1. *The price function $\pi_i^\star$ defined as*

$$\pi_i^\star(\xi_i) := \left[\nabla_{\xi_i}(G \circ H)(\xi_1^\star, \xi_2^\star)\right] \xi_i, \quad \forall \xi_i \in \mathcal{U}_i \tag{5.13}$$

is efficient.

Proof. *Let $C_i^\star$ denote the cost function of power plant i defined as per (5.7) when using the price function $\pi_i^\star$. Note that this function is convex since J_i is convex and $\pi_i^\star$ is linear. Hence, $C_i^\star$ has a unique minimum $\bar{\xi}_i$ in $\mathcal{U}_i$ characterized by the Karush-Kuhn-Tucker (KKT) conditions*

$$g_{ij}(\bar{\xi}_i) \leq 0, \; \forall j \tag{5.14a}$$

$$\mu_j \geq 0, \; \forall j \tag{5.14b}$$

$$\mu_j^T g_{ij}(\bar{\xi}_i) = 0, \; \forall j \tag{5.14c}$$

$$\nabla C_i^\star(\bar{\xi}_i) + \sum_{j=1}^{m_i} \mu_j^T \nabla g_{ij}(\bar{\xi}_i) = 0. \tag{5.14d}$$

Now, note that the minimum $(\xi_1^\star, \xi_2^\star)$ *of problem (5.11) must satisfy the corresponding set of necessary KKT conditions, which are identical to (5.14), except for (5.14d), which is replaced by*

$$\nabla J(\xi_i^\star) + \nabla_{\xi_i} [G \circ H] (\xi_1^\star, \xi_2^\star) + \sum_j \mu_j^T \nabla g_{ij}(\xi_i^\star) = 0.$$

By definition, $\nabla C_i^\star(\xi) = \nabla J_i(\xi) + \nabla_{\xi_i} [J \circ H] (\xi_1^\star, \xi_2^\star)$. *Thus,* $\xi_i^\star$ *is a solution of (5.14) as well with the same multipliers* $\{\mu_j\}_{j=1}^{m_i}$ *and, in turn, the unique minimizer of* $C_i^\star$. □

Price function $\pi_i^\star$ has the advantage of being linear in the decision variable ξ_i. However it also requires the price designer to compute the optimal decisions $\xi_1^\star$, $\xi_2^\star$, i.e., *to solve problem (5.11)*. This is undesirable because cost functions J_1, J_2 and coupling functions H_1 and H_2 are typically not available to the leader a priori, and must be elicited from the subsystems.

If we assume that these subsystems are strategic and solely interested in minimizing their individual costs, with no regard to solving problem (5.11), we should expect them to misreport information if this can result in a price that is advantageous to them. This would lead the designer to compute incorrect values for $\xi_1^\star$ and $\xi_2^\star$ in (5.13) and, in turn, to use an inefficient price function. Such strategic misreporting has been observed in practice on many occasions, particularly in the context of power networks. For example, the sudden rise of the Midwest wholesale spot energy prices to $7000/MWh over the course of a few minutes in June 1998, which resulted in local shortages, was widely attributed to "ramping constraint gaming" [12, 20, 13]—a practice in which providers colluded to lie about their generators' instantaneous start-up and turn-off times.

In summary, the only context in which price function $\pi_i^\star$ is acceptable is when the leader has a priori knowledge of subsystems' characteristics, or when subsystems can be trusted to report them truthfully and act as price takers. However, if we expect subsystems to be strategic, it is necessary we can ensure that the price function be efficient even in the presence of incorrect information about the subsystems' characteristics. This naturally raises the question of designing a strategy-proof price function for problem (5.11), i.e., a price function that is efficient for all values of functions J_1 and J_2.

5.3.2 Strategy-Proofness and Mechanism Design

In order to see how strategy-proofness might be achieved, it is instructive to review some ideas from the field of mechanism design in economics, where this problem has been studied in a different context. To this end, consider yet another problem formulation, which is standard for mechanism design (see, e.g., [9]).

5.3.2.1 The VCG Mechanism: A Building Block for Strategy-Proofness

A typical mechanism design problem involves M agents. Each agent i (which can be identified with a generator in the case of load-frequency control) has a cost function v_i which depends on its type $\theta_i \in \mathbb{R}$ and the decision d taken by the central planner. The type θ_i represents all the information that is privately known to the agent, but the functional form of v_i is assumed to be globally known. Note that this is in contrast with problem (5.11) in Sec. 5.3, where a subsystem's private information is the arbitrary convex function J_i. The goal of the central planner (akin to the leader in our case) is to compute the decision d that minimizes the social cost

$$\sum_{i=1}^{M} v_i(\theta_i, d) \tag{5.15}$$

of the agents, even though it does not know the types $\theta := (\theta_1, ..., \theta_M)$. To do this, the planner first queries the agents for their type and receives a report $\hat{\theta} = (\hat{\theta}_1, ..., \hat{\theta}_M)$ that may not be true. Based on this possibly false report, the planner (i) computes his decision $d(\hat{\theta})$ and (ii) rewards each player i by paying him a price $t_i(\hat{\theta})$. Player i thus incurs a net cost of

$$V_i(\theta_i, d(\hat{\theta})) = v_i(\theta_i, d(\hat{\theta})) + t_i(\hat{\theta}) \tag{5.16}$$

when his true type is θ_i and the report received by the planner is $\hat{\theta}$. Assuming that each player acts so as to minimize this net cost, regardless of the other players' reports (i.e., that he plays a dominant strategy), the planner wants to find payments $\{t_i\}_{i=1}^{M}$ that induce the revelation of the true types, i.e., such that

$$V_i(\theta_i, d(\theta_i, \hat{\theta}_{-i})) \leq V_i(\theta_i, d(\hat{\theta})), \forall \hat{\theta} \tag{5.17a}$$

$$d(\theta) \text{ minimizes } (5.15), \tag{5.17b}$$

where we have introduced the notation θ_{-i} to designate the $(M-1)$-tuple

$$(\theta_1, \dots, \theta_{i-1}, \dots, \theta_{i+1}, \dots, \theta_M) \ \forall i$$

Conditions (5.17) are the counterparts of incentive conditions (5.7), in the case where truthfulness, instead of a known decision, is to be induced by the planner. It can be shown [9] that when the types θ_is are scalar, any set of price functions

$\{t_i\}_{i=1}^M$ satisfying (5.17) can be written as

$$t_i(\hat{\theta}) = \sum_{j \neq i} v_j(\hat{\theta}_j, \text{opt}(\hat{\theta})) + f_i(\hat{\theta}_{-i}), \tag{5.18}$$

where f_i is an arbitrary function and *opt* is the map which, to a type-report $\hat{\theta}$, associates the maximum of social welfare function (5.15) when $\theta = \hat{\theta}$. Note that this function, and hence price function (5.18) as well, can be directly computed by the planner, since we have assumed that the functional form of $\{v_i\}_{i=1}^M$ was known to the planner. Among all functions satisfying (5.18), the following one, known as the Vickrey–Clarke–Groves (VCG) scheme, is particularly interesting, because it can be given a natural auction interpretation [18]:

$$t_i^{VCG}(\hat{\theta}) = \sum_{j \neq i} v_j(\hat{\theta}_j, \text{opt}(\hat{\theta})) - \min_d \sum_{j \neq i} v_j(\hat{\theta}_j, d). \tag{5.19}$$

In particular, player i receives as a payment his marginal contribution to the optimal global cost $\sum_j v_j(\hat{\theta}_j, \text{opt}(\hat{\theta}))$.

5.3.2.2 Achieving Strategy-Proofness in Nash Equilibrium for Problem (5.11)

The discussion of Sec. 5.3.2.1 shows how one can design mechanisms which ensure that it is in the subsystems' best interest is to report truthful information. With this background, it seems natural to try to design a strategy-proof price function for problem (5.11) in two steps, by first constructing a VCG revelation mechanism to induce each subsystem to report the true value of function J_i, and then using this report to compute an efficient price function, such as $\{\pi_i^\star\}$. However, this forces the leader to pay a price for both decision ξ_i and a description of function J_i. It is also not clear what form the report could take in the first step, since function J_i is infinite-dimensional. We propose a different kind of provably strategy-proof price-based control law, which builds on the intuition provided by the VCG mechanism in a more indirect way but requires us to slightly modify the set-up of Sec. 5.3. First, we allow subsystem i's price function π_i to depend on the decision of both subsystems, instead of just ξ_i. In turn, the resulting net cost function

$$C_i(\xi_i, \xi_{-i}) = J_i(\xi_i) + \pi_i(\xi_i, \xi_{-i})$$

also depends on the decisions of both subsystems. Second, we replace the price-taking assumption (i.e., the assumption that subsystems respond to prices by minimizing C_i, as in (5.7)) by a different model of subsystem rational behavior. We assume that subsystems act noncooperatively and make decisions $\bar{\xi}_1$ and $\bar{\xi}_2$ that constitute a Nash equilibrium, i.e., such that

$$C_i(\bar{\xi}_i, \bar{\xi}_{-i}) < C_i(\xi_i, \bar{\xi}_{-i}) \forall \xi_i \text{ and } \forall i = 1, 2. \tag{5.20}$$

Accordingly, we replace the notion of efficiency introduced in Sec. 5.3 by that of Nash efficiency and say that price functions $\{\pi_i\}_{i=1}^2$ are *Nash efficient* if the optimal solution of problem (5.11) satisfies the Nash equilibrium conditions (5.20). We say they are *Nash strategy-proof* if they are Nash efficient for all convex functions J_1, J_2.

Theorem 5.1. *Price functions $\{\pi_i\}_{i=1}^2$ are smooth and Nash strategy-proof for problem (5.11) if and only if there exist smooth functions $\{F_i : \mathcal{U}_{-i} \to \mathbb{R}\}_{i=1}^2$ such that*

$$\pi_i(\xi_i, \xi_{-i}) = [G \circ H](\xi_1, \xi_2) + F_i(\xi_{-i}) \tag{5.21}$$

for all i and all $(\xi_i, \xi_{-i}) \in int\, \mathcal{U}_i \times int\, \mathcal{U}_{-i}$.

Before providing a proof of Theorem 5.1, we should note the strong resemblance between characterizations (5.21) and (5.18). with the well-known Vickrey–Clarke–Groves mechanism for welfare maximization (see, e.g., [9]). It is worth re-emphasizing that the VCG mechanism generates prices that implement truthfulness in *dominant strategies* (i.e., such that $C_i(\xi_i^\star, \xi_{-i}) < C_i(\xi_i, \xi_{-i})$ for all ξ_i, ξ_{-i}) instead of being a Nash equilibrium. This dominant strategies implementation is achieved by use of the efficient decision rule, i.e., the map that to each value of the privately known types associates the optimal decision. In the setting of problem (5.11), computing this map forces the leader to make use of $(\xi_1^\star, \xi_2^\star)$ which, as explained earlier, does not lead to strategy-proofness. It is because of the impossibility of using the efficient decision rule that we had to settle for Nash strategy-proofness.

Finally, note that just like for the Clarke and Groves payment, we can choose

$$F_i(\xi_{-i}) = -\min_{\xi_i \in \mathcal{U}_i} [G \circ H](\xi_1, \xi_2)$$

when the constraint sets $\mathcal{U}_1, \mathcal{U}_2$ are known to the leader. This results in prices $\pi_i(\xi_i, \xi_{-i})$ that are always nonnegative, i.e., such that subsystems always receive money from the leader.

Proof. *The method of proof is adapted from Theorem 2 in [9]. First, if $\pi_i(\xi_i, \xi_{-i})$ satisfies (5.21), then $C_i(\xi_i, \xi_{-i}) = J_i(\xi_i) + G(H(\xi_1, \xi_2)) + F(\xi_{-i})$ for all (ξ_i, ξ_{-i}). Hence, if there exists $\xi_i \neq \xi_i^\star$ such that $C_i(\xi_i, \xi_{-i}^\star) \leq C_i(\xi_i^\star, \xi_{-i}^\star)$, it holds that*

$$J_i(\xi_i) + J_{-i}(\xi_{-i}^\star) + G(H_i(\xi_i) + H_{-i}(\xi_{-i}^\star)) \leq J_i(\xi_i^\star) + J_{-i}(\xi_{-i}^\star) + G(H(\xi_1^\star, \xi_2^\star))$$

and $(\xi_i^\star, \xi_{-i}^\star)$ cannot be the unique minimum of problem (5.23), which is a contradiction. As a result, $(\xi_1^\star, \xi_2^\star)$ is a Nash equilibrium of the net cost functions C_1, C_2. Conversely, let us assume that $\{\pi_i\}_{i=1}^2$ are smooth and Nash strategy-proof for (5.11). Let $(\bar{\xi}_1, \bar{\xi}_2) \in int\, \mathcal{U}_1 \times int\, \mathcal{U}_2$ be such that there exist convex functions $\bar{J}_1$ and $\bar{J}_2$ for which $(\bar{\xi}_1, \bar{\xi}_2)$ is the optimal decision in problem (5.11).

By the assumption on the price functions, $(\bar{\xi}_1, \bar{\xi}_2)$ is a Nash equilibrium for the game described by the agents' net cost functions $\{\bar{J}_i + \pi_i\}_{i=1}^2$. Hence,

$$\nabla \bar{J}_i(\bar{\xi}_i) + \nabla_{\xi_i} \pi_i(\bar{\xi}_i, \bar{\xi}_{-i}) = 0,$$

since $\bar{\xi}_i$ must be a local minimizer of $\bar{J}_i(.) + \pi_i(., \bar{\xi}_{-i})$ over $\mathcal{U}_i$ and is an interior point by assumption. At the same time, $\bar{\xi}_1, \bar{\xi}_2$ also minimizes the convex *function $\bar{J}_1 + \bar{J}_2 + [G \circ H]$ over $\mathcal{U}_1 \times \mathcal{U}_2$, so it is the unique solution of the corresponding KKT conditions, with Lagrange multipliers of all g_{ij} equal to zero (again, because it is an interior point). Hence, we find that*

$$\nabla_{\xi_i} \pi_i(\bar{\xi}_i, \bar{\xi}_{-i}) = \nabla_{\xi_i} [G \circ H](\bar{\xi}_1, \bar{\xi}_2). \tag{5.22}$$

To finish the proof, it is now enough to show that equality (5.22) holds for all $(\bar{\xi}_1, \bar{\xi}_2) \in int\, \mathcal{U}_1 \times int\, \mathcal{U}_2$, i.e., that every such interior point can be written as the optimal decision of problem (5.11) for some functions $\bar{J}_1$, $\bar{J}_2$. This latter statement clearly holds since, for any given $(\bar{\xi}_1, \bar{\xi}_2) \in int\, \mathcal{U}_1 \times int\, \mathcal{U}_2$, we can define function $\bar{J}$ on $\mathcal{U}_1 \times \mathcal{U}_2$ as

$$\bar{J}_i(\xi) = \frac{1}{2}\xi^T \xi - (\bar{\xi}_i + \nabla_{\xi_i} [G \circ H](\bar{\xi}_1, \bar{\xi}_2))^T \xi.$$

$\bar{J}_i$ is convex and $\bar{\xi}_i$ satisfies

$$\nabla \bar{J}_i(\bar{\xi}_i) = -\nabla_{\xi_i} [G \circ H](\bar{\xi}_1, \bar{\xi}_2),$$

which are the KKT conditions for problem (5.11). □

5.4 Back to Load Frequency Control and Dynamic Prices

So far, we have seen how to build on the VCG mechanism to construct Nash strategy-proof price functions for problem (5.11). In this section, we go back to the motivating example of price-based load frequency control and show how to turn these static price functions into dynamic ones. To this end, consider the following specific linear quadratic instance of the load frequency control problem:

$$\min \quad \frac{1}{2}\sum_{t=0}^{N-1} \left(\sum_{i=1}^{2} x_i(t)^T Q_i x_i(t) + u_i(t)^T R_i u_i(t) + z(t+1)^T Q z(t+1) \right) \tag{5.23a}$$

$$\text{subject to} \quad x_i(t+1) = A_i x_i(t) + B_i u_i(t),\ x_i(0) = \bar{x}_i,\ \forall i, \tag{5.23b}$$

$$z(t+1) = z(t) + M_1 x_1(t) + M_2 x_2(t);\ z(0) = \bar{z}, \tag{5.23c}$$

$$\|u_i(t)\| \leq 1,\ \forall i \tag{5.23d}$$

Introducing

$$\mathbf{x_i}[N] = \begin{bmatrix} \bar{x}_i \\ \vdots \\ x_i(N-1) \end{bmatrix}, \quad \mathbf{z}[N] = \begin{bmatrix} z(1) \\ \vdots \\ z(N) \end{bmatrix},$$

$$\mathbf{u_i}[N] = \begin{bmatrix} u_i(0) \\ \vdots \\ u_i(N-1) \end{bmatrix} \mathbf{A_i}[N] = \begin{bmatrix} I \\ \vdots \\ A_i^N \end{bmatrix},$$

$$\mathbf{B_i}[N] = \begin{bmatrix} 0 & 0 & \dots & \dots & 0 \\ B_i & 0 & & & \vdots \\ A_iB_i & B_i & & & \vdots \\ \vdots & \ddots & \ddots & & \vdots \\ A_i^{N-1}B_i & \dots & A_iB_i & B_i & 0 \end{bmatrix},$$

$$\mathbf{M_i}[N] = \begin{bmatrix} M_i & 0 & \dots & 0 \\ M_i & M_i & 0 & 0 \\ \vdots & \vdots & \ddots & \vdots \\ M_i & \dots & \dots & M_i \end{bmatrix}$$

and

$$\mathbf{Q_i}[N] = I_N \otimes Q_i, \;\; \mathbf{Q}[N] = I_N \otimes Q,$$

$$\mathbf{R_i}[N] = I_N \otimes R_i, \; \mathbf{E}[N] = \begin{bmatrix} I \\ \vdots \\ I \end{bmatrix},$$

we can rewrite (5.23) as

$$\min \frac{1}{2}\left(\mathbf{z}[N]^T\mathbf{Q}[N]\mathbf{z}[N] + \sum_{i=1}^{2} \begin{bmatrix} \mathbf{x_1}[N] \\ \mathbf{u_1}[N] \end{bmatrix}^T \begin{bmatrix} \mathbf{Q_1}[N] & 0 \\ 0 & \mathbf{R_1}[N] \end{bmatrix} \begin{bmatrix} \mathbf{x_1}[N] \\ \mathbf{u_1}[N] \end{bmatrix}\right)$$

$$\text{subject to } \; \mathbf{x}_i[N] = \mathbf{A_i}[N]\bar{x}_i + \mathbf{B_i}[N]\mathbf{u_i}[N],$$
$$\mathbf{z}[N] = \mathbf{E}[N]\bar{z} + \mathbf{M_1}[N]\mathbf{x_1}[N] + \mathbf{M_2}[N]\mathbf{x_2}[N].$$

This is of the form (5.11), with appropriately defined functions J_i, H_i and G, and $\mathcal{U}_i$ a hypercube. In particular,

$$H(\mathbf{u_1}[N], \mathbf{u_2}[N]) = \mathbf{E}[N]\bar{z} + \sum_{i=1}^{2} \mathbf{M_i}(\mathbf{A_i}[N]\bar{x}_i + \mathbf{B_i}[N]\mathbf{u_i}[N])$$

and

$$J(\mathbf{z}[N]) = \frac{1}{2}\mathbf{z}[N]^T\mathbf{Q}[N]\mathbf{z}[N].$$

According to Theorem 5.1, a Nash strategy-proof price function π_i for problem (5.23) must be of the form

$$\pi_i(\mathbf{u_i}[N], \mathbf{u_{-i}}[N]) = [G \circ H](\mathbf{u_1}[N], \mathbf{u_2}[N]) + F_i(\mathbf{u_{-i}}[N])$$

for some function F_i. Among such prices, we are particularly interested in those that can be expressed as functions of $\mathbf{M_i}[N]\mathbf{x_i}[N]$ and $\mathbf{M_{-i}}[N]\mathbf{x_{-i}}[N]$ since, unlike the applied inputs $\mathbf{u_i}[N]$ and $\mathbf{u_{-i}}[N]$, these outputs are easily measured by the utility. A possible choice of price function, then, is $\tilde{\pi}_i$ defined as

$$\tilde{\pi}_i(\mathbf{u_i}[N], \mathbf{u_{-i}}[N]) = \left(\mathbf{E}[N]\bar{z} + \frac{1}{2}\mathbf{M_i}[N]\mathbf{x_i}[N] + \mathbf{M_{-i}}[N]\mathbf{x_{-i}}[N]\right)^T \mathbf{Q}[N]\mathbf{M_i}[N]\mathbf{x_i}[N]. \tag{5.24}$$

In order to obtain a dynamic price scheme we must, as stated in (5.9), decompose π_i as

$$\sum_{t=0}^{N-1} \pi_i^t(\mathbf{M_i}[N]\mathbf{x}_i[N], \mathbf{M_{-i}}[N]\mathbf{x}_{-i}[N]) = \tilde{\pi}_i(\mathbf{u}_i[N], \mathbf{u}_{-i}[N]) \tag{5.25}$$

where functions $\{\pi_i^t\}_{i=1}^2$ satisfy

$$\pi_i^t(\mathbf{M_i}[N]\mathbf{x}_i[N], \mathbf{M_{-i}}[N]\mathbf{x}_{-i}[N]) = f_i(\{M_i x_i(s)\}_{s=0}^t, \{M_{-i} x_{-i}(s)\}_{s=0}^t, t)$$

for all $0 \le t \le N-1$ and some function f_i. Among the multiple choices for such *incremental price functions*, $\{\rho_i^t\}_{i=1}^2$ described as follows are of particular interest:

$$\rho_i^0(x_i(0), x_{-i}(0)) = \frac{1}{2}x_i(0)^T M_i^T Q M_i x_i(0) + x_{-i}(0)^T M_{-i}^T Q M_i x_i(0) + \bar{z}^T Q M_i x_i(0)$$

where

$$\begin{aligned}\rho_i^{t+1} = \rho_i^{t+1}(\{x_i(s)\}_{s=0}^t, \{x_{-i}(s)\}_{s=0}^t) &= \rho_i^t + \bar{z}^T Q M_i x_i(t+1) \\ &+ \frac{1}{2}x_i(t+1)^T M_i^T Q M_i x_i(t+1) + x_{-i}(t+1)^T M_{-i}^T Q M_i x_i(t+1) \\ &+ \sum_{s \le t} \big(x_i(s)^T M_i^T Q M_i x_i(t+1) \\ &+ x_{-i}(t+1)^T M_{-i}^T Q M_i x_i(s) + x_{-i}(s)^T M_{-i}^T Q M_i x_i(t+1)\big)\end{aligned} \tag{5.26}$$

for all $0 < t \le N-1$. Indeed, note that the following equality holds for all t, not necessarily equal to $N-1$:

$$\sum_{s=0}^{t} \rho_i^s = \left(\mathbf{E}[t]\bar{z} + \frac{1}{2}\mathbf{M_i}[t]\mathbf{x_i}[t] + \mathbf{M_{-i}}[t]\mathbf{x_{-i}}[t]\right)^T \mathbf{Q}[t]\mathbf{M_i}[t]\mathbf{x_i}[t].$$

This in particular means that the price functions $\{\sum_{s=0}^t \rho_i^t\}_{i=1}^2$ are Nash strategy-proof for problem (5.23) regardless of the horizon length N. In that sense, it can be said that incremental prices $\{\rho_i^t\}_{i=1}^2$ implement the correct decisions in subgame perfect Nash equilibrium [8]. This is particularly useful in a situation where the horizon is chosen by the plants, e.g., when they must declare, before the beginning of the planning phase, a time during which they will be connected to the network and equations (5.23) will be valid. In contrast, when using the one-shot price functions $\{\tilde{\pi}_i\}_{i=1}^2$ (or even the incremental payments $\{\tilde{\pi}_i^t\}$ defined by

$$\begin{aligned}\tilde{\pi}_i^t &= \tilde{\pi}_i^t(\{x_i(s)\}_{s=0}^t, \{x_{-i}(s)\}_{s=0}^t) \qquad (5.27)\\ &= (N-t)\Big[\frac{1}{2}x_i(t)^T M_i^T Q M_i x_i(t) + x_{-i}(t)^T M_{-i}^T Q M_i x_i(t) + \bar{z}^T Q M_i x_i(t)\\ &+ \sum_{s<t}\big(x_i(s)^T M_i^T Q M_i x_i(t) + x_{-i}(t)^T M_{-i}^T Q M_i x_i(s) + x_{-i}(s)^T M_{-i}^T Q M_i x_i(t)\big)\Big]\end{aligned}$$

for all $0 \le t \le N-1$ and thus satisfy (5.25)), there is an opportunity for power plants to misrepresent their plans by declaring a time horizon N that is different from the true time N' that they intend to stay connected. Prices $\tilde{\pi}_i$ and $\tilde{\pi}_i^t$ would then be computed with this declared value N and fail to guarantee that $\mathbf{u}_1^\star[N']$ and $\mathbf{u}_2^\star[N']$ are chosen over the true horizon of operation, when plants play a Nash strategy over $[0, N'-1]$.

5.5 Discussion of Models of Rational Behavior and Future Work

The model of rational behavior (Nash equilibrium playing over the full planning horizon $[0, N-1]$) used so far has the disadvantage that decision $\bar{u}_i[N](s)$ made by agent i at step s depends on the horizon over which it is acting strategically. As a result, the decision made by the plants do not depend causally on the prices, even when the utility uses incremental price functions. For example, when using $\{\rho_i^t\}_{i=1}^2$, the decision made at s by agent i depends on the full payment $\sum_{t=0}^{N-1}\rho_i^t$, not just on the sequence $\{\rho_i^t\}_{t=0}^s$.

While such a noncausal relation is acceptable for planning problems like (5.23), where one can realistically imagine the utility announcing how prices will be computed for all time and plants using this information to decide how to act in advance, one can also think of scenarios (different from the original load frequency control problem of [2]) where agents will *not* plan ahead for a predetermined horizon, but simply react to past and present posted prices, thus making the relation between prices and decision causal, as pictured in the block diagram of Fig. 5.3. This may happen, for example, for smaller and inconsistent generators owned by private users, which may connect to the grid only sporadically.

We can think of two approaches for investigating situations in which agents are constrained to act causally with respect to published prices.

- One can use a different model of rational behavior that is compatible with causality and try to determine price functions that induce the desired outcome under such assumptions. This is the approach taken in some recent works on dynamic mechanism design in economics, in which mechanisms have been designed for systems described by Markovian decision processes that implement efficient de-

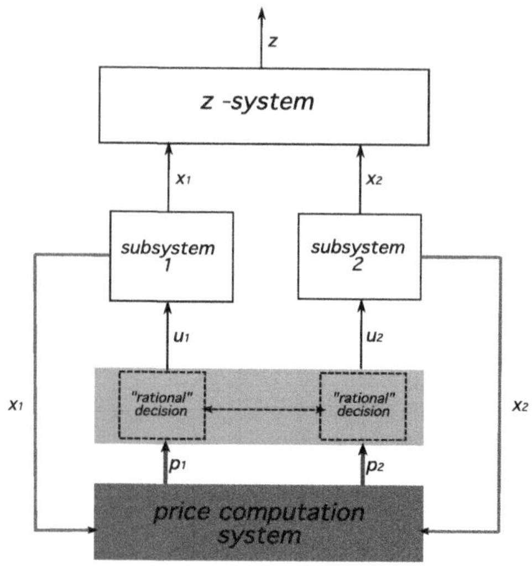

Fig. 5.3 System (5.23) controlled by incremental price functions $\{\rho_i^t\}$. Note that all subsystems are causal, in accordance with the standard practice of block diagrams. See text for details.

cisions in so called *Markov perfect equilibrium* [4] or can be shown to be *interim* incentive compatible [1]. However, both mechanisms use incremental price functions whose computation relies on the value function of optimal control problem (5.23) and thus depend on the agents' cost functions J_1 and J_2.

- One could take the price-computation block of Fig. 5.3 as given (and determined according to an arbitrary model of rational behavior, which, although believed to be invalid and/or incompatible with causality, leads to easily computable prices such as $\{\rho_i^t\}_{i=1}^2$) and study the robustness of/performance degradation in the closed-loop system when the rational decision subsystems vary in a set of operators, possibly under the additional constraint of causality. This, in principle, could be done with the tools of robust control theory [6] and presents two advantages. First, this robustness analysis could also incorporate other sources of uncertainty such as measurement errors and degradation in the channels between subsystems and the price-computation block (or even in that block itself if, e.g., the utility has uncertainty on the H function). Second, it would also allow us to study the properties of mechanisms under a departure of strict Nash-playing behavior. This seems like a valuable endeavor, as the question as to whether strategic agents actually play a Nash equilibrium in practical noncooperative settings is still largely open in economics.

We intend to explore this second, control theory-based, approach in our future work, as a complement to the direction taken by dynamic mechanism design methods.

References

1. D. Bergemann and J. Valimaki, Efficient dynamic auctions, *Cowles Foundation Discussion Paper* 1584, http://cowles.econ.yale.edu/P/cd/d15b/d1584.pdf, 2006.
2. A. W. Berger and F. C. Schweppe, Real Time Pricing to Assist in Load Frequency Control, *IEEE Transactions on Power Systems*, vol. 4, no. 3, pp. 920-926, 1989.
3. A. Berger, Market-clearing prices in coupled, dynamic systems, *Automatica*, vol. 27, no. 4, pp. 743-747, 1991.
4. R. Cavallo, D.C. Parkes, and S. Singh, Efficient Online Mechanisms for Persistent, Periodically Inaccessible Self-Interest Agents, *Harvard University Technical Report*, http://www.eecs.harvard.edu/econcs/pubs/persistent.pdf, 2007.
5. M. Crow, Power System Deregulation, *IEEE Potentials*, pp. 8-9, December 2001- January 2002.
6. G. E. Dullerud and F. Paganini, *A Course in Robust Control Theory*, Springer-Verlag, New York, 1999.
7. B. Daryanian, R. E. Bohn and R. D.Tabors, An experiment in real time pricing for control of electric thermalstorage systems, *IEEE Transactions on Power Systems*, Vol. 6, issue 4, Nov. 1991, pp. 1356-1365.
8. D. Fudenberg and D. Levine, Subgame perfect equilibria of finite- and infinite-horizon games, *Journal of Economic Theory*, vol. 31, no. 2, pp. 251-268, 1983.
9. M. Jackson, Mechanism Theory, in *Encyclopedia of Life Support Systems*, Ulrich Derigs (Ed.), EOLSS Publishers: Oxford UK, 2003.
10. R. Johari and J. N. Tsitsiklis. Efficiency loss in a network resource allocation game. Mathematics of Operations Research, 29(3):407-435, 2004.
11. A. Jokic, *Priced-based optimal control of electrical power systems*, Ph.D. dissertation, Eindhoven University of Technology, September 2007.
12. D. A. Lusan, Z. Yu, and F. T. Sparrow. Market gaming and market power mitigation in deregulated electricity markets. In *Proceedings of the IEEE Power Engineering Society*, vol. 1, pp. 839-843, 1999.
13. S. S. Oren and A. M. Ross, Can we prevent the gaming of ramp constraints?, *Decision Support Systems*, 40, pp. 461-471, 2005.
14. A. Ozdaglar and R. Srikant, Incentives and Pricing in Communication Networks, in *Algorithmic Game Theory, (Noam Nisan, Tim Roughgarden, Eva Tardos, and Vijay Vazirani, Eds.)*, Cambridge University Press, 2007.
15. A. Rantzer, Distributed Control using Decompositions and Games, *Semi-plenary talk at the IEEE Conference on Decision and Control*, 2008.
16. F. C. Schweppe, M. C. Caramanis, R. D. Tabors, and R. E. Bohn, *Spot Pricing of Electricity*, Kluwer Press, 1988.
17. R. Srikant, *The Mathematics of Internet Congestion*, Birkhäuser, 2003.
18. W. Vickrey, Counterspeculation, Auctions, and Competitive Sealed-Tenders, *Journal of Finance*, Vol. 16, pp. 837, 1961.
19. S. L. Waslander, K. Roy, R. Johari, C. J. Tomlin, Lump-Sum Markets for Air Traffic Flow Control With Competitive Airlines, *Proceedings of the IEEE Publication* Volume 96, Issue 12, pp. 2113-2130, 2008.
20. R. Wilson, Architecture of Power Markets, *Econometrica*, vol. 70, no. 4, pp. 1299-1340, 2002.

Chapter 6
Recursive Bargaining with Dynamic Accumulation

Francesca Flamini

Abstract We study a bargaining game (á la Rubinstein) in which parties are allowed to invest part of an available surplus. Therefore, in addition to the standard problem of how to divide a surplus for their own consumption, parties face the additional problem of how much to invest, knowing that the level of investment affects the surplus available in the next period. We provide an algorithm to solve the game when the number of bargaining stages is finite but tends to infinity. We show that the equilibrium investment and consumption shares become independent of the capital stock. The convergence of the equilibrium demands is affected by the elasticity of substitution and parties' patience.

6.1 Introduction

We study a classic bargaining game (á la Rubinstein [9]) with the crucial difference that parties are allowed to invest part of the surplus available. Therefore, in addition to the standard problem of how to divide the surplus for their own consumption, parties face the additional problem of how much to invest. Since the level of investment affects the future capital stock and consequently, the surplus available in the following bargaining stage, the model aims to capture the dynamic nature of long-term negotiations. That is, current agreements affect future bargaining possibilities. For instance, trade talks and negotiations over environmental issues typically take place in a series of rounds in which initial agreements are revisited and revised.

We focus on a finite-stage game: the number of bargaining stages n, is finite, although each stage can be protracted potentially forever. Moreover, we study the asymptotic game in which the number of bargaining stages tends to infinity. We solve the game recursively, for any arbitrary n, and show that when the number

Francesca Flamini
Department of Economics, University of Glasgow, Glasgow, G12 8RT, UK
e-mail: f.flamini@lbss.gla.ac.uk

R. Johansson and A. Rantzer (Eds.): Distributed Decision Making and Control, LNCIS 417, pp. 131–144.
springerlink.com

of bargaining stages n increases, parties invest independently of n for most of the game (only towards the final stages is there a deadline effect). The convergence of the equilibrium demands is also fairly fast and the speed of convergence can be increased when parties are more impatient and their elasticity of substitution is higher, since in this case they invest little in any bargaining stage.

Generally, bargaining games focus on negotiations over consumption levels and therefore parties do not have the ability to affect future bargaining possibilities (see Muthoo [7] and Osborne and Rubinstein [8] for recent reviews of bargaining games and their applications). However, recent studies have attempted to incorporate the investment problem within bargaining games [7, 10, 3]. Muthoo's focus is on parties who share an infinite number of surpluses of the same size (the steady-state stationary subgame perfect equilibria) [7]. As a result the investment problem is simplified since parties need to invest as much as it is necessary so as to have surpluses of the same size. In this sense the problem of how much parties invest in a strategic framework remains open. Sorger extended the Nash bargaining solution to a dynamic accumulation game [10]. In each period, parties can reach an agreement (threat points are endogenized as well); however, the bargaining process is simplified in that it is given by the solution of Nash products, a cooperative perspective. Flamini [3], instead, considered different noncooperative bargaining procedures with dynamic accumulation, including the alternating-offer procedure considered in this chapter, although in a setting with an infinite number of bargaining stages. The main difference with [3] is that in this chapter, we use backward induction to identify the equilibrium. We do not need to impose any functional form on the consumption and investment rules, as these are simply derived by solving the game. We can also look at the asymptotic game in which the number of bargaining stages tend to infinity and study the convergence of the equilibrium demands.

The chapter is organized as follows. In the next section we present the model. In Sec. 6.3, we provide the algorithm to solve the model. The main features of the equilibrium are presented in Sec. 6.4. Some final remarks are made in Sec. 6.5.

6.2 The Model

We consider a two-player bargaining game in which bargaining and production stages alternate (and each stage can start only after the other has taken place). We focus on the case in which there are n bargaining stages, with n $(= 2, 3, \ldots)$, which tends to infinity. Time is discrete, $t = 0, 1, \ldots$. In the first period, at $t = 0$, the surplus available is k_0, by assumption, and a bargaining stage starts. In the subsequent periods, during the production stage, the surplus is generated according to the production function $F(k_t) = k_t$, where k_t is the capital stock at period t, with $t = 0, 1, \ldots$.[1] Production takes time and τ indicates the interval of time in which the surplus is

[1] To simplify the numerical analysis we assume that there is a constant gross rate of return equal to 1. The analysis can be generalized to a constant gross rate equal to G (as long as G is sufficiently small, so that for the asymptotic game the transversality condition holds).

generated. At the beginning of the game or once the surplus is generated, a bargaining stage begins and players attempt to divide the surplus according to a classic infinitely repeated, alternating-offer bargaining procedure [9].[2] In our framework, a proposal by player i consists of a pair $({}_ix_t,\ {}_iI_t)$, where ${}_iI_t$ is the investment level proposed by i and ${}_ix_t$ is the share demanded by i over the remaining surplus. The subscript t indicates the dependence of the proposal $({}_ix_t,\ {}_iI_t)$ on capital at time t, denoted by k_t, that is, the state variable in the model. If there is an acceptance, the bargaining stage ends and the proposer's current per period utility is $u_i({}_ix_t,\ {}_iI_t)$ with

$$u_i({}_ix_t,\ {}_iI_t) = \frac{{}_ic_t^{1-\eta}}{1-\eta} \text{ for } \eta < 1 \tag{6.1}$$

where ${}_ic_t = {}_ix_t(k_t - {}_iI_t)$ is the level of consumption.[3] The output available at the next bargaining stage (at $t+1$) is $F(k_{t+1})$, where k_{t+1} is the capital stock in the next period and it is given by the investment level ${}_iI_t$, that is $k_{t+1} = {}_iI_t$.[4] Regardless of whether the proposal at t has been successful the responder at t becomes the next proposer. If there is a rejection, after an interval of time Δ, the rejecting player can make a counter offer. We assume that the capital stock remains unchanged.[5] In perpetual disagreement, parties consume ${}_ic_t = 0$. Players' time preference is represented by the common discount rate h. Since intervals of time have different lengths, there are two distinct discount factors in our model: the between-cake discount factor $\alpha = exp(-h\tau)$, which takes into account that production takes time, and the between-cake discount factor $\delta = exp(-h\Delta)$, which takes into account that there is an interval of time between a rejection and a new proposal. This discounting structure has been introduced first by Muthoo [6]. It is reasonable to assume that a bargaining round is quicker than a production stage, since a counter offer can be made fairly quickly (that is, $\alpha < \delta$,; see [6]).

The focus is on Markov perfect equilibria (MPE), where the Markov strategies specify players' actions for each time period t as a function of the state of the system at the beginning of that period, k_t.

The timeline in a specific example of this game is represented in Fig. 6.1 below. In this example it is assumed that there are only three bargaining stages ($n = 3$) with g rejections only in the first stage. Hence, in the first stage after g rejections (or $g+1$ rounds), a player, say j (with $j = 1,2$) accepts the proposal $({}_ix_0,\ {}_iI_0)$ made

[2] In a standard bargaining game, the interdependence between players' strategies is related to a single division (parties makes offers and counter-offers until they find an agreement); in our framework it has also a dynamic component, since current actions affect future payoffs (via the investment level).

[3] The case of $\eta \geq 1$ is excluded to simplify the analysis. The analysis could be extended to the case of $\eta \geq 1$, as long as shares are normalized so that even when players do not consume any positive share their payoffs are bounded.

[4] We assume that there is maximum depreciation. However, the analysis can be easily extended to smaller level of capital depreciation.

[5] Alternatively, production takes place again and therefore the capital stock depreciates. In this case, the qualitative results we show in the chapter are unaffected, since the equilibrium is characterized by no delays even when the capital stock does not shrink after a rejection.

by his opponent, where the subscript 0 indicates the dependence of the demand on the initial capital stock, k_0. In the next stage, production takes place and the surplus for the second bargaining stage is given by $k_1 = I_0$. Since players alternate in making offers and there are no rejections in the second/third bargaining stage, then the successful proposer at the beginning of the second bargaining stage is player j, with the division $({}_jx_1, {}_jI_1)$, while at the beginning of the third bargaining stage, player i, after production, makes a successful proposal, say $({}_ix_2, {}_iI_2)$. This is the end of the game in this example. Then, player i's sum of discounted utilities is as follows[6]:

$$\frac{\delta^g}{1-\eta}\Big(\sum_{t=0}^{1}(\alpha^{2t}\left({}_ix_{2t}(k_{2t} - {}_iI_{2t})\right)^{1-\eta}) + \alpha\left((1 - {}_jx_1)(k_1 - {}_jI_1)\right)^{1-\eta}\Big)$$

Similarly, for player j,

$$\frac{\delta^g}{1-\eta}\Big(\sum_{t=0}^{1}\alpha^{2t}\left(1 - {}_ix_{2t})(k_{2t} - {}_iI_{2t})\right)^{1-\eta} + \alpha\left({}_jx_1(k_1 - {}_jI_1)\right)^{1-\eta}\Big)$$

6.3 Algorithm to Identify the Solution

Despite the fact that each bargaining stage can last potentially forever, when the number of bargaining rounds n is finite the game can be solved using backward induction, since it is known that the last stage is the Rubinsteinian game, which has a unique solution with immediate agreement. For any n, at the last stage, in equilibrium, players will invest nothing[7] and divide the surplus for their own consumption as in the classic Rubinsteinian game [9], but with CES utilities. This implies that if there are only two bargaining stages, at the beginning of the first bargaining stage the value of the continuation game (that is, what players will obtain in the last stage) is already known (aside from the size of the surplus, which is defined in the first stage). Hence, we can easily solve the problem at the beginning of the game. In this section, we will first focus on this case ($n = 2$). Once this is solved we will iterate the method for n bargaining stages (with $n > 2$).

Assume that there are only two bargaining stages ($n = 2$). As already pointed out, the last bargaining stage is the Rubinsteinian game (players will invest nothing). Then, at the first bargaining stage, a proposer will make an acceptable offer if it is such that the responder is weakly better off in accepting it rather than rejecting it and making a counter offer (when the constraint is binding it is called an *indifference condition*, since the responder is indifferent with respect to accepting or rejecting the

[6] The general formula for players' utility is much more complex, since there may be many rounds in a single bargaining stage, due to delays, and many bargaining stages or surpluses to be shared. However, the elements it contains are, in essence, the same as those outlined in this simple example. To simplify we omit the general formula, but it is available upon request.

[7] Or disinvest if the depreciation rate is less than 1.

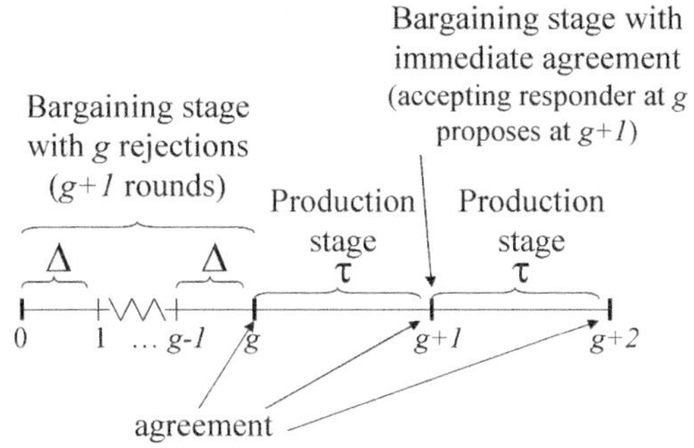

Fig. 6.1 Timeline for a game with three bargaining stages with g (no, respectively) rejections in the first (second/third, respectively) stage.

offer; see, for instance, Muthoo [7]).[8] In other words, at the beginning of the game, the problem of a proposer, say i, is given by the following Lagrangian function L_i:

$$
\begin{aligned}
L_i = \max_{x_i, I_i} & \frac{[x_i(k_0 - I_i)]^{1-\eta}}{1-\eta} + \frac{\alpha}{1-\eta}\left(\frac{\delta^{\frac{1}{1-\eta}}}{1+\delta^{\frac{1}{1-\eta}}} I_i\right)^{1-\eta} \\
& + m_i\left(\frac{[(1-x_i)(k_0 - I_i)]^{1-\eta}}{1-\eta} + \frac{\alpha}{1-\eta}\left(\frac{I_i}{1+\delta^{\frac{1}{1-\eta}}}\right)^{1-\eta}\right) \\
& - m_i\delta\left(\frac{[x_j(k_0 - I_j)]^{1-\eta}}{1-\eta} + \frac{\alpha}{1-\eta}\left(\frac{\delta^{\frac{1}{1-\eta}}}{1+\delta^{\frac{1}{1-\eta}}} I_j\right)^{1-\eta}\right)
\end{aligned}
\tag{6.2}
$$

where m_i are the (non-negative) Kuhn–Tucker multipliers, with $i, j = 1, 2$ and $i \neq j$. The first row of the maximisation problem is player i's current payoff after implementing the agreement (x_i, I_i), plus his discounted utility for receiving the Rubinsteinian share, as a responder, over the surplus I_i. The rest of the Lagrangian includes player j's indifference condition. In particular, the expression in parentheses in the second line is player j's sum of discounted utility for accepting the proposal (x_i, I_i) (or player j's current per-period payoff followed by his discounted utility for receiving the Rubinsteinian share, as a proposer, over the surplus I_i), while the expression in parentheses in the third line is player j's sum of discounted utility after making the counter offer (x_j, I_j). The indifference condition says that the former has to equal the latter multiplied by the discount factor δ, since there has been a rejection.[9]

[8] As for the Rubinsteinian game delays are not profitable in the first bargaining game. Within a dynamic framework, players can always invest a sufficiently high amount as to avoid costly rejections. Instead, there can be strategic delays in bargaining games, when these can resolve some uncertainty [1] or when more than two players are involved [2].

[9] For readers more familiar with dynamic programming, the problem can generally be written using a Bellman equation [4]. In this case, each proposer's value function is given by the maximization of the current per period utility, assuming players reach an agreement, plus the discounted value of

The first-order conditions (FOCs) with respect to the controls imply that

$$x_i = \frac{1}{1+m_i^{\frac{1}{\eta}}} \tag{6.3}$$

$$I_i = \varphi_i k_0, \text{ where } \varphi_i = \frac{1}{1+\frac{\left(1+m_i^{\frac{1}{\eta}}\right)\left(1+\delta^{\frac{1}{1-\eta}}\right)^{\frac{1-\eta}{\eta}}}{\alpha^{\frac{1}{\eta}}(\delta+m_i)^{\frac{1}{\eta}}}} \tag{6.4}$$

Using the FOCs (6.3) and (6.4) in the indifference conditions (or the FOCs with respect to the multipliers), i.e.,

$$\left(\frac{[(1-x_i)(k_0-I_i)]^{1-\eta}}{1-\eta}+\frac{\alpha}{1-\eta}\left(\frac{I_i}{1+\delta^{\frac{1}{1-\eta}}}\right)^{1-\eta}\right)=$$

$$=\delta\frac{[x_j(k_0-I_j)]^{1-\eta}}{1-\eta}+\delta\frac{\alpha}{1-\eta}\left(\frac{\delta^{\frac{1}{1-\eta}}}{1+\delta^{\frac{1}{1-\eta}}}I_j\right)^{1-\eta}$$

we obtain

$$\left(m_i^{\frac{1}{\eta}}\left(1+\delta^{\frac{1}{1-\eta}}\right)^{\frac{1-\eta}{\eta}}\right)^{1-\eta}+\alpha\left(\frac{\alpha^{\frac{1}{\eta}}(\delta+m_i)^{\frac{1}{\eta}}}{1+\delta^{\frac{1}{1-\eta}}}\right)^{1-\eta}= \tag{6.5}$$

$$=\delta\left(\left(1+\delta^{\frac{1}{1-\eta}}\right)^{\frac{1-\eta}{\eta}}\right)^{1-\eta}+\alpha\delta^2\left(\frac{\alpha^{\frac{1}{\eta}}(\delta+m_i)^{\frac{1}{\eta}}}{1+\delta^{\frac{1}{1-\eta}}}\right)^{1-\eta}$$

for $i, j = 1,2$ and $i \neq j$. The focus is on symmetric solutions.[10] Then, $m_i = m$, $\varphi_i = \varphi$ for $i = 1,2$ and the system becomes:

$$\left(1+\delta^{\frac{1}{1-\eta}}\right)^{\frac{1-\eta}{\eta}}\left(m^{\frac{1-\eta}{\eta}}-\delta\right)+\alpha^{\frac{1}{\eta}}(1-\delta^2)(\delta+m)^{\frac{1-\eta}{\eta}}=0 \tag{6.6}$$

By necessity, the following must hold:

$$m^{\frac{1-\eta}{\eta}}<\delta$$

---

the continuation game (based on the Rubinsteinian shares for the last bargaining stage). In addition an offer is accepted if the responder's sum of discounted utility in case of an acceptance is not smaller than his discounted utility in the case where he rejects the offer and makes a counter-offer. Note that this is a constrained optimization in which the constraint has a recursive structure, since it includes the opponent's value function.

[10] We solved system (6.5) numerically for a range of discount factors, and no asymmetric solutions could be identified (we excluded the unreasonable range of small discount factors with $\alpha < \delta$ [6].

If not, the left hand side of (6.6) would be always positive for any α and δ in (0,1) and therefore by the principle of complementary slackness, the multiplier m would be zero, in contradiction with the assumption of

$$m^{\frac{1-\eta}{\eta}} \geq \delta > 0$$

Then, for

$$m^{\frac{1-\eta}{\eta}} < \delta$$

Eq. (6.6) becomes

$$\left(1+\delta^{\frac{1}{1-\eta}}\right)^{\frac{1-\eta}{\eta}}\left(\delta - m^{\frac{1-\eta}{\eta}}\right) = \alpha^{\frac{1}{\eta}}(1-\delta^2)(\delta+m)^{\frac{1-\eta}{\eta}} \tag{6.7}$$

It is still possible that corner solutions exist (that is, $m^{\frac{1-\eta}{\eta}} < \delta$ and the left hand side of (6.6) is positive); however, this requires sufficiently impatient parties. Note that i t is straightforward to check that for δ which tends to 1, there are only interior solutions, i.e.,

$$0 < m^{\frac{1-\eta}{\eta}} < \delta$$

In the next proposition we derive an algorithm to solve the iteration process when the number of bargaining stages is n (with $n \geq 3$). We focus on interior solutions since as already shown for the two-stage game this is the only solution for sufficiently patient players.

Proposition 6.1. *When the number of bargaining stages is n (with $n \geq 3$), the subgame perfect equilibrium proposal is given by the following shares:*

$$x_n = \frac{1}{1+m_n^{1/\eta}} \tag{6.8}$$

$$\varphi_n = \frac{a_n}{a_n+b_n} \tag{6.9}$$

where

$$a_n = (\alpha(c_{n-1}+d_{n-1}m_n))^{\frac{1}{\eta}} \tag{6.10}$$

$$b_n = 1+m_n^{1/\eta} \tag{6.11}$$

$$c_n = ((1-x_n)(1-\varphi_n))^{1-\eta} + \alpha\varphi_n^{1-\eta}d_{n-1} \tag{6.12}$$

$$d_n = (x_n(1-\varphi_n))^{1-\eta} + \alpha\varphi_n^{1-\eta}c_{n-1} \tag{6.13}$$

and

$$c_2 = ((1-x_2)(1-\varphi_2))^{1-\eta} + \alpha\left(\frac{1}{1+\delta^{\frac{1}{1-\eta}}}\varphi_2\right)^{1-\eta} \tag{6.14}$$

$$d_2 = (x_2(1-\varphi_2))^{1-\eta} + \alpha\left(\frac{\delta^{\frac{1}{1-\eta}}}{1+\delta^{\frac{1}{1-\eta}}}\varphi_2\right)^{1-\eta} \tag{6.15}$$

and the multiplier m_n is the solution of the following equation

$$\frac{m_n^{\frac{1-\eta}{\eta}} - \delta}{b_n^{1-\eta}}(1-\varphi_n)^{1-\eta} + \alpha\varphi_n^{1-\eta}(d_{n-1} - \delta c_{n-1}) = 0 \tag{6.16}$$

Proof. *We numerically solve (6.7) for m. The solution is named m_2. The subscript 2 indicates that the solution is related to the game with two bargaining stages. Then, the MPE demand (x_2, φ_2) is given by (6.3) and (6.4).*

Let j be the first mover at $t = 0$ when there are three bargaining stages $(n = 3)$. Then, the problem is

$$\max_{x_j, I_j} \frac{[x_j(Gk_0 - I_j)]^{1-\eta}}{1-\eta} + \frac{\alpha}{1-\eta} I_j^{1-\eta} c_2$$

where c_2 is defined in (6.14). The immediate agreement condition implies that the expression below must be zero:

$$\frac{[(1-x_j)(k_0 - I_j)]^{1-\eta}}{1-\eta} + \frac{\alpha}{1-\eta} I_j^{1-\eta} d_2 - \delta\left(\frac{[x_j(k_0 - I_i)]^{1-\eta}}{1-\eta} + \frac{\alpha}{1-\eta} I_j^{1-\eta} c_2\right)$$

where d_2 is defined in (6.15). Then, the first order condition implies that x_j and I_j have the same linear structure as in (6.3) and (6.4), in particular (6.8) and (6.9). Also the indifference condition can be written as in (6.16), which is one equation in one unknown, the multiplier m_n. A solution to (6.16) defines the solution to (x_n, φ_n), given (6.10)–(6.15) with $n = 3$.

Similarly, for a game with n bargaining stages (with $n > 3$), x_j and I_j are given by (6.8) and (6.9) with a_n and b_n as in (6.10)–(6.13). The indifference condition can be written as in (6.16). This defines the multiplier m_n. That is, a solution to (6.16) defines the solution to (x_n, φ_n). □

Proposition 6.1 allows us to solve the dynamic game for any number of bargaining stages. In the next section, we characterize the solution.

6.4 Features of the Equilibrium

In this section we (numerically) solve the game by using the algorithm in Proposition 6.1 with the aim of highlighting the main features of the solution. First of all, the rules which specify the investment and consumption levels in equilibrium are simple, as shown in the following remark.

Remark 6.1. The investment and consumption levels in equilibrium are linear functions of the capital stock.

Proof. *The first order conditions (6.3) and (6.4) and the indifference condition (6.7) imply that the multiplier, the consumption and the investment shares are independent of the capital stock. By iteration, this can be proved also for $n > 2$ (see (6.16)).* □

Numerical examples of the equilibrium consumption and investment shares are depicted in Figs. 6.2 and 6.3.[11] Figure 6.2 (6.3, respectively) represents the equilibrium consumption (investment, respectively) shares for different elasticities of substitution (η equal to 1/3, 1/2 and 2/3) when the game has n bargaining stages (with $n = 1, 2, \ldots$) and the discounting structure is $(\alpha, \delta) = (0.8, 0.9)$.[12] Figures 6.2 and 6.3 show a second feature of the solution, namely, in equilibrium, investment and consumption shares decrease with the elasticity of substitution (which is inversely related to η). This is due to the curvature of the utility function in (6.1). As η increases the utility in (6.1) is higher and has a higher curvature. As a result parties prefer more consumption smoothing, hence they invest more (as shown in Fig. 6.3). A higher investment today implies higher consumption levels for both players in the future. The current proposer can then exploit a trade-off between current and future consumption: By investing more today he is able to extract a relative higher surplus today for his own consumption (as shown in Fig. 6.2). This explains why when η increases, x and φ increase.

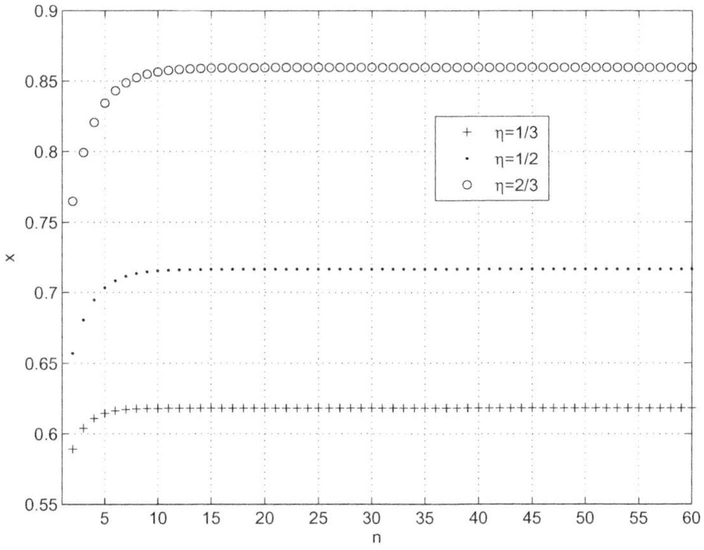

Fig. 6.2 Equilibrium shares x over the number of bargaining stages n for different levels of η.

[11] For any given discount structure (α, δ) and η, we find a unique sequence of demands, one demand for each bargaining stage. That is, we searched across a range of starting values when using the algorithm and found a unique sequence.

[12] Similar graphs with different sets of discounting factors (α, δ) are omitted for simplicity.

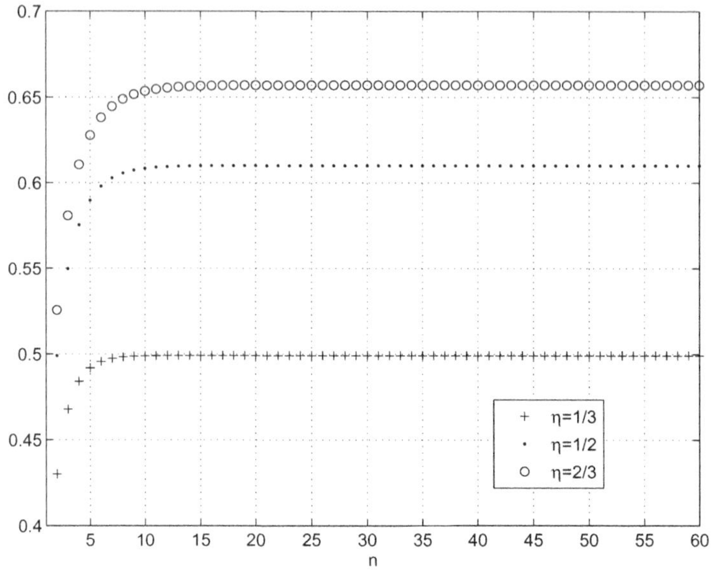

Fig. 6.3 Equilibrium investment shares φ over the number of bargaining stages n for different η.

A third feature of the solution, again shown in Figs. 6.2 and 6.3, is that the equilibrium demands converge as the number of bargaining stages increases. That is, the equilibrium demands (x_n, φ_n) are independent of n as long as n is sufficiently large. Figures 6.2 and 6.3 show that if the number of bargaining stages is small, say less than 10, then consumption and investment shares vary in every stage (since the deadline effect is strong). However, for n larger than 10, players' investment and consumption plans are (almost) unchanged, independently of n, except when they are approaching the last stages. For instance, with 60 bargaining stages ($n = 60$), the equilibrium demands appear to be very similar for the first 50 bargaining stages, then share consumed and invested decline when the deadline effects become dominant.

We further explore the convergence of the equilibrium demands in the following proposition.

Proposition 6.2. *The speed of convergence of the MPE demands in the asymptotic game increases with the elasticity of substitution and parties' impatience.*

This is shown in Tables 6.1–6.2.[13] The former indicates the equilibrium demands when parties are more impatient (the discount factors are $(\alpha, \delta) = (0.8, 0.9)$, in Fig. 6.1 and $(\alpha, \delta) = (0.9, 0.95)$ in Fig. 6.2). The tables also show the bargaining stage where the demands converge, named cr. In other words, if the game has n bargaining stages, with $n > cr$, then the equilibrium demands are unchanged in the first $n - cr$

[13] Similar tables with different sets of discount factors are omitted for simplicity.

stages (the lower the *cr* the higher the convergence speed). Both Tables 6.1 and 6.2 show that convergence is quicker when η is smaller (or the elasticity of substitution is higher). By comparing Tables 6.1 and 6.2, one can show that for any given level of η, convergence increases when parties are more impatient (*cr* is lower in Table 6.1, where parties are more impatient).

The intuition is that parties who are more impatient discount future payoffs more strongly, hence, they invest less (and consume more). As a result, the MPE demands converge more quickly in a game with impatient parties, since even in games with a small n, parties invest little. Similarly, since as explained above, the elasticity of substitution affects negatively the investment level (i.e., the smaller the η, the less parties invest), convergence is quicker when η is smaller.

Table 6.1 The equilibrium demands for discounting (0.8, 0.9), when the number of bargaining stages tends to infinity; *cr* indicates the bargaining stage where demands converge.

	$\eta = 1/3$	$\eta = 1/4$	$\eta = 1/2$	$\eta = 2/3$	$\eta = 3/4$
x	0.5824	0.6182	0.7167	0.8595	0.9381
φ	0.4025	0.4992	0.6100	0.6571	0.6666
cr	24	31	44	58	59

Table 6.2 The equilibrium demands for discounting (0.9, 0.95), when the number of bargaining stages tends to infinity; *cr* indicates the bargaining stage where demands converge.

	$\eta = 1/3$	$\eta = 1/4$	$\eta = 1/2$	$\eta = 2/3$	$\eta = 3/4$
x	0.6184	0.5808	0.7200	0.8639	0.9416
φ	0.7200	0.6506	0.7910	0.8185	0.8239
cr	63	47	87	101	113

Finally, we show that bargaining can be efficient when counter offers can be made very quickly (in accordance with Lockwood and Thomas [5][14] and Flamini [3]).

Proposition 6.3. *At the limit of Δ that tends to zero, players invest and consume efficiently.*

Proof. *To simplify the analysis, we focus on the case of $\eta = 1/2$ (although numerically it can be shown for any $\eta < 1$). Then we can solve analytically (6.7), and the multiplier for the two-stage game is*

[14] Lockwood and Thomas [5] showed that the level of cooperation among players tends to the efficient level in the limit as players become patient, although their framework is quite different from ours, since players cannot bargain.

$$m_2 = \frac{\delta[(1+\delta^2) - \alpha^2(1-\delta)]}{(1+\delta^2) + \alpha^2(1-\delta)}$$

Moreover, at the limit for Δ that tends to 0, the multiplier m tends to 1 and therefore parties share equally the surplus not invested and invest a share equal to

$$\varphi = \frac{\alpha^2}{1+\alpha^2} \tag{6.17a}$$

It can be show that the noncooperative solution coincides with the efficient level of consumption and investment. Indeed, a social planner would maximize the following expression:

$$\sum_{i=1}^{2}\sum_{t=0}^{1} u_i({}_ix_t, {}_tI_i) \tag{6.18}$$

In the last bargaining stage, since $n = 2$, the social planner would split the remaining surplus among the symmetric players (without investing any positive amount). Then, in the first bargaining stage the problem is to maximize

$$\max_{x_i, I_i} \frac{[x_i(k_0 - I_i)]^{1-\eta}}{1-\eta} + \frac{[(1-x_i)(k_0 - I_i)]^{1-\eta}}{1-\eta} + \frac{2\alpha}{1-\eta}\left(\frac{1}{2}I_i\right)^{1-\eta}$$

The FOCs imply that at the first stage $x_i = 1/2$ and

$$I_i = \frac{\alpha^{1/\eta}}{1+\alpha^{1/\eta}} k_0$$

Hence, the socially optimal solution coincides with the noncooperative solution for $\eta = 1/2$. Similarly, with three bargaining stages ($n = 3$), at the limit for δ that tends to 1 it can be shown (see Proposition 6.1) that m_{3s} tends to 1 and therefore x_{3s} tends to $1/2$ while the share invested tend to

$$_{I}\varphi_{3s} = \frac{\alpha^2}{1+2\alpha^2}$$

which again coincides with the socially optimal solution. The problem can be iterated for any n and indeed any η, although we need to rely on numerical solutions. For instance, using Proposition 6.1 for the asymptotic game (n tends to infinity) with $\eta = 1/2$, $(\alpha, \delta) = (0.9, 0.99)$, the shares invested are 0.809, and the proposers obtain shares equal to 0.547. When patience increases, to $\delta = 0.9999$ and $\alpha = 0.9$, the share consumed decreases to 0.500 and the share invested increases to 0.810, which is the social optimum. To see this, we need to solve problem (6.18) for an infinite horizon. Since players are symmetric, a utilitarian social planner would split the surplus not invested equally between parties. The investment problem, instead is reduced to a standard intertemporal saving/consumption problem [11] and, can be shown that the social optimal is $\varphi = \alpha^{1/\eta}$ (which is 0.81 for $\eta = 1/2$ and $\alpha = 0.9$).

□

The intuition is that the frictions in bargaining create a hold-up problem. That is, a higher investment today will lead to a stream of benefits that the proposer cannot fully exploit (since he has to reach a compromise with the other party), hence proposers tend to underinvest. Only when the bargaining is frictionless is bargaining is efficient.

6.5 Conclusions

Bargaining games with dynamic accumulation are almost unexplored in the literature but they are interesting because they capture an important feature of real-life negotiations: Current agreements affect future bargaining possibilities. In this chapter, we analyze a finite-stage model where the number of bargaining stages could be arbitrarily high.

We showed that, in equilibrium, parties maintain the same offer in each bargaining stage independently of the capital stock for sufficiently long games, only closer to end of the game, their investment shares drop and so do their consumption shares (deadline effect).

The equilibrium consumption and investment levels follow a simple rule: they are linear functions of the state variable, that is, the capital stock. In dynamic games often the attention is on linear Markov strategies, to simplify the analysis, however, in our model the linear rules are derived by solving the game.

It can be shown that our qualitative results are robust and do not depend on the procedure chosen (as long as we exclude extreme procedures such as the ultimatum, since in this case the incentives in the game are strongly modified). Although the alternating-offer procedure seems realistic in many negotiations, other procedures could be considered, for instance, a random proposer mechanism. Such a change would not modify the features of the equilibrium, making the results of our model fairly robust.

Acknowledgements The author would like to thank the participants of the LCCC workshop, the anonymous referee and the editors for useful comments and suggestions.

References

1. Admati, A.R., Perry, M.: Strategic delay in bargaining. Review of Economic Studies **54**, 345–364 (1987)
2. Cai, H.: Delay in multilateral bargaining under complete information. Journal of Economic Theory **93**, 260–276 (2000)
3. Flamini, F.: Dynamic Bargaining. Mimeograph. University of Glasgow, Glasgow, Scotland (2009)
4. Ljungqvist, L., Sargent, T.: Recursive Macroeconomic Theory. MIT Press, Cambridge, MA (2000)

5. Lockwood, B., Thomas, J.: Gradualism and irreversibility. Review of Economic Studies **69**, 339–356 (2002)
6. Muthoo, A.: Bargaining in a long run relationship with endogenous termination. Journal of Economic Theory **66**, 590–598 (1995)
7. Muthoo, A.: Bargaining Theory with Applications. Cambridge University Press, Cambridge, UK (1999)
8. Osborne, M., Rubinstein, A.: Bargaining and Markets. Academic Press, San Diego, CA (1990)
9. Rubinstein, A.: Perfect equilibrium in a bargaining game. Econometrica **50**, 97–109 (1982)
10. Sorger, G.: Recursive Nash bargaining over a productive asset. Journal of Economic Dynamic and Control **30**, 2637–2659 (2006)
11. Stokey, N., Lucas, R.: Recursive Methods in Economic Dynamics. Harvard University, Cambridge, MA (1989)

Part II

ADAPTATION AND LEARNING IN AUTONOMOUS SYSTEMS

ADAPTATION AND LEARNING IN AUTONOMOUS SYSTEMS

Concepts of adaptation and learning have been addressed by several different research communities, with different mathematical tools and models. Adaptive methods in control engineering date back to the 1950s, but the most active theory development on adaptive control was done in the 1970–80s. More recently, many new industrial applications have become feasible. A parallel development of theory has taken place in the community of machine learning, with roots in computer science and artificial intelligence. Similarly, in game theory and economics there is active research on learning algorithms. The second part of the book has a focus on adaptation and learning, in particular for autonomous systems, such as robots and unmanned vehicles. The choice of application area limits the scope, yet it leaves room for cross-disciplinary investigation and interaction and it has industrial relevance.

Chapter 7
Distributed Nonlinear Estimation for Diverse Sensor Devices

Andrea Simonetto and Tamás Keviczky

Abstract Distributed linear estimation theory has received increased attention in recent years due to several promising, mainly industrial applications. Distributed nonlinear estimation, however, is still a relatively unexplored field despite the need for such a theory in numerous practical problems with inherent nonlinearities. This work presents a unified way of describing distributed implementations of three commonly used nonlinear estimators: the extended Kalman filter (EKF), the unscented Kalman filter (UKF) and the particle filter. Leveraging on the presented framework, we propose new distributed versions of these methods, in which the nonlinearities are locally managed by the various sensors, whereas the different estimates are merged based on a weighted average consensus process. We show how the merging mechanism can handle sensors running different filters, which is especially useful when they are endowed with diverse local computational capabilities. Numerical simulations of the proposed algorithms are shown to outperform the few published ones in a localization problem via range-only measurements. Quality and effectiveness are investigated in a heterogeneous filtering scenario as well. As a special case, we also present a way to manage the computational load of distributed particle filters using graphical processing unit (GPU) architectures.

Andrea Simonetto
Delft Center of Systems and Control, Delft Technical University, Mekelweg 2, 2628 CD Delft, The Netherlands
e-mail: a.simonetto@tudelft.nl

Tamás Keviczky
Delft Center of Systems and Control, Delft Technical University, Mekelweg 2, 2628 CD Delft, The Netherlands
e-mail: t.keviczky@tudelft.nl

R. Johansson and A. Rantzer (Eds.): Distributed Decision Making and Control, LNCIS 417, pp. 147–169.
springerlink.com

7.1 Introduction

In the past few years, the large number of potential applications of sensor networks have increased the interest in distributed estimation techniques, where sensors compute estimates locally and communicate with others to improve their quality. So far, research has been mainly focused on state estimation of linear dynamical systems. References [1, 7, 37, 28] give a comprehensive overview of the field.

There are many situations in which such a framework cannot be applied, due to nonlinearities either in the dynamical system or in the sensing equation or both. An important example of these scenarios is the localization of a moving object via range-only measurements. This particular problem arises in applications such as indoor robot localization [36, 20, 24], underwater sensor networks [9, 11, 21] and space exploration [29], among others.

Although there are a few cases in the literature in which distributed nonlinear estimation has been addressed, a clear performance evaluation is still missing. For instance, in [28] a distributed extended Kalman filter is suggested, but not implemented, while in [8] and [15] distributed unscented Kalman filters and distributed particle filters are presented, respectively, without extensive analysis.

Our aim in this contribution is to propose effective distributed algorithms for nonlinear estimation. We introduce a unified framework, in which we design versions of distributed extended Kalman filters, distributed unscented Kalman filters and distributed particle filters that show better performance in simulation compared to the aforementioned literature. The core of our framework is a merging mechanism based on a weighted consensus procedure similar to the one developed in [39].

Furthermore, we will show how the mechanism can handle different filters implemented on different sensors. This is especially practical when the sensor devices have different computational capabilities and we want to exploit their resources efficiently. In this respect, the proposed framework can be used to tailor the composition of various filters to the diverse sensor devices in the network.

Finally, we focus on a special case of filter and sensor computing architectures and show how the computational load of distributed particle filters can be managed efficiently using Graphical Processing Units (GPUs).

It is well known that in nonlinear, non-Gaussian problems, particle filters usually outperform Kalman filter-type methods but suffer from very high computational requirements when using many particles. This drawback could be mitigated by reliance on general purpose GPU applications, which are making massive parallelization of algorithms possible by distributing computing tasks on multiple cores, leading to significant improvements in computational time. These trends are reflected by the rise of reasonably priced graphical processing units featuring thousands of GPU cores that compete with and supersede the best traditional desktop processing architectures. It is not far fetched to imagine thousands of particles running on each one of the thousand cores.

This chapter is organized as follows: a description of the distributed estimation problem and our problem formulation is presented in Sec. 7.2. Section 7.3 and Sec. 7.4 contain our main contributions on distributed nonlinear estimation, while Sec.

7.5 explores the distribution of computations on different cores for particle filters. In Sec. 7.6, the algorithms are evaluated in a simulated localization problem via range-only measurements for an Autonomous Underwater Vehicle. Conclusions are drawn in Sec. 7.7.

7.2 Problem Formulation

7.2.1 Notation

The state variable is denoted by $x \in \mathbb{R}^n$, $z \in \mathbb{R}^{n_z}$ the measurement, $w \in \mathbb{R}^{n_w}$ and $v \in \mathbb{R}^{n_v}$ two independent zero mean Gaussian noise terms; $\hat{x}$ represents the estimate of x; $x(k)$ is the state at time instant k; $z_i(k)$ is the measurement vector of sensor i at time k. For all the other variables the same notation holds. If $a_i(k)$ is a generic vector variable associated with sensor i at time k, and $i = 1,\dots,N$, then $\mathbf{a}(k) = (a_1^T(k),\dots,a_N^T(k))^T$ is a stacked vector of all the sensor variables.

7.2.2 Distributed Estimation

Let the nonlinear time-invariant dynamical model of the system with state x be

$$x(k+1) = f(x(k), w(k)) \tag{7.1}$$

where f is a nonlinear function. Let the process be observed by N sensors each with some processing and communication capability. The sensors are labeled $i = 1,\dots,N$ and form the set $\mathcal{V}$. The communication topology is modeled as an undirected graph $\mathcal{G} = (\mathcal{V}, \mathcal{E})$, where an edge (i, j) is in $\mathcal{E}$ if and only if node i and node j can exchange messages. Let the graph be connected, let the sensor clocks be synchronized and assume perfect communication (no delays or packet losses). The nodes with which node i communicates are called *neighbors* and are contained in the set $\mathcal{N}_i$. Note that node i is not included in the set $\mathcal{N}_i$. We define $\mathcal{J}_i = \mathcal{N}_i \cup \{i\}$ and $N_i = |\mathcal{J}_i|$. Let the measurement equation for each sensor $z_i(k)$ be

$$z_i(k) = g_i(x(k)) + v_i(k), \qquad i = 1,\dots,N \tag{7.2}$$

where $v_i(k)$ is a Gaussian noise term and $g_i(\cdot)$ nonlinear functions. We consider a distributed computation setting, where each sensor computes its version of the state estimate locally, and $\hat{x}_i(k)$ denotes the local estimate of sensor i at time k. This problem set-up leads to the following distributed estimation problem. Allowing communication only within the neighborhood, estimate N different copies of the state $\hat{x}_i(k)$ so that the following requirements are satisfied:

(R1) each $\hat{x}_i(k)$ is an unbiased estimate of $x(k)$ at each time step k;

(R2) as $k \to \infty$, all the local estimates $\hat{x}_i(k)$ converge to the same value.

In this chapter, we are interested in algorithms that do not exchange raw measurements among the neighbors but communicate only computed quantities with aggregated information content (such as local estimates). This implies that the structure of our algorithms will consist of N local filters and a distributed merging mechanism, which aggregates the different estimates.

Remark 7.1. The information exchange graph $\mathcal{G}$ is assumed to be time-invariant. However, under weak technical conditions, we could extend the algorithms to situations in which the graph $\mathcal{G}$ is time-varying; see for instance [39].

Remark 7.2. Since the noise process is not assumed to be bounded, Requirement R2 should be interpreted in an almost sure sense.

7.2.3 Distributed Localization

In this chapter, we will use a distributed localization problem via range-only measurements to demonstrate and analyze the proposed algorithms. This problem is a good example of a distributed nonlinear estimation problem where the process is locally unobservable by the individual sensors. Let x_{p} be the position of a moving agent and b_i, $i = 1, \dots, N$, the positions of the fixed range sensors. The measurement equation can then be written as

$$z_i(k) = ||x_{\mathrm{p}}(k) - b_i|| + v_i(k), \qquad i = 1, \dots, N \tag{7.3}$$

Since the state (agent position) is locally unobservable, the sensors have to communicate with each other. In the following section we will present some possibilities for combining their estimates.

7.3 Consensus Algorithms

We want the distributed estimation algorithms that we are interested in to operate in such a way that the local estimates eventually converge to the same value (R2). This could be fulfilled by applying an appropriate consensus algorithm that merges the different estimates, assuming that R1 is satisfied by the choice of a suitable local nonlinear filter.

Consensus algorithms can be represented by linear maps between some local variables. For example, the estimates $\hat{x}_j(k)$ with $j \in \mathcal{J}_i$, and their weighted nodal average $\bar{x}_i(k)$, where

$$\bar{x}_i(k) = \sum_{j \in \mathcal{J}_i} [W]_{ij} \hat{x}_j(k)$$

can be represented in matrix-vector notation as

$$\bar{x}(k) = (W \otimes I_N)\hat{x}(k) \tag{7.4}$$

Equation (7.4) constitutes a consensus iteration. Running this iteration τ times corresponds to repeating the update

$$\begin{cases} \bar{x}(k) = (W \otimes I_N)\hat{x}_{t-1}(k) \\ \hat{x}_t(k) = \bar{x}(k) \end{cases}$$

with $\hat{x}_0(k) = \hat{x}(k)$ from $t = 1$ to τ. The τth iteration can be compactly computed as $(W^\tau \otimes I_N)\hat{x}(k)$. The matrix W is required to satisfy

$$\lim_{\tau \to \infty} W^\tau = \frac{1}{N}\mathbf{1}_N\mathbf{1}_N^T \tag{7.5}$$

so that the consensus iterations do not only converge but also give the mean of the initial values (which is given the name *average-consensus*, as a particular instance of the more general χ-consensus [12]). This property has been used in the last few years [30] as a means of averaging the different estimates $\hat{x}_i(k)$ without requiring N (see also in [27, 26, 13]).

A typical form for W is

$$W = I_N - \varepsilon L$$

where L is the weighted Laplacian associated with the graph $\mathcal{G}$ and ε is a positive constant, which has to be less than 1 to ensure convergence.

In its typical implementation, a consensus algorithm merges the different $\hat{x}_i(k)$ according to

$$\mathcal{A}[\bar{x}, i] = \mathcal{A}[\hat{x}, i] + \frac{\varepsilon}{N_i - 1} \sum_{j \in \mathcal{N}_i} (\mathcal{A}[\hat{x}, j] - \mathcal{A}[\hat{x}, i]) \tag{7.6}$$

delivering

$$\bar{x} = \mathcal{A}[\bar{x}, i] = \frac{1}{N} \sum_{j \in \mathcal{V}} \mathcal{A}[\hat{x}, j]$$

for all i as $\tau \to \infty$. Often, only a few consensus iterations are taken ($\tau \ll \infty$), or even simply one, for instance in [28]. This reduces significantly the communication among sensors, but causes the convergence property to be lost. Detailed analysis of such interleaved schemes, where local estimation problems are solved followed by a finite number of consensus iterations repeatedly, is still an open problem. The reader is referred to [23] for a stability/convergence proof applied to a particular case.

We propose to use a different merging mechanism, based on a weighted version of the typical algorithm (7.6), similarly to [39]. We make use of the following lemma:

Lemma 7.1 ([39]). *Given a set of independent and unbiased estimates, $\mathcal{A}[\hat{x}, i]$, with associated covariance matrices, $\mathcal{A}[P, i]$, where $i \in \mathcal{V}$, the following weighted averaging:*

$$\hat{x} = \left(\sum_{j\in\mathcal{V}} \mathcal{A}[P^{-1}, j]\right)^{-1} \sum_{j\in\mathcal{V}} \mathcal{A}[P^{-1}, j]\, \mathcal{A}[\hat{x}, j]$$
$$P^{-1} = \sum_{j\in\mathcal{V}} \mathcal{A}[P^{-1}, j]$$

gives the linear minimum-variance unbiased estimate of x.

The weighted averaging given in Lemma 7.1 can be seen as

$$\hat{x} = Y^{-1}Z$$
$$P^{-1} = NY$$

where $Z = (1/N)\sum_{j\in\mathcal{V}} \mathcal{A}[Z, j]$, $Y = (1/N)\sum_{j\in\mathcal{V}} \mathcal{A}[Y, j]$, $\mathcal{A}[Z, i] = \mathcal{A}[P^{-1}, i]\mathcal{A}[\hat{x}, i]$, $\mathcal{A}[Y, i] = \mathcal{A}[P^{-1}, i]$. Therefore Z and Y can be calculated using standard averaging. Hence a consensus iteration can be implemented in the form

$$\mathcal{A}[\bar{Z}, i] = \frac{1}{N_i} \sum_{j\in\mathcal{J}_i} \mathcal{A}[Z, j]$$
$$\mathcal{A}[\bar{Y}, i] = \frac{1}{N_i} \sum_{j\in\mathcal{J}_i} \mathcal{A}[Y, j] \tag{7.7}$$

which has iteration matrix W, where $[W]_{ij} = 1/N_i$ if and only if $j \in \mathcal{J}_i$, and $[W]_{ij} = 0$ otherwise. Furthermore, W can be proven to satisfy (7.5).

Note that in general the assumption of independence in Lemma 7.1 will not hold. The different $\hat{x}_i(k)$ local state estimates will be correlated since the sensors are observing the same model. This implies that the optimality of the merged estimate will be lost. We can still show however that the merged estimate will be unbiased.

Lemma 7.2. *Given a set of possibly dependent but unbiased estimates, $\mathcal{A}[\hat{x}, i]$, with associated covariance matrices, $\mathcal{A}[P, i]$, where $i \in \mathcal{V}$, the weighted average*

$$\hat{x} = \left(\sum_{j\in\mathcal{V}} \mathcal{A}[P^{-1}, j]\right)^{-1} \sum_{j\in\mathcal{V}} \mathcal{A}[P^{-1}, j]\, \mathcal{A}[\hat{x}, j]$$
$$P^{-1} = \sum_{j\in\mathcal{V}} \mathcal{A}[P^{-1}, j]$$

will also be an unbiased estimate of x.

Proof The expected value of $\hat{x}$ can be written as

$$\mathbb{E}[\hat{x}] = \mathbb{E}\left[\left(\sum_{j\in\mathcal{V}} \mathcal{A}[P^{-1}, j]\right)^{-1} \sum_{j\in\mathcal{V}} \mathcal{A}[P^{-1}, j]\, \mathcal{A}[\hat{x}, j]\right]$$
$$= \left(\sum_{j\in\mathcal{V}} \mathcal{A}[P^{-1}, j]\right)^{-1} \sum_{j\in\mathcal{V}} \mathcal{A}[P^{-1}, j]\, \mathbb{E}[\mathcal{A}[\hat{x}, j]]$$

from which the claim follows. □

Although the merged estimate will not be optimal, constructing an optimal one would require computing the cross-covariances among all the estimates (i.e., $\mathbb{E}[\hat{x}_i\hat{x}_j^T]$, for all i, j), which leads to adopting a solution that is closer to the centralized one. Moreover, our numerical simulation studies [34] indicate that the presented merge algorithm often leads to more accurate localization solutions than do standard consensus algorithms. Thus, finding a solution that is close to the centralized one seems less relevant in practice.

In our numerical experiments we typically employ $\tau = 1$ iteration per update step. This implies that further analytical studies regarding asymptotic convergence for $k \to \infty$ (R2) and the consistency of the filters are in order. However, from a practical point of view, one could validate at least the consistency property using standard tests on recorded data, such as in [3] and [38].

Algorithm 7.1 shows the method for each sensor i.

Algorithm 7.1 Merge algorithm

1: Input: $\hat{x}_j(k)$, $P_j(k)$, $j \in \mathcal{J}_i$
2: Merge:

$$\bar{x}_i(k) = \left(\sum_{j\in\mathcal{J}_i} P_j^{-1}(k)\right)^{-1} \sum_{j\in\mathcal{J}_i} P_j^{-1}(k)\hat{x}_j(k)$$
$$\bar{P}_i^{-1}(k) = \sum_{j\in\mathcal{J}_i} P_j^{-1}(k)$$

3: Output: $\bar{x}_i(k)$, $\bar{P}_i(k)$

7.4 Distributed Nonlinear Estimation

Given the premises of the previous sections, our general distributed nonlinear estimation framework will consist of N possibly different local filters, connected to the merge/consensus algorithm as depicted in Fig. 7.1. This setting unifies and simplifies the distributed implementation of three of the most widely used nonlinear filters, namely the extended Kalman filter (EKF), unscented Kalman filter (UKF) and particle filter (PF), which will be presented in the next sections. First, the distributed EKF formulation of [28] will be reviewed, the differences from the new proposed solution will pointed out. Then, a distributed UKF algorithm will be proposed relying on the same merge mechanism, and finally, a novel distributed particle filter scheme will be shown and compared with [16, 15]. We note that each local filter can be chosen to be tailored to the computational capabilities of the particular sensor device. The benefits of this aspect will be illustrated in Sec. 7.6.2.

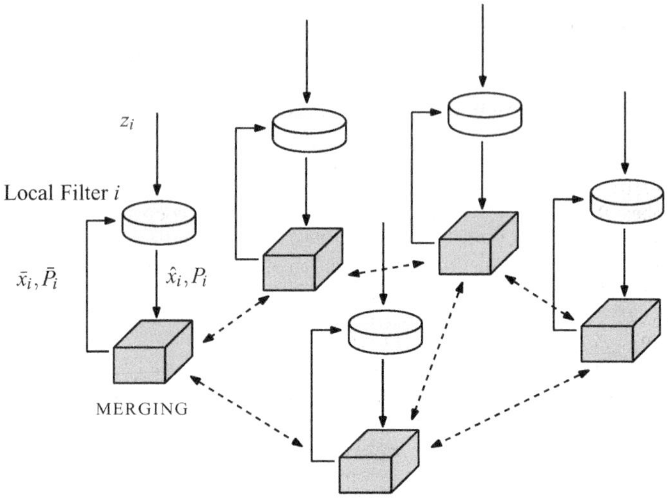

Fig. 7.1 A unified setting for distributed nonlinear estimation. Each cylinder represents a sensor in the network. The sensors can communicate within their neighborhoods and exchange information. This is shown via the grey boxes.

7.4.1 Distributed Extended Kalman Filters

Let F_i, G_i and H_i be, respectively:

$$F_i = \left.\frac{\partial f(x(k), w(k))}{\partial x(k)}\right|_{(\bar{x}_i(k),0)}$$

$$G_i = \left.\frac{\partial f(x(k), w(k))}{\partial w(k)}\right|_{(\bar{x}_i(k),0)}$$

$$H_i = \left.\frac{\partial g(x(k))}{\partial x(k)}\right|_{(\bar{x}_i(k))}$$

Let $Q = \mathbb{E}[w(k)w^T(k)]$, $R_i = \mathbb{E}[v_i(k)v_i^T(k)]$. Define the weighted predicted observation vector $\mathcal{A}[o,i]$, the weighted true observation vector $\mathcal{A}[y,i]$ and their nodal aggregates

$$\mathcal{A}[o,i] = H_i^T \mathcal{A}[R,i]^{-1} g(F_i\bar{x}_i(k)), \quad \mathcal{A}[\bar{o},i] = \sum_{j\in\mathcal{J}_i} \mathcal{A}[o,j]$$

$$\mathcal{A}[y,i] = H_i^T \mathcal{A}[R,i]^{-1} z_i(k+1), \quad \mathcal{A}[\bar{y},i] = \sum_{j\in\mathcal{J}_i} \mathcal{A}[y,j]$$

whose differences $o_i - y_i$ and $\mathcal{A}[\overline{o},i] - \mathcal{A}[\overline{y},i]$ represent the mismatch between the prediction and the measurements. Define the inverse covariance matrix S_i and its nodal aggregate:

$$\mathcal{A}[S,i] = H_i^T \mathcal{A}[R,i]^{-1} H_i, \quad \mathcal{A}[\overline{S},i] = \sum_{j \in \mathcal{J}_i} \mathcal{A}[S,j]$$

The distributed extended Kalman filter (DEKF) algorithm of [28] implements a typical consensus iteration. For each sensor it consists of the following two steps:
Step 1 (Prediction):

$$\begin{aligned} \check{x}_i(k+1) &= F_i \bar{x}_i(k) \\ \check{P}_i(k+1) &= F_i \bar{P}_i(k) F_i^T + G_i Q G_i^T \end{aligned} \tag{7.8}$$

Step 2 (Update):

$$\begin{aligned} (\bar{P}_i(k+1))^{-1} &= \left(\check{P}_i(k+1)\right)^{-1} + \mathcal{A}[\overline{S},i] \\ \bar{x}_i(k+1) &= \check{x}_i(k+1) + \\ &\quad \bar{P}_i(k+1)\left(\mathcal{A}[\overline{y},i] - \mathcal{A}[\overline{o},i]\right) + \varepsilon \bar{P}_i(k+1) \cdot \\ &\quad \sum_{j \in \mathcal{N}_i} \left(\check{x}_j(k+1) - \check{x}_i(k+1)\right) \end{aligned} \tag{7.9}$$

Our proposed version has a different update step and makes use of Algorithm 7.1:
Step 1 (Prediction): same as in (7.8)
Step 2 (Modified update):

$$\begin{aligned} (P_i(k+1))^{-1} &= \left(\check{P}_i(k+1)\right)^{-1} + \mathcal{A}[S,i] \\ \hat{x}_i(k+1) &= \check{x}_i(k+1) + P_i(k+1)\left(\mathcal{A}[y,i] - \mathcal{A}[o,i]\right) \end{aligned} \tag{7.10}$$

Step 3 (Merge):

$$(\bar{x}_i(k+1), \bar{P}_i(k+1)) = \text{MERGE}_{j \in \mathcal{J}_i}(\hat{x}_j(k+1), P_j(k+1))$$

Note that the consensus algorithm in (7.9) has a form similar to (7.6). In addition, y_i, o_i and S_i are used in the modified update step rather than $\overline{y}_i$, $\overline{o}_i$ and $\overline{S}_i$. The reason is to reduce the amount of data that the sensors send to each other. Hence, the final algorithm consists of a modified local prediction-update step, followed by a merge step.

7.4.2 Distributed Unscented Kalman Filters

The unscented Kalman filter performs a statistical linearization through the use of a weighted statistical linear regression process. Instead of approximating the dynamical model by a Taylor series expansion (such as in EKF), the UKF deterministically extracts so-called sigma points from the noise Gaussian terms and passes these through the model. From the transformation of these points it computes the state estimate's Gaussian distribution. The UKF is generally more accurate than the EKF, although it is more computationally expensive. Let

$$(\hat{x}_i(k+1), P_i(k+1)) = \text{UKF}_i(\bar{x}_i(k), \bar{P}_i(k), z_i(k+1))$$

be the local UKF, whose formulation can be found for example in [22].

Our proposed implementation of a distributed UKF algorithm is shown in Algorithm 7.2. Note the general structure of a local nonlinear filter and the merging step. The reader is referred to [8] for an alternative approach, involving exchange of raw measurements.

Algorithm 7.2 Distributed UKF algorithm

1: Given: $\bar{x}_i(0) = \hat{x}_i(0)$, $\bar{P}_i(0) = P_i(0)$
2: **while** new data exists **do**
3: Input: $\bar{x}_i(k), \bar{P}_i(k)$
4: Local UKF:
$$(\hat{x}_i(k+1), P_i(k+1)) = \text{UKF}_i(\bar{x}_i(k), \bar{P}_i(k), z_i(k+1))$$
5: Merge:
$$(\bar{x}_i(k+1), \bar{P}_i(k+1)) = \text{MERGE}_{j \in \mathcal{J}_i}(\hat{x}_j(k+1), P_j(k+1))$$
6: Output: $\bar{x}_i(k+1), \bar{P}_i(k+1)$
7: **end while**

7.4.3 Distributed Particle Filters

While standard particle filters have been studied intensively (a comprehensive overview can be found in [2]), distributed particle filters are a rather unexplored research field [35].

Since the field is rather new, the terminology is not always coherent. We will refer to methods that use central units to collect data from the sensors and compute the state estimate as quasi-distributed particle filter algorithms. This class has been studied in the literature starting from the work of [4], where three algorithms have been presented, including more recent papers on hierarchical distributed particle filters [21]. We will refer to particle filters that do not require centralized data collection as the standard class of distributed particle filters (DPFs).

Although references on DPFs date back to the work of [31] and [10], detailed analysis and evaluation studies on their properties have been initiated only recently. The main reasons are the following:

- Particle filters cannot easily share quantities based on their measurements: they cannot send particles among themselves because these are weighted with dif-

ferent measurements and thus they are noncompatible. Moreover, sending raw measurement data can lead to a heavy communication burden [10].
- Particle filters cannot easily be implemented in parallel; in general, the resampling step needs the whole particle population knowledge.

Furthermore, it is crucial that the distributed algorithms be consistent with the centralized counterparts. For SIR filters this is naturally ensured by the relation

$$p(z(k)|x(k)) = \prod_{i=1}^{N} p(z_i(k)|x(k)),$$

which states that each sensor can agree on the same $p(z(k)|x(k))$, provided all the sensors have full information. Although this statement is true, we must make sure that the sensors speak the same language, i.e., either

- they are sharing raw measurements, or
- they have exactly the same particle population, or
- they have the same representation of $p(z(k)|x(k))$ as a function of $x(k)$.

The first situation above, even though not ideal, is exploited in [31], where a query-answer protocol is used in order to decrease the communication cost. The basic idea is that each sensor keeps in memory the entire time-evolution of its particles and all the measurements and sends some of the particles to the neighbors. The neighbors decide whether some measurements, at some time instant, are valuable for them and they reply with the data. This protocol is valuable when the sensors have enough data to run accurate particle filters on their own, and they need extra information only in some special cases. A typical application is localization in a building: When a sensor has a clear view of the object to be localized, it can run its own particle filter with no extra information. On the contrary, when the object is hidden behind a wall, it needs some data from other sensors that can see the object.

A completely different approach is studied in [10] where the sensors are supposed to share the very same particle population. An algorithm is developed in the framework of parametric modeling for this purpose. Here each $p(z_i(k)|x(k))$ is approximated by a parametric model $\mathcal{F}_i(x(k), \theta_i(k))$. The model parameters $\theta_i(k)$ are estimated at each node from the particles, and then they are disseminated, instead of the data being disseminated directly. Although interesting, this method imposes strict limitations on the distributed algorithms. For instance, in some cases the sensors have to be synchronized, and in general, the communication graph among them has to have a specified structure: either a chain, a ring or a tree.

The recent work of [14] also proposes to share the same particle population among the sensors, but the particles are drawn from an a priori PDF scaled via an adaptation mechanism that makes it closer to the a posteriori PDF through prefiltering.

Another way to implement a consistent version of DPF is to guarantee that all the sensors have the same representation of $p(z(k)|x(k))$. Since this is not easy to ensure, usually, this condition is relaxed by requiring that the representation of the proposal distribution be the same. There are several reasons behind this choice. First, the sensors have different sets of measurements and the particles are not compatible among the sensors, thus they cannot be shared. Second, even if the sensors agree upon a common, continuous $p(z(k)|x(k))$, in the resampling stage they have to increase the number of particles in order to capture the important features of it. However, this last computation should be avoided to limit the computational load.

For these reasons, recent research on DPF is focused on guaranteeing that the PDF from which the samples are drawn be the same among the sensors. This idea is exploited in the papers [32, 33, 17, 16, 15], where the authors use different models. In particular, the first three papers focus on a Gaussian Mixture Model, or GMM, which can be written as

$$q(x(k+1)|x(k),z(k+1)) = \sum_{c=1}^{C} \lambda_c(k)\mathcal{N}(\mu_c(k),\Sigma_c(k))$$

where C, λ_c, μ_c, and Σ_c are parameters of the model and they represent the chosen number of Gaussians, their relative importance, their mean and their covariance, respectively. This representation has the drawbacks that, first, the sensors have to agree upon several variables if $C \gg 1$, and second, the local representation is built via an iterative optimization scheme, which requires time and may lead to local minima (see [32] for further details).

The use of a mono-modal Gaussian distribution generated via an unscented transformation (UT) can instead have accurate results while keeping the algorithm as simple as possible, as pointed out in [16, 15].

In fact, even if different representations are involved, these approaches can both fit naturally in the framework of a consensus process and they can be generalized in the context of Gaussian proposal distributions [18]. The key concept is to use a Gaussian proposal distribution such as

$$q(x(k+1)|x(k),z(k+1)) = \mathcal{N}(\mu(k),\Sigma(k))$$

where $\mathcal{N}$ represents a normal distribution with mean $\mu(k)$ and covariance $\Sigma(k)$.

The pair $(\mu(k),\Sigma(k))$ is calculated by propagating $(\hat{x}(k),P(k))$ via an unscented Transformation, UT, as in [16, 15]

$$(\mu(k),\Sigma(k)) = \mathrm{UT}(\hat{x}(k),P(k))$$

whereas the couple $(\hat{x}(k),P(k))$ can be approximated via consensus using the local couples $(\hat{x}_i(k),P_i(k))$. These can be computed by each local particle filter (at the preceding time instant) in the following way:

$$\hat{x}_i(k) = \sum_{j=1}^{m} \omega_{i,k}{}^j x_i(k)^j \tag{7.11}$$

$$P_i(k) = \sum_{j=1}^{m} \omega_{i,k}{}^j (x_i(k)^j - \hat{x}_i(k))(x_i(k)^j - \hat{x}_i(k))^T \tag{7.12}$$

where j represents the particle index, m the number of particles, $x_i(k)^j$ and $\omega_{i,k}{}^j$, respectively, the state and the weight of the jth particle for the ith sensor at time k. In the formulation of [16], the global couple $(\hat{x}(k), P(k))$ is approximated via a consensus algorithm in the form of (7.4):

$$\bar{x}_i(k) = \hat{x}_i(k) + \frac{\varepsilon}{N_i - 1} \sum_{j \in \mathcal{N}_i} (\hat{x}_j(k) - \hat{x}_i(k))$$

$$\bar{P}_i(k) = P_i(k) + \frac{\varepsilon}{N_i - 1} \sum_{j \in \mathcal{N}_i} (P_j(k) - P_i(k))$$

Therefore, after the consensus iteration each local PF has the new proposal distribution:

$$q(x(k+1)|x(k), z_i(k+1)) = \mathcal{N}(\mu_i(k), \Sigma_i(k))$$

with

$$(\mu_i(k), \Sigma_i(k)) = \mathrm{UT}(\bar{x}_i(k), \bar{P}_i(k))$$

In our formulation, we will use the MERGE algorithm instead of the standard consensus algorithm, thus:

$$(\bar{x}_i(k), \bar{P}_i(k)) = \mathrm{MERGE}_{j \in \mathcal{J}_i}(\hat{x}_j(k), P_j(k))$$

Algorithm 7.3 presents the proposed modified method. Note that PF indicates the local particle filter.

Remark 7.3. Note that in [16] the choice of τ for the consensus algorithm is left to the user as a parameter. We will assume $\tau = 1$ to compare it with our scheme.

7.5 Distributed Computation of Particle Filters on GPUs

Particle filters suffer from higher computational demand than Kalman filters and extensions, especially when a high number of particles is important. However, their number should be tunable and increasable to obtain satisfactory results. The aim of this section is to devise a method to distribute the number of particles on different processing cores and yet obtain a comparable accuracy as if they were on the same core. In this section we focus on a single sensor and its processing cores, our aim being to distribute the computation itself. We stress the fact that in contrast to the available literature, e.g., [6, 5, 19, 25], we will look for algorithms that do not involve

Algorithm 7.3 Distributed PF algorithm

1: Given: $\bar{x}_i(0) = \hat{x}(0)$, $\bar{P}_i(0) = P(0)$
2: **while** new data exists **do**
3: Input: $\bar{x}_i(k)$, $\bar{P}_i(k)$, $z_i(k+1)$
4: LOCAL PF{

a. Propagation of $(\bar{x}_i(k), \bar{P}_i(k))$:

$$(\mu_i(k), \Sigma_i(k)) = \mathrm{UT}(\bar{x}_i(k), \bar{P}_i(k))$$

b. Local PF with proposal distribution $\mathcal{N}(\mu_i(k), \Sigma_i(k))$:

$$(x_i(k+1)^j, \omega_{i,k+1}{}^j) = \mathrm{PF}_i(\mu_i(k), \Sigma_i(k), z_i(k+1))$$

c. Compute $(\hat{x}_i(k+1), P_i(k+1))$ via (7.11) - (7.12)

}

5: Merge:

$$(\bar{x}_i(k+1), \bar{P}_i(k+1)) = \mathrm{MERGE}_{j \in \mathcal{J}_i}(\hat{x}_j(k+1), P_j(k+1))$$

6: Output: $\bar{x}_i(k+1)$, $\bar{P}_i(k+1)$
7: **end while**

extensive communication among the cores and, in particular, do not make use of centralized information.

Even if recent developments in general purpose multicore GPU architectures make the study of this question very relevant and interesting, distributing particle filter computations is not a trivial task. The main reason, as pointed out in the previous sections, is that usually the resampling step requires us to have knowledge of all the particles.

We propose in the following a rather simple yet effective idea, which leads to distributed particle filter computations. First, we assume that all the cores have the same set of measurements, that is $z_i(k+1)$ for the cores of sensor i. For this reason, the particles are compatible from core to core and can be shared within each sensor. Of course, there is no point in sharing all the subsets of particles among the different cores, since this would simply lead to the nondistributed version. However, inspired by some ideas in [31], it makes much more sense to share only a few representative particles. If each of the C cores is running m particles and they share only those $n \ll m$ with higher weights, then the communication is reduced and the accuracy will still approach that of a particle filter with mC particles. A simulation example is shown in the next section to support this observation, while Algorithm 7.4 describes the main steps of the method.

We note that the proposed algorithm can also be cast in the framework of distributed estimation presented earlier, if we assume that each local filter has access to different computing cores.

Algorithm 7.4 Distributed computation particle filter

1: Given: $x(0)^j = x(0)$
2: **while** new data exists **do**
3: Input: $x(k)^j$, $z_i(k+1)$
4: Draw samples for each j:
$$x(k+1)^j \sim p(x(k+1)|x(k)^j)$$
5: Calculate the weights:
$$\omega_{k+1}{}^j = p(z_i(k+1)|x(k+1)^j)$$
6: Share n particles with high weights with the neighbors.
7: Calculate the local state estimate.
8: Resample m particles.
9: Output: $x(k+1)^j$
10: **end while**

7.6 Numerical Evaluation and Comparison

Now we present a realistic test case to analyze the proposed algorithms. We consider the localization via range-only measurements of an autonomous underwater vehicle (AUV), which can represent a scaled model of many existing underwater robotic platforms; see for example [11]. We define the error $e_i(k)$ of sensor i at time k, as the distance between the true position at that time and the one estimated by the sensor i. Let the mean error e_{m} be defined as:

$$e_{\mathrm{m}} = \frac{1}{NT} \sum_{i=1}^{N} \sum_{k=0}^{T} e_i(k)$$

where T is the final time of simulation. Let the average error be the mean error averaged on a number of different simulations.

7.6.1 The Mobile Agent Model

The state of the AUV is chosen as $x = (x_{\mathrm{p}}^T, s_{\mathrm{p}}^T)^T$, where $x_{\mathrm{p}} \in \mathbb{R}^3$ is the position and $s_{\mathrm{p}} \in \mathbb{R}^3$ is the velocity. The dynamical equations are:

$$\begin{aligned} x_{\mathrm{p}}(k+1) &= x_{\mathrm{p}}(k) + s_{\mathrm{p}}(k)\Delta t \\ s_{\mathrm{p}}(k+1) &= s_{\mathrm{p}}(k) + \frac{\Delta t}{m}\left(\hat{u} - \alpha \left\|s_{\mathrm{p}}(k)\right\| s_{\mathrm{p}}(k)\right) \\ \hat{u} &= u + n_u \end{aligned}$$

where m is the mass of the vehicle, α is a drag coefficient, and n_u is a noise term. We assume that we have $N = 25$ sensors (7.3) sparsely distributed at varying heights from a plane surface. The different heights simulate an uneven seafloor. We take $\Delta t = 1$ s, $T = 130$ s, $m = 1$ kg, $\alpha = 1$ kg/s. We define an open-loop control sequence of magnitude $||u|| = 0.5$ N and varying direction, while we select $\mathrm{std}(n_u) = (0.05, 0.05, 0.025)^T$ N. We assume that the measurement error in Eq. (7.2) is $\mathrm{std}(v_i) = 0.1$ m, for all the sensors. We consider 500 particles for the DPF. A schematic representation of the simulation test case is shown in Fig. 7.2.

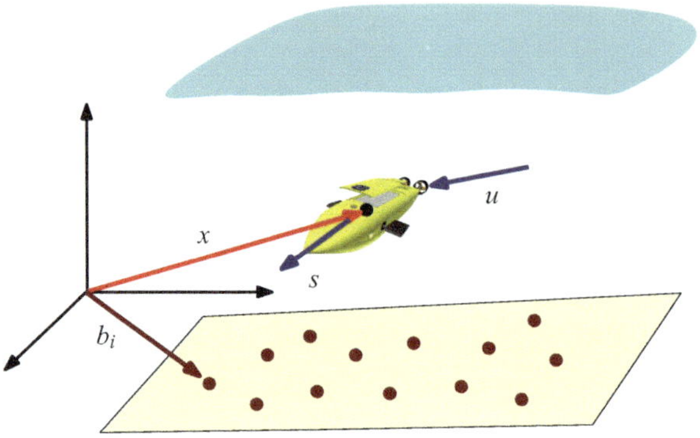

Fig. 7.2 Schematic representation of the test case.

7.6.2 Simulation Results

In the first scenario we consider, each sensor is assumed to run the same type of local filter. We collect the results for 2500 different simulation runs, varying randomly the position and the communication range of the sensors. Figure 7.3 depicts the average error of the proposed algorithms versus the second smallest eigenvalue of the communication graph Laplacian (also called the algebraic connectivity). The second smallest eigenvalue, denoted as λ_2, or its normalized counterpart, $\lambda_2/\lambda_{2,\max}$, dictate the convergence rate of the consensus algorithm [27]. Values of $\lambda_2/\lambda_{2,\max}$ close to 0 represent graphs that are not highly connected, leading to more distributed estimation problems. Values of $\lambda_2/\lambda_{2,\max}$ near 1 correspond to highly connected graphs, thus estimation problems close to the centralized case. Here, $\lambda_{2,\max}$ is the maximum over the graphs from the 2500 simulations. In Fig. 7.3 a dot at the coordinate (ϕ, ψ), corresponds to an average error of ψ for graphs with $\lambda_2/\lambda_{2,\max} \in (\phi - 0.05, \phi + 0.05)$. The shaded areas show the standard deviation of these errors. Note that the DEKF estimations are not depicted here because they do not converge. The DUKF is shown without the standard deviation, to make the graph more readable; its value is on the order of 0.3 m.

The results show that the proposed DPF outperforms the DPF found in the literature. This is because the MERGE algorithm delivers estimates closer to the minimum-variance one than the literature, e.g. [16], where simple averaging algorithms are implemented. This also means that a given set of particles will charac-

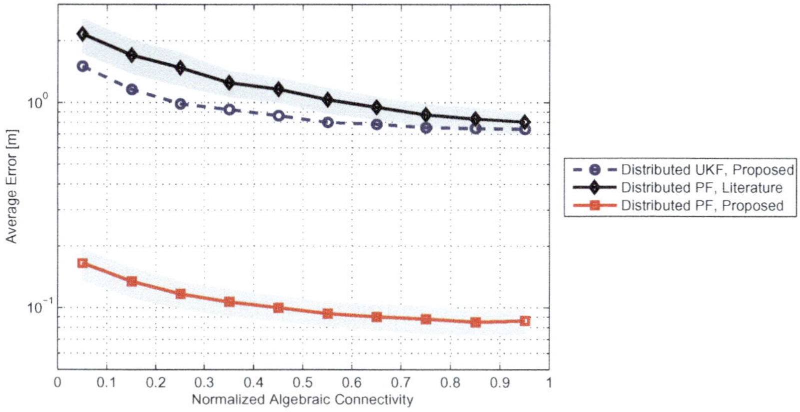

Fig. 7.3 Comparison between the proposed algorithms and that of the literature with respect to the normalized algebraic connectivity of the information exchange graph. The average error is computed as the mean error of the sensor averaging 2500 different simulations. The shaded areas are the standard deviations of the bold lines. The DUKF's standard deviation is not depicted.

terize the a posteriori distribution better, since the trace of our covariance is smaller than in [16]. A limitation of our procedure is that this "small covariance" could cause an impoverishment of the particle diversity, which may lead to a loss in accuracy. This has not been detected in our simulations but it is a topic of further investigation.

Our results show also that in this simulation study, the DUKF has similar average error as the DPF reported in the literature. This is important because the DUKF is less computationally expensive than the DPF, which is crucial in the context of designing fast yet accurate algorithms.

As a second test scenario, we fix the graph topology with a normalized algebraic connectivity of 0.6, and we vary the types of filters embedded on each sensor. We collect data from 1500 simulation runs, with a varying number of PFs, UKFs and EKFs present in the sensor network and their physical location. Figure 7.4 summarizes our results. The curves represent different numbers of particle filters. Since the overall number of sensors is fixed ($N = 25$), one can compute the number of EKFs present in the network from knowledge of the number of PFs and UKFs.

We can observe that even with a relatively small number of more accurate filters (for example 1 PF and 5 UKFs), the distributed estimation converges. This was not the case in the first test scenario, where the local estimates were diverging using the EKFs alone. This is a very interesting observation that seems to support having many cheap devices and only a very few expensive ones.

We may also notice that by increasing the number of UKFs, the accuracy improves initially quite noticeably but eventually deteriorates. For a low number of UKFs, this trend can be explained by the merging mechanism. Let n_{UKF} be the number of UKFs and n_{PF} the number of the particle filters, and let the covariances of the filters be P_{PF}, P_{UKF}, and P_{EKF} respectively. Assume, for simplicity and without loss of generality, a scalar state vector. The average error is then related to the merge

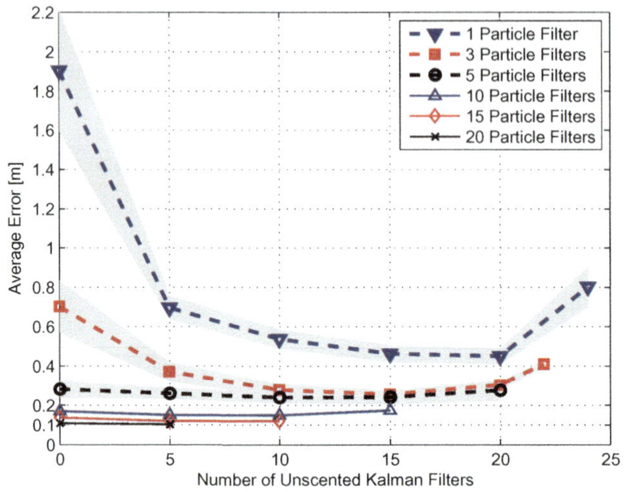

Fig. 7.4 Results for different filters running on different sensors. The vverage error is computed as the mean error of the sensor averaging 1500 different simulations. The shaded areas are the standard deviations of the bold lines. Since the number of sensors is fixed ($N = 25$), one can compute the number of EKFs present in the network from the knowledge of the number of PFs and UKFs.

covariance, whose expression is

$$\bar{P} = \frac{P_{\mathrm{PF}} P_{\mathrm{UKF}} P_{\mathrm{EKF}}}{25 P_{\mathrm{PF}} P_{\mathrm{UKF}} + P_{\mathrm{UKF}}(P_{\mathrm{EKF}} - P_{\mathrm{PF}}) n_{\mathrm{PF}} + P_{\mathrm{PF}}(P_{\mathrm{EKF}} - P_{\mathrm{UKF}}) n_{\mathrm{UKF}}}$$

which for a given n_{PF} is proportional to

$$\bar{P} \propto \frac{1}{\alpha + n_{\mathrm{UKF}}}$$

with the constant $\alpha > 0$. This explains the initial decrease of the average error for an increasing number of UKFs.

The deteriorating performance for an increasing number of UKFs is still under investigation. One potential reason is the suboptimality of the merging mechanism and the “small covariance” problem. In fact, the proposed algorithm may lead to a smaller covariance estimate than the actual one, which leads to optimistic filters [3], and this could cause a bad selection of the sigma points for the UKFs. Adding more EKFs increases the average error and brings the estimated covariance closer to the real one, which improves performance.

7.6.3 Distributed Computation Particle Filters

In our third and final test scenario, we assume to have access to multiple cores on the sensors. In order to make the analysis specific to the distribution of the computation, we consider a centralized unit that collects all the measurements from all the 25 sensors. This centralized unit is equipped with $C = 16$ cores to compute the estimate via particle filters. We stress that this setting is not restrictive, meaning that we do not need the knowledge of all the measurements, and one could still use the framework we proposed earlier assuming multiple cores on each sensor. However, by choosing a centralized unit, we will avoid the effects of the distributed estimation and consensus problem using local measurements, which could lead to bias in our conclusions. Thus, we assume that each core of the centralized unit can have access to all the sensor measurements and can communicate with its neighbors. Figure 7.5 shows the computing core configuration.

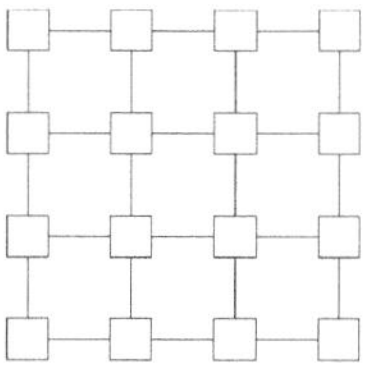

Fig. 7.5 Core configuration: the cores are represented by the squares, while the lines indicate possible communication paths.

The simulation parameters are the same as before, only the noise of the sensors is made 10 times bigger to stress the effect of the particle number, i.e., $\text{std}(v_i) = 1$ m. Five different cases are considered:

- Case 1: $m = 2$, $n = 1$;
- Case 2: $m = 4$, $n = 1$;
- Case 3: $m = 8$, $n = 1$;
- Case 4: $m = 16$, $n = 1$;
- Case 5: $m = 32$, $n = 1$;

where m is the number of particles in each core and n is the number of particles that each core sends to its neighbors. The results of 200 simulations for each case are summarized in Fig. 7.6, where the left bar on each case (from 1 to 5) represents the average error of a centralized PF with m particles. The right bar represents the average error of the distributed computations particle filters on all cores. The bar on the far right represents the average error for a centralized PF with $m = 500$ particles and it illustrates the performance that could be obtained with a high number of particles, indicating an approximate lower bound. The vertical interval lines represent the standard deviation.

As this numerical example illustrates, the distributed computations particle filters outperforms a standard centralized PF, which does not even converge for $m = 2$. This can be seen more easily when the number of particles is low, since both approaches hit the same accuracy limit when many particles are used. Moreover, we

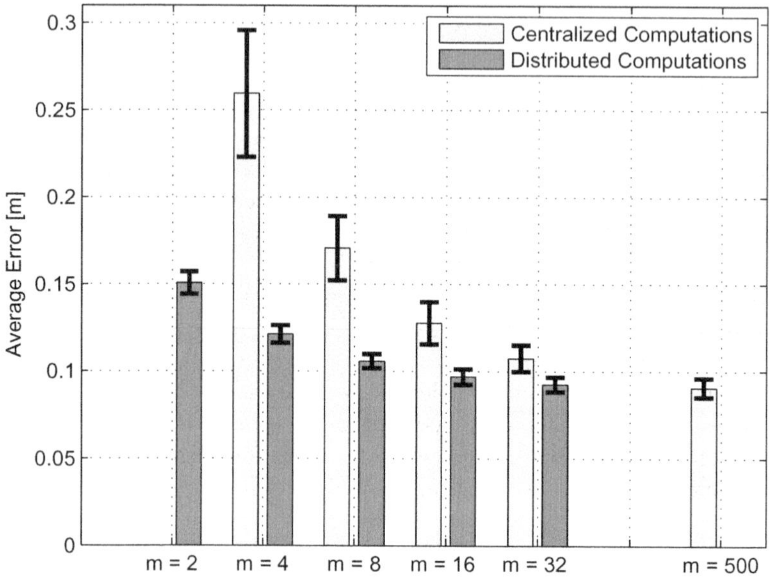

Fig. 7.6 Comparison among different particle filters. The left bar in each case (from 1 to 5) represents the average error of a centralized PF with m particles. The right bar represents the average error of the distributed computations particle filters on all cores. The vertical interval lines represent the standard deviation.

can observe that even using $n = 1$, the accuracy of the distributed approach is closer to that of a centralized filter with mC particles than one with m particles. Further improvements are expected by allowing more communication among the cores, for example using $n = 2$. This is encouraging, since using the distributed computations particle filters decreases computation time by a factor of at least $(1+4n/m)/C$. This speedup estimate follows from the fact that each core is at most dealing with $m+4n$ particles, where 4 is the maximum number of connections with neighboring cores. This result comes at the price of little communication among the cores. However, since the interconnection among them is rather sparse, it scales linearly with the number of cores, and it is not affected by data loss or network reconfiguration.

7.7 Conclusions

We proposed an effective scheme to distribute the nonlinear estimation problem among different sensing units. We applied the methods to a localization problem with range-only measurements, designing distributed extended Kalman filters, distributed unscented Kalman filters and distributed particle filters. The proposed algorithms outperform those found in the literature based on a simulated benchmark.

Furthermore, we showed how to implement different filters on different sensors tailoring the estimators to the specific device, which can be useful to exploit the sensor capabilities to their maximum. Finally, we proposed a way to decrease the computational load of particle filters, distributing the particles on different cores. Our initial promising results suggest that this can be particularly valuable given the recent interest in GPU architectures.

As future work we plan to implement the scheme in a real robotic testbed, which at this writing is under development. Moreover, we will extend the formulation to multi-robot settings, in which some of the sensors are moving with the robots themselves. Convergence, consistency and optimality of the merging mechanisms represent important future research directions.

References

1. Alriksson, P.: State estimation for distributed and hybrid systems. Ph.D. thesis, Department of Automatic Control, Lund University, Sweden (2008)
2. Arulampalam, M., Maskell, S., Gordon, N., Clapp, T.: A tutorial on particle filters for online nonlinear/non-gaussian bayesian tracking. IEEE Trans. Signal Processing **50**(2), 174 – 188 (2002)
3. Bar-Shalom, Y., Rong Li, X., Kirubarajan, T.: Estimation with Applications to Tracking and Navigation. Wiley Inter-Science, New York (2001)
4. Bashi, A.S., Jilkov, V.P., Li, X.R., Chen, H.: Distributed implementations of particle filters. In: Proc. Int. Conf. Information Fusion, pp. 1164–1171. Cairns, Australia (2003)
5. Bolić, M., Djurić, P.M., Hong, S.: Resampling algorithms and architectures for distributed particle filters. IEEE Trans. Signal Processing **53**, 2442–2450 (2005)
6. Brun, O., Teuliere, V., Garcia, J.: Parallel particle filtering. Journal of Parallel and Distributed Computing **62**, 1186–1202 (2002)
7. Carli, R., Chiuso, A., Schenato, L., Zampieri, S.: Distributed Kalman filtering based on consensus strategies. IEEE J. Selected Areas in Communications **26**, 622–633 (2008)
8. Cattivelli, F.S., Sayed, A.H.: Distributed nonlinear Kalman filtering with applications to wireless localization. In: Proc. 35th IEEE Int. Conf. Acoustics, Speech, and Signal Processing, pp. 3522–3525. Dallas, TX, USA (2010)
9. Chandrasekhar, V., Seah, W.K., Choo, Y.S., Ee, H.V.: Localization in underwater sensor networks—Survey and challenges. In: Proc. 1st ACM Int. Workshop on Underwater Networks, pp. 33–40. Los Angeles, CA, USA (2006)
10. Coates, M.: Distributed particle filters for sensor networks. In: Proc. Third Int. Symp. Information Processing in Sensor Networks, pp. 99 – 107. Berkeley, CA, USA (2004)
11. Corke, P., Detweiler, C., Dunbabin, M., Hamilton, M., Rus, D., Vasilescu, I.: Experiments with underwater robot localization and tracking. In: Proc. IEEE Int. Conf. Robotics and Automation, pp. 4556–4561. Roma, Italy (2007)
12. Cortés, J.: Distributed algorithms for reaching consensus on general functions. Automatica **44**, 726 – 737 (2008)
13. Fagnani, F., Zampieri, S.: Randomized consensus algorithms over large scale networks. IEEE J. Selected Areas in Communications **26**, 634–649 (2008)
14. Farahmand, S., Roumeliotis, S.I., Giannakis, G.B.: Particle filter adaptation for distributed sensors via set membership. In: Proc. 35th IEEE Int. Conf. Acoustics, Speech, and Signal Processing, pp. 3374–3377. Dallas, TX, USA (2010)
15. Gu, D., Hu, H.: Target tracking by using particle filter in sensor networks. Int. J. Robotics and Automation **24**(3), 169 176 (2009)

16. Gu, D., Sun, J., Hu, Z., Li, H.: Consensus based distributed particle filter in sensor network. In: Proc. 2008 IEEE Int. Conf. Information and Automation. Zhangjiajie, China (2008)
17. Gu, D.: Distributed particle filter for target tracking. In: Proc. 2007 IEEE Int. Conf. Robotics and Automation, pp. 3856–3861. Roma, Italy (2007)
18. Guo, D., Wang, X., Chen, R.: New sequential monte carlo methods for nonlinear dynamics systems. Statistics and Computing **15**, 135 – 147 (2005)
19. Hendeby, G., Hol, J.D., Karlsson, R., Gustafsson, F.: A graphics processing unit implementation of the particle filter. Tech. rep., Automatic Control, Linkoping University (2007)
20. Hossain, A., Nguyen Van, H., Jin, Y., Soh, W.: Indoor localization using multiple wireless technologies. In: Proc. IEEE Int. Conf. Mobile Adhoc and Sensor Systems, pp. 1–8. Los Alamitos, CA, USA (2007)
21. Huang, Y., Liang, W., Yu, H., Xiao, Y.: Target tracking based on a distributed particle filter in underwater sensor networks. Wireless Communication and Mobile Computing **8**, 1023 – 1033 (2008)
22. Julier, S.J., Uhlmann, J.K.: Unscented filtering and nonlinear estimation. Proc. IEEE **93**(3), 401 – 422 (2004)
23. Kamgarpour, M., Tomlin, C.: Convergence properties of a decentralized Kalman filter. In: Proc. 47th IEEE Conf. Decision and Control, pp. 3205–3210. Cancun, Mexico (2008)
24. Kim, B.K., Jung, H.M., Yoo, J.B., Lee, W.Y., Park, C.Y., Ko, Y.W.: Design and implementation of cricket-based location tracking system. Proc. World Academy of Science, Engineering and Technology **28**, 96 – 100 (2008)
25. Lozano, O.M., Otsuka, K.: Real-time visual tracker by stream processing. simultaneous and fast 3D tracking of multiple faces in video sequences by using a particle filter. J. Signal Processing Systems **57**(2), 285–295 (2009)
26. Olfati-Saber, R., Fax, J.A., Murray, R.: Consensus and cooperation in networked multi-agent systems. Proc. IEEE **95**(1), 215–233 (2007)
27. Olfati-Saber, R., Murray, R.: Consensus problems in networks of agents with switching topology and time-delays. IEEE Trans. Automatic Control **49**(9), 1520 – 1533 (2004)
28. Olfati-Saber, R.: Distributed Kalman filtering for sensor networks. In: Proc. 46th IEEE Conf. Decision and Control 2007 (CDC2007), pp. 5492–5498. New Orleans, LA, USA (2007)
29. Rekleitis, I., Bedwani, J., Dupuis, E.: Autonomous planetary exploration using lidar data. In: Proc. 2009 IEEE Int. Conf. Robotics and Automation (ICRA2009), pp. 3025–3030. Kobe, Japan (2009)
30. Ren, W., Beard, R.: Distributed Consensus in Multi-Vehicle Cooperative Control: Theory and Applications. Springer-Verlag London (2008)
31. Rosencrantz, M., Gordon, G., Thrun, S.: Decentralized sensor fusion with distributed particle filters. In: Proc. Conf. Uncertainty in Artificial Intelligence. Acapulco, Mexico (2003)
32. Sheng, X., Hu, Y., Ramanathan, P.: Distributed particle filter with GMM approximation for multiple targets localization and tracking in wireless sensor network. In: Proc. Fourth Int. Symp. Information Processing in Sensor Networks, pp. 181–188 (2005)
33. Sheng, X., Hu, Y.: Distributed particle filters for wireless sensor network target tracking. In: Proc. IEEE Int. Conf. Acoustics, Speech, and Signal Processing (2005)
34. Simonetto, A., Keviczky, T., Babuška, R.: Distributed nonlinear estimation for robot localization using weighted consensus. In: Proc. IEEE Int. Conf. Robotics and Automation, pp. 3026–3031. Anchorage, AK, USA (2010)
35. Simonetto, A., Keviczky, T.: Recent developments in distributed particle filters: Towards fast and accurate algorithms. In: Proc. 1st IFAC Workshop on Estimation and Control of Networked Systems, pp. 138–143. Venice, Italy (2009)
36. Smith, A., Balakrishnan, H., Goraczko, M., Priyantha, N.: Tracking moving devices with the cricket location system. In: Proc. 2nd Int. Conf. Mobile Systems, Applications and Services (MobiSys '04), p. 190 202. Boston, MA, USA (2004)
37. Speranzon, A., Fischione, C., Johansson, K.H., Sangiovanni-Vincentelli, A.: A distributed minimum variance estimator for sensor networks. IEEE Journal on Selected Areas in Communications **26**(4), 609 – 621 (2008)

38. van der Heijden, F.: Consistency checks for particle filters. IEEE Trans. Pattern Analysis and Machine Intelligence **1**, 140 – 145 (2006)
39. Xiao, L., Boyd, S., Lall, S.: A scheme for robust distributed sensor fusion based on average consensus. In: Proc. Fourth Int. Conf. Information Processing in Sensor Networks, pp. 63–70. Los Angeles, CA, USA (2005)

Chapter 8
Performance Prediction in Uncertain Multi-Agent Systems Using $\mathcal{L}_1$ Adaptation-Based Distributed Event-Triggering

Xiaofeng Wang and Naira Hovakimyan

Abstract This chapter studies the impact of communication constraints and uncertainties on the performance of multi-agent systems, while closing the local loops with embedded $\mathcal{L}_1$ adaptive controllers. A communication and adaptation co-design scheme is proposed that helps to predict system performance. With this scheme, an agent locally determines its broadcast time instants using distributed event-triggering. The embedded $\mathcal{L}_1$ adaptive controller enables each agent to compensate for the local uncertainties and disturbances. Performance bounds are derived on the difference between the signals of the ideal model (in the absence of uncertainties and with perfect communication) and the real system operating with the proposed co-design scheme, which can be arbitrarily reduced subject only to hardware limitations. These results can be used for design guidelines in safety-critical applications.

8.1 Introduction

This chapter examines the performance of multi-agent systems in the presence of communication constraints and modeling uncertainties. We propose a co-design scheme to bound the derivation of the real system from its ideal model. The proposed scheme includes distributed event-triggering, used for determining broadcast time, and embedded $\mathcal{L}_1$ adaptive controller for compensation of uncertainties. By introducing an intermediate system (called "desired system") as a bridge, the design

---

Xiaofeng Wang
Department of Mechanical Science and Engineering, University of Illinois at Urbana-Champaign, 1206 West Green Street, Urbana, IL 61801, USA
e-mail: wangx@illinois.edu

Naira Hovakimyan
Department of Mechanical Science and Engineering, University of Illinois at Urbana-Champaign, 1206 West Green Street, Urbana, IL 61801, USA,
e-mail: nhovakim@illinois.edu

R. Johansson and A. Rantzer (Eds.): Distributed Decision Making and Control, LNCIS 417, pp. 171–193.
springerlink.com

of the broadcast event and the $\mathcal{L}_1$ controller can be decoupled. Embedded $\mathcal{L}_1$ adaptive controllers are designed to ensure closeness between the real system and the desired system, and broadcast-triggering events are designed for closeness between the desired system and the ideal model.

A multi-agent system contains a number of agents that can be geographically distributed and each of these agents is controlled by an embedded processor. Information is exchanged among these processors through communication networks. Control decisions are made based on this information in order to achieve certain global objectives, such as formation [8], consensus [18], flocking [17], coverage maximization [6]. A survey on the relevant work in multi-agent systems can be found in [15].

Most of these results assume that agents are communicating with their neighbors at will and the system dynamics are completely known. These two assumptions, however, may not be practical in reality. In practice, communication networks, especially wireless networks, are digital, which means that the information is transmitted in a discrete-time manner rather than continuous-time. Moreover, all real networks have bandwidth limitations that may cause delays in message delivery. Such limitations may severely degrade the overall system performance [13]. So it is important to effectively manage the available limited network bandwidth. Also, the uncertainties and the disturbances are unavoidable in implementation of control systems. They can lead to loss of stability and/or performance. As a result, the real system may completely deviate from the original design. Therefore an important control objective is to ensure that the discrepancy between the real system and its ideal mathematical model can be quantified and reduced in a predictable manner during the entire operation of the system, including the transient phase.

To solve the communication issue, we focus on event-triggering. Event-triggering has been widely studied for its benefits in saving the communication and/or computational resource. This is done by having agents broadcast their states only in case of occurrence of local events. There are several results on implementation of event-triggering in embedded real-time systems [21, 27], networked control systems (NCSs) [26, 14], network utility maximization (NUM) [22], consensus algorithms [7] , etc. Instead of just focusing on stability, as was done in the prior work, this chapter places emphasis on the performance of the real system both in transient and steady state. We prove that using the event-triggering scheme in this chapter, along with the $\mathcal{L}_1$ adaptation, the performance of the real system can be rendered arbitrarily close to the ideal model in the presence of uncertainties and disturbances.

We notice that disturbances in NCSs have been addressed by studies that resort to input-to-state stability (ISS) of the systems [16, 27, 25, 24]. However, ISS ensures only a peak bound without guarantees for the frequency content of the signals, i.e., robustness. Using the $\mathcal{L}_1$ adaptive control technique [10] in each agent, the closeness between the real system and the ideal one can be adjusted by tuning the adaptation gain, the bandwidth of the low-pass filter and the thresholds of the local events in a decoupled way. As a result, in the presence of uncertainties and disturbances the system with its fast adaptation can compensate for the effect of disturbances, retaining the desired transient performance bounds. This may largely improve the prediction of the performance of the real systems. Simulation results verify our claims. The results in this chapter can be used as design guidelines for safety-critical applications, including air traffic control and collision avoidance in multi-agent systems.

The chapter is organized as follows. Section 8.2 formulates the problem. The $\mathcal{L}_1$ adaptive control structure is introduced in Sec. 8.3. Event-triggering and adaptation co-design are presented in Sec. 8.4. Simulation results are presented in Sec. 8.5. Section 8.6 concludes the chapter. Finally, the proofs are presented in Sec. 8.7.

8.2 Problem Formulation

Notation: We denote by $\mathbb{R}^n$ the n-dimension real vector space and by $\mathbb{R}^+$ the real positive numbers. We also define $\mathbb{R}_0^+ = \mathbb{R}^+ \cup \{0\}$ and $\mathcal{N}$ to be the set of natural numbers. We use $\|\cdot\|_1$, $\|\cdot\|_2$, and $\|\cdot\|_\infty$ for the 1-norm, 2-norm and ∞-norm of a vector, respectively. Further, $\|\cdot\|_{\mathcal{L}_1}$ and $\|\cdot\|_{\mathcal{L}_\infty}$ are the $\mathcal{L}_1$ and $\mathcal{L}_\infty$ norms of a function, respectively. The truncated $\mathcal{L}_\infty$ norm of a function $x : [0,\infty) \to \mathbb{R}^n$ is defined as $\|x\|_{\mathcal{L}_\infty^{[0,\tau]}} \triangleq \sup_{0\le t\le \tau} \|x(t)\|_\infty$. The symbol $\mathbb{I}_{n\times n}$ is used to represent an n by n identity matrix; when it is clear in context, we use simply $\mathbb{I}$ instead. Given a vector $x \in \mathbb{R}^n$, $[x]_k \in \mathbb{R}$ denotes the kth entry of x and for a matrix A, $[A]_{pq} \in \mathbb{R}$ denotes the entry at the pth row and the qth column. For a strictly proper stable system $C(s)$ and an input signal $r(s)$, the system output $y(s) = C(s)r(s)$ satisfies $\|y\|_{\mathcal{L}_\infty^{[0,\tau]}} \le \|C(s)\|_{\mathcal{L}_1} \|r\|_{\mathcal{L}_\infty^{[0,\tau]}}$ [11].

Consider a multi-agent system consisting of N agents (or "subsystems"). Let $\mathcal{N} = \{1,\ldots,N\}$ represent the set of agents. The state equation of agent i is

$$\begin{aligned} \dot{x}_i(t) &= A_i x_i(t) + B_i(u_i(t) + \theta_i(t)\phi_i(x_i(t))) \\ x_i(0) &= x_i^0, \end{aligned} \tag{8.1}$$

where $x_i : \mathbb{R}_0^+ \to \mathbb{R}^n$ is the state of agent i, $u_i : \mathbb{R}_0^+ \to \mathbb{R}^m$ is the control input of agent i, $x_i^0 \in \mathbb{R}^n$ is agent i's initial condition, $A_i \in \mathbb{R}^{n\times n}, B_i \in \mathbb{R}^{n\times m}$ are known matrices with appropriate dimensions, $\text{rank}(B_i) = m \le n$, (A_i, B_i) is controllable, $\phi_i : \mathbb{R}^n \to \mathbb{R}^l$ is a known function and $\theta_i : \mathbb{R}_0^+ \to \Omega_i$ is an unknown mapping from t to a known compact set $\Omega_i = \{\theta_i \in \mathbb{R}^{m\times l} \mid |[\theta_i]_{pq}| \le b^i_{pq}, \forall p = 1,\ldots,m,\ \forall q = 1,\ldots,l\}$. We assume that $\phi_i(x_i)$ is Lipschitz and bounded at $x_i = 0$, i.e.

$$\|\phi_i(y_1) - \phi_i(y_2)\|_\infty \le L_i \|y_1 - y_2\|_\infty \tag{8.2}$$

$$\|\phi_i(0)\|_\infty \le M_i \tag{8.3}$$

hold for any $y_1, y_2 \in \mathbb{R}^n$ in a compact set, and $\theta_i(t)$ is differentiable with bounded first derivative, i.e., there exists a positive constant $d_i^\theta \in \mathbb{R}^+$, such that

$$\sqrt{\text{tr}\left(\dot{\theta}_i^T(t)\dot{\theta}_i(t)\right)} \le d_i^\theta. \tag{8.4}$$

Remark 8.1. In this chapter, we use a simple form of the uncertainty with globally Lipschitz nonlinearity to illustrate the co-design idea. In fact, $\mathcal{L}_1$ adaptive controllers can handle much more complex uncertainties and disturbances, including unmodeled dynamics, unmatched uncertainties [10, 9]. An extension of the proposed scheme for handling interconnected nonlinearities is reported in [23].

Since (A_i, B_i) is controllable, there exist positive definite matrices $P_i, Q_i \in \mathbb{R}^{n\times n}$ and feedback gain $K_i \in \mathbb{R}^{m\times n}$ such that

$$P_i(A_i + B_iK_i) + (A_i + B_iK_i)^T P_i = -Q_i. \tag{8.5}$$

For notational convenience, we define

$$A_i^m = A_i + B_iK_i, \tag{8.6}$$

and $A \in \mathbb{R}^{nN\times nN}$, $B \in \mathbb{R}^{nN\times mN}$, $f : \mathbb{R}_0^+ \times \mathbb{R}^{nN} \to \mathbb{R}^{mN}$, respectively, by

$$A = \begin{pmatrix} A_1 & \cdots & 0 \\ \vdots & \ddots & \vdots \\ 0 & \cdots & A_N \end{pmatrix}, \; B = \begin{pmatrix} B_1 & \cdots & 0 \\ \vdots & \ddots & \vdots \\ 0 & \cdots & B_N \end{pmatrix}, \; f(t,x) = \begin{pmatrix} \theta_1(t)\phi_1(x_1) \\ \vdots \\ \theta_N(t)\phi_N(x_N) \end{pmatrix} \tag{8.7}$$

Therefore, the state equation of the overall system can be written as:

$$\dot{x}(t) = Ax(t) + B(u(t) + f(t,x(t))) \tag{8.8}$$

where $x = (x_1^T, x_2^T, \cdots, x_N^T)^T$ and $u = (u_1^T, u_2^T, \cdots, u_N^T)^T$.

The ideal system dynamics for agent i are given by

$$\begin{aligned} \dot{x}_i^{\text{ideal}}(t) &= A_i x_i^{\text{ideal}}(t) + B_i\left(K_i x_i^{\text{ideal}}(t) + g_i\left(x_i^{\text{ideal}}(t), \{x_j^{\text{ideal}}(t)\}_{j\in\mathcal{N}_i}\right) + r_i(t)\right) \\ x_i^{\text{ideal}}(0) &= x_i^0, \end{aligned} \tag{8.9}$$

where $\mathcal{N}_i \subseteq \mathcal{N}$ is agent i's neighboring set, $g_i : \mathbb{R}^n \times \mathbb{R}^{n|\mathcal{N}_i|} \to \mathbb{R}^m$ is a known control strategy and $r_i : \mathbb{R}_0^+ \to \mathbb{R}^m$ is a known reference signal for agent i. The function g_i can be any control algorithm proposed for the "ideal" multi-agent systems ("ideal" means continuous (perfect) communication and no uncertainty). A survey of these algorithms can be found in [15].

Ideally, if we knew the function $\theta_i(t)$, we could include $-\theta_i(t)\phi_i(x_i)$ in the control input to cancel the existing nonlinearity. Then we could design an appropriate feedback strategy to fulfill the control objective. In that case, the ideal controller for agent i would be

$$u_i^{\text{ideal}}(t) = K_i x_i^{\text{ideal}}(t) + g_i\left(x_i^{\text{ideal}}(t), \{x_j^{\text{ideal}}(t)\}_{j\in\mathcal{N}_i}\right) + r_i(t) - \theta_i(t)\phi_i(x_i^{\text{ideal}}(t)). \tag{8.10}$$

We denote the overall ideal state and input by $x^{\text{ideal}} = ((x_1^{\text{ideal}})^T, \cdots, (x_N^{\text{ideal}})^T)^T$ and $u = ((u_1^{\text{ideal}})^T, \cdots, (u_N^{\text{ideal}})^T)^T$, respectively.

The ideal controller, however, is not practical for two reasons. First $\theta_i(t)$ is unknown. Second, agent i cannot continuously receive its neighbors' state information due to the communication constraints. In practice, agent i can only continuously detect its state $x_i(t)$ and discretely receive data packets broadcast by its neighbors in $\mathcal{N}_i$; furthermore, agent i can also broadcast its state information to its neighbors in a discrete manner. The broadcast state of agent i is denoted by $\hat{x}_i(t)$. Let $t_i[k]$ be the time instant when agent i releases the kth broadcast. Then $\hat{x}_i(t)$ is deemed to be constant over the time interval $[t_i[k], t_i[k+1])$, and the control input of agent i is computed based on $x_i(t)$ and $\{\hat{x}_j(t)\}_{j\in\mathcal{N}_i}$.

To address these issues, we develop a distributed, in contrast to centralized, $\mathcal{L}_1$ adaptive control scheme. We use event-triggering to define the broadcast release time instant $t_i[k]$. Agent i broadcasts its state information to its neighbors only when a "local" event occurs. In other words, agent i broadcasts its states when it detects occurrence of the event using only its local information. The event in this chapter corresponds to certain local measurement error exceeding a prespecified threshold. The local $\mathcal{L}_1$ adaptive controller compensates for the modeling uncertainties and component failures. The objective is to ensure that the internal signals in the overall system in (8.1) can be rendered close to those in the ideal system in (8.9) via appropriately adjusting the design parameters in a *decoupled* way.

To study the closeness between the real system and the ideal model, we need to design feedback and communication schemes addressing issues raised by both uncertainties and communication constraints. To decouple the design procedure, we introduce an intermediate system called *desired system* as a bridge to establish the relation between the real and ideal systems. The ith desired subsystem is defined by

$$\begin{aligned}
\dot{x}_i^{\text{des}}(t) &= A_i x_i^{\text{des}}(t) + B_i(u_i^{\text{des}}(t) + \theta_i(t)\phi_i(x_i^{\text{des}}(t))) \\
u_i^{\text{des}}(t) &= K_i x_i^{\text{des}}(t) - \theta_i(t)\phi_i(x_i^{\text{des}}(t)) + \chi_i^{x_i}(t) \\
x_i^{\text{des}}(0) &= x_i^0,
\end{aligned} \tag{8.11}$$

where $x_i^{\text{des}} : \mathbb{R}_0^+ \to \mathbb{R}^n$ and $u_i^{\text{des}} : \mathbb{R}_0^+ \to \mathbb{R}^m$ are the state and the input of the ith desired subsystem, respectively, and

$$\chi_i^{x_i}(t) \triangleq g_i\left(x_i(t), \{\hat{x}_j(t)\}_{j\in\mathcal{N}_i}\right) + r_i(t). \tag{8.12}$$

It takes two steps to show the closeness between the real and the ideal systems. The first step is to show the closeness between the ith real subsystem and the ith desired subsystem. This step is achieved by the design of the $\mathcal{L}_1$ adaptive controller, which compensates for modeling uncertainties and external disturbances. The second step considers the closeness between the desired system and the originally posed ideal model, which mainly focuses on the communication constraints. The introduction of the desired system, achievable by the $\mathcal{L}_1$ adaptive controller, is the key step in the decoupling between the design of the feedback strategy and the design of the communication protocol.

8.3 $\mathcal{L}_1$ Adaptive Control Structure

This section considers the first step for handling the uncertainties. We treat each agent as an "independent" system and associate with it its desired subsystem, given by (8.11). In this desired subsystem, $\chi_i^{x_i}(t)$ plays the role of an external reference input, which is dependent upon the state of agent i. The desired subsystem is used only in analysis and not synthesis; therefore the dependence of $\chi_i^{x_i}(t)$ upon x_i is not leading to circular arguments. The $\mathcal{L}_1$ adaptive controller is used to close the feedback loop of agent i, given by Eq. (8.1). The $\mathcal{L}_1$ adaptive controller for agent i has the following structure:

$$u_i(t) = K_i x_i(t) + v_i(t), \tag{8.13}$$

where K_i is the nominal feedback gain and $v_i : \mathbb{R}_0^+ \to \mathbb{R}^m$ is the adaptive feedback. For the adaptive feedback, we first need to introduce the identifier:

$$\dot{x}_i^{\text{iden}}(t) = A_i^m x_i^{\text{iden}}(t) + B_i(v_i(t) + \hat{\theta}_i(t)\phi_i(x_i(t))), \quad \dot{x}_i^{\text{iden}}(0) = x_i^0,$$

where $x_i^{\text{iden}} : \mathbb{R}_0^+ \to \mathbb{R}^n$ is the state of the identifier, and $\hat{\theta}_i : \mathbb{R}_0^+ \to \mathbb{R}^{m\times l}$ is updated according to the adaptive law: For any $p \in \{1,\ldots,m\}$ and $q \in \{1,\ldots,l\}$,

$$\begin{aligned} [\dot{\hat{\theta}}_i]_{pq}(t) &= \Gamma_i \mathbf{Proj}_{b_{pq}^i}\left([\hat{\theta}_i]_{pq}(t), -[\tilde{x}_i^T P_i B_i]_p [\phi_i(x_i)]_q\right) \\ \hat{\theta}_i(0) &\in \Omega_i. \end{aligned} \tag{8.14}$$

In Eq. (8.14), $\Gamma_i \in \mathbb{R}^+$ is the adaptation gain, and the operator $\mathbf{Proj}_{b_{pq}^i} : \mathbb{R}\times\mathbb{R} \to \mathbb{R}$ is given by

$$\mathbf{Proj}_{b_{pq}^i}(a,b) = \begin{cases} b - bh(a), & \text{if } h(a) > 0 \text{ and } b\dot{h}(a) > 0 \\ b, & \text{otherwise,} \end{cases} \tag{8.15}$$

where $h(a) = \frac{a^2 - a_{\max}^2}{\varepsilon_a a_{\max}^2}$ and $a_{\max} = \frac{b_{pq}^i}{\sqrt{1+\varepsilon_a}}$ with some $\varepsilon_a \in (0,1)$. The $\mathbf{Proj}_{b_{pq}^i}$ operator ensures that $\hat{\theta}_i(t)$ remains in Ω_i [19].

The adaptive signal v_i is defined by

$$v_i(s) = -C_i(s)\hat{\eta}_i(s) + \chi_i^{x_i}(s), \tag{8.16}$$

where $C_i(s)$ is an $m\times m$ diagonal matrix defined by $C_i(s) = c_i(s)\mathbb{I}_{m\times m}$; $c_i(s)$ is a strictly proper, stable low-pass filter satisfying $c_i(0) = 1$; and $v_i(s)$, $\hat{\eta}_i(s)$, $\chi_i^{x_i}(s)$ are the Laplace transforms of the signals $v_i(t)$, $\hat{\eta}_i(t) \triangleq \hat{\theta}_i(t)\phi_i(x_i(t))$, $\chi_i^{x_i}(t)$, respectively. The structure of the embedded $\mathcal{L}_1$ controller in subsystem i is shown in Fig. 8.1.

The results from [3] can be elaborated to prove that subject to a mild assumption on $\chi_i^{x_i}(t)$, the distance between the trajectories of agent i and its desired subsystem, and also the corresponding control signals, can be uniformly bounded on arbitrary

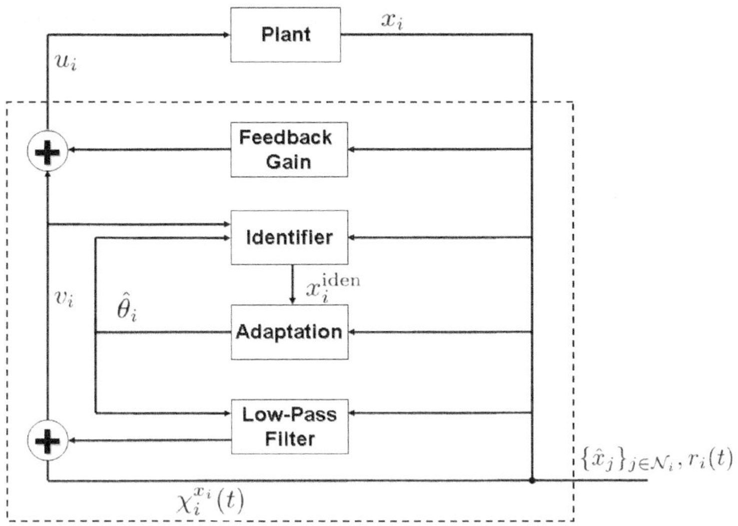

Fig. 8.1 $\mathcal{L}_1$ adaptive controller for subsystem i.

time interval $[0,\tau]$ for any $\tau > 0$. The assumption imposed on $\chi_i^{x_i}(t)$ in Lemma 8.1 will be verified later in the proof of Lemma 8.2.

Lemma 8.1. *Consider the desired subsystem in equation* (8.11) *and the subsystem in Eq.* (8.1) *with the controller in Eqs.* (8.13)–(8.16). *Assume that Eqs.* (8.2), (8.3), (8.4), (8.5) *hold. If there exists a positive constant* ρ_{χ_i} *such that*

$$\|\chi_i^{x_i}(t)\|_\infty \leq \rho_{\chi_i}$$

for any $t \in [0,\tau]$, *and* $C_i(s)$ *is chosen to verify*

$$\lambda_i \triangleq \|G_i(s)\|_{\mathcal{L}_1} \theta_i^{\max} L_i < 1, \tag{8.17}$$

where L_i *is defined in* (8.2) *and*

$$G_i(s) \triangleq H_i(s)(\mathbb{I} - C_i(s)) \tag{8.18}$$

$$H_i(s) \triangleq (s\mathbb{I} - A_i^m)^{-1} B_i \tag{8.19}$$

$$\theta_i^{\max} \triangleq \max_{\theta_i \in \Omega_i} \|\theta_i\|_\infty, \tag{8.20}$$

then the following inequalities hold for any $\tau > 0$:

$$\|x_i^{\text{des}} - x_i\|_{\mathcal{L}_\infty^{[0,\tau]}} \leq \varepsilon_i \tag{8.21}$$

$$\|u_i^{\text{des}} - u_i\|_{\mathcal{L}_\infty^{[0,\tau]}} \leq \delta_i, \tag{8.22}$$

where $\varepsilon_i, \delta_i \in \mathbb{R}_0^+$ *are defined by*

$$\varepsilon_i \triangleq \frac{\theta_i^{\max} M_i \|G_i(s)\|_{\mathcal{L}_1}}{1-\lambda_i} + \frac{\|C_i(s)\|_{\mathcal{L}_1}}{1-\lambda_i} \frac{\psi_i}{\sqrt{\Gamma_i}} + \frac{\lambda_i}{1-\lambda_i} \|H_i(s)\|_{\mathcal{L}_1} \|\chi_i^{x_i}\|_{\mathcal{L}_\infty^{[0,\tau]}} \tag{8.23}$$

$$\begin{aligned}\delta_i \triangleq & (\|K_i\|_\infty + \|C_i(s)\|_{\mathcal{L}_1} \theta_i^{\max} L_i)\varepsilon_i + \frac{\|C_i(s)(c_i^o H_i(s))^{-1} c_i^o\|_{\mathcal{L}_1} \psi_i}{\sqrt{\Gamma_i}} \\ & + \|C_i(s) - \mathbb{I}\|_{\mathcal{L}_1} \theta_i^{\max}(L_i \|H_i(s)\|_{\mathcal{L}_1} \|\chi_i^{x_i}\|_{\mathcal{L}_\infty^{[0,\tau]}} + M_i)\end{aligned} \tag{8.24}$$

$$\psi_i \triangleq \sqrt{\frac{4\,(\theta_i^{\max})^2 + 4\dfrac{\sigma_{\max}(P_i)}{\sigma_{\min}(Q_i)} d_i^\theta \max_{\theta_i \in \Omega_i} \sqrt{\mathrm{tr}\,(\theta_i^T \theta_i)}}{\sigma_{\min}(P_i)}} \tag{8.25}$$

and $c_i^o \in \mathbb{R}^{m\times n}$ *ensures that* $c_i^o H_i(s)$ *has full rank, stable zeros, and relative degree 1.*

Remark 8.2. According to Eq. (8.23), if we increase the cut-off frequency and the value of the adaptation gain Γ_i, ε_i will reduce to zero. However, notice that setting $C_i(s) = \mathbb{I}$, the $\mathcal{L}_1$ adaptive controller degenerates into model reference adaptive controller (MRAC), yielding unbounded δ_i, because the term $C_i(s)(c_i^o H_i(s))^{-1} c_i^o$ is then improper [4].

8.4 Local Event Design

This section studies the impact of communication constraints on the closeness of the real system and the ideal system. As mentioned in Sec. 8.2, we use event-triggering to determine the broadcast time instants. It is, therefore, important to know how to design these events. There are several requirements on the events. First, the event for each agent should be detectable by that agent. Second, the event for an individual agent should not continuously occur. Third, the events need to be designed such that the resulting event-triggered system is close to the ideal system. In the following discussion, we design local events fulfilling these requirements and provide performance bounds between the real and the ideal systems.

It is shown in Sec. 8.3 that the real system can be close to the desired system in Eq. (8.11). So we just need to design the events such that the desired system is close to the ideal system. Let $g : \mathbb{R}^{nN} \times \mathbb{R}^{nN} \to \mathbb{R}^{mN}$ be defined by

$$g(x,\hat{x}) = \begin{pmatrix} g_1(x_1, \{\hat{x}_j\}_{j\in\mathcal{N}_1}) \\ \vdots \\ g_n(x_N, \{\hat{x}_j\}_{j\in\mathcal{N}_N}) \end{pmatrix} \tag{8.26}$$

According to Eqs. (8.9) and (8.11), the error dynamics between the desired system and the overall ideal system is

$$\dot{x}^{\text{des}}(t) - \dot{x}^{\text{ideal}}(t) = A_m(x^{\text{des}}(t) - x^{\text{ideal}}(t)) + B(g(x(t),\hat{x}(t)) - g(x^{\text{ideal}}(t),x^{\text{ideal}}(t))) \tag{8.27}$$

where $x^{\text{des}} = ((x_1^{\text{des}})^T, \cdots, (x_N^{\text{des}})^T)^T$, $A_m \triangleq \text{diag}\{A_1^m, \cdots, A_n^m\} \in \mathbb{R}^{nN \times nN}$ and B, g are defined in equations (8.7) and (8.26), respectively. Let

$$H(s) \triangleq (s\mathbb{I} - A_m)^{-1}B \tag{8.28}$$

Then, Eq. (8.27) implies

$$\|x^{\text{des}} - x^{\text{ideal}}\|_{\mathcal{L}_\infty^{[0,\tau]}} \leq \|H(s)\|_{\mathcal{L}_1} \|g(x(t),\hat{x}(t)) - g(x^{\text{ideal}}(t),x^{\text{ideal}}(t))\|_{\mathcal{L}_\infty^{[0,\tau]}} \tag{8.29}$$

for all $\tau > 0$.

To obtain a bound on $\|x^{\text{des}} - x^{\text{ideal}}\|_{\mathcal{L}_\infty}$, we need to focus on g. Let us first define two types of errors:

$$e(t) \triangleq x(t) - x^{\text{des}}(t) \tag{8.30}$$

$$\hat{e}(t) \triangleq x(t) - \hat{x}(t), \tag{8.31}$$

where $e : \mathbb{R}_0^+ \to \mathbb{R}^{nN}$ is the difference between the real state and the state of the desired system in equation (8.11), and $\hat{e} : \mathbb{R}_0^+ \to \mathbb{R}^{nN}$ is the difference between the real state and the broadcast state. Notice that $x(t) = x^{\text{des}}(t) + e(t)$ and $\hat{x}(t) = x^{\text{des}}(t) + e(t) - \hat{e}(t)$.

Assumption 8.1. *Assume that g is locally Lipschitz, i.e., there exist positive constants $a, b \in \mathbb{R}^+$ such that*

$$\|g(x, x - \hat{e}) - g(x^{\text{ideal}}, x^{\text{ideal}})\|_\infty \leq a\|x - x^{\text{ideal}}\|_\infty + b\|\hat{e}\|_\infty \tag{8.32}$$

hold for all $x, \hat{e}, x^{\text{ideal}}$ in a compact set.

Subject to Assumption 8.1, we conclude from inequalities (8.29) and (8.32) that

$$\|x^{\text{des}} - x^{\text{ideal}}\|_{\mathcal{L}_\infty^{[0,\tau]}} \leq \|H(s)\|_{\mathcal{L}_1} \left(a\|x^{\text{des}} - x^{\text{ideal}}\|_{\mathcal{L}_\infty^{[0,\tau]}} + a\|e\|_{\mathcal{L}_\infty^{[0,\tau]}} + b\|\hat{e}\|_{\mathcal{L}_\infty^{[0,\tau]}} \right)$$

Assume that

$$\|H(s)\|_{\mathcal{L}_1} a < 1 \tag{8.33}$$

Then, inequality (8.33) implies

$$\|x^{\text{des}} - x^{\text{ideal}}\|_{\mathcal{L}_\infty^{[0,\tau]}} \leq \frac{\|H(s)\|_{\mathcal{L}_1}}{1 - a\|H(s)\|_{\mathcal{L}_1}} \left(a\|e\|_{\mathcal{L}_\infty^{[0,\tau]}} + b\|\hat{e}\|_{\mathcal{L}_\infty^{[0,\tau]}} \right) \tag{8.34}$$

The preceding inequality indicates that the bound on $\|x^{\text{des}} - x^{\text{ideal}}\|_{\mathcal{L}_\infty}$ depends on the two types of errors $\hat{e}$ and e, with $\hat{e}$ generated due to limited communication resource and e coming from the uncertainty. We need to find bounds on these two errors. Let us first take a look at $\hat{e}$. Arbitrarily choosing a positive constant $\rho_i \in \mathbb{R}^+$, it is easy to see that if agent i uses the violation of the inequality

$$b\|x_i(t) - x_i(t_i[k])\|_\infty \leq \rho_i \tag{8.35}$$

to determine the next broadcast release time $t_i[k+1]$, then $\|x_i(t) - x_i(t_i[k])\|_\infty \leq \rho_i/b$ holds for all $t \in [t_i[k], t_i[k+1]]$. Notice that $\hat{x}_i(t) = x_i(t_i[k])$ for all $t \in [t_i[k], t_i[k+1]]$. Therefore, for $\hat{e}_i = x_i - \hat{x}_i$, we have

$$\|\hat{e}_i(t)\|_\infty \leq \frac{\rho_i}{b} \tag{8.36}$$

for all $t \in [t_i[k], t_i[k+1]]$ and all $k \in \mathcal{N}$, which implies that the preceding inequality holds for all $t \geq 0$. Therefore, $\|\hat{e}\|_{\mathcal{L}_\infty} \leq \rho_i/b$.

The bound on e is provided in Lemma 8.1. However, to apply Lemma 8.1, we need to verify the assumption that $\chi_i^{x_i}(t)$ is bounded. The following lemma helps us to show this by providing bounds on $\|x^{\text{ideal}}\|_{\mathcal{L}_\infty}$ and $\|x\|_{\mathcal{L}_\infty}$.

Lemma 8.2. *Consider the system in equation* (8.1) *with the controller in equations* (8.13)–(8.16). *Suppose that Assumption 8.1 and Eqs.* (8.2), (8.3), (8.4), (8.5), (8.17), (8.33) *hold and* $\|r\|_{\mathcal{L}_\infty}$ *is bounded. If for any* $i \in \mathcal{N}$,

$$\frac{\lambda_i}{1-\lambda_i}\frac{a\|H(s)\|_{\mathcal{L}_1}}{1-a\|H(s)\|_{\mathcal{L}_1}} < 1 \tag{8.37}$$

holds, where λ_i, a, $H(s)$ *are defined in equation* (8.17), (8.32), (8.28), *respectively, and inequality* (8.35) *holds, then the following inequalities hold*

$$\|x^{\text{ideal}}\|_{\mathcal{L}_\infty} \leq \xi_1 \tag{8.38}$$

$$\|x\|_{\mathcal{L}_\infty} \leq \xi_2 \tag{8.39}$$

for all $t \geq 0$, *where* $\xi_1, \xi_2 \in \mathbb{R}^+$ *are defined by*

$$\xi_1 \triangleq \frac{\|H(s)\|_{\mathcal{L}_1}(\|g(0,0)\|_\infty + \|r\|_{\mathcal{L}_\infty}) + \|(s\mathbb{I} - A_m)^{-1}x(0)\|_{\mathcal{L}_\infty}}{1 - a\|H(s)\|_{\mathcal{L}_1}} \tag{8.40}$$

$$\xi_2 \triangleq \frac{\xi_1 + \zeta + \|H(s)\|_{\mathcal{L}_1}\left(\max_{i\in\mathcal{N}}\rho_i - \xi_1 a + \max_{i\in\mathcal{N}}\left\{\frac{\lambda_i}{1-\lambda_i}\right\}\left(\max_{i\in\mathcal{N}}\rho_i + \|g(0,0)\|_\infty + \|r\|_{\mathcal{L}_\infty}\right)\right)}{1 - \max_{i\in\mathcal{N}}\left\{\frac{1}{1-\lambda_i}\right\}\|H(s)\|_{\mathcal{L}_1} a} \tag{8.41}$$

$$\zeta \triangleq \max_{i\in\mathcal{N}}\left\{\frac{\theta_i^{\max} M_i \|G_i(s)\|_{\mathcal{L}_1}}{1-\lambda_i} + \frac{\|C_i(s)\|_{\mathcal{L}_1}}{1-\lambda_i}\frac{\psi_i}{\sqrt{\Gamma_i}}\right\}. \tag{8.42}$$

Remark 8.3. Inequality (8.37) puts constraints on the low-pass filter and the ideal system. It requires $a\|H(s)\|_{\mathcal{L}_1} < 1$, which ensures the boundedness of the ideal system. It also requires $\lambda_i < 1$, which is needed for the stability of $\mathcal{L}_1$ adaptive controller. Note that $\lambda_i < 1$ can be always satisfied by adjusting the bandwidth of the low-pass filter $C_i(s)$. Inequality (8.37) in fact places a small-gain condition on the system, which may be conservative. An alternative way is to use Lyapunov method [21, 27] that may provide less conservative requirements on the low-pass filter bandwidth and the ideal system, while still achieving the desired system performance.

Since $\|x\|_{\mathcal{L}_\infty} \le \xi_2$, we can easily obtain, by (8.12) and (8.32), that

$$\|\chi^x\|_{\mathcal{L}_\infty} \le a\xi_2 + \max_{i\in\mathcal{N}} \rho_i + \|g(0,0)\|_\infty + \|r\|_{\mathcal{L}_\infty} \triangleq \rho_\chi, \tag{8.43}$$

where

$$\chi^x = (\chi_1^{x_1}, \cdots, \chi_N^{x_N})^T. \tag{8.44}$$

With this bound on $\|\chi^x\|_{\mathcal{L}_\infty}$, we are able to present the main result.

Theorem 8.1. *Assume that the hypotheses in Lemma 8.2 hold. Then, the following inequalities hold:*

$$\|x - x^{\text{ideal}}\|_{\mathcal{L}_\infty} \le \xi_3 \tag{8.45}$$

$$\|u - u^{\text{ideal}}\|_{\mathcal{L}_\infty} \le \xi_4, \tag{8.46}$$

where $\xi_3, \xi_4 \in \mathbb{R}^+$ *are defined by*

$$\xi_3 \triangleq \frac{\zeta + \|H(s)\|_{\mathcal{L}_1}\left(\max_{i\in\mathcal{N}} \rho_i + \max_{i\in\mathcal{N}}\left\{\frac{\lambda_i}{1-\lambda_i}\right\}\rho_\chi\right)}{1 - a\|H(s)\|_{\mathcal{L}_1}} \tag{8.47}$$

$$\xi_4 \triangleq \max_{i\in\mathcal{N}} \delta_i + \left(\|K\|_\infty + \max_{i\in\mathcal{N}}\{\theta_i^{\max} L_i\}\right)\frac{\|H(s)\|_{\mathcal{L}_1}}{1 - a\|H(s)\|_{\mathcal{L}_1}}\left(a\max_{i\in\mathcal{N}} \varepsilon_i + \max_{i\in\mathcal{N}} \rho_i\right) + a\xi_3 + \max_{i\in\mathcal{N}} \rho_i \tag{8.48}$$

and $\rho_\chi, \varepsilon_i, \delta_i, \zeta, \xi_2 \in \mathbb{R}^+$ *are defined in equations* (8.43), (8.21), (8.22), (8.42) *and* (8.41), *respectively.*

Remark 8.4. The inequality in (8.35) defines the time instants for event-triggering. Each agent can use the violation of this inequality to trigger the broadcast. At the time instant when the broadcast is released, $\|e_i(t)\|_\infty$ becomes zero and inequality (8.35) is trivially satisfied. Then, agent i does not broadcast until the next time when (8.35) is violated. Notice that this inequality is only related to each agent's local information, so it can be detected by agents locally.

Remark 8.5. The event-triggering literature is rich on defining various events. In general, the event can be described by a logic rule, when a certain function of the current and broadcast states/ouputs is equal to zero (for example, in our case it is the function $\|x_i(t) - \hat{x}_i(t)\|_\infty - \rho_i/b$). However, for different control purposes, such functions might be different. In [21, 27], the threshold on the error is set as a function of the current and/or broadcast states to ensure asymptotic stability; in [5], the event is designed to enforce the system output to be less than a constant for limit circles; in [28, 20, 12], such functions are derived to establish the optimality in estimation. The event in this chapter is simple in its structure and can be directly related to communication constraints, i.e., hardware specifications. With the $\mathcal{L}_1$ adaptive controller, this definition of *the event* leads to *uniform* performance bounds for the distributed system, the tuning of which can be related to hardware specifications and pursued in *decoupled* way.

Remark 8.6. Note that when the cut-off frequency of $C_i(s)$ increases, λ_i reduces to zero and ζ, ξ_2 decrease. Therefore, according to Eq. (8.45), by reducing the event threshold ρ_i and increasing the adaptation gain Γ_i as well as the cut-off frequency of $C_i(s)$, the bound on $\|x - x^{\text{ideal}}\|_{\mathcal{L}_\infty}$, given by ξ_3, can be arbitrarily close to zero. Note that reducing ρ_i will result in frequent communication, which requires a great amount of communication resource. The performance bounds, therefore, can be directly related to the hardware specifications.

Theorem 8.1 provides uniform performance bounds for the signals in the closed-loop system using the proposed event-triggering scheme. Next we derive the bounds on the transmission periods generated by this scheme.

Corollary 8.1. *Assume that the hypotheses in Theorem 8.1 hold. Then for any $i \in \mathcal{N}$, there exists a positive constant T_i such that*

$$t_i[k+1] - t_i[k] \geq T_i. \tag{8.49}$$

8.5 Simulations

Simulation results are presented in this section to illustrate the proposed co-design scheme. We adopt the example from [2], where N agents move in a two-dimensional plane. The control objective is to move agent i to a prespecified position, assuming that agent i can only receive the broadcast state of the two other agents.

The ith agent's dynamics are given as

$$\begin{aligned} \ddot{x}_i &= u_i^x + \theta_i^x(t)\phi_i(X_i) \\ \ddot{y}_i &= u_i^y + \theta_i^y(t)\phi_i(X_i), \end{aligned}$$

where (x_i, y_i) is agent i's relative position to its destination, X_i is the state of agent i defined by

$$X_i = \begin{pmatrix} x_i & \dot{x}_i & y_i & \dot{y}_i \end{pmatrix}^T, \text{ and } \theta_i^x, \theta_i^y : \mathbb{R}_0^+ \to \mathbb{R}^4$$

are time-varying uncertainties, satisfying $\left|[\theta_i^S(t)]_p\right| \leq 5$ for any $p = 1, \ldots, 4$ and

$\|\dot{\theta}_i^S(t)\|_1 \leq 50, S \in \{x, y\}$; and $\phi_i(X_i) = (\tanh(x_i),\ \tanh(\dot{x}_i),\ \tanh(y_i),\ \tanh(\dot{y}_i))^T$

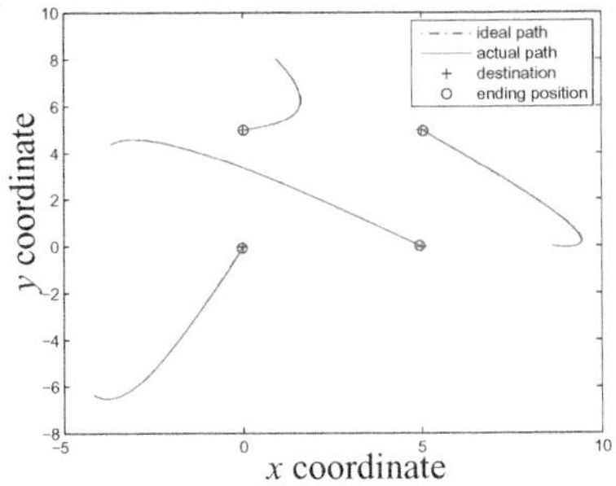

Fig. 8.2 The ideal path and the actual path of 4 agents.

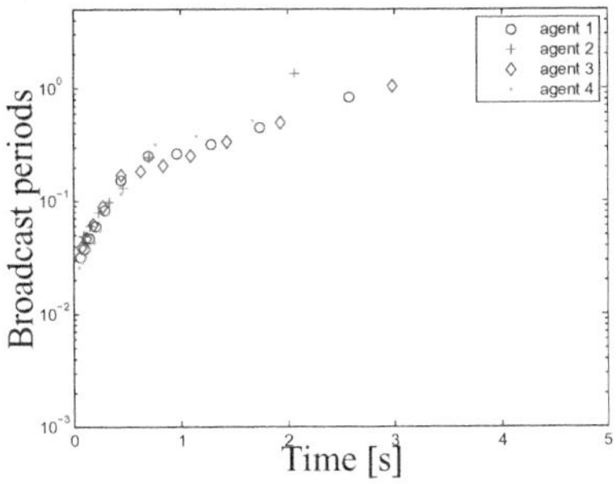

Fig. 8.3 Broadcast periods generated by agents.

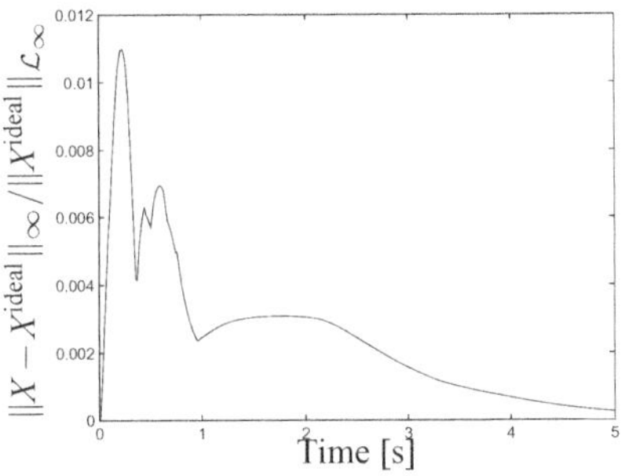

Fig. 8.4 The difference in the overall states of two systems, $\|X - X^{\text{ideal}}\|_\infty / \|X^{\text{ideal}}\|_{\mathcal{L}_\infty}$.

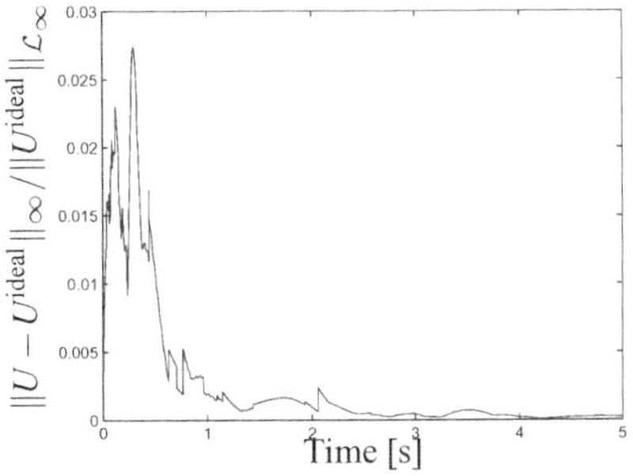

Fig. 8.5 The difference in the overall inputs of two systems, $\|U - U^{\text{ideal}}\|_\infty / \|U^{\text{ideal}}\|_{\mathcal{L}_\infty}$.

The ideal controller for agent i, U_i^{ideal}, is

$$U_i^{\text{ideal}} = K_i X_i^{\text{ideal}} + \sum_{j \in \mathcal{N}_i} K_{ij} X_j^{\text{ideal}} - \begin{pmatrix} \theta_i^x(t) \\ \theta_i^y(t) \end{pmatrix} \phi_i(X_i^{\text{ideal}})$$

where the feedback gain

$$K_i = \begin{pmatrix} -3.45 & -4.34 & 0 & 0 \\ 0 & 0 & -3.45 & -4.34 \end{pmatrix}$$

$$K_{ij} = \begin{pmatrix} 0.16 & 0.16 & 0 & 0 \\ 0 & 0 & 0.16 & 0.16 \end{pmatrix}, \quad \forall\, i \text{ and } j \neq i$$

is obtained using the distributed linear quadratic regulator approach from [2]. The initial state is randomly selected. In the simulations, we use $\|\hat{e}(t)\|_\infty = 1$ to trigger the broadcast. For the embedded $\mathcal{L}_1$ controller, we set $\Gamma_i = 10^5$, $C_i(s) = (30/s + 30) \cdot \mathbb{I}_{4\times 4}$. It is easy to verify that inequalities (8.17), (8.33) and (8.37) hold.

We first set $N = 4$ and run the event-triggered adaptive control system for 5 seconds. The results are plotted in Figs. 8.2–8.5. Figure 8.2 plots agents' ideal motion path and the actual path, which are almost identical. Figures 8.4 and 8.5 show the difference of the overall states and control inputs in these two systems, respectively. We compute the relative errors

$$\frac{||X - X^{\text{ideal}}||_{\infty}}{||X^{\text{ideal}}||_{\mathcal{L}_\infty}}, \qquad \frac{||U - U^{\text{ideal}}||_{\infty}}{||U^{\text{ideal}}||_{\mathcal{L}_\infty}}$$

of the difference in states and inputs, respectively. It shows that the errors in the states and the inputs are less than 1.2% and 3% of the $||X_{\text{ideal}}||_{\mathcal{L}_\infty}$ and $||U_{\text{ideal}}||_{\mathcal{L}_\infty}$, respectively. These errors can be further reduced by deceasing the thresholds in the local events and increasing the local adaptation gain in each agent.

Figure 8.3 plots the broadcast periods of the agents that are defined by $t_i[k+1] - t_i[k]$. It shows that the periods vary a lot. These results indicate that the communication resource might be largely reduced by using event-triggering. Also note that the broadcast periods are getting larger, as the states approach the equilibrium. This is because the plant is linear time-invariant, and the threshold in the event is a constant. In this case, when the state approaches its equilibrium, less information is required to keep the state inside a small neighborhood of the equilibrium. More detailed discussion on this relationship can be found in [1].

We then add transmission delays into the system to examine the robustness of the scheme to the delays. The initial condition is the same as that in the first experiment. The delay in each transmission is a random variable, uniformly distributed over $[0, 0.3]$. Simulation results are shown in Figs. 8.6 – 8.9. The difference in signals becomes a little larger compared with the nondelay case, but it is still acceptably small. It implies that for this specific example, the scheme is robust to the transmission delays.

Finally, we vary the number of agents from 3 to 80 to examine the scalability of this scheme. We compute the normalized $\mathcal{L}_\infty$-norms of the errors in the states and the inputs of two systems, defined by

$$\frac{||X - X^{\text{ideal}}||_{\mathcal{L}_\infty}}{||X^{\text{ideal}}||_{\mathcal{L}_\infty}} \text{ and } \frac{||U - U^{\text{ideal}}||_{\mathcal{L}_\infty}}{||U^{\text{ideal}}||_{\mathcal{L}_\infty}}$$

respectively. Figures 8.10–8.11 plot the number of agents versus $\mathcal{L}_\infty$ norm of the error signals and shows that the errors between the states/inputs of these two systems remain on the same level as the number of agents increases. It implies that in this simulation, the gap between the real system and the ideal system scales well with respect to the number of agents.

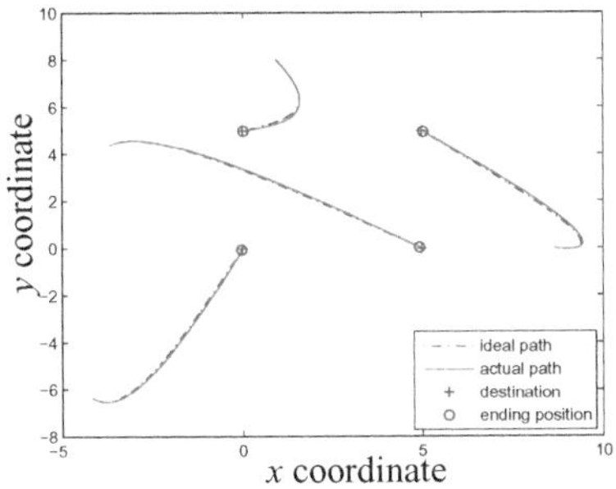

Fig. 8.6 The ideal path and the actual path of 4 agents when transmission delays exist.

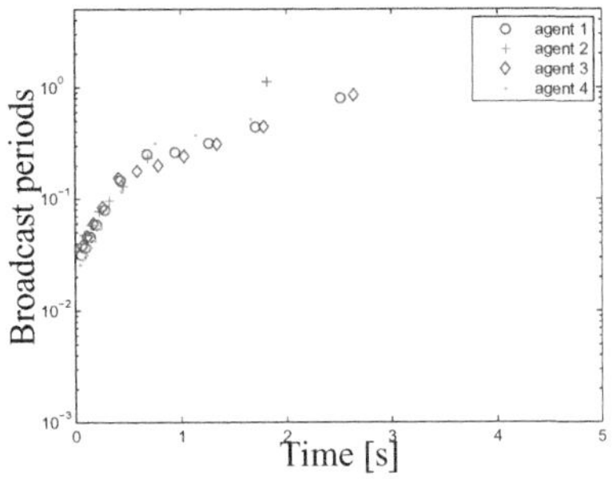

Fig. 8.7 Broadcast periods generated by agent 1 when transmission delays exist.

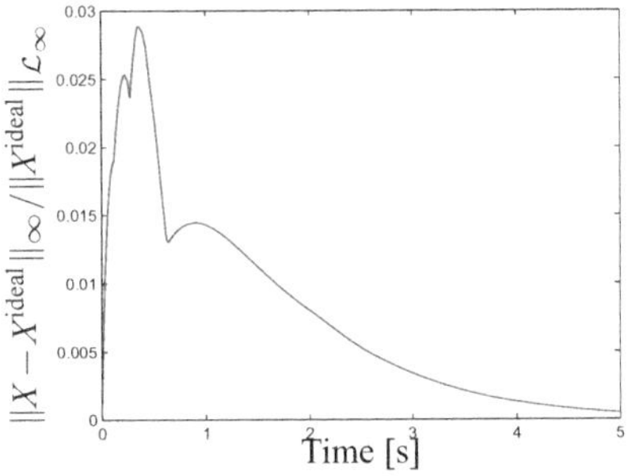

Fig. 8.8 The difference in the overall states of two systems, $||X - X^{\mathrm{ideal}}||_\infty / ||X^{\mathrm{ideal}}||_{\mathcal{L}_\infty}$ when transmission delays exist.

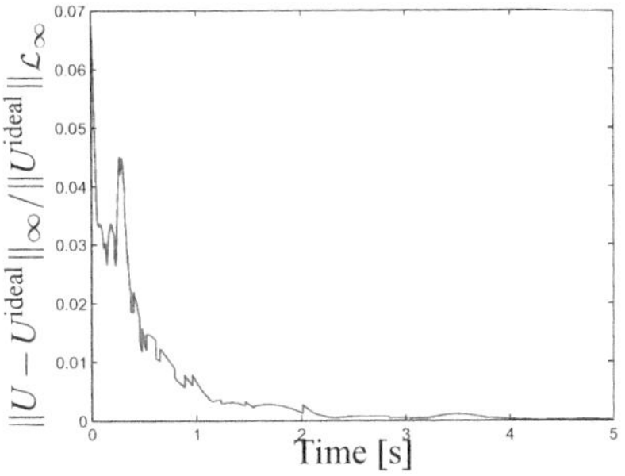

Fig. 8.9 The difference in the overall inputs of two systems, $||U - U^{\mathrm{ideal}}||_\infty / ||U^{\mathrm{ideal}}||_{\mathcal{L}_\infty}$ when transmission delays exist.

8.6 Conclusions

This chapter presented an event-triggering and $\mathcal{L}_1$ adaptation co-design scheme that can be used to predict the performance of networked control systems in the presence of uncertainties and communication constraints. Performance bounds are derived on the difference between the signals in the ideal system and the real system. These bounds can be arbitrarily reduced by deceasing the thresholds in local events and increasing the local adaptation gains, as well as the bandwidths of the embedded low-pass filters. Simulation results show that this co-design scheme seems to be robust to the transmission delays and scale well with the number of agents. Rigorous proofs of these features, and also other features like quantization and delays, will be reported in forthcoming publications.

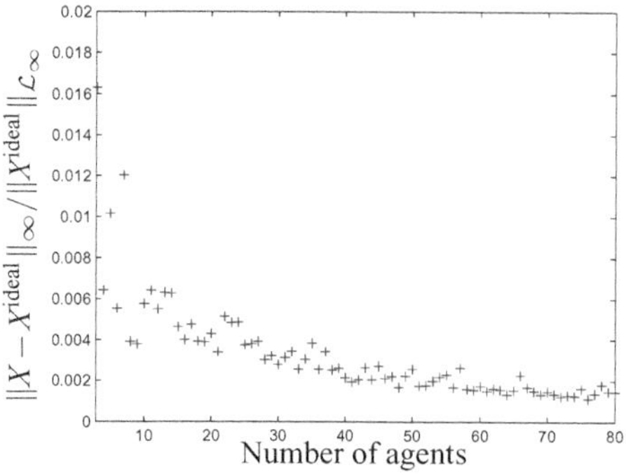

Fig. 8.10 The number of agents versus $\|X - X^{\text{ideal}}\|_{\mathcal{L}_\infty}/\|X^{\text{ideal}}\|_{\mathcal{L}_\infty}$

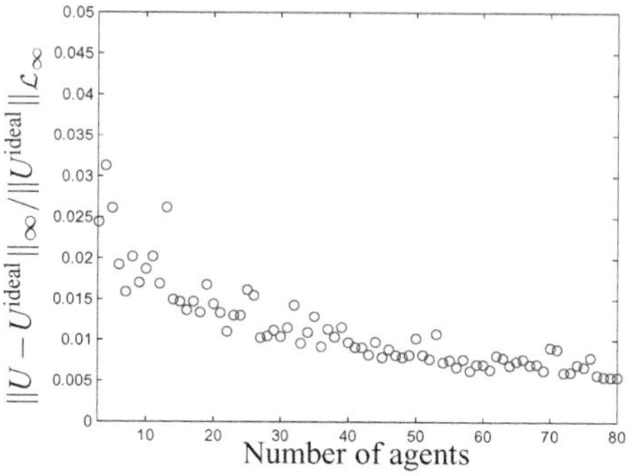

Fig. 8.11 The number of agents versus $\|U - U^{\text{ideal}}\|_{\mathcal{L}_\infty}/\|U^{\text{ideal}}\|_{\mathcal{L}_\infty}$

8.7 Proofs

8.7.1 Proof of Lemma 8.1

Proof. *We first show the boundedness of* $\tilde{x}_i = x_i^{\text{iden}} - x_i$. *From Eqs.* (8.1) *and* (8.14), *we have*

$$\dot{\tilde{x}}_i = A_i^m \tilde{x}_i + B_i \tilde{\theta}_i^T(t)\phi_i(x_i) \tag{8.50}$$

where $\tilde{\theta}_i(t) = \hat{\theta}_i(t) - \theta_i(t)$.

Consider $\dot{V}_i$ *with* $V_i(\tilde{x}_i, \tilde{\theta}_i) = \tilde{x}_i^T P_i \tilde{x}_i + \Gamma_i^{-1}\text{tr}(\tilde{\theta}_i^T\tilde{\theta}_i)$ *and* P_i *defined in Eq.* (8.5)*:*

$$\begin{aligned}
\dot{V}_i &= 2\tilde{x}_i^T P_i \dot{\tilde{x}}_i + 2\Gamma_i^{-1}\text{tr}(\tilde{\theta}_i^T \dot{\tilde{\theta}}_i) \\
&= -\tilde{x}_i^T Q_i \tilde{x}_i + 2\tilde{x}_i^T P_i B_i \tilde{\theta}_i(t)\phi_i(x_i) + 2\Gamma_i^{-1}\text{tr}\left(\tilde{\theta}_i^T \dot{\hat{\theta}}_i - \tilde{\theta}_i^T \dot{\theta}_i\right).
\end{aligned}$$

Since the adaptive law from (8.14) *implies*

$$\tilde{x}_i^T P_i B_i \tilde{\theta}_i(t)\phi_i(x_i) + \Gamma_i^{-1}\text{tr}\left(\tilde{\theta}_i^T \dot{\hat{\theta}}_i\right) \le 0$$

and $\theta_i(t) \in \Omega_i$, $\sqrt{\text{tr}\left(\dot{\theta}_i^T(t)\dot{\theta}_i(t)\right)} \le d_i^\theta$, *we have*

$$\begin{aligned}
\dot{V}_i &\le -\tilde{x}_i^T Q_i \tilde{x}_i - 2\Gamma_i^{-1}\text{tr}\left(\tilde{\theta}_i^T \dot{\theta}_i\right) \\
&\le -\tilde{x}_i^T Q_i \tilde{x}_i - 2\Gamma_i^{-1}\sqrt{\text{tr}\left(\tilde{\theta}_i^T \tilde{\theta}_i\right)}\sqrt{\text{tr}\left(\dot{\theta}_i^T \dot{\theta}_i\right)} \\
&\le -\sigma_{\min}(Q_i)\|\tilde{x}_i\|_2^2 + 4\Gamma_i^{-1} d_i^\theta \max_{\theta_i \in \Omega_i} \sqrt{\text{tr}\left(\theta_i^T \theta_i\right)}.
\end{aligned}$$

Since $V(0) = \tilde{\theta}_i^T(0)\Gamma^{-1}\tilde{\theta}_i(0) \leq 4\theta_i^{\max 2}\Gamma_i^{-1}$, *the inequality above implies*

$$\|\tilde{x}_i(t)\|_\infty \leq \|\tilde{x}_i(t)\|_2 \leq \frac{\psi_i}{\sqrt{\Gamma_i}}, \quad \forall t \geq 0. \tag{8.51}$$

We now consider the bound on $\|x_i^{\text{des}} - x_i\|_{\mathcal{L}_\infty^{[0,\tau]}}$. *Let* $\tilde{x}_i^{\text{des}} = x_i^{\text{des}} - x_i$. *According to equations* (8.1) *and* (8.11)*, the error dynamics are*

$$\begin{aligned}
\tilde{x}_i^{\text{des}}(s) &= H_i(s)\left(-C_i(s)\hat{\eta}_i(s) + \eta_i(s)\right) \\
&= H_i(s)\left(-C_i(s)\hat{\eta}_i(s) + C_i(s)\eta_i(s) - C_i(s)\eta_i(s) + \eta_i(s)\right) \\
&= H_i(s)C_i(s)(\eta_i(s) - \hat{\eta}_i(s)) + H_i(s)(\mathbb{I} - C_i(s))\eta_i(s) \\
&= C_i(s)H_i(s)(\eta_i(s) - \hat{\eta}_i(s)) + H_i(s)(\mathbb{I} - C_i(s))\eta_i(s),
\end{aligned}$$

where $\eta_i(s)$, $\hat{\eta}_i(s)$ *are the Laplace transforms of* $\theta_i(t)\phi_i(x_i)$ *and* $\hat{\theta}_i(t)\phi_i(x_i)$, *respectively. Therefore,*

$$\begin{aligned}
\|\tilde{x}_i^{\text{des}}\|_{\mathcal{L}_\infty^{[0,\tau]}} &\leq \|C_i(s)H_i(s)(\eta_i(s) - \hat{\eta}_i(s))\|_{\mathcal{L}_\infty^{[0,\tau]}} + \|H_i(s)(\mathbb{I} - C_i(s))\eta_i(s)\|_{\mathcal{L}_\infty^{[0,\tau]}} \\
&\leq \|C_i(s)\|_{\mathcal{L}_1}\|H_i(s)(\eta_i(s) - \hat{\eta}_i(s))\|_{\mathcal{L}_\infty^{[0,\tau]}} + \|G_i(s)\|_{\mathcal{L}_1}\|\eta_i(s)\|_{\mathcal{L}_\infty^{[0,\tau]}}
\end{aligned} \tag{8.52}$$

From (8.50)*, we have*

$$\tilde{x}_i(s) = H_i(s)(\hat{\eta}_i(s) - \eta_i(s)). \tag{8.53}$$

This implies

$$\|H_i(s)(\hat{\eta}_i(s) - \eta_i(s))\|_{\mathcal{L}_\infty^{[0,\tau]}} \leq \|\tilde{x}_i\|_{\mathcal{L}_\infty^{[0,\tau]}} \leq \frac{\psi_i}{\sqrt{\Gamma_i}} \tag{8.54}$$

where the last inequality is obtained by using (8.51).

Consider $\|\eta_i(s)\|_{\mathcal{L}_\infty^{[0,\tau]}}$. *From* (8.2) *and* (8.3)*, we know*

$$\begin{aligned}
\|\eta_i(s)\|_{\mathcal{L}_\infty^{[0,\tau]}} &= \sup_{0 \leq t < \tau} \|\theta_i(t)\phi_i(x_i(t))\|_\infty \\
&\leq \theta_i^{\max}(L_i\|x_i\|_{\mathcal{L}_\infty^{[0,\tau]}} + M_i) \\
&\leq \theta_i^{\max}(L_i\|\tilde{x}_i^{\text{des}}\|_{\mathcal{L}_\infty^{[0,\tau]}} + L_i\|x_i^{\text{des}}\|_{\mathcal{L}_\infty^{[0,\tau]}} + M_i)
\end{aligned} \tag{8.55}$$

Based on equation (8.11)*, we have*

$$\|x_i^{\text{des}}\|_{\mathcal{L}_\infty^{[0,\tau]}} \leq \|H_i(s)\|_{\mathcal{L}_1}\|\chi_i^{x_i}\|_{\mathcal{L}_\infty^{[0,\tau]}} \leq \|H_i(s)\|_{\mathcal{L}_1}\|\chi_i^{x_i}\|_{\mathcal{L}_\infty^{[0,\tau]}}$$

Applying this inequality into (8.55) *implies*

$$\|\eta_i(s)\|_{\mathcal{L}_\infty^{[0,\tau]}} \leq \theta_i^{\max}(L_i\|\tilde{x}_i^{\text{des}}\|_{\mathcal{L}_\infty^{[0,\tau]}} + L_i\|H_i(s)\|_{\mathcal{L}_1}\|\chi_i^{x_i}\|_{\mathcal{L}_\infty^{[0,\tau]}} + M_i) \tag{8.56}$$

Combining inequalities (8.52)*,* (8.54) *and* (8.56) *yields*

$$\|\tilde{x}_i^{\text{des}}\|_{\mathcal{L}_\infty^{[0,\tau]}} \leq \frac{\|C_i(s)\|_{\mathcal{L}_1}\psi_i}{(1-\lambda_i)\sqrt{\Gamma_i}} + \frac{\lambda_i}{1-\lambda_i}\|H_i(s)\|_{\mathcal{L}_1}\|\chi_i^{x_i}\|_{\mathcal{L}_\infty^{[0,\tau]}} + \frac{\|G_i(s)\|_{\mathcal{L}_1}\theta_i^{\max}M_i}{1-\lambda_i}.$$

The bound on $u_i^{\text{des}} - u_i$ *can be obtained by studying* (8.13) *and* (8.11)*:*

$$\begin{aligned} u_i^{\text{des}}(s) - u_i(s) &= K_i x_i^{\text{des}}(s) - \eta_i^{\text{des}}(s) - K_i x_i(s) + C_i(s)\hat{\eta}_i(s) \\ &= K_i\tilde{x}_i^{\text{des}} - \eta_i^{\text{des}}(s) + C_i(s)\eta_i^{\text{des}}(s) - C_i(s)\eta_i^{\text{des}}(s) \\ &\quad + C_i(s)\eta_i(s) - C_i(s)\eta_i + C_i(s)\hat{\eta}_i(s) \end{aligned}$$

where $\eta_i^{\text{des}}(s)$ *is the Laplace transform of* $\theta_i(t)\phi_i(x_i^{\text{des}}(t))$*. This inequality implies*

$$\begin{aligned} &\|u_i^{\text{des}}(s) - u_i(s)\|_{\mathcal{L}_\infty^{[0,\tau]}} \qquad (8.57)\\ &\leq \|K_i\|_\infty\|\tilde{x}_i^{\text{des}}\|_{\mathcal{L}_\infty^{[0,\tau]}} + \|C_i(s)\eta_i^{\text{des}}(s) - \eta_i^{\text{des}}(s)\|_{\mathcal{L}_\infty^{[0,\tau]}} \\ &\quad + \|C_i(s)\eta_i(s) - C_i(s)\eta_i^{\text{des}}(s)\|_{\mathcal{L}_\infty^{[0,\tau]}} + \|C_i(s)\hat{\eta}_i(s) - C_i(s)\eta_i\|_{\mathcal{L}_\infty^{[0,\tau]}} \\ &\leq \|K_i\|_\infty\|\tilde{x}_i^{\text{des}}\|_{\mathcal{L}_\infty^{[0,\tau]}} + \|C_i(s) - \mathbb{I}\|_{\mathcal{L}_1}\theta_i^{\max}(L_i\|H_i(s)\|_{\mathcal{L}_1}\|\chi_i^{x_i}\|_{\mathcal{L}_\infty^{[0,\tau]}} + M_i) \\ &\quad + \|C_i(s)\|_{\mathcal{L}_1}\theta_i^{\max}L_i\|\tilde{x}_i^{\text{des}}\|_{\mathcal{L}_\infty^{[0,\tau]}} + \|C_i(s)\hat{\eta}_i(s) - C_i(s)\eta_i\|_{\mathcal{L}_\infty^{[0,\tau]}}. \end{aligned}$$

To obtain the bound on $\|C_i(s)\hat{\eta}_i(s) - C_i(s)\eta_i\|_{\mathcal{L}_\infty^{[0,\tau]}}$*, consider* (8.53)*. Since* $c_i^o H_i(s)$ *has full rank, we have*

$$C_i(s)(c_i^o H_i(s))^{-1}c_i^o\tilde{x}_i(s) = C_i(s)(\hat{\eta}_i(s) - \eta_i(s)).$$

Because $c_i^o H_i(s)$ *has stable zeros and relative degree 1, the preceding inequality implies*

$$\begin{aligned} \|C_i(s)(\hat{\eta}_i(s) - \eta_i(s))\|_{\mathcal{L}_\infty^{[0,\tau]}} &\leq \|C_i(s)(c_i^o H_i(s))^{-1}c_i^o\|_{\mathcal{L}_1}\|\tilde{x}_i(s)\|_{\mathcal{L}_\infty^{[0,\tau]}} \\ &\leq \|C_i(s)(c_i^o H_i(s))^{-1}c_i^o\|_{\mathcal{L}_1}\frac{\psi_i}{\sqrt{\Gamma_i}}. \end{aligned}$$

Applying this bound into (8.57) *yields* (8.24). □

8.7.2 Proof of Lemma 8.2

Proof. *We first show that* $\|x^{\text{ideal}}\|_{\mathcal{L}_\infty}$ *is bounded. From equation* (8.9)*, we have*

$$\begin{aligned} \|x^{\text{ideal}}\|_{\mathcal{L}_\infty^{[0,\tau]}} \leq \|H(s)\|_{\mathcal{L}_1}(\|g(x^{\text{ideal}}, x^{\text{ideal}})\|_{\mathcal{L}_\infty^{[0,\tau]}} + \|r\|_{\mathcal{L}_\infty^{[0,\tau]}}) \\ + \|(s\mathbb{I} - A_m)^{-1}x(0)\|_{\mathcal{L}_\infty^{[0,\tau]}}. \end{aligned} \qquad (8.58)$$

Notice that equation (8.32) *implies*

$$\|g(x^{\text{ideal}},x^{\text{ideal}})\|_{\mathcal{L}_\infty^{[0,\tau]}} \leq a\|x^{\text{ideal}}\|_{\mathcal{L}_\infty^{[0,\tau]}} + \|g(0,0)\|_\infty. \tag{8.59}$$

Applying this inequality into inequality (8.58), *with the condition in* (8.33), *we have*

$$\|x^{\text{ideal}}\|_{\mathcal{L}_\infty^{[0,\tau]}} \leq \frac{\|H(s)\|_{\mathcal{L}_1}(\|g(0,0)\|_\infty + \|r\|_{\mathcal{L}_\infty}) + \|(s\mathbb{I}-A_m)^{-1}x(0)\|_{\mathcal{L}_\infty^{[0,\tau]}}}{1-a\|H(s)\|_{\mathcal{L}_1}} \triangleq \xi_1.$$

Next, we show that $\|x\|_{\mathcal{L}_\infty}$ is bounded, using a contradiction argument. It is obvious that $\|x(0)\|_\infty \leq \xi_2$. Suppose that inequality (8.39) *does not hold. Then there must exist a time instant $\tau \geq 0$ and a positive constant $\bar{\xi} > \xi_2$ such that*

$$\|x(\tau)\|_\infty = \bar{\xi} > \xi_2 \tag{8.60}$$

$$\|x(t)\|_\infty \leq \bar{\xi}, \quad \forall t \in [0,\tau]. \tag{8.61}$$

As the hypotheses in Lemma 8.1 are satisfied, we obtain

$$\begin{aligned}\|x_i - x_i^{\text{des}}\|_{\mathcal{L}_\infty^{[0,\tau]}} \leq \varepsilon_i = {} & \frac{\theta_i^{\max} M_i \|G_i(s)\|_{\mathcal{L}_1}}{1-\lambda_i} \\ & + \frac{\|C_i(s)\|_{\mathcal{L}_1}}{1-\lambda_i}\frac{\psi_i}{\sqrt{\Gamma_i}} + \frac{\lambda_i}{1-\lambda_i}\|H_i(s)\|_{\mathcal{L}_1}\|\chi_i^{x_i}\|_{\mathcal{L}_\infty^{[0,\tau]}}\end{aligned} \tag{8.62}$$

for all $i \in \mathcal{N}$, which implies

$$\begin{aligned}\|e\|_{\mathcal{L}_\infty^{[0,\tau]}} & \leq \zeta + \max_{i\in\mathcal{N}}\left\{\frac{\lambda_i}{1-\lambda_i}\|H_i(s)\|_{\mathcal{L}_1}\|\chi_i^{x_i}\|_{\mathcal{L}_\infty^{[0,\tau]}}\right\} \\ & \leq \zeta + \max_{i\in\mathcal{N}}\left\{\frac{\lambda_i}{1-\lambda_i}\right\}\|H(s)\|_{\mathcal{L}_1}\|\chi^x\|_{\mathcal{L}_\infty^{[0,\tau]}}\end{aligned} \tag{8.63}$$

and ζ, χ^x are defined in equations (8.42), (8.44), *respectively.*

Since Assumption 8.1 holds, we have inequality (8.33), *which with inequality* (8.33) *implies the satisfaction of inequality* (8.34). *Combining inequalities* (8.34) *and the inequality above yields*

$$\begin{aligned}\|x - x^{\text{ideal}}\|_{\mathcal{L}_\infty^{[0,\tau]}} & \leq \|x - x^{\text{des}}\|_{\mathcal{L}_\infty^{[0,\tau]}} + \|x^{\text{des}} - x^{\text{ideal}}\|_{\mathcal{L}_\infty^{[0,\tau]}} \\ & \leq \|e\|_{\mathcal{L}_\infty^{[0,\tau]}} + \frac{\|H(s)\|_{\mathcal{L}_1}}{1-a\|H(s)\|_{\mathcal{L}_1}}\left(a\|e\|_{\mathcal{L}_\infty^{[0,\tau]}} + b\|\hat{e}\|_{\mathcal{L}_\infty^{[0,\tau]}}\right) \\ & \leq \frac{\|H(s)\|_{\mathcal{L}_1}}{1-a\|H(s)\|_{\mathcal{L}_1}} b\|\hat{e}\|_{\mathcal{L}_\infty^{[0,\tau]}} + \frac{1}{1-a\|H(s)\|_{\mathcal{L}_1}}\|e\|_{\mathcal{L}_\infty^{[0,\tau]}}\end{aligned}$$

Applying inequalities (8.35) *and* (8.63) *into the preceding inequality implies*

$$\|x - x^{\text{ideal}}\|_{\mathcal{L}_\infty^{[0,\tau]}} \leq \frac{\|H(s)\|_{\mathcal{L}_1} \max_{i\in\mathcal{N}} \rho_i}{1 - a\|H(s)\|_{\mathcal{L}_1}} + \frac{1}{1 - a\|H(s)\|_{\mathcal{L}_1}} \left(\zeta + \max_{i\in\mathcal{N}} \left\{\frac{\lambda_i}{1-\lambda_i}\right\} \|H(s)\|_{\mathcal{L}_1} \|\chi^x\|_{\mathcal{L}_\infty^{[0,\tau]}}\right) \tag{8.64}$$

Notice that

$$\begin{aligned}
\|\chi^x\|_{\mathcal{L}_\infty^{[0,\tau]}} &= \|g(x,\hat{x}) + r\|_{\mathcal{L}_\infty^{[0,\tau]}} \\
&\leq \|g(x,\hat{x}) - g(0,0)\|_{\mathcal{L}_\infty^{[0,\tau]}} + \|g(0,0)\|_\infty + \|r\|_{\mathcal{L}_\infty^{[0,\tau]}} \\
&\leq a\|x\|_{\mathcal{L}_\infty^{[0,\tau]}} + b\|\hat{e}\|_{\mathcal{L}_\infty^{[0,\tau]}} + \|g(0,0)\|_\infty + \|r\|_{\mathcal{L}_\infty^{[0,\tau]}} \\
&\leq a\|x\|_{\mathcal{L}_\infty^{[0,\tau]}} + \max_{i\in\mathcal{N}} \rho_i + \|g(0,0)\|_\infty + \|r\|_{\mathcal{L}_\infty^{[0,\tau]}},
\end{aligned}$$

where the last inequality comes from inequality (8.32).

Applying the preceding inequality into inequality (8.64) *yields*

$$\|x\|_{\mathcal{L}_\infty^{[0,\tau]}} \leq \xi_2, \tag{8.65}$$

which contradicts equation (8.60). *Therefore* $\|x\|_{\mathcal{L}_\infty} \leq \xi_2$ *holds.* □

8.7.3 Proof of Theorem 8.1

Proof. *We first consider the bound on* $\|x - x^{\text{ideal}}\|_{\mathcal{L}_\infty}$. *Since the inequality*

$$\|x - x^{\text{ideal}}\|_{\mathcal{L}_\infty^{[0,\tau]}} \leq \|x - x^{\text{des}}\|_{\mathcal{L}_\infty^{[0,\tau]}} + \|x^{\text{des}} - x^{\text{ideal}}\|_{\mathcal{L}_\infty^{[0,\tau]}} \tag{8.66}$$

holds, the bound on $\|x - x^{\text{ideal}}\|_{\mathcal{L}_\infty}$ *can be easily obtained by applying inequalities* (8.21), (8.34) *and* (8.43) *into the preceding inequality.*

We now consider the bound on $\|u - u^{\text{ideal}}\|_{\mathcal{L}_\infty}$. *By equations* (8.9) *and* (8.11), *we have*

$$\begin{aligned}
u^{\text{des}}(t) - u^{\text{ideal}}(t) = K(x^{\text{des}}(t) - x^{\text{ideal}}(t)) + f(t, x^{\text{des}}(t)) - f(t, x^{\text{ideal}}(t)) \\
+ g(x(t), \hat{x}(t)) - g(x^{\text{ideal}}(t), x^{\text{ideal}}(t)),
\end{aligned}$$

which together with inequalities (8.2), (8.32) *implies*

$$\begin{aligned}
\|u^{\text{des}} - u^{\text{ideal}}\|_{\mathcal{L}_\infty^{[0,\tau]}} \leq \|K\|_\infty \|x^{\text{des}} - x^{\text{ideal}}\|_{\mathcal{L}_\infty^{[0,\tau]}} + \max_{i\in\mathcal{N}}\{\theta_i^{\max} L_i\} \|x^{\text{des}} - x^{\text{ideal}}\|_{\mathcal{L}_\infty^{[0,\tau]}} \\
+ a\|x - x^{\text{ideal}}\|_{\mathcal{L}_\infty^{[0,\tau]}} + b\|\hat{e}\|_{\mathcal{L}_\infty^{[0,\tau]}}.
\end{aligned}$$

Applying inequalities (8.34), (8.35), *and* (8.45) *into the preceding inequality yields*

$$\|u^{\text{des}} - u^{\text{ideal}}\|_{\mathcal{L}_\infty^{[0,\tau]}} \leq (\|K\|_\infty + \max_{i\in\mathcal{N}}\{\theta_i^{\max} L_i\}) \frac{\|H(s)\|_{\mathcal{L}_1}\left(a\|e\|_{\mathcal{L}_\infty^{[0,\tau]}} + \max_{i\in\mathcal{N}} \rho_i\right)}{1 - a\|H(s)\|_{\mathcal{L}_1}}$$
$$+ a\xi_3 + \max_{i\in\mathcal{N}} \rho_i.$$

Since the hypotheses in Lemma 8.1 are satisfied, inequalities (8.21) *and* (8.22) *hold. We therefore have*

$$\|u - u^{\text{ideal}}\|_{\mathcal{L}_\infty^{[0,\tau]}} \leq \|u - u^{\text{des}}\|_{\mathcal{L}_\infty^{[0,\tau]}} + \|u^{\text{des}} - u^{\text{ideal}}\|_{\mathcal{L}_\infty^{[0,\tau]}} \leq \xi_4.$$

□

8.7.4 Proof of Corollary 8.1

Proof. *Since the hypotheses in Theorem 8.1 hold, we have* $\|u - u^{\text{ideal}}\|_{\mathcal{L}_\infty} \leq \xi_4$. *By equation* (8.10), $\|u^{\text{ideal}}\|_{\mathcal{L}_\infty}$ *is bounded and therefore* $\|u\|_{\mathcal{L}_\infty}$ *is also bounded, i.e., there exists a positive constant* ρ_u *such that* $\|u\|_{\mathcal{L}_\infty} \leq \rho_u$. *Also note that by Lemma 8.2,* $\|x\|_{\mathcal{L}_\infty} \leq \xi_2$.

Consider $\frac{d}{dt}\|x_i(t) - x_i(t_i[k])\|_2$ *for* $t \geq t_i[k]$.

$$\begin{aligned}
\rho_i^k = \frac{d}{dt}\|x_i(t) - x_i(t_i[k])\|_2 &\leq \|\dot{x}_i(t)\|_2 \\
&\leq \|A_i x_i(t) + B_i(u_i(t) + \theta_i(t)\phi_i(x_i(t)))\|_2 \\
&\leq \|A_i\|_2\|x_i(t)\|_2 + \|B_i\|_2\|u_i(t)\|_2 + \|B_i\|_2\|\theta_i(t)\|_2\|\phi_i(x_i)\|_2 \\
&\leq \|A_i\|_2 n\|x_i(t)\|_\infty + \|B_i\|_2 m\|u_i(t)\|_\infty + \|B_i\|_2 \max_{\theta_i\in\Omega_i}\|\theta_i\|_2 l\|\phi_i(x_i)\|_\infty \\
&\leq \|A_i\|_2 n\xi_2 + \|B_i\|_2 m\rho_u + \|B_i\|_2 \max_{\theta_i\in\Omega_i}\|\theta_i\|_2 l(L_i\xi_2 + M_i) = \rho_{\dot{x}_i}
\end{aligned}$$

Solving this differential inequality with the initial condition, $\|x_i(t_i[k]) - x_i(t_i[k])\|_2 = 0$, *implies for any* $t \geq t_i[k]$,

$$\|x_i(t) - x_i(t_i[k])\|_\infty \leq \|x_i(t) - x_i(t_i[k])\|_2 \leq \rho_{\dot{x}_i}(t - t_i[k]).$$

Since $t_i[k+1]$ *is the time instant when*

$$\|x_i(t_i[k+1]) - x_i(t_i[k])\|_\infty = \frac{\rho_i}{b}$$

holds, we have

$$\frac{\rho_i}{b} = \|x_i(t_i[k+1]) - x_i(t_i[k])\|_\infty \leq \rho_{\dot{x}_i}(t_i[k+1] - t_i[k]),$$

which means

$$t_i[k+1] - t_i[k] \geq \frac{\rho_i}{b\rho_{\dot{x}_i}} = T_i.$$

□

Acknowledgements This research was supported by AFOSR under Contract No. FA9550-09-1-0265.

References

1. Anta, A., Tabuada, P.: Self-triggered stabilization of homogeneous control systems. In: Proc.American Control Conference (ACC2008), pp. 4129–4134. Seattle, WA (2008)
2. Borrelli, F., Keviczky, T.: Distributed LQR design for identical dynamically decoupled systems. IEEE Trans. Automatic Control **53**(8), 1901–1912 (2008)
3. Cao, C., Hovakimyan, N.: Design and analysis of a novel $\mathcal{L}_1$ adaptive control architecture with guaranteed transient performance. IEEE Trans. Automatic Control **53**(2), 586–590 (2008)
4. Cao, C., Hovakimyan, N.: Stability margin of $\updownarrow_1$ adaptive control architecture. IEEE Trans. Automatic Control **55**(2), 480–487 (2010)
5. Cervin, A., Åström, K.: On limit cycles in event-based control systems. In: Proc. IEEE Conf. Decision and Control (CDC2007), pp. 3190–3195. New Orleans, LA (2007)
6. Cortes, J., Martinez, S., Karatas, T., Bullo, F.: Coverage control for mobile sensing networks. IEEE Trans. Robotics and Automation **20**(2), 243–255 (2004)
7. Dimarogonas, D.V., Johansson, K.H.: Event-triggered control for multi-agent systems. In: Proc. 48th IEEE Conf. Decision and Control (CDC2009), pp. 7131–7136. Shanghai, China (2009)
8. Egerstedt, M., Hu, X.: Formation constrained multi-agent control. IEEE Trans. Robotics and Automation **17**(6), 947–950 (2001)
9. Gregory, I., Xargay, E., Cao, C., Hovakimyan, N.: Flight test of $\mathcal{L}_1$ adaptive controller on the NASA AirSTAR flight test vehicle. In: Proc. AIAA Guidance, Navigation and Control Conference. Toronto, Canada (2010)
10. Hovakimyan, N., Cao, C.: $\mathcal{L}_1$ Adaptive Control Theory: Guaranteed Robustness with Fast Adaptation. SIAM, Philadelphia, PA (2010)
11. Khalil, H.: Nonlinear Systems. Prentice Hall, Upper Saddle River, NJ (2002)
12. Li, L., Lemmon, M., Wang, X.: Optimal event triggered transmission of information in distributed state estimation problems. In: American Control Conference (ACC2010). Baltimore, MD (2010)
13. Lian, F.L., Moyne, J., Tilbury, D.: Network design consideration for distributed control systems. IEEE Trans. Control Systems Technology **10**(2), 297–307 (2002)
14. Mazo, M., Tabuada, P.: On event-triggered and self-triggered control over sensor/actuator networks. In: Proc. 47th IEEE Conf. Decision and Control (CDC2008), p. 435 440. Cancun, Mexico (2008)
15. Murray, R.: Recent research in cooperative control of multivehicle systems. Journal of Dynamic Systems, Measurement, and Control **129**, 571 (2007)
16. Nesić, D., Teel, A.: Input-output stability properties of networked control systems. IEEE Trans. Automatic Control **49**, 1650–1667 (2004)
17. Olfati-Saber, R.: Flocking for multi-agent dynamic systems: Algorithms and theory. IEEE Trans. Automatic Control **51**(3), 401 (2006)
18. Olfati-Saber, R., Murray, R.: Consensus problems in networks of agents with switching topology and time-delays. IEEE Trans. Automatic Control **49**(9), 1520–1533 (2004)
19. Pomet, J., Praly, L.: Adaptive nonlinear regulation: estimation from the Lyapunov equation. IEEE Trans. Automatic Control **37**(6), 729–740 (1992)

20. Rabi, M., Johansson, K., Johansson, M.: Optimal stopping for event-triggered sensing and actuation. In: Proc. IEEE Conf. Decision and Control (CDC2008), pp. 3607–3612. Cancun, Mexico (2008)
21. Tabuada, P.: Event-triggered real-time scheduling of stabilizing control tasks. IEEE Trans. Automatic Control **52**(9), 1680–1685 (2007)
22. Wan, P., Lemmon, M.: An event-triggered distributed primal-dual algorithm for network utility maximization. In: Proc. 48th IEEE Conf. Decision and Control (CDC2009), pp. 5863–5868. Shanghai, China (2009)
23. Wang, X., Hovakimyan, N.: Performance prediction in uncertain networked control systems using $\mathcal{L}_1$-adaptation-based distributed event-triggering. In: Proc. IEEE Conf. Decision and Control (CDC2010), p. 7570 7575. Atlanta, GA (2010)
24. Wang, X., Lemmon, M.: Finite-gain $\mathcal{L}_2$ stability in distributed event-triggered networked control systems with data dropouts. In: European Control Conference (ECC2009). Budapest, Hungary (2009)
25. Wang, X., Lemmon, M.: Self-triggered feedback systems with state-independent disturbances. In: Proc. American Control Conference (ACC2009), pp. 3842–3847. St. Louis, MO (2009)
26. Wang, X., Lemmon, M.: Event-triggering in distributed networked systems with data dropouts and delays. In: Hybrid Systems: Computation and Control (HSCC2009), pp. 366–380. San Francisco, CA (2009)
27. Wang, X., Lemmon, M.: Self-triggered feedback control systems with finite-gain $\mathcal{L}_2$ stability. IEEE Trans. Automatic Control **54**(3), 452–467 (2009)
28. Xu, Y., Hespanha, J.: Optimal communication logics in networked control systems. In: Proc. IEEE Conf. Decision and Control (CDC2004), pp. 3527–3532. Paradise Island, Bahamas (2004)

Chapter 9
Weight Determination by Manifold Regularization

Henrik Ohlsson and Lennart Ljung

Abstract A new type of linear kernel smoother is derived and studied. The smoother, referred to as *weight determination by manifold regularization*, is the solution to a regularized least squares problem. The regularization avoids overfitting and can be used to express prior knowledge of an underlying smooth function. An interesting property of the kernel smoother is that it is well suited for systems governed by the semi-supervised smoothness assumption. Several examples are given to illustrate this property. We also discuss why these type of techniques can have a potential interest for the system identification community.

9.1 Introduction

A central problem in many scientific areas is linking certain observations to each other and building *models* for how they relate. In loose terms, the problem could be described as relating y to φ in

$$y = f_0(\varphi) \tag{9.1}$$

where φ is a vector of observed variables, a *regressor* vector, and y is a characteristic of interest, an *output*. In system identification φ could be observed past behavior of a dynamical system and y the predicted next output.

Observations are often imperfect or noisy, and we are therefore led to consider

Henrik Ohlsson
Division of Automatic Control, Department of Electrical Engineering, Linköping University, SE-583 37 Linköping, Sweden
e-mail: ohlsson@isy.liu.se

Lennart Ljung
Division of Automatic Control, Department of Electrical Engineering, Linköping University, SE-583 37 Linköping, Sweden
e-mail: ljung@isy.liu.se

R. Johansson and A. Rantzer (Eds.): Distributed Decision Making and Control, LNCIS 417, pp. 195–214.
springerlink.com

$$y = f_0(\varphi) + e, \quad e \sim \mathcal{N}(0, \sigma^2). \tag{9.2}$$

Assume now that a set of observations, $\{(\varphi_t, y_t)\}_{t=1}^{N_e}$, of how f_0 transforms φ is available. $f_0 : \mathcal{R}^{n_\varphi} \to \mathcal{R}$ is itself unknown. The conventional approach within system identification is to make use of a parametric expression $f(\varphi_t, \theta)$, which is we hope is flexible enough to imitate the transformation f_0. $f(\varphi_t, \theta)$ is adjusted to the observations by choosing θ as

$$\hat{\theta} = \arg\min_{\theta} \sum_{t=1}^{N_e} l(y_t - f(\varphi_t, \theta)). \tag{9.3}$$

$l : \mathcal{R} \to \mathcal{R}$ is here a measure of how well the model predicts the estimation data $\{(\varphi_t, y_t)\}_{t=1}^{N_e}$ e.g.—l being chosen as a norm.

There are a number of parametric expressions and of varying flexibility, and to choose a model structure just flexible enough is crucial when Eq. (9.3) is used. Let e.g.

$$f(\varphi, \theta) = \theta \tag{9.4}$$

and $l(\cdot) = (\cdot)^2$ in (9.3). $\hat{\theta}$ and $f(\varphi, \hat{\theta})$ then become the mean of the observed outputs $\sum_{t=1}^{N_e} y_t / N_e$. This model has of course very good predictive abilities if f_0 is constant but has otherwise rather limited abilities to produce satisfying predictions.

The other extreme, and of particular interest in this chapter, would be to use a parameter for each of the φ_t we will work with. Let $\mathcal{D}$ denote that set of regressors. $\mathcal{D}$ is typically larger than the set of regressors in the estimation set. (If nothing else, we have occasion to compute the response value $f_0(\varphi)$ at new points). Let Θ be a parameter vector of the same size as the number of elements in $\mathcal{D}$:

$$\text{card}(\mathcal{D}) = \dim \Theta \tag{9.5}$$

We can then associate each parameter value in Θ with a response value $f_0(\varphi_t)$ for any $\varphi_t \in \mathcal{D}$: for convenience denote the elements of Θ by f_t. This particular model hence takes the form

$$f(\varphi_t, \Theta) = f_t, \quad \forall \varphi_t \in \mathcal{D}. \tag{9.6}$$

Remark 9.1 (Nonparametric Model). Somewhat misleading, a model for which the number of parameters grows with the number of estimation data is called a *nonparametric model*. The model given in (9.6) is hence a nonparametric model.

9.2 Supervised, Semi-Supervised and Unsupervised Learning

Before continuing it is useful to introduce the notions of *supervised*, *semi-supervised* and *unsupervised learning*.

The term *supervised learning* is used for algorithms for which the construction of $f(\varphi, \hat{\theta})$ is "supervised" by the measured information in y. In contrast to this, *unsupervised learning* has only the information of the regressors $\{\varphi_t, t = 1, \ldots, N_e\}$. In unsupervised classification, e.g. [10], the classes are constructed by various clustering techniques. *Manifold learning*, e.g. [26, 25] deals with unsupervised techniques to construct a manifold in the regressor space that houses the observed regressors.

Semi-supervised algorithms are less common. In semi-supervised algorithms, both (y, φ)-pairs and φs, for which no output has been observed, are used to construct the model $f(\varphi, \theta)$. This is particularly interesting if extra effort is required to obtain y. Thus, costly (y, φ)-pairs are supported by less costly regressors to improve the result. It is clear that unsupervised and semi-supervised algorithms are of interest only if the regressors have a pattern that is unknown a priori.

Semi-supervised learning is an active area within classification and machine learning (see [4, 33] and references therein). The main reason that semi-supervised algorithms are not often seen in regression and system identification may be that it is less clear when regressors alone can be of use. We will try to bring some clarity to this through this chapter. Generally, it could be said that regression problems having regressors constrained to rather limited regions in the regressor space may be suitable for a semi-supervised regression algorithm. It is also important that regressors be available and comparably "cheap" to get as opposed to the (y, φ)-pairs.

9.3 Cross Validation and Regularization

Let us now return to the model given in (9.6). If we let $l(\cdot) = (\cdot)^2$ again and $\mathcal{D} = \{\varphi_1, \dots, \varphi_{N_e}\}$, the criterion of fit (9.3) now takes the form

$$\hat{\Theta} = (\hat{f}_1, \hat{f}_2, \dots, \hat{f}_{N_e}) = \arg \min_{f_1, f_2, \dots, f_{N_e}} \sum_{t=1}^{N_e} (y_t - f_t)^2 = (y_1, y_2, \dots, y_{N_e}) \tag{9.7}$$

$f(\varphi_t, \hat{\Theta})$ hence "succeeds" to perfectly fit the estimation data. If there were no measurement noise polluting the observations, this would be good. However, with measurement noise present, obtaining a perfect fit is not desirable and termed *overfitting*. Overfitting is a problem for very flexible models and to chose a model structure just flexible enough to imitate f_0 (and not flexible enough to being able to imitate the noise) would be ideal.

There are a number of approaches to discovering what is "just flexible enough." Most approaches can be seen belonging to either *cross validation* or *regularization*.

In *cross validation*, a new data set $\{(\varphi_t, y_t)\}_{t=1}^{N_v}$ is utilized to avoid overfitting. The data set

$$\{(\varphi_t, y_t)\}_{t=1}^{N_v}$$

is denoted the *validation data set*. Since measurement noise e of the validation data set is impossible to predict, the best scenario possible would be perfectly predicting the outcome of the deterministic part of (9.2) i.e., $f_0(\varphi)$. Therefore, for a number of candidate structures $f_i(\varphi, \hat{\theta}_i)$, $i = 1, \dots, m$ ($\hat{\theta}$ found using (9.3)), a model is chosen by

$$\arg \min_{f_i(\varphi,\hat{\theta}_i), i=1,\dots,m} \sum_{t=1}^{N_v} l(y_t - f_i(\varphi_t, \hat{\theta}_i)). \tag{9.8}$$

To evaluate (9.8) we need to compute predictions for f_0 at regressors not included in the estimation data set i.e., $f_i(\varphi_t, \hat{\theta}_i)$, $t = 1, \dots, N_v$. For the model $f(\varphi_t, \theta) = \theta$, see (9.4), this is straightforward. $f(\varphi_t, \hat{\theta})$, $t = 1, \dots, N_v$ are simply equal to $\sum_{t=1}^{N_e} y_t/N_e$. For the model given in (9.6), however, it is not trivial and we will discuss this in the next section.

In *regularization*, a cost on flexibility is added to the criterion of fit. $f(\varphi_t, \theta)$ is now adjusted to the observations by choosing θ as

$$\hat{\theta} = \arg\min_{\theta} \sum_{t=1}^{N_e} l(y_t - f(\varphi_t, \theta)) + \lambda J(\theta, \varphi_t) \tag{9.9}$$

rather than using (9.3). $J(\theta, \varphi_t)$ serves as a cost on flexibility and is often used to penalize nonsmooth estimates. λ is seen as a design parameter and regulates the trade-off between fit to the estimation data and smoothness. Choosing the "just flexible enough" model structure is now transformed to choosing the right λ value.

For the model proposed in (9.6), a suitable regularizer is

$$J(\Theta, \varphi_t) = \sum_{t=1}^{N_e} \left(f_t - \sum_{s=1}^{N_e} \frac{k(\varphi_t, \varphi_s) f_s}{\sum_{r=1}^{N_e} k(\varphi_t, \varphi_r)} \right)^2 \tag{9.10}$$

$k : \mathcal{R}^{n_\varphi} \times \mathcal{R}^{n_\varphi} \rightarrow \mathcal{R}$ is a *kernel*. The regularizer (9.10) makes sure that close-by regressors are transformed in a similar way by $f(\varphi, \Theta)$ and therefore reassures smoothness. This remedies the overfitting problem that (9.6) was previously suffering from.

There are a number of different kernels that could be of interest to use in (9.10). Some of the most interesting kernels are the *squared exponential*, *KNN* and *LLE* kernels. The details of these kernels are outlined in Appendix 9.10.

9.4 Generalization

For most practical purposes it is not enough to find a model $f(\varphi, \theta)$ that well imitates f_0 at $\{(\varphi_t, y_t)\}_{t=1}^{N_e}$. Generalization to nonobserved data is often of more importance. This is denoted the model's ability to *generalize* to unseen data.

Let φ_* be an unseen regressor, i.e.$\varphi_* \neq \varphi_t, t = 1, \dots, N_e$,. For the simple model (9.4),

$$f(\varphi, \hat{\theta}) = \sum_{t=1}^{N_e} y_t/N_e \tag{9.11}$$

generalization is trivial since the prediction does not depend on the regressor. The estimate for $f(\varphi_*)$, $\varphi_* \neq \varphi_t$, $t = 1, \dots, N_e$, is simply taken as $\sum_{t=1}^{N_e} y_t/N_e$.

For the model (9.6),

$$f(\varphi_t, \hat{\Theta}) = \hat{f}_t, \quad t = 1, \ldots, N_e \tag{9.12}$$

with

$$\hat{\Theta} = \arg \min_{f_1, f_2, \ldots, f_{N_e}} \sum_{t=1}^{N_e} (y_t - f_t)^2 + \lambda \sum_{t=1}^{N_e} \left(f_t - \sum_{s=1}^{N_e} \frac{k(\varphi_t, \varphi_s) f_s}{\sum_{r=1}^{N_e} k(\varphi_t, \varphi_r)} \right)^2 \tag{9.13}$$

generalization is a bit more involved. The most natural way is to introduce a new parameter f_* for the estimate of $f_0(\varphi_*)$ and let the smoothness implied by the regularization give an estimate

$$f(\varphi_*, \hat{\Theta}) = \hat{f}_* \tag{9.14}$$

with

$$\hat{\Theta} = \arg \min_{f_1, f_2, \ldots, f_{N_e}, f_*} \sum_{t=1}^{N_e} (y_t - f_t)^2 + \lambda \sum_{\varphi_t \in \mathcal{D}} \left(f_t - \sum_{\varphi_s \in \mathcal{D}} \frac{k(\varphi_t, \varphi_s) f_s}{\sum_{\varphi_r \in \mathcal{D}} k(\varphi_t, \varphi_r)} \right)^2. \tag{9.15}$$

$\mathcal{D}$ now contains both the estimation data and φ_*, i.e.$\mathcal{D} = \{\varphi_1, \varphi_2, \ldots, \varphi_{N_e}, \varphi_*\}$. Since Eq. (9.15) is quadratic in the optimization variables, an explicit solution can be computed. Introduce first the notation

$$\mathbf{J} \triangleq [I_{N_e \times N_e} \; 0_{N_e \times 1}], \; \mathbf{y} \triangleq [y_1 y_2 \ldots y_{N_e}]^T,$$

$$\hat{\mathbf{f}} \triangleq [\hat{f}_1 \, \hat{f}_2 \ldots \hat{f}_{N_e} \, \hat{f}_*]^T, \; \bar{k}(\varphi_t, \varphi_s) \triangleq \frac{k(\varphi_t, \varphi_s)}{\sum_{\varphi_r \in \mathcal{D}} k(\varphi_t, \varphi_r)},$$

$$\mathbf{K} \triangleq \begin{bmatrix} \bar{k}(\varphi_1, \varphi_1) & \bar{k}(\varphi_1, \varphi_2) & \ldots & \bar{k}(\varphi_1, \varphi_{N_e}) & \bar{k}(\varphi_1, \varphi_*) \\ \bar{k}(\varphi_2, \varphi_1) & \bar{k}(\varphi_2, \varphi_2) & & \bar{k}(\varphi_2, \varphi_{N_e}) & \bar{k}(\varphi_2, \varphi_*) \\ \vdots & & \ddots & & \vdots \\ \bar{k}(\varphi_{N_e}, \varphi_1) & \bar{k}(\varphi_{N_e}, \varphi_2) & \ldots & \bar{k}(\varphi_{N_e}, \varphi_{N_e}) & \bar{k}(\varphi_{N_e}, \varphi_*) \\ \bar{k}(\varphi_*, \varphi_1) & \bar{k}(\varphi_*, \varphi_2) & \ldots & \bar{k}(\varphi_*, \varphi_{N_e}) & \bar{k}(\varphi_*, \varphi_*) \end{bmatrix}$$

Equation (9.15) can then be written as

$$(\mathbf{y} - \mathbf{J}\hat{\mathbf{f}})^T (\mathbf{y} - \mathbf{J}\hat{\mathbf{f}}) + \lambda (\hat{\mathbf{f}} - \mathbf{K}\hat{\mathbf{f}})^T (\hat{\mathbf{f}} - \mathbf{K}\hat{\mathbf{f}}) \tag{9.16}$$

which expands into

$$\hat{\mathbf{f}}^T \left(\lambda (\mathbf{I} - \mathbf{K})^T (\mathbf{I} - \mathbf{K}) + \mathbf{J}^T \mathbf{J} \right) \hat{\mathbf{f}} - 2\hat{\mathbf{f}}^T \mathbf{J}^T \mathbf{y} + \mathbf{y}^T \mathbf{y} \tag{9.17}$$

Setting the derivative with respect to $\hat{\mathbf{f}}$ to zero and solving gives

$$\hat{f}_* = \mathbf{e}_* \left(\lambda (\mathbf{I} - \mathbf{K})^T (\mathbf{I} - \mathbf{K}) + \mathbf{J}^T \mathbf{J} \right)^{-1} \mathbf{J}^T \mathbf{y}, \quad \mathbf{e}_* \triangleq [0_{1 \times N_e} 1] \tag{9.18}$$

It is straightforward to do the generalization for more than one unobserved regressor at a time. $\mathcal{D}$ used in (9.15) then indexes all regressors, the N_e estimation regressors and the regressors for which we seek an estimate of the function value but have not observed the output. We will refer to the method outlined in (9.18) to as *weight determination by manifold regularization* (WDMR [20], see also [17, 18]). The reason for the name will become clear later.

Proposition 9.1 (Linear Kernel Smoother). *The estimate given in* (9.18) *can be rewritten in the form*

$$f(\varphi_*, \hat{\Theta}) = \sum_{t=1}^{N_e} w_t y_t \tag{9.19}$$

and is therefore a linear estimator, since it is linear in the estimation outputs. The estimate given in (9.18) *is also a kernel smoother since it is constructed using kernels. These two combined make WDMR a linear kernel smother (see, e.g.,[8], p. 129). For WDMR w_t is given by*

$$w_t = \mathbf{e}_* \left(\lambda (\mathbf{I} - \mathbf{K})^T (\mathbf{I} - \mathbf{K}) + \mathbf{J}^T \mathbf{J}\right)^{-1} \mathbf{J}^T \mathbf{e}_t^T \tag{9.20}$$

with $\mathbf{e}_t = [0_{1\times t-1} \; 1 \; 0_{1\times N_e - t}]$. *The expression for constructing the weights w in* (9.19) *is referred to as the equivalent kernel in literature (see, e.g., [8] p. 170).*

Notice that the resulting estimates coming from estimating the function-value of unobserved regressors one-by-one and all at the same time will not be the same. This is a property of semi-supervised regression approaches. The regularization will make sure that the estimated f_t varies smoothly on regressor-dense regions. We will return to this property later and discuss when it can be useful.

9.5 WDMR and the Nadaraya–Watson Smoother

In the linear kernel smoother WDMR the kernel was used to provide a smoothness prior. A kernel can also be used to obtain an estimate for $f_0(\varphi_*)$, via

$$f(\varphi_*) = \sum_{t=1}^{N_e} \frac{k(\varphi_*, \varphi_t) y_t}{\sum_{r=1}^{N_e} k(\varphi_*, \varphi_r)} \tag{9.21}$$

which also is a linear kernel smoother. This is referred to as the *Nadaraya–Watson smoother* or *estimator* [13, 29]. It may seem a bit overcomplicated to, as in WDMR, use a kernel as, for example, smoothness prior, but in the end we nevertheless end up with a linear kernel smoother. What is achieved by using a kernel in WDMR compared to using the kernel direct in the Nadaraya–Watson smoother as in Eq. (9.21)?

Remark 9.2. Note that the Nadaraya–Watson smoother weights are used together with noisy observations $\{y_t\}_{t=1}^{N_e}$ to obtain an estimate $f(\varphi_*)$. To reduce the influence of noise, y_t in (9.21) could itself be replaced by an estimate of $f_0(\varphi_t)$ by using (9.21) a second time, i.e.,

$$f(\varphi_t) = \frac{1}{\sum_{r=1}^{N_e} k(\varphi_t,\varphi_r) + k(\varphi_t,\varphi_*)} \Big(\sum_{s=1}^{N_e} k(\varphi_t,\varphi_s) y_s + k(\varphi_t,\varphi_*) f(\varphi_*) \Big) \tag{9.22}$$

If all noisy observations are replaced, we obtain the system of equations

$$f(\varphi_t) = \sum_{\varphi_s \in \mathcal{D}} \frac{k(\varphi_t,\varphi_s) f(\varphi_s)}{\sum_{\varphi_r \in \mathcal{D}} k(\varphi_t,\varphi_r)}, \quad \forall \varphi_t \in \mathcal{D}, \mathcal{D} = \{\varphi_1,\varphi_2,\dots,\varphi_{N_e},\varphi_*\} \tag{9.23}$$

which takes a familiar form (see the regularization in (9.15)). The regularization in WDMR represents that we wish to obtain an estimate satisfying this system of equations.

Remark 9.3. The resulting estimates coming from estimating the function value at unobserved regressors one-by-one and all at the same time will not be the same for WDMR. This is a property of *semi-supervised* regression approaches. The regularization will make sure that the estimated f_ts vary smoothly on regressor-dense regions. The Nadaraya–Watson smoother is not a semi-supervised approach, and estimating one function-value at a time or all at the same time would give the same result.

To start to examine the advantages and disadvantages of WDMR compared to the Nadaraya–Watson smoother, let us look at an example.

Example 9.1. Nadaraya-Watson Smoother versus WDMR

Let us now consider a standard test example from [14], "the Narendra-Li example":

$$x_{t+1} = \left(\frac{x_t}{1+x_t^2} + 1 \right) \sin(z_t) \tag{9.24a}$$

$$z_{t+1} = z_t \cos(z_t) + x_t \exp\left(-\frac{x_t^2+z_t^2}{8} \right) + \frac{u_t^3}{1+u_t^2+0.5\cos(x_t+z_t)} \tag{9.24b}$$

$$y_t = \frac{x_t}{1+0.5\sin(z_t)} + \frac{z_t}{1+0.5\sin(x_t)} + e_t \tag{9.24c}$$

This dynamical system was simulated with 2000 samples using a random binary input, giving input-output data $\{y_t, u_t, t = 1,\dots,2000\}$. A separate set of 200 validation data was also generated with a sinusoidal input. To get an idea of the complexity of the system, see Fig. 9.1, which shows these validation data. were also generated with a sinusoidal input. The chosen regression vector was

$$\varphi_t = \begin{bmatrix} y_{t-1} & y_{t-2} & y_{t-3} & u_{t-1} & u_{t-2} & u_{t-3} \end{bmatrix}^T \tag{9.25}$$

Let us use the squared exponential kernel (see Appendix 9.10) and apply the Nadaraya–Watson smoother and WDMR ($\lambda = 0.0001$ was used in (9.20)) to es-

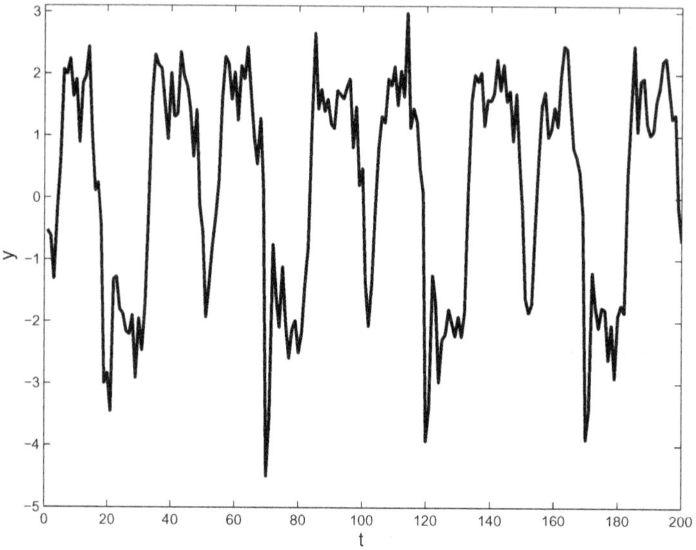

Fig. 9.1 Validation data for the Narendra-Li example.

timate the function values at the validation regressors. The result is given in Table 9.1. A length scale (see Appendix 9.10 for the definition of length scale) of 0.6 and 0.7 gave the best performing Nadaraya–Watson smoother and WDMR, respectively. The table also give the fit for a neural network (a single layer sigmoid network with 23 units in the System Identification Toolbox [11] gave the best performance) and the prediction given by guessing that the next f_0 value will equal the previous observation.

The direct usage of the squared exponential kernel in the Nadaraya–Watson smoother does very well compared to the neural network and guessing that the next f_0 value will be equal the previous observation. However, WDMR does even better. As mentioned earlier (see Remark 9.2), WDMR has a hierarchical scheme for denoising the observations. One may therefore wonder if enlarging the bandwidth/length scale in the Nadaraya–Watson smoother would have the same denoising effect. Figure 9.2 shows that doing so is not that easy and that enlarging the bandwidth/length scale in fact does not help the Nadaraya–Watson smoother. It is also interesting to examine what happens if the measurement noise changes. Table 9.2 shows the result of an experiment where the noise level was decreased in three steps. We see that as noise level decreased, the difference in performance between the Nadaraya-Watson smoother and WDMR disappeared.

So far we have only applied WDMR to a batch of data. We claimed earlier that applying WDMR to a batch of data is not the same as applying it to the regressors one-by-one. Unfortunately, the advantage seen for the batch tends to disappear when the latter is done. WDMR and the Nadaraya–Watson smoother then perform similarly. There are many possible reasons for this. One is that in the batch setting,

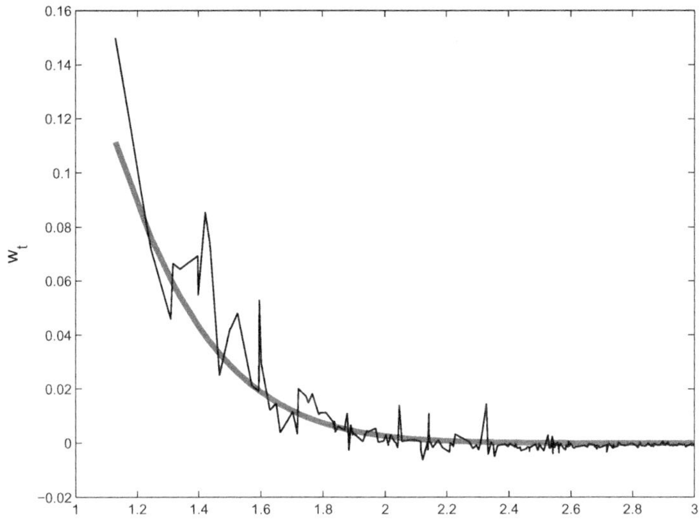

Fig. 9.2 w_i of (9.20) (thin black line) plotted as a function of $|\varphi_t - \varphi_*|, t = 1 \ldots, N_e$ and φ_* being one of the validation regressors. The thick grey line shows the corresponding weights of the Nadaraya–Watson smoother.

Table 9.1 Mean fit over 20 noise and input realization data for the Nadaraya-Watson smoother and WDMR using a squared exponential kernel, a neural network and the estimate obtained by simply taking the previous output as an estimate for the next function value.

Algorithm	Mean fit (%)
Nadaraya-Watson (squared exponential, $l = 0.6$)	68
WDMR (squared exponential, $l = 0.7$, $\lambda = 10^{-4}$)	71
Neural network (23 units)	66
Last measurement	47

Table 9.2 The Nadaraya–Watson (NW) smoother and WDMR's performance for three different noise levels. Both algorithms used a squared exponential kernel and were tuned for each of the noise levels for optimal performance.

Algorithm	Fit ($\sigma^2 = 0.5$)	Fit ($\sigma^2 = 0.1$)	Fit ($\sigma^2 = 0.05$)	Fit ($\sigma^2 = 0.01$)
Nadaraya–Watson ($l = 0.8, 0.6, 0.6, 0.5$)	56	69	72	74
WDMR ($l = 0.8, 0.7, 0.7, 0.7$, $\lambda = 10^{-4}, 10^{-4}, 0.8 \cdot 10^{-4}, 0.5 \cdot 10^{-4}$)	60	71	73	74

the regularization in WDMR ensures that the estimate varies smoothly over regions of regressors. So-called boundary effects have however been observed. This means that the estimates for regressors at the end of dense regions often are worse than estimates for regressors surrounded by many other regressors. Boundary effects are a known issue of kernel smoothers, see e.g.[8, p. 168]. This supports that batch would do better than one-by-one. Unfortunately, this means that WDMR is less interesting for other than batch and nonlinear finite impulse response (FIR) models. In the next section, we will only discuss a batch approach.

9.6 The Semi-Supervised Smoothness Assumption

WDMR does not only have good denoising properties, it can also provide desirable properties when it comes to problems of regressors confined to limited regions e.g.manifolds, in the regressor space. Let us illustrate this by a pictorial example.

Consider the five regressors shown on the left of Fig. 9.3. For four of the regressors the output has been observed and their outputs written next to them. One regressor output is unknown. To estimate the function value at that regressor, we could use the Nadaraya–Watson smoother and compute the average of the two closest regressors' outputs, which would give an estimate of 2.5. Let us now add the information that the regressors and the outputs were sampled from a continuous-time process and that the value of the regressor was evolving along the curve shown on the right of Fig. 9.3. Knowing this, a better estimate of the function value would probably be 1. Knowing that the regressors are restricted to a certain region in the regressor space can hence make us reconsider our estimation strategy.

In regression we are interested in finding estimates for the conditional distribution $p(f|\varphi)$. For the regressors without observed output to be useful, it is required that the regressor distribution $p(\varphi)$ bring information concerning the conditional

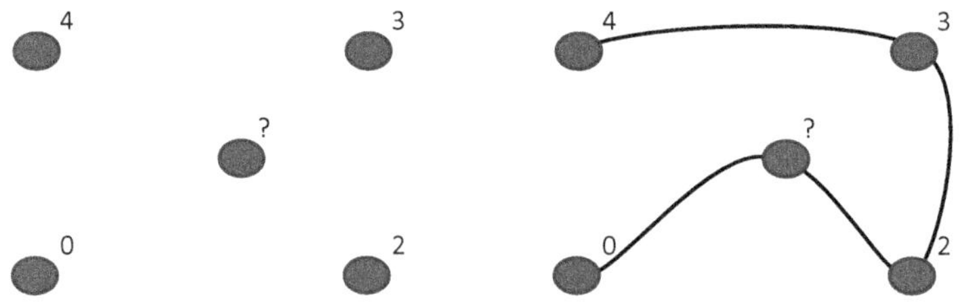

Fig. 9.3 Illustration of the information value of regressor manifolds: The left side shows five regressors, four with measured outputs and one with unknown output marked by '?'. If these regressors have no known underlying structure or ordering, one would interpolate the unknown output from its neighbors to 2.5 or so. On the other hand, if a one-dimensional manifold can be associated with the regressors along the indicated line on the right hand side, the regressors have a certain ordering. Then it is natural to interpolate the unknown value from its new neighbors to 1.

$p(f|\varphi)$. We saw from the pictorial example that one situation for which this is the case is that in which we make the assumption that the sought function value changes continuously along high-density areas in the regressor space. This assumption is referred to as the *semi-supervised smoothness assumption* [4]:

Assumption 9.1 (Semi-Supervised Smoothness). *If two regressors φ_1, φ_2 in a high-density region are close, then $f_0(\varphi_1)$ and $f_0(\varphi_2)$ should be closed.*

"High density region" is a somewhat loose term: In many cases it corresponds to a manifold in the regressor space, such that the regressors for the application in question are confined to this manifold. That two regressors are "close" then means that the distance between them along the manifold (the geodesic distance) is small.

In classification, this smoothness assumption is interpreted as meaning the class labels should be the same in the high density regions. In regression, we interpret this as a slowly varying function along high density regions. Note that in regression, it is common to assume that the function value varies smoothly in the regressor space; the semi-supervised smoothness assumption is less conservative since it only assumes smoothness in the high density regions in the regressor space. Two regressors could be close in the regressor space metric, but far apart along the high-density region (the manifold): think of the region being a spiral in the regressor space.

One may discuss how common it is in system identification that the regressors are constrained to a manifold. The input signal part of the regression vector should according to identification theory be "persistently exciting," which is precisely the opposite of being constrained. However, in many biological applications and in DAE (differential algebraic equation) modeling such structural constraints are frequently occurring.

9.6.1 A Comparison between the Nadaraya-Watson Smoother and WDMR Using the KNN Kernel

To illustrate the advantage of WDMR under the semi-supervised smoothness assumption, we continue to discuss the previous pictorial example. We now add 5 regressors with unobserved output to the 5 previously considered. Hence, we have 10 regressors, 4 with observed outputs and 6 with unobserved outputs, and we desire an estimate of the output marked with a question mark in Fig. 9.4. The left of Fig. 9.4 shows how the Nadaraya–Watson smoother solves the estimation problem if the KNN kernel (see Appendix 9.10) is used. The kernel will cause the searched function value to be similar to the observed outputs of the K closest regressors. On the right of Fig. 9.4, WDMR with the KNN kernel is used. This kernel grants estimates of the K closest regressors (observed or unobserved output) to be similar. Since the regressors closest to the regressor for which we search the function value, are unobserved, information is propagated from the observed regressors towards the one for which we search a function value estimate along the chain of unobserved regressors. The shaded regions in both the left and right part of the figure symbol-

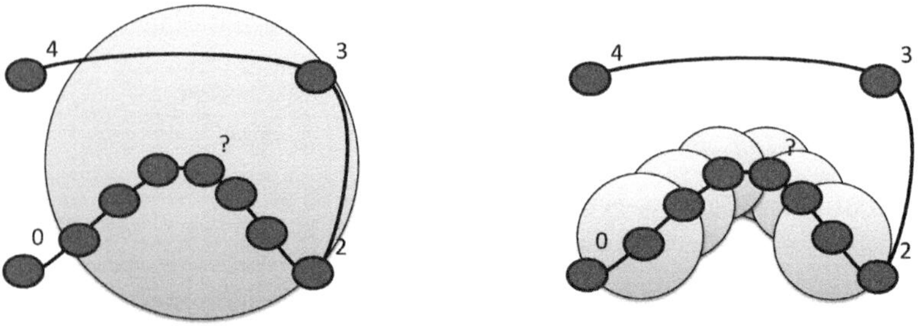

Fig. 9.4 An illustration of the difference of using the Nadaraya-Watson smoother (left part of the figure) and WDMR (right part of the figure) with the KNN kernel.

ize the way information is propagated using the Nadaraya–Watson smoother and WDMR. In the left part of the figure we will therefore obtain an estimate equal to 2.5 while in the right we get an estimate equal to 1.

The ability of WDMR to account for manifolds in the regressor space and the semi-supervised smoothness assumption is a rather unique property of a kernel smoother and the reason it is called "weight determination by manifold regularization."

9.7 Related Approaches

Semi-supervised learning has been around since the 1970s (some earlier attempts exist). *Fisher's linear discriminant rule* was then discussed under the assumption that each of the class conditional densities was Gaussian. *Expectation maximization* was applied using both regressor output pairs and regressors to find the parameters of the Gaussian densities [9]. During the 1990s, interest in semi-supervised learning increased, mainly due to its application to text classification, see e.g. [16]. The first usage of the term *semi-supervised learning*, as it is used today, was not until 1992 [12].

The boost in the area of manifold learning in the 1990s brought with it a number of semi-supervised methods. Semi-supervised manifold learning is a type of semi-supervised learning used for nonlinear dimensionality reduction. Most of the algorithms are extensions of unsupervised manifold learning algorithms [2, 31, 15, 24, 23, 20, 32]. Another interesting contribution is found in the developments by Rahimi in [21]. A time series of regressors, some with measured outputs and some not, are considered there. The series of estimates best fitting the given outputs and at the same time satisfying some temporal smoothness assumption is then computed.

Most of the references above are to semi-supervised classification algorithms. They are however relevant since most semi-supervised classification methods can, with minor modifications, be applied to regression problems. The modification or the application to regression problems are however almost never discussed or exemplified. For more historical notes on semi-supervised learning, see [4].

Similar methods to WDMR have also been discussed; see e.g. [7, 19, 31, 3, 2, 28]. In [31], manifold learning is discussed with construction of a semi-supervised version of the manifold learning technique known as *locally linear embedding* (LLE, [25]), which coincides with a particular choice of kernel in (9.15). Combining LLE with system identification was also discussed in [19]. In [7], graph-based semi-supervised methods for classification were studied, deriving an objective function similar to (9.15). References [3, 28] discuss a classification method called *label propagation* which is an iterative approach converging to (9.15). In [2], support vector machines were extended to work under the semi-supervised smoothness assumption. There is also a huge literature on kernel smoothers; see e.g. [8].

9.8 Examples

The following two examples are regression problems for which the semi-supervised smoothness assumption is motivated. Data presented and analyzed here were collected for the present article.

9.8.1 Example 1—Functional Magnetic Resonance Imaging

Functional magnetic resonance imaging (fMRI) is a technique that measures brain activity [30]—fMRI measurements reveal the degree of oxygenation in the blood via the blood oxygenation level dependent (BOLD) response. The degree of oxygenation reflects the neural activity in the brain, so fMRI is indirect measure of brain activity.

Measurements of brain activity can be acquired with fMRI as often as once a second and are given as an array, each element giving a scalar measure of the average activity in a small volume element of the brain. These volume elements are commonly called *voxels* (short for volume pixel) and they can be as small as 1 mm^3. The fMRI measurements were heavily affected by noise.

We considered measurements from an $8 \times 8 \times 2$ array covering parts of the visual cortex gathered with a sampling period of 2 seconds. To remove noise, data were prefiltered by applying a spatial and temporal filter with a squared exponential kernel. The filtered fMRI measurements at each time t were vectorized into the regression vector φ_t. fMRI data were acquired during 240 seconds (giving 120 samples, since the sampling period was 2 seconds) from a subject that was instructed to look away from a flashing checkerboard covering 30% of the field of view. The flashing checkerboard moved around and caused the subject to look to the left, right, up and down. The gaze direction was seen as the output. The output was chosen to

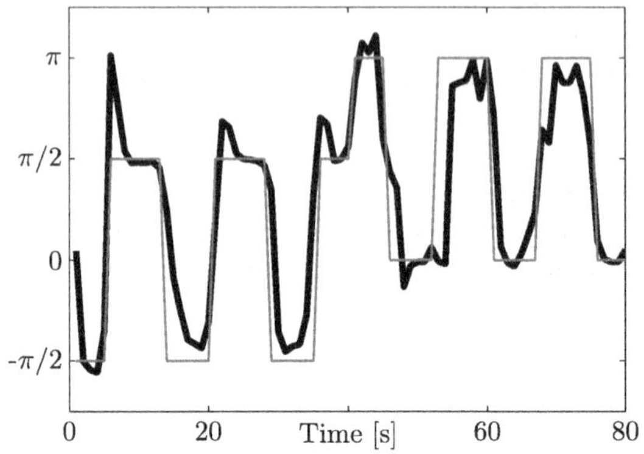

Fig. 9.5 WDMR applied to brain activity measurements (fMRI) of the visual cortex in order to tell what direction the subject in the MR scanner was looking. Thin grey line shows direction in which the subject was looking and thick black line the estimated direction by WDMR.

0 when the subject was looking to the right, $\pi/2$ when looking up, π when looking to the left and $-\pi/2$ when looking down.

Gaze direction is described by its angle, a scalar. The fMRI data should hence be constrained to a one-dimensional closed manifold residing in the 128 dimensional regressor space (since the regressors can be parameterized by the angle). If we assume that the semi-supervised smoothness assumption holds, WDMR therefore seems like a good choice.

The 120 regressors with observed output were separated into two sets, a training set consisting of 80 regressors and a test set consisting of 40 regressors. The training set was further divided into an estimation set and a validation set, both the same size. The estimation set and the regressors of the validation set were used in WDMR. The estimated outputs of the validation regressors were compared to the measured outputs and used to determine the design parameters. λ in (9.15) was chosen as 0.8 and K (using the kernel determined by LLE, see Appendix 9.10) as 6. The tuned WDMR regression algorithm was then used to predict the direction in which the person was looking. The results from applying WDMR to the 40 regressors of the test set are shown in Fig. 9.5.

The result is satisfactory but it is not clear to what extent the one-dimensional manifold has been found. The number of regressors with unobserved output used are rather low and it is therefore not surprising that the Nadaraya–Watson smoother with the KNN kernel can be shown to perform almost as well as WDMR in this example. One would expect that adding more regressors with unobserved output would improve the result obtained by WDMR. The estimates of the Nadaraya–Watson smoother would however stay unchanged since the Nadaraya–Watson smoother is a supervised method and therefore not affected by regressors with unobserved output.

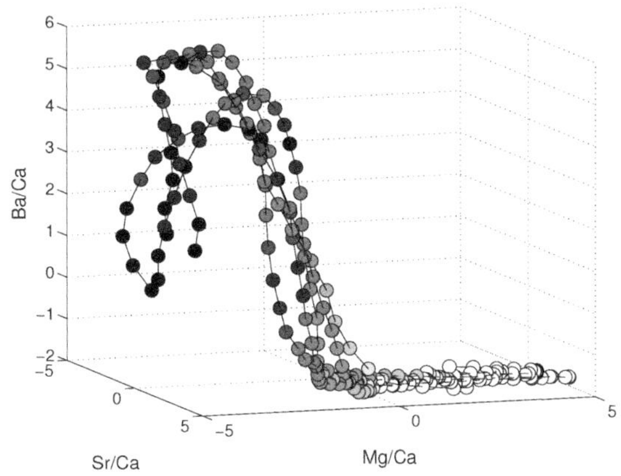

Fig. 9.6 A plot of the [Sr]/[Ca], [Mg]/[Ca] and [Ba]/[Ca] concentration ratio measurements from five shells. Lines connect measurements (ordered chronologically) coming from the same shell. The temperatures associated with the measurements were color coded and are shown as different grey scales on the measurement points.

9.8.2 Example 2—Climate Reconstruction

There exist a number of climate recorders in nature from which past temperature can be extracted. However, only a few natural archives are able to record climate fluctuations with high enough resolution so that seasonal variations can be reconstructed. One such archive is a bivalve shell. The chemical composition of a shell of a bivalve depends on a number of chemical and physical parameters of the water in which the shell was composed. Of these parameters, water temperature is probably the most important one. It should therefore be possible to estimate the water temperature for the time the shell formed from measurements of the shell's chemical composition. This feature would, for example, give climatologists the ability to estimate past water temperatures by analyzing ancient shells.

In this example, we analyzed 10 shells grown in Belgium. Since the temperature in the water for these shells had been monitored, this data set provides an excellent means of testing how well one could predict water temperature from chemical composition measurements. Chemical composition measurements had been taken along the growth axis of the shells and paired up with temperature measurements. Between 30 and 52 measurement were provided from each shell, corresponding to a time period of a couple of months. The 10 shells were divided into an estimation set and a validation set. The estimation set consisted of 6 shells (a total of 238 regressors with observed output) grown in Terneuzen in Belgium. Measurements from 5 of these shells are shown in Fig. 9.6. The figure shows measurements of the relative concentrations of [Sr]/[Ca], [Mg]/[Ca] and [Ba]/[Ca] ([Pb]/[Ca] is also measured but not shown in the figure). As seen in the figure, measurements are highly restricted to a small region in the measurement space. Also, the water temperature (grey level-coded in Fig. 9.6) varies smoothly in the high density regions. This feature together with the fact that it is a biological process generating data, motivates the semi-supervised smoothness assumption as we try to estimate water temperature from chemical composition measurements (four-dimensional regressors).

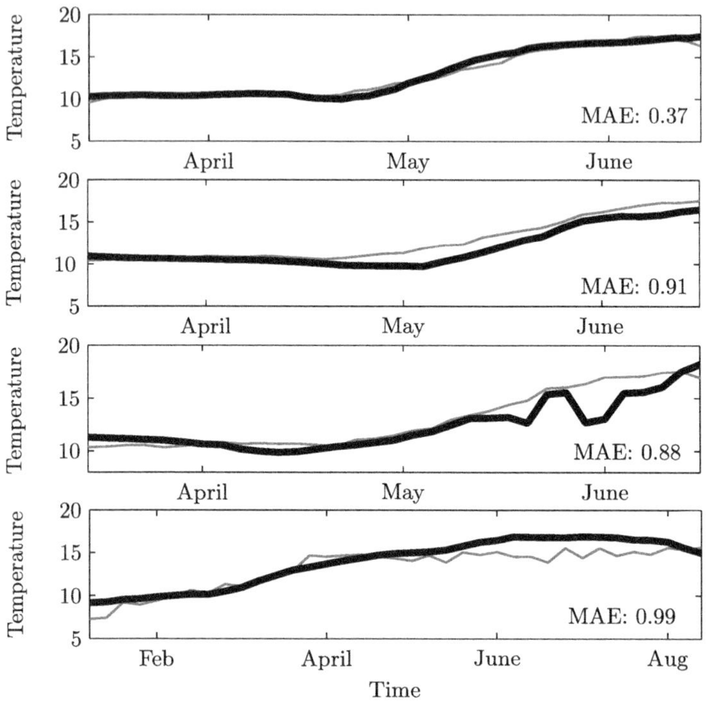

Fig. 9.7 Water temperature estimations using WDMR for validation data (thick line) and measured temperature (thin line). From top to bottom figure, shells from: Terneuzen, Breskens, Ossenisse, Knokke.

The 4 shells in the validation set came from four different sites (Terneuzen, Breskens, Ossenisse, Knokke) and from different time periods. The estimated temperatures for the validation data obtained by using WDMR with the kernel determined by LLE (see Appendix 9.10) are shown in Fig. 9.7. For comparison purposes, we mention that the Nadaraya–Watson smoother using the LLE kernel had a mean absolute error (MAE) nearly twice as high as WDMR.

A more detailed discussion of this example is presented in [1]. The data sets used were provided by Vander Putten et al. [27] and Gillikin et al. [5, 6].

9.9 Conclusion

This chapter presents and discusses a novel linear kernel smoother, that of weight determination by manifold regularization. The regression method is of particular interest when regressors are confined to limited regions in the regressor space and under the semi-supervised smoothness assumption. Examples of this type of problem were given.

9.10 Appendix—Kernels

This section presents kernels referred to in the chapter. The convention that $k(\varphi_i, \varphi_j) = 0$ if $i = j$ is always used. See Chapter 4 in [22] for more on kernels.

9.10.1 The KNN Kernel

Define the *K-nearest neighbor kernel* as

$$k(\varphi_i, \varphi_j) \triangleq \begin{cases} \frac{1}{K}, & \text{if } \varphi_j \text{ is one of the } K \text{ closest neighbors,} \\ 0, & \text{otherwise.} \end{cases} \tag{9.26}$$

9.10.2 The Squared Exponential Kernel

Define the *squared exponential kernel* (sometimes called a Gaussian kernel) as

$$k(\varphi_i, \varphi_j) \triangleq e^{-\|\varphi_i - \varphi_j\|^2 / 2\mathfrak{l}^2}. \tag{9.27}$$

$\mathfrak{l}$ is a parameter of the kernel and denoted the *length scale.*

9.10.3 The LLE Kernel

Local linear embedding (LLE), [25] is a technique to find lower dimensional manifolds to which an observed collection of regressors belong. A brief description of it is as follows:

Let $\{\varphi_i, i = 1, \ldots, N\}$ belong to $U \subset R^{n_\varphi}$ where U is an unknown manifold of dimension n_z. A coordinatization z_i ($z_i \in R^{n_z}$) of U is then obtained by first minimizing the cost function

$$\varepsilon(l) = \sum_{i=1}^{N} \left\| \varphi_i - \sum_{j=1}^{N} l_{ij} \varphi_j \right\|^2 \tag{9.28a}$$

under the constraints

$$\begin{cases} \Sigma_{j=1}^{N} l_{ij} = 1, \\ l_{ij} = 0 \text{ if } \|\varphi_i - \varphi_j\| > C_i(\kappa) \text{ or if } i = j. \end{cases} \tag{9.28b}$$

Here, $C_i(\kappa)$ is chosen so that only κ weights l_{ij} become nonzero for every i. κ is a design variable. It is also common to add a regularization to (9.28a) to prevent degenerate solutions.

Then, for the determined l_{ij}, find z_i by minimizing

$$\sum_{i=1}^{N} \left\| z_i - \sum_{j=1}^{N} l_{ij} z_j \right\|^2 \tag{9.29}$$

with respect to $z_i \in R^{n_z}$ under the constraint

$$\frac{1}{N} \sum_{i=1}^{N} z_i z_i^T = I_{n_z \times n_z}$$

z_i will then be the coordinate for φ_i in the lower dimensional manifold. Define now the *LLE kernel* as

$$k(\varphi_i, \varphi_j) \triangleq l_{ij} \tag{9.30}$$

where l_{ij} is defined in (9.28).

Note that the LLE kernel is invariant to translation, rotation and rescaling of the regressors φ. By using the LLE kernel in (9.18) we hence assume that the map f_0 is a linear combination of some coordinates that are invariant to translation, rotation and rescaling of the regressors.

Acknowledgements This work was supported by the Strategic Research Center MOVIII, funded by the Swedish Foundation for Strategic Research, SSF, and CADICS, a Linnaeus center funded by the Swedish Research Council.

References

1. Bauwens, M., Ohlsson, H., Barbé, K., Beelaerts, V., Dehairs, F., Schoukens, J.: On climate reconstruction using bivalve shells: Three methods to interpret the chemical signature of a shell. In: Proc. 7th IFAC Symposium on Modelling and Control in Biomedical Systems, p. 407412. Aalborg, Denmark (2009)
2. Belkin, M., Niyogi, P., Sindhwani, V.: Manifold regularization: A geometric framework for learning from labeled and unlabeled examples. Journal of Machine Learning Research **7**, 2399–2434 (2006)
3. Bengio, Y., Delalleau, O., Le Roux, N.: Label propagation and quadratic criterion. In: O. Chapelle, B. Schölkopf, A. Zien (eds.) Semi-Supervised Learning, pp. 193–216. MIT Press, Cambridge, MA (2006)
4. Chapelle, O., Schölkopf, B., Zien, A. (eds.): Semi-Supervised Learning. MIT Press, Cambridge, MA (2006)
5. Gillikin, D.P., Dehairs, F., Lorrain, A., Steenmans, D., Baeyens, W., André, L.: Barium uptake into the shells of the common mussel (*Mytilus edulis*) and the potential for estuarine paleo-chemistry reconstruction. Geochimica et Cosmochimica Acta **70**(2), 395–407 (2006)
6. Gillikin, D.P., Lorrain, A., Bouillon, S., Willenz, P., Dehairs, F.: Stable carbon isotopic composition of *Mytilus edulis* shells: Relation to metabolism, salinity, $\delta^{13}C_{DIC}$ and phytoplankton. Organic Geochemistry **37**(10), 1371–1382 (2006)

7. Goldberg, A.B., Zhu, X.: Seeing stars when there aren't many stars: Graph-based semi-supervised learning for sentiment categorization. In: Proc. TextGraphs: The First Workshop on Graph Based Methods for Natural Language Processing (TextGraphs'06), pp. 45–52. Association for Computational Linguistics, Morristown, NJ, USA (2006)
8. Hastie, T., Tibshirani, R., Friedman, J.: The Elements of Statistical Learning: Data Mining, Inference, and Prediction. Springer Series in Statistics. Springer, New York (2001)
9. Hosmer, D.W.J.: A comparison of iterative maximum likelihood estimates of the parameters of a mixture of two normal distributions under three different types of sample. Biometrics **29**(4), 761–770 (1973)
10. Kohonen, T.: Self-Organizing Maps, *Springer Series in Information Sciences*, vol. 30. Springer, Berlin (1995)
11. Ljung, L.: The System Identification Toolbox: The Manual. The MathWorks Inc. 1st ed. 1986, 7th ed. 2007, Natick, MA, USA (2007)
12. Merz, C.J., St. Clair, D.C., Bond, W.E.: SeMi-supervised adaptive resonance theory (SMART2). In: Proc. Int. Joint Conf. Neural Networks (IJCNN), vol. 3, pp. 851–856 (1992)
13. Nadaraya, E.A.: On estimating regression. Theory of Probability and Its Applications **9**, 141–142 (1964)
14. Narendra, K.S., Li, S.M.: Neural networks in control systems. In: P. Smolensky, M.C. Mozer, D.E. Rumelhard (eds.) Mathematical Perspectives on Neural Networks, chap. 11, pp. 347–394. Lawrence Erlbaum Associates, Hillsdale, NJ, USA (1996)
15. Navaratnam, R., Fitzgibbon, A.W., Cipolla, R.: The joint manifold model for semi-supervised multi-valued regression. Proc. IEEE 11th Int. Conf. Computer Vision (ICCV 2007) pp. 1–8 (2007)
16. Nigam, K., McCallum, A., Thrun, S., Mitchell, T.: Learning to classify text from labeled and unlabeled documents. In: Proc. 15th National/10th Conf. Artificial Intelligence/Innovative Applications of Artificial Intelligence (AAAI '98/IAAI '98), pp. 792–799. American Association for Artificial Intelligence, Menlo Park, CA, USA (1998)
17. Ohlsson, H.: Regression on manifolds with implications for system identification. Licentiate thesis no. 1382, Linköping University, Linköping, Sweden (2008)
18. Ohlsson, H., Ljung, L.: Semi-supervised regression and system identification. In: X. Hu, U. Jonsson, B. Wahlberg, B. Ghosh (eds.) Three Decades of Progress in Control Sciences. Springer, New York (2010). To appear
19. Ohlsson, H., Roll, J., Glad, T., Ljung, L.: Using manifold learning for nonlinear system identification. In: Proc. 7th IFAC Symposium on Nonlinear Control Systems (NOLCOS2007), p. 706711. Pretoria, South Africa (2007)
20. Ohlsson, H., Roll, J., Ljung, L.: Manifold-constrained regressors in system identification. In: Proc. 47th IEEE Conf. Decision and Control (CDC2008), p. 13641369. Cancun, Mexico (2008)
21. Rahimi, A., Recht, B., Darrell, T.: Learning to transform time series with a few examples. IEEE Trans. Pattern Analysis and Machine Intelligence **29**(10), 1759–1775 (2007)
22. Rasmussen, C.E., Williams, C.K.I.: Gaussian Processes for Machine Learning. The MIT Press, Cambridge, MA (2005)
23. de Ridder, D., Duin, R.P.: Locally linear embedding for classification (2002). Technical Report, PH-2002-01, Pattern Recognition Group, Dept. of Imaging Science & Technology, Delft University of Technology, Delft, The Netherlands.
24. de Ridder, D., Kouroptеva, O., Okun, O., Pietikinen, M., Duin, R.: Supervised locally linear embedding. In: O. Kaynak, E. Alpaydin, E. Oja, L. Xu (eds.) Artificial Neural Networks and Neural Information Processing – ICANN/ICONIP 2003, *Lecture Notes in Computer Science*, vol. 2714, pp. 333–341. Springer Berlin / Heidelberg (2003)
25. Roweis, S.T., Saul, L.K.: Nonlinear dimensionality reduction by locally linear embedding. Science **290**(5500), 2323–2326 (2000)
26. Tenenbaum, J.B., de Silva, V., Langford, J.C.: A global geometric framework for nonlinear dimensionality reduction. Science **290**(5500), 2319–2323 (2000)

27. Vander Putten, E., Dehairs, F., André, L., Baeyens, W.: Quantitative in situ microanalysis of minor and trace elements in biogenic calcite using infrared laser ablation - inductively coupled plasma mass spectrometry: A critical evaluation. Analytica Chimica Acta **378**(1), 261–272 (1999)
28. Wang, F., Zhang, C.: Label propagation through linear neighborhoods. IEEE Trans. Knowledge and Data Engineering **20**(1), 55–67 (2008)
29. Watson, G.S.: Smooth regression analysis. Sankhyā Ser. **26**, 359–372 (1964)
30. Weiskopf, N., Sitaram, R., Josephs, O., Veit, R., Scharnowski, F., Goebel, R., Birbaumer, N., Deichmann, R., Mathiak, K.: Real-time functional magnetic resonance imaging: Methods and applications. Magnetic Resonance Imaging **25**, 989–1003 (2007)
31. Yang, X., Fu, H., Zha, H., Barlow, J.: Semi-supervised nonlinear dimensionality reduction. In: Proceedings of the 23rd international conference on Machine learning (ICML '06), pp. 1065–1072. ACM, New York, NY, USA (2006)
32. Zhao, L., Zhang, Z.: Supervised locally linear embedding with probability-based distance for classification. Computers & Mathematics with Applications **57**(6), 919–926 (2009)
33. Zhu, X.: Semi-supervised learning literature survey. Tech. Rep. 1530, Computer Sciences, University of Wisconsin, Madison, WI (2005)

Chapter 10
Dynamic Coverage and Clustering: A Maximum Entropy Approach

Carolyn Beck, Srinivasa Salapaka, Puneet Sharma and Yunwen Xu

Abstract We present a computational framework we have recently developed for solving a large class of dynamic coverage and clustering problems, ranging from those that arise in the deployment of mobile sensor networks to the identification of ensemble spike trains in computational neuroscience applications. This framework provides for the identification of natural clusters in an underlying dataset, while addressing inherent tradeoffs such as those between cluster resolution and computational cost. More specifically, we define the problem of minimizing an instantaneous *coverage* metric as a combinatorial optimization problem in a Maximum Entropy Principle framework, which we formulate specifically for the dynamic setting. Locating and tracking dynamic cluster centers is cast as a control design problem that ensures the algorithm achieves progressively better coverage with time.

Carolyn Beck
Coordinated Science Lab, University of Illinois at Urbana-Champaign & Department of Industrial and Enterprise Systems Engineering, University of Illinois, 117 Transportation Building MC-238, 104 S. Mathews Ave., Urbana, IL 61801, USA
e-mail: beck3@illinois.edu

Srinivasa Salapaka
Coordinated Science Lab, University of Illinois at Urbana-Champaign & Department of Mechanical Science and Engineering, 362c Mechanical Engineering Building, 1206 West Green Street, Urbana, IL 61801
e-mail: salapaka@illinois.edu

Puneet Sharma
Integrated Data Systems Department, Siemens Corporate Research, 755 College Road East, Princeton, NJ 08540, USA
e-mail: sharma.puneet@gmail.com

Yunwen Xu
Coordinated Science Lab, University of Illinois at Urbana-Champaign & Department of Industrial and Enterprise Systems Engineering, University of Illinois, 1308 W Main Street Urbana, IL 61801-2307, USA
e-mail: xu27@illinois.edu

R. Johansson and A. Rantzer (Eds.): Distributed Decision Making and Control, LNCIS 417, pp. 215–243.
springerlink.com

10.1 Introduction

There has been a recent growing interest in the development of algorithms for deployment of mobile resources that continuously *cover* a set of mobile sites in a region, that is, algorithms related to determining clusters in an ensemble of moving objects. Such algorithms have numerous applications, such as in developing automatic deployment and tracking techniques for surveillance and military applications [4, 8], in clustering of the spatio-temporal dynamics of brain signals [18, 17, 1] and in routing traffic and detecting traffic jams by clustering traffic flow [30].

Clustering algorithms for *static data* have been widely studied in various areas, such as the minimum distortion problem in data compression [10], facility location assignments [5], optimal quadrature rules and discretization of partial differential equations [6], pattern recognition [33], drug discovery [28] and recently in coarse quantization [7, 16, 22]. However, the focus on clustering *dynamically evolving data* is new and many issues pertaining to quantification, analysis and design for coverage remain unsolved.

Although the static problems focus on seemingly unrelated goals, they, in fact, share a number of fundamental common attributes. They are typically formulated under a class of combinatorial optimization problems that searches for an optimal partition of the underlying domain, as well as an optimal assignment of elements, from a finite *resource* set to each *cell* in the partition. Since both the number of partitions as well as the number of element associations to partitions are combinatorial, these problems are computationally intractable (i.e., NP-hard [11]), which rules out exhaustive search methods. The cost functions in these optimization problems contain numerous local minima, and therefore it is crucial to design algorithms that do not get trapped in local minima [11].

The complexity of the combinatorial problem addressed in this chapter is further compounded by the *dynamics* associated with each data point, i.e., the moving objects. These moving objects could be, for example, mobile threat locations in a battlefield scenario, ensemble neuronal spike trains in brain signal data, formations of unmanned vehicles, or naturally occuring swarms or flocks. The task at hand is to design a dynamic control law for mobile resources such that they continuously identify and track cluster centers of groups of moving objects. Thus, locations of each data-point in the static case are replaced by velocity and/or acceleration fields in the dynamic case. From a naive computational viewpoint, the dynamic problem can be regarded as a time-indexed set of static problems. Adding dynamics also introduces new complexities to the notions of coverage due to the dynamic nature of cluster sizes, number of clusters, and relative distances between the individual elements.

Problems related to dynamic coverage were considered in [4, 8, 3, 30], in which distributed implementation approaches are used. Distributed schemes aim to address the issue of large impractical or nonviable computational efforts required for centralized schemes; however, these algorithms are prone to converge to one of the many local minima typically present in these problems. As a consequence, the performance of distributed algorithms is extremely sensitive to initial placement of the resource locations. At present, there is little research that addresses the development of tractable algorithms for dynamic problems of a nondistributed nature, that is, algorithms that aim simultaneously to attain global solutions and maintain low computational expense.

In this chapter, we present a general framework based on the maximum entropy principle (MEP) to formulate and solve dynamic clustering problems, which addresses both coverage and tracking. The framework we propose, which we refer to as the dynamic maximum entropy (DME) framework, adapts the notion of coverage to the dynamic setting and resolves the inherent trade-off between resolution of the clusters and computational cost, while avoiding shallow local minima. The algorithms we propose are hierarchical in that they progressively seek finer subclusters from larger clusters. An important feature of the proposed framework is its ability to detect natural cluster centers in the underlying data set, without the need to initialize or define the clusters a priori. The computational complexity of these algorithms is reduced by exploiting structure in the problem formulation. The algorithms, as they proceed, become more "local", that is, the computation of clusters becomes less sensitive to distant sites. This feature is exploited to make these algorithms scalable and computationally efficient. Our simulation results, employing algorithms based in the DME framework, demonstrate improvements in computation time and illustrate the flexibility of this framework.

We note that an MEP-based approach has been developed in the static setting in the context of vector quantization (VQ) problems in the data-compression community. The resulting algorithm, known as the *deterministic annealing* (DA) algorithm, has been shown to avoid local minima, provide provably better clustering results than the popular k-means algorithms, and converge faster than other heuristics designed to avoid local minima such as the simulated annealing algorithm [21]. The DME framework we propose inherits the advantages of the DA algorithm and, further, addresses tracking as well as resolution issues originating from cluster dynamics.

This chapter is organized as follows. We discuss the general setting of the dynamic coverage problem in Sec. 10.2 and highlight key issues and challenges that must be resolved so such problems can be solved. We also provide an overview of the DA algorithm developed in [21] for clustering static data in this section and analyze those features that are relevant to the dynamic problem. The proposed DME framework for solving the clustering problem for dynamic data is presented in Sec. 10.3. Scalability issues are discussed in Sec. 10.4. Implementation and simulation results for a variety of data sets are presented in Sec. 10.5. Conclusions and directions for future work are discussed in Sec. 10.6.

10.2 Dynamic versus Static Clustering

Although the static clustering problem has been extensively studied, in this chapter we focus on the scenario where the problem is to detect and track a group of moving objects (henceforth referred to as *sites*) in a given area; that is, the *dynamic clustering problem*.

10.2.1 Problem Formulation

The mobile sites we study may move in clusters and change cluster associations, and the clusters themselves can split and rejoin over time. This problem can be posed as a coverage problem, where the aim is to successively identify representative object locations (in the sequel referred to as *resource locations*, or *cluster centers*) such that the resources provide adequate *coverage* of the moving sites at all times. The number of resources is assumed to be far fewer than the number of moving sites. Each resource can then be thought of as a cluster center.

For notational convenience, we consider a domain $\Omega \subset \mathbb{R}^2$ with N mobile sites and M resources, where $N \gg M$, on a time horizon in $[0,\infty)$. (Note that the approach we develop in this chapter is applicable to domains $\Omega \subset \mathbb{R}^k$, for any $k \in \mathbb{N}$; however, we will restrict our discussion herein to $k = 2$ for ease of exposition.) The location of the ith mobile site ($i \le N$) and the jth resource ($j \le M$) at time instance $t \in [0,\infty)$ is represented by $x_i(t) = [\xi_i(t)\ \eta_i(t)]^T \in \mathbb{R}^2$ and $y_j(t) = [\rho_j(t)\ \omega_j(t)]^T \in \mathbb{R}^2$, respectively. We will sometimes simply use x_i and y_j in place of $x_i(t)$ and $y_j(t)$, where the time dependence is assumed and is clear from the context. The dynamics are given by the continuously differentiable velocity fields, $\phi_i(x,y,t) \in \mathbb{R}^2, i \le N$, for the ith site and $u_j(t) \in \mathbb{R}^2, j \le M$, for the jth resource, where x and y represent the locations of all sites and resources, respectively. More precisely, we have a domain Ω with N sites $\{x_i\}$ and M resource locations $\{y_j\}$, whose dynamics are given by

$$\begin{aligned} \dot{x}(t) &= \phi(x(t), y(t), t), \quad x(0) = x_0 \\ \dot{y}(t) &= u(t), \quad y(0) = y_0 \\ &\Updownarrow \\ \dot{\zeta} &= f(\zeta, t), \end{aligned} \tag{10.1}$$

where $x(t) = [x_1(t)\ x_2(t)\ \cdots\ x_N(t)]^T$, $y(t) = [y_1(t)\ y_2(t)\ \cdots\ y_M(t)]^T$, $\phi(t) = [\phi_1(t)\ \phi_2(t)\ \cdots\ \phi_N(t)]^T$, $u(t) = [u_1(t)\ u_2(t)\ \cdots\ u_M(t)]^T$, and $\zeta(t) = [x^T\ y^T]^T$. This system can be viewed as a control system where the control field u is to be determined for the M mobile resources such that a notion of coverage is maintained. We summarize the main challenges and our objectives in addressing dynamic coverage problems in the following:

- One of the main challenges in dynamic coverage problems is fundamental: quantifying the performance objectives. We adopt the concept of *distortion* and its variants from the data compression literature (which deals with static coverage problems) as a metric for coverage and modify this to make it suitable to a dynamic setting. Distortion, in a static coverage problem, is a measure of the (weighted) average distance of the site locations to their nearest resource location. For a static problem ($\phi(t) \equiv 0, u(t) \equiv 0$), the distortion measure is given

Fig. 10.1 The clustering solution in (a) identifies 4 coarse clusters centers $y_j, 1 \leq j \leq 4$ (shown by stars) in the underlying data $x_i, 1 \leq i \leq 37$ (shown by squares). On the other hand, the solution in (b) identifies 7 finer clusters at a higher resolution on the same underlying data. The solution in (b) has lower distortion D than in (a), but at the expense of higher computation time.

by

$$D(x,y) = \sum_{x_i \in \Omega} p_i \left\{ \min_{1 \leq j \leq M} d(x_i, y_j) \right\}, \tag{10.2}$$

where p_i represents the weight or relative importance of the site x_i and $d(x_i,y_j)$ is a metric defined on Ω, which represents a distance function; this is typically given by a squared Euclidean distance function $d(x_i,y_j) = \|x_i - y_j\|^2$. Thus, for a given set of site locations $\{x_i\}$, $1 \leq i \leq N$, the set of resource locations $\{y_j\}$, $1 \leq j \leq M$ that achieves lower distortion results in better coverage.

- A second challenge arises in defining clusters. This challenge primarily stems from the fact that clusters are hierarchical in nature. Each cluster can be thought of as smaller subclusters (Fig. 10.1), and in the limiting case comprising every moving site can be thought of as a distinct cluster. On the other hand, the entire domain can be thought of as a cluster. Thus, we consider assigning a notion of *resolution* for a cluster. Iterative algorithms that identify finer and finer (higher resolution) clusters progressively reduce the coverage cost function, but at the expense of increased computation.

  Modeling the coverage function and defining clusters is even more challenging in the dynamic case, since the coverage function must adjust to and appropriately reflect the expanding, shrinking, splitting and/or coalescing components of clusters, adding further variability to the control design.
- Computational complexity is a third important challenge confronting the solution of coverage problems, arising from the inherent nonconvex nature of the distortion function (10.2). The dynamic distortion function contains numerous local minima, just as it does in the static case. Consequently, the problem necessitates designing algorithms that do not get trapped at local minima, while also avoiding expensive divide-and-search strategies. This complexity is further aggravated

by the additional time component in the dynamic setting. These challenges are addressed by the framework proposed in this section. More specifically we formulate a coverage problem and propose algorithms that determine the resource velocity fields $u(t)$, such that they *track* each cluster over time. If a cluster splits, the resource locations mimic this behavior, thereby maintaining coverage.

10.2.2 The Static Resource Allocation Approach

Dynamic data may be viewed as a time-indexed series of static data, thus the simplest approach conceptually is to perform static clustering periodically. However, if the elapsed time between two successive clustering events is small, the algorithm is unnecessarily expensive because the spatial clustering from previous time steps is not exploited for clustering at current and future time steps. On the other hand, if the time period is long the clustering obtained at the previous time step does not provide adequate tracking during the interval to the current time step. In short, such a simplistic approach ignores the spatio-temporal aspects of clustering moving data. Nevertheless, we provide a brief overview of this approach to provide insights into the challenges that are inherited by the dynamic problems. In this *frame-by-frame approach*, at each instant of time t, we solve the following static resource allocation problem:

Given N sites $x_i(t), 1 \leq i \leq N$, in a domain Ω with relative weights p_i, find the set of M resource locations $y_j(t), 1 \leq j \leq M$, that minimizes the distortion (at fixed time t); that is, find

$$\underset{y_j,\ 1\leq j\leq M}{\arg\min} \left(\sum_{i=1}^{N} p_i \left\{ \min_{1\leq j\leq M} d(x_i,y_j) \right\} \right) \tag{10.3}$$

Again, $d(x_i,y_j)$ represents an appropriate distance function between the resource location y_j and the site x_i, for example, $d(x_i,y_j) = \|x_i - y_j\|^2 + \|\phi_i - u_j\|^2$, where velocity terms as well as location terms are included in the Euclidean distance function. Minimizing (10.3) is akin to finding a velocity field for the resources such that the coverage condition is satisfied at time t.

This problem is equivalent to finding an *optimal* partition of the domain space Ω at time t into M cells ($R_j, j = 1 \ldots M$) and assigning to each cell R_j a resource location y_j that minimizes the partition cost, given by $\sum_j \sum_{x_i \in R_j} d(x_i,y_j) p_i$. Solving this partitioning problem at a fixed time t is equivalent to the vector quantization problem in data compression, which is addressed by the DA algorithm [21, 20].

We now describe relevant features of the DA algorithm [19], which provide a basis for our DME algorithm.

10.2.3 The Deterministic Annealing Algorithm: Clustering in the Static Setting

One of the main objectives of the DA algorithm is to avoid local minima. This algorithm can be viewed as a modification of Lloyd's algorithm [15, 10], in which the initial step consists of randomly choosing resource locations and then successively iterating between the steps of: (1) forming Voronoi partitions and (2) moving the resource locations to respective centroids of cells until the sequence of resource locations converges. The solution depends substantially on the initial placement of resources, as in successive iterations the locations are influenced only by nearby points of the domain and are virtually independent of distant points. As a result, Lloyd's algorithm typically converges to local minima.

The DA algorithm overcomes the local influence of domain elements by allowing each element $x_i \in \Omega$ to be associated with every resource location y_j through a weighting parameter $p(y_j|x_i)$ (without loss of generality, $\sum_j p(y_j|x_i) = 1$ for every $1 \leq i \leq N$) [21, 20, 26, 27]. Thus, the DA algorithm eliminates the hard partitions of Lloyd's algorithm and seeks to minimize a modified distortion term given by

$$D(x,y) = \sum_{i=1}^{N} p_i \sum_{j=1}^{M} d(x_i,y_j)p(y_j|x_i). \tag{10.4}$$

Note that the instantaneous weighting term $p(y_j|x_i)$ (at time t) can also be viewed as an association probability between the mobile site x_i and the mobile resource y_j. The choice of weights $\{p(y_j|x_i)\}$ determines the trade-off between decreasing effects of local influence and deviation of the distortion (10.4) from the original cost function (10.2). For instance, a uniform weighting function, where $p(y_j|x_i) = 1/M$ for all $i \leq N, j \leq M$, makes the minimization of (10.4) with respect to y_j independent of initial placement of y_j; however, the corresponding distortion function is considerably different from the cost function in (10.2). At the other extreme, setting $p(y_j|x_i) = 1$ when $d(x_i,y_j) = \min_k d(x_i,y_k)$, but otherwise setting $p(y_j|x_i) = 0$, makes the distortion term in (10.4) identical to the cost function (10.2) but retains the "local influence effect" when minimizing with respect to y_j.

The MEP provides a systematic way to determine a weighting function that achieves a specific feasible value of distortion, and thereby achieves a prespecified tradeoff in the above context [12, 13]. More specifically, we seek a weight distribution $p(y|x)$ that maximizes the Shannon entropy [25]

$$H(y|x) = -\sum_{i=1}^{N} p_i \sum_{j=1}^{M} p(y_j|x_i)\log(p(y_j|x_i)), \tag{10.5}$$

at a given (feasible) level of coverage $D(x,y) = D_0$. The entropy term quantifies the level of randomness in the distribution of association weights, so maximizing this term causes the weights to be maximally noncommittal toward any single cluster. In this framework, we first determine the weighting functions by maximizing the unconstrained Lagrangian $L = H(y|x) - \beta(D(x,y) - D_0)$ with respect to $\{p(y_j|x_i)\}$, where $H(y|x)$ and $D(x,y)$ are given by (10.5) and (10.4) respectively, D_0 is the desired value of distortion and β is a Lagrange multiplier. This is equivalent to the minimization problem:

$$\min_{\{p(y_j|x_i)\}} \underbrace{D(x,y) - TH(y|x)}_{\triangleq F}, \tag{10.6}$$

where the Lagrange multiplier, denoted by $T = 1/\beta$, and the term F are called *temperature* and *free energy*, respectively, analogously to quantities in statistical physics [14]. The MEP theorem (see [12, 13]) gives an explicit solution for the weights, given by the Gibbs distribution

$$p(y_j|x_i) = \frac{\exp\{-\beta d(x_i,y_j)\}}{\sum_{k=1}^{M} \exp\{-\beta d(x_i,y_k)\}} \tag{10.7}$$

On substituting (10.7) into the Lagrangian (10.6) we have

$$F(x,y) = -\frac{1}{\beta}\sum_{i=1}^{N} p_i \log \sum_{k=1}^{M} \exp\{-\beta d(x_i,y_k)\}. \tag{10.8}$$

This function plays an important role in representing the coverage function (as will be shown later). The resource locations $\{y_j\}$ are specified by setting $\partial F/\partial y_j = 0$, which yields

$$y_j = \sum_{i=1}^{N} p(x_i|y_j)x_i, \ \forall j = 1,2,\ldots,M, \text{ with } p(x_i|y_j) = \frac{p_i p(y_j|x_i)}{\sum_{m=1}^{N} p_m p(y_j|x_m)} \tag{10.9}$$

where $p(x_i|y_j)$ denotes the posterior association weight calculated using Bayes's rule. The above equations clearly convey the "centroid" aspect of the solution.

The temperature variable $T = 1/\beta$ is fixed by the constraint value D_0 of the distortion. A simple sensitivity analysis shows that lower values of D_0 correspond to lower values of the temperature variable [13]. Clearly, for small values of β (i.e., large values of T) in (10.6), we are mainly maximizing the entropy. Thus a choice of weights corresponding to a high value (near infinity) of T leads to algorithms that are insensitive to the initial allocation of resource locations, since their subsequent locations are affected almost equally by all sites. As β is increased, we trade entropy for the reduction in distortion, and as β approaches infinity (i.e., T approaches zero), we minimize distortion D directly to obtain a *hard* (nonrandom) solution. An *annealing* process is incorporated where the minimization problem (10.6) is repeatedly solved at different values $\beta = \beta_k$, where $\beta_{k+1} > \beta_k$. Solving the implicit equation (10.9) to determine y_j entails the most computationally expensive step for each value of β_k. This equation is solved by evolving the following dynamical system to convergence:

$$y_j^k(n+1) = \sum_{i=1}^{N} p(x_i|y_j^k(n))x_i, \quad 1 \leq j \leq M, \; n \geq 0, \tag{10.10}$$

where $y_j^k(n)$ represents the value of the estimate of y_j (when the temperature value is given by $\beta = \beta_k$) at the nth step of this iterative procedure. At each k, the initial value $y_j^k(0)$ is set to the final value for $k-1$, that is $y_j^k(0) = y_j^{k-1}(\infty)$. Note that the iterative process (10.10) is equivalent to a Newton descent method and, accordingly, this procedure inherits the convergence properties of a Newton descent method.

The annealing process itself exhibits a *phase transition* property. That is, there are critical values of β at which the number of *distinct* resource locations abruptly change. At $\beta = 0$, there is only one resource at the weighted centroid of all site locations x_i. As the parameter β is increased, there exists some critical value β_c such that the number of distinct resource locations that minimize free energy jumps from $m = 1$ for $\beta < \beta_c$ to some $m > 1$ for $\beta > \beta_c$. This critical value, β_c, can be computed explicitly from the distribution of the sites x_i and resource locations y_j. As β is increased further, successive critical values are reached that also satisfy the phase transition condition. At each such critical value, there is an increase in the number of the distinct resource locations. A detailed version of the implementation steps is presented [21].

10.2.4 Properties of the DA Algorithm

The features developed for the static coverage algorithm form the basis for our DME framework. In this formulation, various cluster attributes, such as relative size and shape, arise naturally, thereby preemting the need for new characterization variables. The weights $\{p(x_i|y_j)\}$ characterize soft clusters; a high value of $p(x_i|y_j)$ implies the site x_i predominantly belongs to jth cluster, which is covered by the resource at location y_j. The weight $p(x_i|y_j)$, when viewed as a function of x_i for a fixed y_j, determines the shape of the jth cluster. Cluster mass is characterized by the set of weights $\{p(y_j)\}$. Since $p(y_j) = \sum_i p_i p(y_j|x_i)$, this represents the *total mass* associated with the resource y_j. That is, $p(y_j)N$ is an estimate of the number of sites that determine the jth cluster. In this work, we assume $\min_j\{p(y_j)\} > \nu > 0$ to avoid modeling degenerate (or zero-mass) clusters. As the annealing parameter $\beta \to \infty$, the weights $\{p(y_j|x_i)\}$ become binary valued and the resulting clusters become "hard".

Notation: We define matrices

$$P_x \triangleq \mathrm{diag}(p(x_i)) \in \mathbb{R}^{N\times N}, \; P_{y|x} \triangleq [p(y_j|x_i)]$$

and

$$P_{x|y} \triangleq [p(x_i|y_j)] \in \mathbb{R}^{N\times M}, \quad P_y \triangleq \mathrm{diag}(p(y_j)) \in \mathbb{R}^{M\times M}$$

. In this notation, $P_{x|y}$ and P_y contain information about the relative shapes and sizes of the clusters. Also note that

$$P_x P_{y|x} = P_{x|y} P_y = P_{xy} = [p(x_i, y_j) = p_i p(y_j|x_i)]$$

The expression for cluster centers, given by (10.9), can be written concisely as $y = \check{P}_{x|y}^T x$ where

$$\check{P}_{x|y} \triangleq (I_2 \otimes P_{x|y})$$

I_2 is the 2×2 identity matrix, and $\otimes$ represents the matrix Kronecker product. Similarly, we define matrices

$$\check{P}_{y|x} \triangleq (I_2 \otimes P_{y|x}), \quad \check{P}_x \triangleq (I_2 \otimes P_x), \quad \check{P}_y \triangleq (I_2 \otimes P_y), \text{ and } \check{P}_{xy} \triangleq (I_2 \otimes P_{xy})$$

The *radius* of each cluster can be inferred from the magnitudes of the vectors $x_i - y_{c[i]}$ $(1 \leq i \leq N)$, where $y_{c[i]}$ denotes the resource location closest to the site x_i. We define

$$\bar{x} \triangleq x - \check{P}_{y|x} \check{P}_{x|y}^T x$$

which determines the weighted average distance of x_i from the cluster centers $\{y_j\}$; that is,

$$\bar{x}_i = x_i - \sum_j p(y_j|x_i) y_j$$

Additional important features of the DA algorithm are summarized in the following.

Theorem 10.1. *The following properties hold for the Deterministic Annealing algorithm:*

1. Centroid property [21]: $\lim_{\beta \to 0} y_j = \sum_{i=1}^N p_i x_i$.
2. Phase transition Property [21]: *The resource locations $\{y_j\}$ given by (10.9) give a local minimum for free energy F at every value of β except at* critical temperatures *when $\beta = \beta_c$, given by $\beta_c^{-1} = 2\lambda_{max}(C_{x|y_j})$ for some $1 \leq j \leq M$, where*

$$C_{x|y_j} = \sum_{i=1}^N p(x_i|y_j)(x_i - y_j)(x_i - y_j)^T \tag{10.11}$$

   and $\lambda_{max}(\cdot)$ represents the largest eigenvalue. Moreover, the number of distinct locations in $\{y_j(\beta)\}$ for $\beta > \beta_c$ is greater than for $\beta < \beta_c$.
3. Sensitivity-to-temperature property [29]: *If the parameter value β is far from the critical values β_c, that is the minimum eigenvalue of $(I - 2\beta C_{x|y_j}) \geq \Delta$ for some $\Delta > 0$ and $1 \leq j \leq M$, then*

$$\left\|\frac{dy}{d\beta}\right\| \triangleq \left(\sum_j \left\|\frac{dy_j}{d\beta}\right\|^2\right)^{\frac{1}{2}} \leq \frac{c(\beta)}{\Delta}$$

where $c(\beta)$ monotonically decreases to zero with β and is completely determined by β and the size of the space Ω.

Sketch of Proof: Proofs for properties 1 and 2 can be found in [21]). Herein we provide a sketch of a proof for property 3; full details can be found in [29].

We obtain $dy_j/d\beta$ by differentiating (10.9) with respect to β. Premultiplying $dy_j/d\beta$ by

$$p(y_j)\frac{dy_j}{d\beta}^T$$

and summing over j gives

$$\underbrace{\sum_j p(y_j)\frac{dy_j}{d\beta}^T\left[I-2\beta C_{x|y_j}\right]\frac{dy_j}{d\beta}}_{T_1}$$

$$= \underbrace{\sum_i p_i \sum_j p(y_j|x_i)(x_i-y_j)^T\frac{dy_j}{d\beta}\left\{\sum_k p(y_k|x_i)d(x_i,y_k)-d(x_i,y_j)\right\}}_{T_2}$$

$$- \underbrace{2\beta\sum_i p_i \sum_j p(y_j|x_i)(y_j-x_i)^T\frac{dy_j}{d\beta}\sum_k p(y_k|x_i)(y_k-x_i)^T\frac{dy_k}{d\beta}}_{T_3}.$$

Since T_3 is nonnegative, and $T_1+T_3=T_2$, we have $T_2 \geq T_1$, which in turn implies

$$T_2 \geq \min_j\{p(y_j)\}\Delta\sum_j \frac{dy_j}{d\beta}^T\frac{dy_j}{d\beta} = \min_j\{p(y_j)\}\Delta\left\|\frac{dy}{d\beta}\right\|^2. \tag{10.12}$$

Also, since $\sum_k p(y_k|x_i)=1$, T_2 can be rewritten as

$$T_2 = \sum_i p_i \sum_j\sum_k p(y_j|x_i)p(y_k|x_i)\left[d(x_i,y_k)-d(x_i,y_j)\right](y_j-x_i)^T\frac{dy_j}{d\beta}. \tag{10.13}$$

To obtain an upper bound on T_2, we first obtain a bound on the association weights,

$$p(y_j|x_i) = \frac{e^{-\beta(d(x_i,y_j)-d(x_i,y_k))}}{1+\sum_{m\neq k}e^{-\beta(d(x_i,y_m)-d(x_i,y_k))}} \leq e^{-\beta(d(x_i,y_j)-d(x_i,y_k))}. \tag{10.14}$$

Since $\|x-y_j\| \leq 2w_1 \stackrel{\triangle}{=} 2\text{ diameter}(\Omega)$ and

$$|e^{-\beta(d(x_i,y_m)-d(x_i,y_k))}(d(x_i,y_m)-d(x_i,y_k))| \leq \frac{e^{-1}}{\beta},$$

(since $\theta e^{-\gamma\theta} < \frac{e^{-1}}{\gamma}$ for $\gamma > 0$), we can infer a bound on T_2 given by

$$T_2 \leq \frac{e^{-1}}{\beta}(2w_1)\sum_j \left\| \frac{dy_j}{d\beta} \right\| \leq \left(\frac{e^{-1}}{\beta}(2w_1)\sqrt{M} \right) \left\| \frac{dy}{d\beta} \right\| \tag{10.15}$$

From (10.12) and (10.15), we have that

$$\min_j\{p(y_j)\}\Delta\|\frac{dy}{d\beta}\|^2 \leq |T_2| \leq \left(\frac{e^{-1}}{\beta}(2w_1)\sqrt{M} \right) \left\| \frac{dy}{d\beta} \right\|,$$

which implies that

$$\left\| \frac{dy}{d\beta} \right\| \leq \frac{c(\beta)}{\Delta}, \text{ where } c(\beta) = \left(\frac{e^{-1}}{\nu\beta}(2w_1)\sqrt{M} \right)$$

is completely determined by the value of β and the bound of the size of the space Ω. Here ν is a lower bound on $\min_{y_j}\{p(y_j)\}$, the minimum cluster size and w_1 is the diameter of space Ω. □

The centroid property implies that for small values of β the DA algorithm places all resources at the weighted centroid of the sites, that is, at $\beta = 0$ the cost function (10.6) achieves the global minimum with $p(y_j|x_i) = 1/M$ and with all resource locations $\{y_j\}$ being placed at the centroid of the data set. The annealing process deforms the free energy F from the entropy function at $\beta = 0$ to the distortion function at $\beta = \infty$, which allows us to track the evolution of the global minimum as β is increased.

The phase transition property guarantees that the resource locations obtained at noncritical β values are local minima of F. In this sense, the performance of this algorithm is as good as or better than related algorithms such as Lloyd's algorithm. As $\beta \to \infty$, the DA algorithm essentially converges to Lloyd's algorithm, with the choice of initial placement of locations (through the annealing process) designed to achieve better minima. The phase transition property also obviates the hierarchical nature of the algorithm. When the value of β crosses a critical threshold value, β_c, the number of *distinct* resource locations jumps; this is referred to as a *splitting* process and is used explicitly to identify *natural* clusters in the data.

The sensitivity-to-temperature property implies that in the static algorithm—the rate of change of the resource locations between two critical temperatures—can be bounded above. The condition on the minimum eigenvalue of $I - 2\beta C_{x_i|y_j}$ for $1 \leq j \leq M$ being bounded away from zero corresponds to non-critical temperature values. The bounds given in this theorem are conservative; better bounds can be obtained by imposing additional assumptions on the data [29]. This property has important consequences—namely the rate at which we change temperature values, that is the *cooling rate*, can be high—and typically we increase β geometrically, that is

$$\beta_k = \gamma^k \beta_0 \quad \text{for some } \gamma > 1. \tag{10.16}$$

It is this property that forms the basis for extending the MEP-based framework to the dynamic setting.

Remark 10.1. Recall that the static problem is NP-hard. The cooling law for the annealing process in the DA algorithm is typically geometric, as is indicated in (10.16), and therefore avoiding local minima comes at a cost of only a few iteration steps (in comparison, other annealing procedures such as simulated annealing require much slower ($O(\log N)$) cooling rates [9]). The MEP-based framework has additional flexibility, for example, we have proposed adoptings constraints that reflected computational expense and obtained weight functions that accommodated for this constraint. The resulting scalable algorithms were 75% – 80% more efficient in terms of computation time than the original algorithm.

10.3 The Dynamic Maximum Entropy Framework

In this section, we present our general framework for formulating the dynamic clustering problem and determining computationally efficient algorithms that resolve issues of numerous local minima, quantifying coverage and cluster resolution, spatio-temporal smoothening, achieving trade-offs between computational cost and resolution and tracking dynamically evolving clusters. As noted earlier, using a frame-by-frame approach is not computationally viable as it requires multiple iterations at each time step.

We use free energy as the metric for coverage, which together with an annealing process provides a basis for an algorithm that is independent of the initial choice of resource locations and avoids local minima. The instantaneous cluster center z^c, which is determined solely by site locations x_i (see (10.9)), is given by the recursion $z^c = \check{P}^{cT}_{x|y} x$, where $P^c_{x|y} = [p(x_i|z^c_j)] \in \mathbb{R}^{N\times M}$ (see Sec. 10.2.4 for details on notation). Note we represent the cluster center by z^c to distinguish it from the instantaneous resource location y. Since cluster centers, shapes, and mass are specified by

$$z^c = \check{P}^{cT}_{x|y} x,\ \check{P}^c_{x|y} \text{ and } \check{P}^c_y$$

respectively (see Sec. 10.2.4), the cluster-drift, intra-cluster, and inter-cluster-interaction dynamics are quantified by ϕ, $\dot{\check{P}}^c_{x|y}$, and $\dot{\check{P}}^c_y$, respectively.

In adapting the static concepts for the dynamic setting we must determine appropriate cooling rates (as in (10.16)) relative to the time scales of the site dynamics. For example, if the rate of cooling is much faster than the given dynamics of the sites, the resulting algorithm is similar to the frame-by-frame approach. However, from the sensitivity-to-temperature property in Theorem 10.1, we know the resource locations are insensitive to temperatures between two successive critical temperatures. Critical temperatures are indicative of splits and resolution, thus decreasing temperature values matter only for forcing these splits or obtaining higher resolution. Note that in the dynamic setting, the cluster splits at time t are still identified by critical temperatures given by $\beta_c^{-1} = 2\lambda_{\max}(C_{x(t)|y_j(t)})$ (once the resource locations $y_j(t)$ are at cluster centers). However, unlike the static case, these splitting conditions can be reached due to dynamics of x and y, in addition to decreasing values of β.

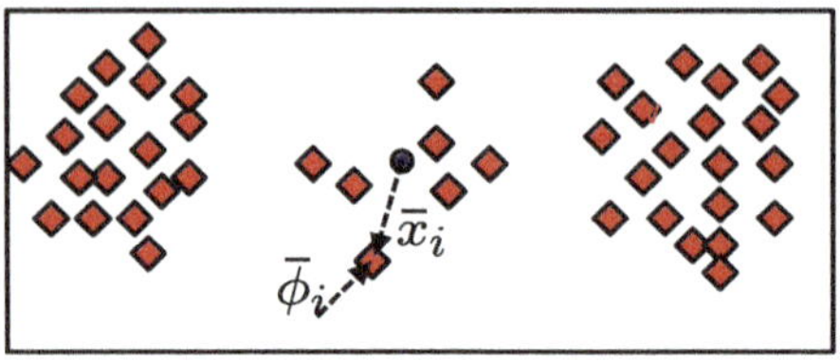

Fig. 10.2 The condition on the velocity field ϕ for consistent clusters. The solid circle represents the weighted cluster centroid x_i^c that is closest to the ith site with velocity ϕ_i, i.e., $\bar{\phi}_i = p_i\phi_i$. The condition $\bar{\phi}^T\bar{x}^c < 0$ implies that the angle between $\bar{x}_i^c$ and $\bar{\phi}_i$ is obtuse (in an average sense) and hence the individual sites are directed toward the cluster and not away from it.

This interpretation of the sensitivity-to-temperature property allows us to separate the dynamic clustering problem into two subproblems: (1) tracking cluster centers; and (2) monitoring splitting conditions. In our implementations, we split the resource locations only after they have reached the cluster centers. We monitor the cluster splits that result due to site dynamics or the cooling law. The resulting algorithm, as well as the incorporation of user-specified decisions on splits and resolutions, are presented later in this section.

Our main tracking results are briefly summarized in the following.

1. Under the assumption that the site dynamics given by $\phi(\zeta,t)$ in (10.1) are continuously differentiable, we provide a control law $u(\zeta)$ such that the resource locations asymptotically track the cluster centers (see Theorem 10.3).
2. We show this control law is nonconservative, that is, if $u(\zeta)$ is not bounded then there *does not exist any* Lipschitz control law that achieves asymptotic tracking for the system given by (10.1) (see Theorem 10.4).
3. If we further assume the clusters are *consistent* (that is, the average cluster size is nonincreasing and clusters are separated), then the prescribed control that achieves asymptotic tracking is *bounded* (see Theorem 10.5).

The results stated in 1 and 2 require only mild assumptions on the site dynamics. For the result in 3 we require the clusters to be consistent, that is, the distance from a site location x_i to the closest cluster center z_j^c does not increase with time. We impose this constraint, albeit in an average sense, by assuming that ϕ satisfies

$$\bar{\phi}^T\bar{x}^c \leq 0, \text{ where } \bar{\phi} = \check{P}_x^c\phi \text{ and } \bar{x}^c = x - \check{P}_{y|x}^c\check{P}_{x|y}^{cT}x$$

Since the cluster center $z^c = \check{P}_{x|y}^{cT}x$, $\bar{x}^c$ denotes the average weighted vector from each site location x_i to these cluster centers (with the nearest cluster center weighted heavily for large β), consequently, $\bar{\phi}^T\bar{x}^c \leq 0$ implies that velocity ϕ is such that magnitudes of these vectors are nonincreasing (see Fig. 10.2). We first elaborate on properties of the free energy term and its time derivative, which form the basis for addressing both the problems of tracking cluster centers and monitoring splitting conditions.

10.3.1 The Free Energy Term

The crux of the DA algorithm for the static setting is that it replaces the notion of distortion in (10.2) by the free energy in (10.8) as a metric for coverage. In order to achieve the objective of tracking the cluster centers without resorting to the frame-by-frame approach, we formulate a control problem, where we design the velocity $u(t)$ for the resource locations such that the time derivative $\dot{F}$ of the free energy function is nonpositive along the trajectories of $x(t)$ and $y(t)$. Such a formulation not only addresses the drawbacks of the frame-by-frame clustering approach, but also preserves the advantages of the DA algorithm found with static data. Moreover, the condition under which control authority is lost, that is, where $\partial F/\partial y = 0$, provides explicit connections to the splitting conditions and cooling laws encountered in the static setting. We summarize the properties of free energy and its time derivative in the following theorem. For a proof of these properties see [29].

Theorem 10.2. *Let F given by (10.8) be the free energy for the sites and resources x_i, $1 \le i \le N$, and y_j, $1 \le j \le M$, whose dynamics are defined by (10.1). Then*

1. Positivity: $F(\zeta) + \frac{1}{2\beta}\log M > 0$ *for all ζ in* $\left(\mathbb{R}^2\right)^{N+M}$.
2. Structured derivative: *The derivative of the free energy term has the following structure:*

$$\dot{F} = 2\zeta^T \Gamma(\zeta) f(\zeta), \ \ \Gamma = \begin{pmatrix} \check{P}_x & -\check{P}_{xy} \\ -\check{P}_{xy}^T & \check{P}_y \end{pmatrix}. \tag{10.17}$$

   The matrix Γ is a symmetric positive semidefinite matrix for all ζ, which can be decomposed as $\alpha(I-W)$ where $\alpha > 0$, I is the identity matrix and W is a symmetric doubly stochastic matrix with spectral radius $\rho(W) = 1$.
3. Loss of dynamic control authority at cluster centers: *The derivative $\dot{F}$ becomes independent of the control, that is, $\partial(\dot{F})/\partial u = 0$ (or equivalently the partial derivative $\partial F/\partial y = 0$) only at those time instants t_c when the resource locations $y_j(t_c)$ are coincident with the cluster centers, that is, only when $y_j(t_c) = \sum_{i=1}^{N} p(x_i(t_c)|y_j(t_c))x_i(t_c)$ for $1 \le j \le M$.*

10.3.2 Control Design: Tracking Cluster Centers

The resource locations coincide with the cluster centers only when $y_j - \sum_i p(x_i|y_j)x_i$ is zero for each j, that is, when

$$\bar{y} \stackrel{\triangle}{=} y - \check{P}_{x|y}^T x = 0$$

For design of u, we transform the coordinates $\zeta = (x,y)$ to $\bar{\zeta} = (x,\bar{y})$, in which the dynamics in (10.1) and $\dot{F}$ given by (10.17) are rewritten as

$$\left.\begin{aligned}\dot{x}(t) &= \phi,\\ \dot{\bar{y}}(t) &= u - \dot{\check{P}}_{x|y}^{T} x - \check{P}_{x|y}^{T}\phi\end{aligned}\right. \Longleftrightarrow \dot{\bar{\zeta}} = \bar{f}(\bar{\zeta},t) \text{ and } \dot{F} = \bar{x}^T\bar{\phi} + \bar{y}^T\check{P}_y\bar{u} \tag{10.18}$$

where $\bar{u} = u - \check{P}_{x|y}^{T}\phi$. We exploit the affine dependence of $\dot{F}$ in (10.18) on $\bar{u}$ to make $\dot{F}$ nonpositive analogous to control design based on control Lyapunov functions [24, 31, 32]. More specifically we choose $\bar{u}$ from sets of the form:

$$\bar{\mathcal{U}}(\alpha) = \{\bar{u} : (\mathbb{R}^2)^{N+M} \to (\mathbb{R}^2)^M \mid \bar{u}(\bar{\zeta}) = -[K_0 + \frac{\alpha(\bar{\zeta}) + \theta(\bar{\zeta})}{\bar{y}^T\check{P}_y\bar{y}}]\bar{y},\ \bar{y} \neq 0\} \tag{10.19}$$

for some $\theta(\bar{\zeta}) \geq 0$ and $K_0 > 0$, which are parameterized by functions

$$\alpha(\cdot) : (\mathbb{R}^2)^{N+M} \to \mathbb{R}$$

The following theorem establishes that the assumption of continuously differentiable ϕ in (10.1) (which ensures $\dot{F}$ is of bounded variation) is adequate for the control design to achieve asymptotic tracking of clusters. We first state the following lemma, which is used in the proof of the theorem; a proof for this lemma can be found in [29].

Lemma 10.1. *For a nonnegative function $g : \mathbb{R} \to \mathbb{R}$ of bounded variation, if $\int_0^\infty g(\tau)d\tau < \infty$, then $\lim_{t\to\infty} g(t) = 0$.*

Theorem 10.3 (Tracking). *For the site-resource dynamics given by (10.18), if*

$$u = \bar{u} + \check{P}_{x|y}^{T}\phi$$

where $\bar{u} \in \bar{\mathcal{U}}(\bar{x}^T\bar{\phi})$, then $\dot{F} \leq 0$ for all $t \geq 0$ and the resource locations asymptotically track the cluster centers, that is, $\lim_{t\to\infty}\bar{y}(t) = 0$.

Proof. *For $\bar{u} \in \bar{\mathcal{U}}(\bar{x}^T\bar{\phi})$, it is straightforward to show that $\dot{F}$ given by (10.18) reduces to $\dot{F} = -K_0\bar{y}^T\check{P}_y\bar{y} - \theta(\bar{\zeta})$ for some $\theta(\bar{\zeta}) \geq 0$ and $K_0 > 0$. Therefore $\dot{F} \leq 0$. Since $F(t)$ (which denotes the time-dependence of free energy $F(\zeta(t))$) is bounded below (from Theorem 10.2(1)) and $\dot{F} \leq 0$, this implies $F(t) \to F_\infty$ for some $|F_\infty| < \infty$ as $t \to \infty$. Thus,*

$$\int_0^\infty |\dot{F}(\tau)|d\tau = -\int_0^\infty \dot{F}(\tau)d\tau = F(0) - F_\infty < \infty$$

Therefore, since $\dot{F}$ is of bounded variation, we can deduce that $\lim_{t\to\infty}|\dot{F}(t)| = 0$ from Lemma 10.1. Since

$$\dot{F} = -K_0\bar{y}^T\check{P}_y\bar{y} - \theta(\bar{\zeta}) \Rightarrow K_0\bar{y}^T\check{P}_y\bar{y} \leq |\dot{F}| \text{ and } \check{P}_y \text{ is positive definite}$$

with elements bounded away from zero (since $\min_j\{p(y_j)\} \geq \nu > 0$), we have $\bar{y} \to 0$ as $t \to \infty$. □

Clearly, resource velocities of the form $u = \bar{u} + \check{P}^T_{x|y}\phi$ can take very large values near $\bar{y} \neq 0$. The following theorem emphasizes that this control design is not conservative, that is, if there exists a control design that achieves $\dot{F} \leq 0$ without growing unbounded near $\bar{y} = 0$, then there necessarily exists a bounded element from $\bar{\mathcal{U}}(\bar{x}^T\bar{\phi}) + \check{P}^T_{x|y}\phi$ that guarantees $\dot{F} \leq 0$.

Theorem 10.4 (Lipschitz property of control). *If there exists a* $\hat{u} : \left(\mathbb{R}^2\right)^{N+M} \to \left(\mathbb{R}^2\right)^M$ *such that* $\hat{u}$ *is Lipschitz at* $\bar{\zeta} = 0$ *and*

$$\dot{F} = \bar{x}^T\bar{\phi} + \bar{y}^T\check{P}_y(\hat{u} - \check{P}^T_{x|y}\phi) \leq 0$$

then $u = \bar{u}_S + \check{P}^T_{x|y}\phi$ *is also Lipschitz at* $\bar{\zeta} = 0$*, where*

$$\bar{u}_S(\bar{\zeta}) = -[K_0 + \frac{\bar{x}^T\bar{\phi} + \sqrt{|\bar{x}^T\bar{\phi}|^2 + (\bar{y}^T\check{P}_y\bar{y})^2}}{\bar{y}^T\check{P}_y\bar{y}}]\bar{y} \in \bar{\mathcal{U}}(\bar{x}^T\bar{\phi}). \tag{10.20}$$

i.e., there exists a $\delta > 0$ *and a constant* c_0 *such that* $\|\bar{u}_s(\bar{\zeta})\| \leq c_0\|\bar{\zeta}\|$ *for* $\|\bar{\zeta}\| \leq \delta$.

Theorem 10.4 can be proven in a manner analogous to the proof for Proposition 3.43 in [24]); see [29] for details. Theorems 10.3 and 10.4 are general in the sense that they do not impose conditions other than smoothness on ϕ. We can obtain better bounds on the control effort needed if we assume additional conditions on the cluster dynamics. We use the following lemma, the proof of which can be found in [29].

Lemma 10.2. *Let* $x \in \left(\mathbb{R}^2\right)^N$ *and* y *in* $\left(\mathbb{R}^2\right)^M$ *satisfy*

$$\frac{\|y_{c[i]} - x_i\|^2}{\|y_j - x_i\|^2} \leq \delta < 1,\ \forall 1 \leq j \leq M,\ j \neq c[i]$$

where $y_{c[i]}$ *is the unique closest resource location to* x_i*, that is,*

$$\|y_{c[i]} - x_i\| = \min_{1 \leq j \leq M} \|y_k - x_i\| \text{ for } 1 \leq i \leq N.$$

Then, there exists a constant $\bar{k}_2 < \infty$ *such that*

$$\|\check{P}_x\check{P}_{y|x}\check{P}^T_{x|y} - \check{P}_x\check{P}^c_{y|x}\check{P}^{cT}_{x|y}\| < 3k_2\|\bar{y}\|(1 + k_2\|\bar{y}\|^2 + k_2^2\|\bar{y}\|^2)$$

Theorem 10.5 (Bounded control). *Assume the cluster dynamics are size-consistent, that is*

$$\bar{\phi}^T\bar{x}^c \leq 0 \text{ and } \frac{\|y_{c[i]} - x_i\|^2}{\|y_j - x_i\|^2} \leq \delta < 1\ \forall j,\ 1 \leq j \leq M,\ j \neq c[i],$$

where $y_{c[i]}$ *is the unique closest resource location to* x_i*. Consider the control law*

$$u(\bar{\zeta}) = \check{P}^T_{x|y}\phi + [K_0 + \frac{\alpha + \sqrt{\alpha^2 + (\bar{y}^T\check{P}_y\bar{y})^2}}{\bar{y}^T\check{P}_y\bar{y}}]\bar{y},$$

where $\alpha = \bar{\phi}^T\bar{x} - \bar{\phi}^T\bar{x}^c$ *and* $K_0 > 0$. *Then, for some quadratic function* $c_1(\cdot)$,

$$\lim_{t\to\infty}\bar{y} = 0 \text{ and } \|u(\bar{\zeta})\| \leq (2\nu^{-1}c_1(\|\bar{y}\|)\|\|x\| + 1)\|\phi\| + (K_0+1)\|\bar{y}\|$$

Therefore, if $\|\phi\|, \|x\| \leq c_2 < \infty$, *then* $\|u(\bar{\zeta}(t))\|$ *is bounded.*

Proof. *Note that* $\dot{F} = \bar{\phi}^T\bar{x} + \bar{y}^T P_y(u - \check{P}^T_{x|y}\phi) = \bar{\phi}^T\bar{x}^c + \alpha + \bar{y}^T P_y(u - \check{P}^T_{x|y}\phi)$. *After substituting for u, we obtain*

$$\dot{F} = \bar{\phi}^T\bar{x}^c - K_0\bar{y}^T\check{P}_y\bar{y} - \sqrt{\alpha^2 + (\bar{y}^T\check{P}_y\bar{y})^2} \leq 0$$

By following the same arguments as in the proof for Theorem 10.3, we can show that $\lim_{t\to\infty}\bar{y} = 0$. *From Lemma 10.2 we have*

$$\|\check{P}_x\check{P}_{y|x}\check{P}^T_{x|y} - \check{P}_x\check{P}^c_{y|x}\check{P}^{cT}_{x|y}\| < \|\bar{y}\|c_1(\|\bar{y}\|)$$

for some quadratic function $c_1(\cdot)$. *Since* $\alpha = x^T(\check{P}_x\check{P}_{y|x}\check{P}^T_{x|y} - \check{P}_x\check{P}^c_{y|x}\check{P}^{cT}_{x|y})\phi$, *we have*

$$|\alpha\|\bar{y}\|/\bar{y}^T\check{P}_y\bar{y}| \leq \nu^{-1}c_1(\|\bar{y}\|)\|x\|\|\phi\|$$

Further

$$\|u(\bar{\zeta})\| \leq \|\phi\| + (2|\alpha|\|\bar{y}\|/\bar{y}^T\check{P}_y\bar{y}) + (1 + K_0\|\bar{y}\|)$$

implies $\|u(\bar{\zeta})\|$ *is bounded by*

$$\|\phi\| + 2\nu^{-1}c_1(\|\bar{y}\|)\|x\|\|\phi\| + (K_0+1)\|\bar{y}\| \leq (2\nu^{-1}c_1(\|\bar{y}\|)\|c_2 + 1)c_2 + (K_0+1)\|\bar{y}\|$$

where $\bar{y} \to 0$ *from above.* □

□

This design procedure provides great flexibility, with the free energy F viewed as a control Lyapunov function and u designed to make $\dot{F} \leq 0$. Alternative computationally efficient control laws can be devised by exploiting the properties of Γ. For instance, we can guarantee exponential convergence of $\bar{y} \to 0$ by appropriately choosing the function $\theta(\bar{\zeta})$ in the control design [29].

Note that the size-consistency assumption on cluster dynamics is required only in the interim period when resource locations are far from the cluster centers (i.e., when $\bar{y} \neq 0$). Once the resource locations track the cluster centers, these assumptions are not required, and instead monitoring cluster splits is required.

10.3.3 Cluster Evaluation

As the resource locations begin to track the cluster centers, cluster evaluation is completed based on the following decision choices.

Splitting: The decision to split can be enforced when the parameter β satisfies the condition

$$\beta^{-1} = 2\lambda_{\max}(C_{x(t_c)|y_j(t_c)})$$

If time is fixed at $t = t_c$, this splitting condition implies that y is at a local maxima (or inflection point) of F (i.e., the Hessian of F is no longer positive definite).

The set of resource locations after splitting is determined by the same procedure used as in the static case described in [28]. The new locations are denoted by

$$\mathbf{y}_{new} = q_{\text{split}}(\mathbf{y})$$

Splitting does not pose problems for the dynamic implementation of the algorithm as the computation time to determine $q_{\text{split}}(\mathbf{y})$ is negligible compared to computation of $u(t)$.

Tracking: At time t if $y_j, 1 \leq j \leq M$, are at the cluster centers and the parameter β does not satisfy the splitting condition, then one can continue to track the cluster centers by assigning $y_j(t) = \sum_i p(x_i(t)|y_j(t))x_i(t)$. Alternatively, one can choose to improve coverage resolution by increasing cooling rates, eventually leading to the splitting condition being satisfied, resulting in finer clusters and therefore higher resolution.

10.4 Scalability of the DME Algorithm

The framework presented in this chapter aims at avoiding local minima. As a consequence the computations are global, or centralized, in the sense that the computation of *each* resource location uses the values of *all* the site locations. However, the contribution of distant site locations becomes progressively smaller as the parameter β is increased. In fact, the partitions are hard as $\beta \to \infty$ and consequently the contribution of site locations that are not nearest neighbors goes to zero. In this sense, the computation of resource locations changes from being *truly global* to *truly local*, or distributed, as β is increased from zero to infinity.

In the case of static sites, we have exploited this tendency toward distributedness to develop a scalable algorithm that provides a close approximation to the original algorithm. The basic idea is straightforward: quantify and exploit inherent cluster separations based on cluster interaction information, then apply the MEP-based algorithm separately to each cluster. This approach leads to a hierarchical application of this algorithm. More specifically, at each phase transition, we compute the *level of interaction* between clusters by

$$\varepsilon_{ji} = \sum_{x \in R_i} p(y_j|x)p(x), \tag{10.21}$$

which represents the level of interaction between sites in cluster R_i with cluster R_j. We then determine subsets of strongly interacting clusters based on these ε-values and apply the DA to each subset individually, ignoring the other clusters. Obviously the trade-off in this is that the resulting distortions will be larger than in the original algorithm. We have shown we can bound the difference in the resulting cluster centers found by the scalable algorithm as follows:

$$|y_i - \hat{y}_i| < \varepsilon_i \left(\frac{1 - \sum_{x \in R_i} p(x)}{1 - \varepsilon_i \sum_{x \in R_i} p(x)} \right) \quad \text{with } \varepsilon_i = \sum_{j \neq i} \varepsilon_{ij}, \tag{10.22}$$

where y_i represent the cluster center for cluster R_i determined from the original DA algorithm, and $\hat{y}_i$ represent the cluster center for cluster R_i determined from the scalable algorithm.

In simulations on data sets containing from 5000 to 15,000 static sites, the scalable algorithm has converged to a solution approximately six to seven times faster on average than the DA algorithm, with an increase in the final distortion value obtained of only about 5%. On data sets containing from 15,000 up to 50,000 static sites, the scalable algorithm converges to a solution where the original DA algorithm fails; see [28] for further details.

To extend this approach to the dynamic setting, we first note that clusters can interact even when β values are high, due to site dynamics. Thus we propose to monitor values of the norm of the Hessian $\partial^2 F / \partial y^2$ and estimate the *effective* radius around each cluster center y_j, beyond which the site locations can be ignored.

10.4.1 Incorporating Scalability into the Cost Function

An additional approach for improving scalability in our DME algorithm is to consider incorporating computational cost into the algorithm's objective function. More specifically, we consider computing the association weights $p(x_i, y_j)$ and resource locations y_j such that the entropy term

$$H(x,y) = -\sum_j \sum_i p(x_i, y_j) \log(p(x_i, y_j))$$

is maximized subject to constraining the distortion, or coverage cost, as before by D_0, with

$$D_0 = \sum_j \sum_i p(x_i) p(y_j|x_i) d(x_i, y_j)$$

and also constraining the *cluster interaction* and the *computational cost* by values C_I and C_C, respectively, where

$$C_I = \sum_j \varepsilon_{ji} d(x_i, y_j) M_{ij}, \text{ with } M_{ij} = \begin{cases} 0 & x_i \in R_j \\ 1 & x_i \notin R_j \end{cases} \tag{10.23}$$

and

$$C_C = \sum_j \varepsilon_{ij} N_j^2, \text{ with } N_j = \sum_i (1 - M_{ij}). \tag{10.24}$$

That is, we now solve the modified Lagrangian

$$\max_{y_j, p(y_j|x_i)} \quad H - \beta_1 D_0 - \beta_2 C_I - \beta_3 C_C \tag{10.25}$$

To further improve tractability of this algorithm, we consider a relaxation of the problem where M_{ij} and N_j are approximated by the differentiable functions

$$M_{ij} = 1 - e^{-\frac{d(x_i, y_j)}{\sigma_j}} \text{ and } N_j = \sum_i e^{-\frac{d(x_i, y_j)}{\sigma_j}}. \tag{10.26}$$

Implementation and simulations using these approaches are ongoing.

10.5 Simulations

In this section, we present simulation results for a variety of data sets with different underlying dynamics. These test cases are specifically designed to highlight key features of our DME framework and to allow for performance analysis and discussion.

10.5.1 The Basic Algorithm

We summarize in the following outline the main steps and flow of the the algorithm. This implementation applies to the basic version of the algorithm, having no external user-based directives:

- **Step 0:** Initialize the algorithm: $t = t_0$, $\beta = 0$, and $u = 0$
- **Step 1:** Determine the resource locations (10.9) together with the association probabilities (10.7)
- **Step 2:** Simulate motion of sites x_i under (10.1) and determine resource velocities from (10.1) for time interval Δt
- **Step 3a:** If $\bar{y} = 0$, then
  - Evaluate phase transition condition:
    - Satisfied:
      - If more resolution required, split y_j, redistribute weights, determine new y_j values and return to **Step 2**; otherwise check stopping-time criteria and exit if met, else return to **Step 2**

  - Not satisfied:
    - If more resolution required, increase β and return to **Step 2**; otherwise set y_j to respective centroids and return to **Step 2**

- **Step 3b:** If $\bar{y} \neq 0$ then compute control u using (10.20) and return to **Step 2**

All simulations were carried out in MATLAB. For these simulations the dynamics in (10.1) were completed by discretizing using a fourth-order Runge–Kutta method (RK-4) [2]. In the RK-4 method, the error per step is $O((\Delta t)^5)$, while the total accumulated error is $O((\Delta t)^4)$. The time steps were chosen so that the solution converges. Note that the time required to compute $q_{\text{split}}(y)$ is comparable to that required to compute $u(t)$.

10.5.2 Natural Cluster Identification and Tracking

In this primary simulation, we use a data set with 160 mobile sites. A velocity field $\phi(t,x)$ is assigned to each of the mobile sites such that natural clusters emerge within 8 seconds.

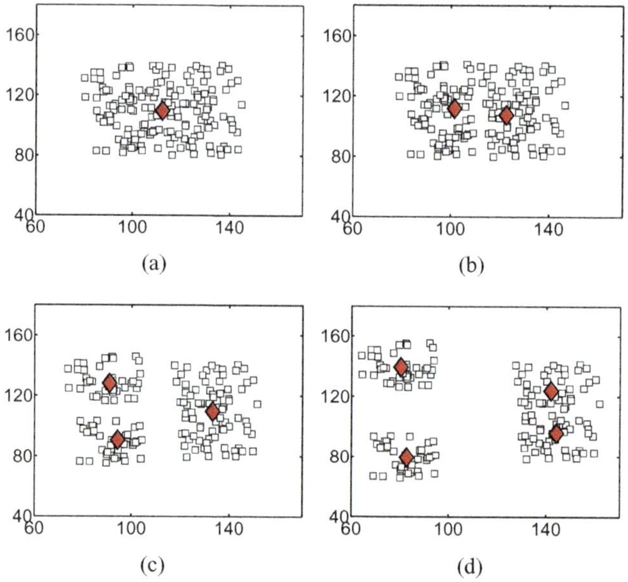

Fig. 10.3 Simulation results showing cluster identification and tracking. Snapshots (a), (b), (c) and (d) show the locations of mobile sites $x_i, 1 \leq i \leq 160$ (shown by squares) and resources $y_j, 1 \leq j \leq M$ (shown by diamonds) at various phase transition instances. All sites are initially concentrated at the center of the domain area and then slowly drift apart. Four natural clusters emerge at the end of the time horizon, shown in (d).

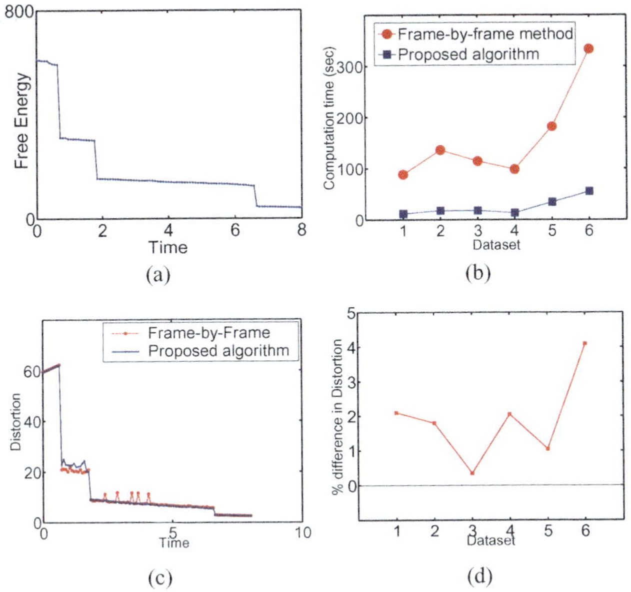

Fig. 10.4 (a) Free energy F with respect to time. The control value u ensures that $dF/dt \leq 0$. Sharp decreases in F are due to phase transitions. Progressively decreasing free energy results in improved coverage and tracking throughout the time horizon. (b) Comparison of the computation times for the frame-by-frame approach and the DME algorithm. (c) Comparison of the distortion achieved by the frame-by-frame method and the proposed framework. (d) Percentage error in final distortion achieved by the proposed DME algorithm versus the frame-by-frame method.

The algorithm is initiated with one resource placed at the centroid of the sites (at $t = 0$) (a diamond denotes this location and squares represent the sites, Fig. 10.3(a)). As the site dynamics evolve, the site locations move according to the equation $\dot{x} = \phi(x,t)$. The DME algorithm progressively updates the association probabilities and resource locations, and determines control values (i.e., the resource velocities from (10.20)) in order to track cluster centers. Figures 10.3(b), (c) and (d) show the locations of the sites and the resources in the interim instants. The number of resource locations increases progressively due to successive phase transitions, and as seen in the figure, the resource locations identify and track natural clusters in the underlying data. At the end of the time horizon, all natural clusters are identified. The algorithm successfully avoids several local minima and provides progressively better coverage. Figure 10.4(a) shows a plot of the coverage function F with respect to time. Note that at each phase transition, there is a sharp decline in F. This decline is due to the fact that at every phase transition, one or more resource locations are added, which results in lower free energy, thereby providing better coverage.

Depending on the distribution of the underlying data, the DME algorithm is five to seven times faster than the frame-by-frame method. A comparison of the instantaneous distortion value $\sum_i p_i \min_j d(x_i, y_j)$ obtained by the two algorithms is presented in Fig. 10.4(c). The proposed algorithm identifies and tracks the clusters hierarchically such that the distortion steadily decreases, with sharp decreases at splits. As seen in the figure, the frame-by-frame method for the same number of resources achieves slightly lower distortion, but with frequent spikes due to the spatio-temporal effects. Moreover, the frame-by-frame method uses five times the computation time required for the proposed algorithm. Figure 10.4(d) shows a comparison of the final distortion achieved by the proposed algorithm and the frame-by-frame method (for the same time step and same number of resources). As is seen in the figure, the proposed algorithm achieves a distortion similar to the frame-by-frame approach (within 0.5% to 4.3%), however it uses considerably less computation time as shown in (b).

In the absence of spatio-temporal smoothening, the clustering solutions obtained at two successive time instants might be considerably disparate, even though the number of natural clusters in the data set remains the same. Figure 10.5 shows the clustering solution obtained by the frame-by-frame approach at two successive time instants. Three clusters are identified by the algorithm at each instant, but at considerably different spatial locations. This occurs because no information from the previous clustering solution is used for determining the solution at the next time instant. On the other hand, the proposed framework overcomes this problem by using a smooth control value everywhere except during cluster splits via phase transition. In order to speed up the frame-by-frame approach, we increase the time step between successive frames. During such an implementation, the resource locations obtained at the previous time instant are used in the interim between successive frames. This results in a clustering solution that deteriorates in the interim because of the lack of new information. Figures 10.5(c) and (d) show the distortion obtained by a three fold and six fold increase in the time steps. As is seen in the figures, the distortion obtained from the frame-by-frame approach deteriorates considerably with respect to the proposed algorithm. Note that even for a six fold increase in the time step, the computation time for the frame-by-frame is slightly higher than that of the proposed algorithm. In most of the applications, such a phenomenon would not be desirable.

Once natural clusters are identified and tracked, the user may desire higher resolution clustering and/or simultaneous tracking. User-based directives are easily incorporated in our DME framework, providing flexibility in the modes of operation. Higher resolution clusters can be achieved by increasing the annealing parameter value until the phase transition condition is satisfied and the eventual splitting of resources is obtained.

10.6 Ongoing Work and Conclusions

In this section we discuss ongoing work and further extensions of the framework presented herein.

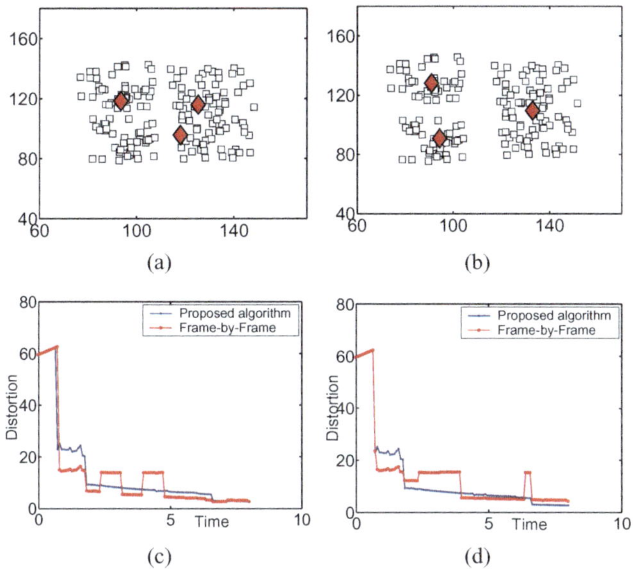

Fig. 10.5 (a) and (b): Clustering results from two successive frames using the frame-by-frame approach. In both instances, three resource locations were identified by the algorithm, but at considerably different positions. Such a solution violates the spatio-temporal requirement of the dynamic clustering algorithm. This happens because none of the clustering information from previous frame is employed in order to determine the solution at future frames. (c) and (d): Comparison of distortion obtained by the proposed algorithm and the frame-by-frame approach under different time steps. (c) Proposed algorithm : $\Delta t = 0.08$ s, frame-by-frame approach: $\Delta t = 0.24$ s. (d) Proposed algorithm : $\Delta t = 0.08$ s, frame-by-frame approach: $\Delta t = 0.48$ s.

10.6.1 General Extensions of the Coverage Problem

Constraints on resources: The resources in the framework presented are assumed to be identical. However, depending on the underlying problem domain, these resources may be heterogeneous, where different constraints apply to different resources; for example, vehicles may be of different sizes with different coverage capacities. Resources can be made nonidentical in our framework by introducing weights λ_j to each resource location y_j. This interpretation yields a modified free energy function,

$$F = -\frac{1}{\beta}\sum_i p_i \log \sum_j \lambda_j e^{-\beta d(x_i,y_j)}. \tag{10.27}$$

Constraints on resources are implemented by specifying constraints on λ_j. For example, constraining the λ_j by constants W_j yields resource locations that have weights in the same ratios as $\{W_j\}$. More details on this formulation for static problems in the context of facility location, drug discovery and vector quantization are presented, respectively, in [23, 28, 21]. The same formulation easily extends to the dynamic case, where static constraints can be incorporated by redefining the free energy in (10.27), and minimizing the appropriate Lagrangian $L = F + \sum_j \mu_j(\lambda_j - W_j)$ with respect to $\{y_j\}$ and $\{\lambda_j\}$.

Inertial forces in vehicle dynamics: In this chapter, we have focused on tracking of cluster centers when velocity fields (one state per direction for each vehicle) are given for sites. Procedures that are applicable when higher order (i.e., multiple-state) differential equations are given are presented in [34]. For example, in the context of vehicle systems, the dynamics of autonomous mobile agents are often controlled by thrust actuators, the control term being the amount of thrust in each direction. The corresponding model for a domain with N mobile agents becomes $x_i = [\xi_i \;\; \eta_i]^T \in \mathbb{R}^2$ and M resource locations $y_j = [\rho_j \;\; \omega_j]^T \in \mathbb{R}^2$ as before, whose dynamics are now given by

$$\ddot{x}(t) = \Upsilon(x(t), \dot{x}(t), y(t), \dot{y}(t), t), \;\; x(0) = x_0, \;\; \dot{x}(0) = \dot{x}_0, \tag{10.28}$$

$$\ddot{y}(t) = u(t), \;\; y(0) = y_0, \; \dot{y}(0) = \dot{y}_0. \tag{10.29}$$

Our task is to determine the accelerations ($\ddot{y}$) for the M mobile resources such that they track cluster centers. However by writing the above second order differential equations for each vehicle into first order vector differential equations, these equations take the same form as (10.1), albeit with additional algebraic structure. Under this scenario, the Euclidean distance metric is of the form

$$d(x_i, y_j) = (x_i - y_j)^2 + \theta(\dot{x}_i - \dot{y}_j)^2,$$

where θ is a constant. Choosing high values of θ thus gives relatively more importance to velocities and hence yields cluster centers for instantaneous headings (see [34]).

10.6.2 Estimating Data Dynamics

In the framework presented, the velocity fields ϕ of sites are assumed to be known. In the case of noisy dynamics, where ϕ is given by perturbations $n(t)$ about a nominal function $\bar{\phi}$, that is $\phi = \bar{\phi}(x, y, t) + n(t)$, and where the measurements of site and resource locations are noisy, control designs based on estimated values of (x, y) have provided satisfactory performance. This is primarily due to the fact that the shapes of the Gibbs distribution functions are insensitive to slight perturbations in x and y. When site dynamics are completely realized by velocity fields, if these fields are not known a priori, we can estimate them using system identification methods, and design $u(t)$ based on the estimated fields. The inherent robustness in the algorithm due to the properties of the Gibbs distribution again yield satisfactory performance.

10.6.3 Robustness to Modeling Uncertainties

The effect of noisy channel transmissions can also be accommodated in our framework, by modifying the instantaneous distortion term so that

$$\tilde{D} = \sum_i \sum_j p_i p(y_j|x_i) d'(x_i, y_j)$$

where $d'(x_i, y_j)$ reflects the probability of the event that x_i was actually sent given that the received message is interpreted as x_k.

Our DME framework can address robustness issues to a variety of transmission uncertainties. We can accommodate resource-dependent site-location uncertainties, for example, where the probability of receiving the location x_k when the location x_i is sent now depends on the receiver at the resource y_j. Additionally, transmission errors in resource locations can also be addressed; this case is analogous to the noisy vector quantization problem addressed in [21].

We can also address communication link failures; to do so, we introduce a binary random variable $\chi_{ij} \in \{0,1\}$, with a given probability distribution $p_{\chi_{ij}}$, where $\chi_{ij} = 1$ (or 0) implies that the ith site is (or is not) in communication with the jth resource. Again, we modify the instantaneous distortion term such that

$$\hat{D} = \sum_i \sum_j p_i p(y_j|x_i) p_{\chi_{ij}}(\chi_{ij}) \chi_{ij} d(x_i, y_j) \tag{10.30}$$

Appending the entropy to include the link failure probability distribution gives $\hat{H} = -\sum_i \sum_j p_i p(y_j|x_i) p_{\chi_{ij}}(\chi_{ij}) \log p(y_j|x_i)$, and the association probabilities then take the form

$$p(y_j|x_i) = \frac{\exp(-\beta(\sum_{\chi_{ij}} p_{\chi_{ij}}(\chi_{ij}) \chi_{ij} d(x_i, y_j)))}{Z_i} \tag{10.31}$$

$$\text{where } Z_i = \sum_j \exp(-\beta(\sum_{\chi_{ij}} p_{\chi_{ij}}(\chi_{ij}) \chi_{ij} d(x_i, y_j))) \tag{10.32}$$

The instantaneous free energy is then rewritten as

$$\hat{F} = -\frac{1}{\beta} \sum_i p_i \log \sum_j [-\beta(\sum_{\chi_{ij}} p_{\chi_{ij}}(\chi_{ij}) \chi_{ij} d(x_i, y_j))] \tag{10.33}$$

This term can now be used as a metric for instantaneous coverage under link failures. An analysis similar to that of the proposed algorithm in Sec. 10.2 can be employed to obtain dynamic clustering.

10.6.4 Conclusions

In this chapter, we proposed the dynamic maximum entropy (DME) framework for formulating and solving the dynamic coverage problem. As shown in the simulations, the proposed framework resolves both the coverage as well as the tracking aspects of the dynamic coverage problem. Using a control-theoretic approach to determine the velocity field for the cluster centers, we achieve progressively better coverage with time, which is shown to be five to seven times faster than the frame-by-frame method. The hierarchical aspect of the proposed algorithm enables us to identify natural clusters in the underlying data and characterize the notion of cluster resolution. Notions of coverage and clusters based on the MEP naturally allow for quantification of inter cluster and intra cluster dynamics, and greatly facilitate the simultaneous design process for the coverage and tracking objectives of the underlying problems.

Acknowledgements This work is partially supported by NSF label grants ECS-0725708 and ECS-0449310 CAREER.

References

1. Blankertz, B., Dornhege, G., Krauledat, M., Müller, K., Curio, G.: The non-invasive berlin brain–computer interface: Fast acquisition of effective performance in untrained subjects. Neuroimage **37**(2), 539–550 (2007)
2. Butcher, J.C.: Numerical Methods for Ordinary Differential Equations. John Wiley, New York (2004)
3. Chudova, D., Gaffney, S., Mjolsness, E., Smyth, P.: Translation-invariant mixture models for curve clustering. In: Proc. 9th ACM SIGKDD, pp. 79–88 (2003)
4. Cortés, J., Martínez, S., Karatas, T., Bullo, F.: Coverage Control for Mobile Sensing Networks. IEEE Trans. Robotics and Automation **20**(2), 243–255 (2004)
5. Drezner, Z.: Facility Location: A Survey of Applications and Methods. Springer Verlag, New York (1995)
6. Du, Q., Faber, V., Gunzburger, M.: Centroidal Voronoi Tessellations: Applications and Algorithms. SIAM Review **41**(4), 637–676 (1999)
7. Elia, N., Mitter, S.: Stabilization of Linear Systems with Limited Information. IEEE Trans. Automatic Control **46**(9), 1384–1400 (2001)
8. Frazzoli, E., Bullo, F.: Decentralized algorithms for vehicle routing in a stochastic time-varying environment. Proc. IEEE Conf. Decision and Control (CDC2004) pp. 3357–3363 (2004)
9. Geman, S., Geman, D.: Stochastic relaxation, gibbs distribution, and the bayesian restoration of images. IEEE Trans. Pattern Analysis and Machine Intelligence **6**, 721–741 (1984)
10. Gersho, A., Gray, R.: Vector Quantization and Signal Compression, 1st edn. Kluwer, Boston, MA (1991)
11. Gray, R., Karnin, E.D.: Multiple local minima in vector quantizers. IEEE Trans. Information Theory **IT-28**, 256–361 (1982)
12. Jaynes, E.T.: Information theory and statistical mechanics. Physical Review **106**(4), 620–630 (1957)
13. Jaynes, E.T.: Probability Theory—The Logic of Science. Cambridge University Press (2003)
14. Landau, L.D., Lifshitz, E.M.: Statistical Physics, Part 1, vol. 3, 3rd edn. Oxford (1980)

15. Lloyd, S.: Least-squares quantization in PCM. IEEE Trans. Information Theory **28**(2), 129–137 (1982)
16. Mitter, S.K.: Control with limited information. Plenary lecture at International Symposium on Information Theory, Sorrento, Italy (2000)
17. Nunez, P.L., Srinivasan, R.: Electric Fields of the Brain: The Neurophysics of EEG. Oxford University Press, USA (2006)
18. Omar, C., Vaidya, M., Bretl, T.W., Coleman, T.P.: Using feedback information theory for closed-loop neural control in brain-machine interfaces. In: 17th Ann. Computational Neuroscience Meeting (CNS). Portland, OR (2008). Invited Session on Methods of Information Theory in Computational Neuroscience
19. Rose, K.: Deterministic annealing, clustering, and optimization. Ph.D. thesis, California Institute of Technology, Pasadena (1991)
20. Rose, K.: Constrained Clustering as an Optimization Method. IEEE Trans. Pattern Anal. Machine Intell. **15**, 785–794 (1993)
21. Rose, K.: Deterministic annealing for clustering, compression, classification, regression and related optimization problems. Proc. IEEE **86**(11), 2210–39 (1998)
22. Salapaka, S.: Combinatorial optimization approach to coarse control quantization. In: Proc. 45th IEEE Conf. Decision and Control (CDC2005), pp. 5234–5239. San Diego, CA (2006)
23. Salapaka, S., Khalak, A.: Locational optimization problems with constraints on resources. In: Proc. 41st Allerton Conference, pp. 1240–1249 (2003)
24. Sepulchre, R., Jankovic, M., Kokotovic, P.: Constructive Nonlinear Control. Springer-Verlag (1997). URL http://www.montefiore.ulg.ac.be/services/stochastic/pubs/1997/SJK97a
25. Shannon, C.E., Weaver, W.: The mathematical theory of communication. Univ. of Illinois Press, Urbana IL (1949)
26. Sharma, P., Salapaka, S., Beck, C.: A maximum entropy based scalable algorithm for resource allocation problems. Proc. American Control Conference (ACC2007) pp. 516–521 (2007)
27. Sharma, P., Salapaka, S., Beck, C.: Entropy based algorithm for combinatorial optimization problems with mobile sites and resources. In: Proc. American Control Conference (ACC2008), pp. 1255–1260 (2008)
28. Sharma, P., Salapaka, S., Beck, C.: A scalable approach to combinatorial library design for drug discovery. Journal of Chemical Information and Modeling **48**(1), 27–41 (2008)
29. Sharma, P., Salapaka, S., Beck, C.: Entropy-based framework for dynamic coverage and clustering problems. IEEE Trans. Automatic Control (Accepted; to Appear 2011)
30. Sheu, J.B.: A fuzzy clustering-based approach to automatic freeway incident detection and characterization. Fuzzy Sets Syst. **128**(3), 377–388 (2002)
31. Sontag, E.D.: A Lyapunov-like characterization of asymptotic controllability. SIAM J. Control Optim. **21**, 462–471 (1983)
32. Sontag, E.D.: A 'universal' construction of Artstein's theorem on nonlinear stabilization. System and Control Letters **13**(2), 117–123 (1989)
33. Therrien, C.W.: Decision, Estimation and Classification: An Introduction to Pattern Recognition and Related Topics, vol. 14, 1st edn. Wiley, New York (1989)
34. Xu, Y., Salapaka, S., Beck, C.: Dynamic maximum entropy algorithms for clustering and coverage control. In: Proc. IEEE Conf. Decision and Control (CDC2010), pp. 1836–1841. Atlanta, GA (2010)

Chapter 11
Transverse Linearization for Underactuated Nonholonomic Mechanical Systems with Application to Orbital Stabilization

Leonid B. Freidovich and Anton S. Shiriaev

Abstract We consider a class of mechanical systems with an arbitrary number of passive (nonactuated) degrees of freedom, which are subject to a set of nonholonomic constraints. We assume that the challenging problem of motion planning is solved giving rise to a feasible desired periodic trajectory. Our goal is either to analyze orbital stability of this trajectory with a given time-independent feedback control law or to design a stabilizing controller. We extend our previous work done for mechanical systems without nonholonomic constraints. The main contribution is an analytical method for computing coefficients of a linear reduced-order control system, solutions of which approximate dynamics that is transversal to the preplanned trajectory. This linear system is shown to be useful for stability analysis and for design of feedback controllers orbitally, exponentially stabilizing forced periodic motions in nonholonomic mechanical systems. We illustrate our approach on a standard benchmark example.

11.1 Introduction

The problem of orbital stabilization of periodic motions in mechanical systems is challenging. However, it naturally appears in various applications. It is well known that mathematically rigorous analysis of orbital stability and design of orbitally stabilizing feedback controllers can be based on the properties of the dynamics that

Leonid Freidovich
Department of Applied Physics and Electronics, Umeå University, SE-901 87 Umeå, Sweden
e-mail: leonid.freidovich@tfe.umu.se

Anton S. Shiriaev
Department of Engineering Cybernetics, Norwegian University of Science and Technology (NTNU), NO-7491 Trondheim, Norway
Department of Applied Physics and Electronics, Umeå University, SE-901 87 Umeå, Sweden
e-mail: anton.shiriaev@itk.ntnu.no

R. Johansson and A. Rantzer (Eds.): Distributed Decision Making and Control, LNCIS 417, pp. 245–258.
springerlink.com

is transverse to the orbit [18, 8, 2, 14]. Defining such dynamics is equivalent to introducing special coordinates in a vicinity of a periodic motion [18] and has a nice geometric interpretation, known as a *moving Poincaré section* [10].

We have recently shown that, under some mild conditions, for mechanical systems with various controlled and uncontrolled forces such transverse dynamics can be computed analytically [15, 13, 14, 6]. Below, we derive an analogous result for a large class of underactuated nonholonomic mechanical systems.

11.2 Class of Systems and Problem Formulation

We follow the formulation of nonholonomic dynamics given in [11]; see also [1, 4, 3, 9] and references therein.

11.2.1 Dynamics of Underactuated Nonholonomic Systems

We consider a class of mechanical systems, dynamics of which can be described by the Euler–Lagrange equations with nonholonomic constraints:

$$\begin{aligned} M(q)\,\ddot{q}+C(q,\dot{q})\,\dot{q}+G(q) &= J(q)^T\lambda+B(q)\,u, \\ J(q)\,\dot{q}=0, \qquad u &= R(q,\dot{q})+Q(q,\dot{q})\,v, \end{aligned} \tag{11.1}$$

where $q \in \mathbb{R}^n$ and $\dot{q} \in \mathbb{R}^n$ are vectors of the generalized coordinates and velocities; $M(q)$ is a positive definite matrix of inertia; $\lambda \in \mathbb{R}^k$ is a vector of Lagrange multipliers related to the constraints defined by a matrix $J(q) \in \mathbb{R}^{k\times n}$ of constant full rank $k < n$. The multiplier λ can be interpreted as a vector of (normalized) amplitudes of the reaction forces necessary for keeping the k not integrable[1] relations $J(q)\,\dot{q} = 0$ invariant along the solutions; $u \in \mathbb{R}^m$ is a vector of control inputs with $0 \le m \le n-k$; $R(q,\dot{q})$ represents the preliminary feedback torque; the matrix $Q(q,\dot{q})$ completes the definition of a preliminary feedback transformation from u to v, it is not necessarily square and may change rank; $B(q)$ is a matrix function of constant full rank m; $G(q)$ represents the generalized forces due to gravity; and $C(q,\dot{q})\dot{q}$ is quadratic in $\dot{q}$ and describes Coriolis and generalized centrifugal forces.

The initial conditions for (11.1) are assumed to satisfy $J(q)\,\dot{q} = 0$ which leads to appropriate definitions of solutions that stay on the $(2n-k)$-dimensional submanifold of the state space

$$\mathcal{M} = \{(q,\dot{q}) : J(q)\,\dot{q} = 0\} \tag{11.2}$$

and the stability of the solutions; see e.g. [4, 3].

[1] If one of the k relations were integrable, it would be called *holonomic* while q would be a vector of not generalized but extensive coordinates. The lack of integrability is actually not essential for our derivations. However, we keep this assumption for the sake of terminological consistency.

Now, suppose that there exists a nontrivial periodic solution of (11.1) with $v \equiv 0$, along which

$$q = q_\star(t) = q_\star(t+T), \quad \dot{q} = \frac{d}{dt} q_\star(t) = \frac{d}{dt} q_\star(t+T) \tag{11.3}$$

with some $T > 0$. The task is to analyze the orbital stability of this solution under a particular choice of time-independent state feedback control law for v and to provide guidelines for designing stabilizing controllers.

As in our previous work, see e.g., [13, 14], we will use a special time-independent reparametrization of the desired orbit, known as *virtual holonomic constraint*; see [19] and references therein.

11.2.2 Target Periodic Motion via Virtual Holonomic Constraints

We assume that over the period the target solution (11.3) can be represented as

$$q_\star(t) = \Phi\big(\theta_\star(t)\big), \quad \dot{q}_\star(t) = \left.\frac{d\Phi(\theta)}{d\theta}\right|_{\theta=\theta_\star(t)} \dot{\theta}_\star(t), \tag{11.4}$$

where the smooth vector function $\Phi(\theta)$ defines the desired synchronization among the generalized coordinates, and $\theta_\star(t)$ may coincide with the target time evolution of one of the generalized coordinates.

Such a reparametrization is always possible along a finite number of pieces of the trajectory. A generalization for the case when a single virtual holonomic constraint for the whole period does not exist or is hard to find can be easily made along the lines of [6], but it would lead to a notion of hybrid transverse linearization, which is more complex. So, for simplicity, we restrict ourselves to the one-piece case.

11.2.3 Problem Formulation

Our goal is to define a procedure for computing a T-periodic linear control system

$$\frac{d}{d\tau} X = \mathcal{A}(\tau) X + \mathcal{B}(\tau) V \tag{11.5}$$

with $X \in \mathbb{R}^{2n-k-1}$ and $V \in \mathbb{R}^m$ such that its solutions, being appropriately initiated, approximate that part of the solution—i.e., the solution of the dynamics in Eq. (11.1)—that is transversal to the target periodic solution (11.3) defined for $v \equiv 0$.

As known, this linear system, if found, should allow us to analyze orbital stability of the nonlinear system and to design orbitally stabilizing feedback control laws.

As in our previous work [14], the procedure consists of finding new $2n-k-1$ independent state variables that vanish along the target motion whenever $v \equiv 0$, computing their dynamics and finding its linearization.

To compute the transverse linearization (11.5) for (11.1), we introduce a moving Poincaré section [10] following the derivations presented in [14]. This will be a family of the surfaces orthogonal to the target periodic motion. Linearizing the dynamics projected onto these surfaces allows one to obtain a family of linear, time-varying system solutions. These allow one to approximate the change in the distance between the orbits of (11.1) and the target orbit defined by (11.3) for an arbitrary choice of v.

Below, we propose two approaches. The first relies on an order reduction procedure, which can be found in many references; see e.g., [11, 4]. It is applicable for quite a restrictive class of nonholonomic systems. However, the problem of planning a feasible motion might be easier to solve in the case when the reduction of the order is possible. The second approach is more general and does not assume knowledge of a order reduction transformation, which for a particular system may be definable only locally.

11.3 Transverse Linearization with a Preliminary Reduction

Let us first try to eliminate invariants from our description, thereby obtaining a system of a lower dimension.

11.3.1 Reduced-Order Dynamics

To simplify the subsequent calculations, one might be able to rewrite the system (11.1) in coordinates for the reduced state space of dimension $2n-k$. There are various procedures that can be used to eliminate the vector of Lagrange multipliers λ. For example, following [4], suppose that the manifold $\mathcal{M}$ in (11.2) contains a neighborhood of the desired periodic trajectory (11.3) where the equation $J(q)\dot{q}=0$ can be written as

$$J(q)\dot{q} = J_a(q)\dot{q}_a + J_c(q)\dot{q}_c = 0 \tag{11.6}$$

with $q=[q_a;\, q_c]$ and $J_c(q)$ being a nonsingular $k \times k$ matrix. Then, one can define a smooth matrix function

$$J_{ac}(q) = \begin{bmatrix} I_{(n-k)\times(n-k)} \\ -\left(J_c(q)\right)^{-1} J_a(q) \end{bmatrix}, \tag{11.7}$$

columns of which span the null space of $J(q)$, and, by substituting the identities

$$\dot{q} = J_{ac}(q)\dot{q}_a, \qquad \ddot{q} = \dot{J}_{ac}(q)\dot{q}_a + J_{ac}(q)\ddot{q}_a \tag{11.8}$$

into (11.1) multiplied by $J_{ac}(q)^T$, obtain the following reduced-order dynamics [4]:

$$\begin{aligned} M_{ac}(q)\,\ddot{q}_a + C_{ac}(q,\dot{q}_a)\,\dot{q}_a + G(q) &= B_{ac}(q)\,u, \\ \dot{q}_c = J_{ca}(q)\,\dot{q}_a, \qquad u &= R_{ac}(q,\dot{q}_a) + Q_{ac}(q,\dot{q}_a)\,v, \end{aligned} \tag{11.9}$$

where $J_{ca}(q) = -\left(J_c(q)\right)^{-1} J_a(q)$, the matrix $M_{ac}(q) = J_{ac}(q)^T M(q) J_{ac}(q)$ is positive definite, the vector $C_{ac}(q,\dot{q}_a)\,\dot{q}_a = C\left(q, J_{ac}(q)\,\dot{q}_a\right) J_{ac}(q)\,\dot{q}_a + M(q)\,\dot{M}(q)\,\dot{q}_a$ is quadratic in $\dot{q}_a$ and the rank of $B_{ac}(q) = J_{ac}(q)^T B(q)$, which is less than or equal to $n-k$, defines the number of independently actuated degrees of freedom.

Let us assume that $v \in \mathbb{R}^m$ and rank $\{B_{ac}(q)\} =$ rank $\{B(q)\} = m$. If $n-k=m$, the system is called *fully actuated*. Otherwise, it is called underactuated, and $n-k-m$ is the nonholonomic underactuation degree.

In the rest of the chapter, we consider only the challenging underactuated case, i.e., we assume that

$$n-k-m \geq 1$$

It is of interest to note that when the simple reduction procedure described above is not applicable, it is often possible to introduce so-called quasicoordinates q_a [11, Chapter III] to obtain the relation (11.8) and, consequently, the reduced-order dynamics in the form of (11.9) and with similar properties. So, for the rest of this section let us assume that such a transformation is available and we do have an equivalent description (11.9), called in [4] as the normal form equations for nonholonomic control systems.

Let us assume that the vector of virtual holonomic constraints in (11.4) is partitioned as

$$\Phi(\theta) = [\Phi_a(\theta);\ \Phi_c(\theta)]$$

according to $q = [q_a;\ q_c]$ and that $\theta_\star(t)$ coincides with the desired time evolution of one of the components of q_a. Substituting $q = \Phi(\theta)$ into the second equation in (11.9), one obtains the identity

$$\Phi_c'(\theta) = J_{ca}\left(\Phi(\theta)\right) \Phi_a'(\theta)$$

which must be satisfied.

On the other hand, following [16], multiplying the first equation in (11.9) by an annihilator $B^\perp$ for $B_{ac}(q)$ and substituting $q = \Phi(\theta)$, one obtains $n-m-k$ differential equations

$$\alpha_j(\theta)\,\ddot{\theta} + \beta_j(\theta)\,\dot{\theta}^2 + \gamma_j(\theta) = 0, \quad j = 1,\ldots,n-m-k \tag{11.10}$$

with coefficients as defined in [16, 14], admitting $\theta = \theta_\star(t)$ as a common solution. Moreover, each of these equations with nonvanishing $\alpha_j(\theta)$ admits the conserved quantity [17]

$$I_j = \dot{\theta}^2 - \Psi_j(\theta,\theta_\star(0))\,\dot{\theta}_\star^2(0) + \int_{\theta_\star(0)}^{\theta(t)} \frac{2\Psi_j(\theta,s)\,\gamma_j(s)}{\alpha_j(s)}\,ds$$
$$\Psi_j(a,b) = \exp\left\{\int_a^b \frac{2\beta_j(\tau)}{\alpha_j(\tau)}\,d\tau\right\}, \tag{11.11}$$

which is zero for $\theta = \theta_\star(t)$ and is equivalent to the Euclidean distance between a solution of (11.10) and the target solution [14].

Introducing the change of coordinates $(q,\dot{q}) \leftrightarrow (\theta,\dot{\theta},y_a,y_c,\dot{y}_a)$ defined by

$$y_a = (n-k-1) \text{ components of } \{q_a - \Phi_a(\theta)\},$$
$$y_c = q_c - \Phi_c(\theta), \qquad y = [y_a;\, y_c] \in \mathbb{R}^{n-1} \tag{11.12}$$

one can easily rewrite (11.9) as

$$\begin{aligned} \alpha_j(\theta)\,\ddot{\theta} + \beta_j(\theta)\,\dot{\theta}^2 + \gamma_j(\theta) &= g_0(\theta,\dot{\theta},y,\dot{y}_a,v), \\ \ddot{y}_a &= g_a(\theta,\dot{\theta},y,\dot{y}_a,v), \\ \dot{y}_c &= g_c(\theta,\dot{\theta},y,\dot{y}_a) \end{aligned} \tag{11.13}$$

such that the right-hand sides are zeros whenever $\theta = \theta_\star(t)$, $\dot{\theta} = \dot{\theta}_\star(t)$, $y = 0$, $\dot{y}_a = 0$ and $v = 0$.

Introducing the vector of transversal coordinates

$$x_\perp = [I_j;\, y;\, \dot{y}_a] \in \mathbb{R}^{2n-k-1} \tag{11.14}$$

with the first component defined by (11.11), it is possible to make another change of coordinates $(\theta,\dot{\theta},y_a,y_c,\dot{y}_a) \leftrightarrow (\psi_j,x_\perp)$ without implicitly defining the variable ψ_j that characterizes the dynamics along the target motion. The first equation in (11.13) can be rewritten as [15]

$$\dot{I}_j = \frac{2\dot{\theta}}{\alpha_j(\theta)}\Big(g_0(\theta,\dot{\theta},y,\dot{y}_a,v) - \beta_j(\theta)\,I_j\Big) \tag{11.15}$$

It is left to linearize the rewritten dynamics with respect to the components of $x_\perp$. The only nontrivial part here is obtaining the coefficients for the variations with respect to I_j; the following formula [5] is useful:

$$g(\theta,\dot{\theta},y,\dot{y}_a,v) = \frac{\partial g}{\partial I_j} I_j + \frac{\partial g}{\partial y} y + \frac{\partial g}{\partial \dot{y}_a}\dot{y}_a + \frac{\partial g}{\partial v} v + \ldots,$$
$$\frac{\partial g}{\partial I_j} := \left.\frac{\dot{\theta}\,\dfrac{\partial g}{\partial \dot{\theta}} - \ddot{\theta}\,\dfrac{\partial g}{\partial \theta}}{2\left(\dot{\theta}^2 + \ddot{\theta}^2\right)}\right|_{\substack{\theta=\theta_\star(\tau)\\ \dot{\theta}=\dot{\theta}_\star(\tau)\\ \ddot{\theta}=\ddot{\theta}_\star(\tau)\\ y=0,\ \dot{y}_a=0}} \tag{11.16}$$

where τ defines the closest point to $(\theta,\dot\theta)$ on the target trajectory and ... denotes small higher-order terms with respect to the distance to this point.

Hence, to compute the coefficients of the linearization, one needs to obtain a few partial derivatives and to know $\theta_\star(t)$, together with its two derivatives, that can be computed solving (11.10). The result is (11.5) with

$$\mathcal{B}(\tau) = \left[\frac{\partial g_0}{\partial v};\ 0_{m\times(n-k-1)};\ 0_{m\times k};\ \frac{\partial g_a}{\partial v}\right]_{\substack{\theta=\theta_\star(\tau)\\ \dot\theta=\dot\theta_\star(\tau)\\ \ddot\theta=\ddot\theta_\star(\tau)\\ y=0,\ \dot y_a=0}},$$

$$\mathcal{A}(\tau) = \left[\begin{array}{ccc} \frac{2\dot\theta}{\alpha_j}\left(\frac{\partial g_0}{\partial I_j} - \beta_j\right) & \frac{2\dot\theta}{\alpha_j}\frac{\partial g_0}{\partial y} & \frac{2\dot\theta}{\alpha_j}\frac{\partial g_0}{\partial \dot y_a} \\ 0_{(n-k-1)\times 1} & 0_{(n-k-1)\times(n-1)} & I_{(n-k-1)} \\ \frac{\partial g_c}{\partial I_j} & \frac{\partial g_c}{\partial y} & \frac{\partial g_c}{\partial \dot y_c} \\ \frac{\partial g_a}{\partial I_j} & \frac{\partial g_a}{\partial y} & \frac{\partial g_a}{\partial \dot y_a} \end{array}\right]_{\substack{\theta=\theta_\star(\tau)\\ \dot\theta=\dot\theta_\star(\tau)\\ \ddot\theta=\ddot\theta_\star(\tau)\\ y=0,\ \dot y_a=0}},$$

where, e.g., the first row in the matrices $\mathcal{A}(\tau)$ and $\mathcal{B}(\tau)$ is defined by (11.16) and (11.15).

We show in Sec. 11.5 how to use the computed transverse linearization for stability analysis and for exponential orbital stabilization. Before that, let us propose a more general approach for its computation.

11.4 Computation of a Transverse Linearization without a Preliminary Reduction of Order

In case the partition of q that corresponds to (11.6) with nonsingular $J_c(q)$ is not known or does not exist, computing the transverse linearization as in the previous section is not possible. However, there is another way. Differentiating the equation of constraint one obtains

$$\dot J(q)\,\dot q + J(q)\,\ddot q = 0$$

while since $M(q)$ is not singular

$$\ddot q = -M(q)^{-1}\left(C(q,\dot q)\,\dot q + G(q) - J(q)^T\lambda - B(q)\,u\right)$$

Now, using the fact that $J(q)$ is of full rank, one can obtain the explicit expression for λ solving

$$\left(J(q)M(q)^{-1}J(q)^T\right)\lambda + \dot J(q)\,\dot q = J(q)M(q)^{-1}\left(B(q)u - C(q,\dot q)\dot q - G(q)\right) \quad (11.17)$$

and rewrite the system (11.1) in an equivalent form,

$$
\begin{aligned}
&M(q)\ddot{q}+C_\lambda(q,\dot{q})\dot{q}+G_\lambda(q)=B_\lambda(q)u,\\
&J(q)\dot{q}=0, \qquad u=R(q,\dot{q})+Q(q,\dot{q})v,
\end{aligned}
\tag{11.18}
$$

where the identity $J(q)\dot{q}=0$ holds provided it is satisfied in the initial moment of time.

It is easy to verify that the expression $C_\lambda(q,\dot{q})\dot{q}$ is still quadratic in $\dot{q}$, inheriting this property from $C(q,\dot{q})\dot{q}$ and $\dot{J}(q)\dot{q}$. As a result it is straightforward to follow the derivations presented in [14], which are similar to the ones presented in the previous section, taking

$$y=(n-1)\text{ components of }\{q-\Phi(\theta)\}$$

and

$$x_\perp^e=[I_j;\, y;\, \dot{y}]\in\mathbb{R}^{2n-1}$$

and obtaining the extended transverse linearization

$$\frac{d}{d\tau}x_\perp^e=\mathcal{A}_e(\tau)x_\perp^e+\mathcal{B}_e(\tau)V$$

It is clear that this system should have at least k uncontrollable states since there are k conserved identities. They, in principle, can be obtained by linearizing the identity $J(q)\dot{q}=0$ expressed in terms of the components of $x_\perp^e$ and computing an appropriate change of coordinates that explicitly contain the conserved quantities. Removing the corresponding extraneous states should allow us to obtain the transverse linearization (11.5) of the correct dimension.

11.5 Orbital Stability and Stabilization

Let us illustrate how to use the computed transverse linearization (11.5) for analysis of orbital stability and stabilization via design of a feedback controller. For simplicity, we will do this only for the case where the reduction of order is possible.

11.5.1 Analysis of Orbital Stability

Suppose someone suggested a stabilizing feedback control law for the system (11.1). It can certainly be expressed in terms of the new coordinates as a feedback control law

$$v=k(\theta,\dot{\theta},y,\dot{y}_a) \tag{11.19}$$

for the system (11.9).

The corresponding expression for V in (11.5) can be computed as

$$V=K(\tau)X, \tag{11.20}$$

where

$$K(\tau) = \left[\frac{\dot{\theta}\dfrac{\partial k}{\partial \dot{\theta}} - \ddot{\theta}\dfrac{\partial k}{\partial \theta}}{2\left(\dot{\theta}^2 + \ddot{\theta}^2\right)}, \left(\frac{\partial k}{\partial y}\right)^T, \left(\frac{\partial k}{\partial \dot{y}_a}\right)^T \right]\Bigg|_{\substack{\theta=\theta_\star(\tau)\\ \dot{\theta}=\dot{\theta}_\star(\tau)\\ \ddot{\theta}=\ddot{\theta}_\star(\tau)\\ y=0,\ \dot{y}_a=0}} \tag{11.21}$$

is obtained via linearization.

The transition matrix for the transverse linearization can be computed as follows:

$$X(T) = \Phi(T)X(0), \tag{11.22}$$

where $\Phi(T)$ is the end-point of the solution of the initial value problem

$$\frac{d}{d\tau}\Phi = \left(\mathcal{A}(\tau) + \mathcal{B}(\tau)K(\tau)\right)\Phi, \quad \Phi(0) = I. \tag{11.23}$$

Theorem 11.1. *Suppose all the eigenvalues of the matrix $\Phi(T)$ defined by (11.23) are inside the open unit circle. Then, the periodic solution defined by $q = q_\star(t)$ for the system (11.1), (11.19) is exponentially orbitally stable.*

Proof of this statement relies on the fact that $\Phi(T)$ is a linearization of the Poincaré first return map [12] computed on the first surface of the explicitly defined Poincaré section, and it follows the ideas of the proof of the main result in [14].

It appears, however, that suggesting a feedback design technique for (11.19) is more involved.

11.5.2 Orbital Stabilization

The obvious idea is to start with designing a stabilizing feedback controller for (11.5) in the form of (11.21). This is a nontrivial task, although it is clearly much simpler than designing a stabilizing regulator for the system (11.1) directly.

In [5], a time-independent law in the form of (11.21) is computed through a numerical search for a planar three-link model of a walking robot with a torso of underactuation degree one. If this is done, one can take

$$v = K\left[I^{(i)};\ y;\ \dot{y}_a\right]$$

and use Theorem 11.1 to conclude orbital exponential stability.

However, if a nonconstant gain $K(\tau)$ is available for stabilization of the linearization, suggesting a formula for (11.19) to recover (11.21) is a nontrivial task. A possible way to proceed is to use the Poincaré section as in [14, 15] or as defined below:

$$U(\tau)=\Big\{(q,\dot q):\ \big(q-q_\star(\tau)\big)^T\dot q_\star(\tau)+\big(\dot q-\dot q_\star(\tau)\big)^T\ddot q_\star(\tau)=0 \\ \text{and}\ \|q-q_\star(\tau)\|^2+\|\dot q-\dot q_\star(\tau)\|^2\le\varepsilon\Big\}\bigcap\mathcal{M} \tag{11.24}$$

where the manifold $\mathcal{M}$ is defined in (11.2) and $\varepsilon > 0$ is small enough to ensure that all $U(\tau)$ are disjoint.

With such a tabular neighborhood at hand, we can formulate our main stabilization result.

Theorem 11.2. *Suppose all the eigenvalues of the matrix $\Phi(T)$ defined by (11.23) are inside the open unit circle. Then, the periodic solution defined by $q = q_\star(t)$ for the system (11.1) is exponentially orbitally stabilized by*

$$v = K(\tau)x_\perp(t) = K(\tau)\,[I_j;\ y;\ \dot y_a] \tag{11.25}$$

where τ is such that $(q(t),\dot q(t)) \in U(\tau)$ with $U(\tau)$ defined in (11.24).

The proof follows from re-examining the proof of Theorem 11.1. The key observation is that variation in τ resulting from variations in $(q,\dot q)$ restricted to keeping a constant value of $x_\perp$ are of the second order in magnitude with respect to the components of $x_\perp$.

11.6 Example—Steering of a Knife-Edge System without Pushing

Here we illustrate how a transverse linearization can be computed for a standard simple example of a knife-edge system.

11.6.1 Equations of Motion

Dynamics of an underactuated knife-edge system can be described in the form of (11.1) as follows, see e.g. [4]

$$\ddot q_1 = \lambda\,\sin(q_3) \tag{11.26}$$

$$\ddot q_2 = -\lambda\,\cos(q_3) \tag{11.27}$$

$$\ddot q_3 = u \tag{11.28}$$

$$\dot q_1\,\sin(q_3) - \dot q_2\,\cos(q_3) = 0 \tag{11.29}$$

where q_1 and q_2 denote the point of contact of the knife-edge system with the horizontal plane, q_3 is the heading angle and u is the steering torque.

The reduced-order dynamics in the form of (11.9) can be obtained by letting

$$q_a = [q_{a1};\ q_{a2}] = [q_1\cos(q_3) + q_2\sin(q_3);\ q_3],$$
$$q_c = -q_1\sin(q_3) + q_2\cos(q_3)$$

and it is given by (see e.g. [4])

$$\begin{aligned}\ddot{q}_{a1} &= -q_{a1}\dot{q}_{a2}^2 + q_c u,\\ \ddot{q}_{a2} &= u,\\ \dot{q}_c &= -q_{a1}\dot{q}_{a2}\end{aligned}$$

In the rest of the chapter, we would like to analyze or to stabilize a periodic motion that exists in the open loop; so, we take

$$u = v, \tag{11.30}$$

which implies that the system we deal with is

$$\begin{aligned}\ddot{q}_{a1} &= -q_{a1}\dot{q}_{a2}^2 + q_c v,\\ \ddot{q}_{a2} &= v,\\ \dot{q}_c &= -q_{a1}\dot{q}_{a2}\end{aligned} \tag{11.31}$$

11.6.2 Target Periodic Solution

To obtain a virtual-holonomic-constraint description (11.4) of a periodic solution that exists in the open loop, we substitute

$$q_{a1} = \phi_a(\theta), \quad q_{a2} = \theta, \quad q_c = \phi_c(\theta), \quad v = 0$$

into (11.31) and obtain

$$\begin{aligned}\left(\phi_c(\theta) - \phi_a'(\theta)\right)\ddot{\theta} - \left(\phi_a(\theta) + \phi_a''(\theta)\right)\dot{\theta}^2 &= 0,\\ \ddot{\theta} &= 0,\\ \phi_a(\theta) &= -\phi_c'(\theta)\end{aligned}$$

Therefore, the coefficients for Eq. (11.10) are

$$\begin{aligned} \alpha_1(\theta) &= \phi_c(\theta) + \phi_c''(\theta), & \alpha_2(\theta) &= 1, \\ \beta_1(\theta) &= \phi_c'(\theta) + \phi_c'''(\theta), & \beta_2(\theta) &= 0, \\ \gamma_1(\theta) &= 0, & \gamma_2(\theta) &= 0 \end{aligned}$$

Hence, the only possible periodic solution (of the second kind, i.e., $\dot{\theta}_\star$ is periodic) is defined by

$$\theta_\star(t) = \omega_0 t + \phi_0$$

and

$$\phi_c'(\theta) + \phi_c'''(\theta) \equiv 0, \qquad \phi_a(\theta) = -\phi_c'(\theta)$$

We assume $\omega_0 \neq 0$ and take

$$\phi_c(\theta) = a_0 \neq 0, \qquad \phi_a(\theta) = 0,$$

which results in a circular motion described in the original coordinates for (11.29) as follows:

$$\begin{gathered} q_{1\star}(t) = -a_0 \sin(\omega_0 t + \phi_0), \;\; q_{2\star}(t) = a_0 \cos(\omega_0 t + \phi_0), \\ q_{3\star}(t) = \omega_0 t + \phi_0, \quad u_\star = 0. \end{gathered} \tag{11.32}$$

11.6.3 Orbital Stabilization

Note that with (11.32) we have reduced (11.11) to

$$I = \dot{\theta}^2 - \omega_0^2 \tag{11.33}$$

and we should introduce the transversal coordinates (11.14), taking for (11.12)

$$y_a = q_{1a}, \qquad y_c = q_c - a_0.$$

The system (11.31) in the new coordinates rewritten in the form (11.13) becomes

$$\begin{aligned} \ddot{y}_a &= -y_a (I + \omega_0^2) + (y_c + a_0) v, \\ \ddot{\theta} &= v, \\ \dot{y}_c &= -y_a \dot{\theta} \end{aligned}$$

and we obtain a time-invariant transverse linearization (11.5) with

$$\mathcal{A} = \begin{bmatrix} 0 & 0 & 0 & 0 \\ 0 & 0 & 0 & 1 \\ 0 & -\omega_0 & 0 & 0 \\ 0 & -\omega_0^2 & 0 & 0 \end{bmatrix}, \quad \mathcal{B} = \begin{bmatrix} 2\omega_0 \\ 0 \\ 0 \\ a_0 \end{bmatrix}$$

It is easy to verify that following the second approach by taking $\theta = q_3, y_1 = q_1 + a_0 \sin(\theta)$ and $y_2 = q_2 - a_0 \cos(\theta)$, after straightforward computations, one obtains the extended transverse dynamics

$$\begin{aligned} \ddot{y}_1 &= -\dot{\theta}\,(\dot{y}_1 \cos(\theta) + \dot{y}_2 \sin(\theta)) + a_0\, v \cos(\theta), \\ \ddot{y}_2 &= \dot{\theta}\,(\dot{y}_1 \cos(\theta) + \dot{y}_2 \sin(\theta)) + a_0\, v \sin(\theta), \\ \ddot{\theta} &= v, \\ 0 &= \dot{y}_1 \sin(\theta) - \dot{y}_2 \cos(\theta), \end{aligned}$$

which can be transformed into the same form since $y_c = -y_1 \sin(\theta) + y_2 \cos(\theta)$ and $y_a = y_1 \cos(\theta) + y_2 \sin(\theta)$ and so results in an equivalent linearization.

Since the pair $(\mathcal{A}, \mathcal{B})$ obtained is not controllable, exponential stabilization of the circular motion (11.32) via a smooth feedback is impossible. It is not hard to verify with similar calculations that allowing an additional control input that corresponds to a pushing force would still result in an uncontrollable linearization. Note that it was shown in [4] that it is not possible to stabilize the equilibria of the knife-edge system (11.29) via continuous feedback by pushing and steering, and so the failure to stabilize limit cycles is intuitively not surprising.

Dropping the uncontrollable equation for y_c one obtains, for the nontrivial case of $\omega_0 \neq 0$ and $a_0 \neq 0$, the controllable pair

$$\mathcal{A}_2 = \begin{bmatrix} 0 & 0 & 0 \\ 0 & 0 & 1 \\ 0 & -\omega_0 & 0 \end{bmatrix}, \quad \mathcal{B}_2 = \begin{bmatrix} 2\omega_0 \\ 0 \\ a_0 \end{bmatrix}$$

for which it is not hard to design a constant-gain feedback control law exponentially stabilizing the corresponding manifold that contains the desired motion. Note that the final value of y_c for the closed-loop system based on the corresponding constant-gain controller depends on the initial conditions.

11.7 Conclusion

We have studied dynamics of underactuated nonholonomic mechanical systems in a vicinity of a forced periodic solution that can be described via a single virtual holonomic constraint. The main result is a systematic procedure for analytical computation of a transverse linearization, which can be used for orbital stability analysis and design of exponentially orbitally stabilizing controllers. The procedure is explained in detail for a simplified case wherein the order of the system can be reduced via an appropriate change of coordinates eliminating all the conserved quantities. Only a suggestion on how to proceed in the general case is given. The technique is illustrated in an example.

References

1. Arnold, V.I., Kozlov, V.V., Neishtadt, A.I.: Mathematical Aspects of Classical and Celestial Mechanics, vol. 3. Springer-Verlag, Berlin (1988)
2. Banaszuk, A., Hauser, J.: Feedback linearization of transverse dynamics for periodic orbits. Systems and Control Letters **26**, 95–105 (1995)
3. Bloch, A., Baillieul, J., Crouch, P., Marsden, J.: Nonholonomic Mechanics and Control. Springer-Verlag, New York (2003)
4. Bloch, A.M., Reyhanoglu, M., McClamroch, N.H.: Control and stabilization of nonholonomic dynamic systems. IEEE Trans. Automatic Control **37**(11), 1746–1757 (1992)
5. Freidovich, L., Shiriaev, A., Manchester, I.: Stability analysis and control design for an underactuated walking robot via computation of a transverse linearization. In: Proc. 17th IFAC World Congress, pp. 10,166–10,171. Seoul, Korea (2008)
6. Freidovich, L., Shiriaev, A.: Transverse linearization for mechanical systems with passive links, impulse effects, and friction forces. In: Proc. 48th IEEE Conf. Decision and Control (CDC2009) / the 28th Chinese Control Conference, pp. 6490–6495. Shanghai, China (2009)
7. Freidovich, L.B., Robertsson, A., Shiriaev, A.S., Johansson, R.: Periodic motions of the Pendubot via virtual holonomic constraints: Theory and experiments. Automatica **44**(3), 785–791 (2008)
8. Hale, J.: Ordinary Differential Equations. Krieger, Malabar (1980)
9. Hussein, I.I., Bloch, A.M.: Optimal control of underactuated nonholonomic mechanical systems. IEEE Trans. Automatic Control **53**(3), 668–682 (2008)
10. Leonov, G.: Generalization of the Andronov–Vitt theorem. Regular and Chaotic Dynamics **11**(2), 281–289 (2006)
11. Neimark, J.I., Fufaev, F.A.: Dynamics of Nonholonomic Systems, vol. 33. A.M.S. Translations of Mathematical Monographs, Providence, RI (1972)
12. Poincaré, H.: Oeuvres completes, vol. 11. Gauthier-Villars, Paris, France (1916-1954)
13. Shiriaev, A., Freidovich, L.: Computing a transverse linearization for mechanical systems with two and more passive degrees of freedom. IEEE Trans. Automatic Control **54**(12), 2882–2888 (2009)
14. Shiriaev, A., Freidovich, L., Gusev, S.: Transverse linearization for controlled mechanical systems with several passive degrees of freedom. IEEE Trans. Automatic Control **55**(4), 893–906 (2010)
15. Shiriaev, A., Freidovich, L., Manchester, I.: Can we make a robot ballerina perform a pirouette? Orbital stabilization of periodic motions of underactuated mechanical systems. Annual Reviews in Control **32**(2), 200–211 (2008)
16. Shiriaev, A., Perram, J., Canudas-de-Wit, C.: Constructive tool for orbital stabilization of underactuated nonlinear systems: Virtual constraints approach. IEEE Trans. Automatic Control **50**(8), 1164–1176 (2005)
17. Shiriaev, A., Perram, J., Robertsson, A., Sandberg, A.: Periodic motion planning for virtually constrained Euler–Lagrange systems. Systems and Control Letters **55**, 900–907 (2006)
18. Urabe, M.: Nonlinear Autonomous Oscillations. Academic Press, New York (1967)
19. Westervelt, E.R., Grizzle, J.W., Chevallereau, C., Choi, J.H., Morris, B.: Feedback Control of Dynamic Bipedal Robot Locomotion. CRC Press, Taylor and Francis Group, Boca Raton, FL (2007)

Part III
DISTRIBUTED MODEL PREDICTIVE CONTROL AND SUPPLY CHAINS

DISTRIBUTED MODEL PREDICTIVE CONTROL AND SUPPLY CHAINS

One of the most important areas of development in control engineering during the past two decades is model predictive control, a technique to use mathematical models for real-time optimization. The range of applications has grown rapidly and the theoretical understanding of the technique has improved. A major bottleneck for the technique is computational complexity. This has created a strong incentive to look for distributed solutions, to simplify computations, communication and analysis. Distributed approaches also open up more large scale applications, for example, that involving several industrial plants connected by supply chains. This third part of the book illustrates this development with both theoretical and application-oriented contributions.

Chapter 12
A Distributed NMPC Scheme without Stabilizing Terminal Constraints

Lars Grüne and Karl Worthmann

Abstract We consider a distributed NMPC scheme in which the individual systems are coupled via state constraints. In order to avoid violation of the constraints, the subsystems communicate their individual predictions to the other subsystems once in each sampling period. For this setting, Richards and How have proposed a sequential distributed MPC formulation with stabilizing terminal constraints. In this chapter we show how this scheme can be extended to MPC without stabilizing terminal constraints or costs. We show theoretically and by means of numerical simulations that under a suitable controllability condition stability and feasibility can be ensured even for rather short prediction horizons.

12.1 Introduction

In this chapter we consider a distributed nonlinear model predictive control (NMPC) algorithm for systems which are coupled via state constraints. NMPC is a controller design method that relies on the online solutions of optimal control problems on finite optimization horizons in each sampling period. In a distributed setting, the solution of this optimal control problem is distributed among the individual systems. This can be done in various ways (see [12, Chapter 6] or [15] for an overview). One way is to formulate the optimization objective in a centralized way and to solve this problem in a distributed way in each sampling period. The necessary splitting of the optimization problem can be obtained in various ways which, under suitable assumptions, guarantee that the performance of the distributed controller is similar to that of a centralized controller; examples can be found, e.g., in [4] or [12, Chapter 6]. The drawback of this method—which is usually called *cooperative control*—is

Lars Grüne and Karl Worthmann,
Mathematical Institute, University of Bayreuth, 95440 Bayreuth, Germany,
e-mail: lars.gruene, karl.worthmann@uni-bayreuth.de

R. Johansson and A. Rantzer (Eds.): Distributed Decision Making and Control, LNCIS 417, pp. 261–287.
springerlink.com

that it requires numerous information exchanges between the individual systems during the iterative optimization procedure in each sampling interval.

A less demanding approach from the communication point of view is noncooperative control, in which some information from the other systems is taken into account when a system performs its optimization but in which the optimization objectives of the individual systems are independent from each other. It is known that for this setting a solution close to the central optimum can no longer be expected; rather, the best one can get is a Nash equilibrium, see [12, Chapter 6]. However, under suitable conditions the resulting closed loop may still be stable and maintain the imposed coupling constraints. This is the situation we investigate in this chapter. More precisely, we consider a specific noncooperative distributed NMPC algorithm proposed by Richards and How [14, 13] in which each system sends information about its predicted future states once in each sampling period. Via a suitable sequential ordering of the individual optimizations it is then ensured that the coupling state constraints are maintained whenever the optimization problems are feasible, i.e., when optimal solutions exist. Clearly, requiring a strict sequential order is a drawback of this approach, which we will attempt to relax in future research. Still, the numerical effort of this scheme is already significantly lower than for a centralized solution of the optimization problem; cf. the discussion after Algorithm 12.3, below.

In a stabilization setting, the optimal control problem to be solved online in the NMPC iteration usually minimizes the distance to the desired equilibrium. Often, additional stabilizing terminal constraints and costs are imposed in order to ensure asymptotic stability of the resulting closed loop. This means that the optimization on the finite horizon in each sampling instant is performed over those trajectories which—at the end of the optimization horizon—end up in the terminal constraint set, which is typically a neighborhood of the equilibrium to be stabilized. These terminal constraints also play a vital role for ensuring both stability and feasibility in the scheme of Richards and How. In certain situations, however, imposing terminal constraints has the significant drawback that rather long optimization horizons are needed in order to ensure the existence of trajectories that end up in the terminal constraint sets. Furthermore, stabilizing terminal constraints may have negative effects on the performance of the scheme; see, e.g., [6, Sec. 8.4]. As we will see in the detailed description in Sec. 12.3, in the distributed setting the terminal constraint formulation has the additional drawback that possible conflicts between the individual systems, i.e., violations of the coupling state constraints, have to be resolved in an initialization step.

The contribution of this chapter is to give sufficient conditions under which we can ensure stability and feasibility without stabilizing terminal constraints. In the nondistributed setting, several approaches for this purpose have been developed, e.g., in [5, 8, 7, 9]. Here, we use the approach developed in [8, 7], which relies on an asymptoptic controllability assumption taking into account the stage cost of the finite horizon optimal control problems. We will develop an extension of this condition to the distributed setting and we will verify that this condition holds for a simple test example of moving agents in a plane where the coupling constraints are formulated in order to prevent collisions between the agents. Numerical simulations for this example illustrate that with this scheme stability can be achieved with short optimization horizons and that this scheme allows us to resolve conflicts between the individual systems once they become "visible," i.e., at the runtime of the system rather than in an initialization step.

The chapter is organized as follows. In Sec. 12.2 we describe the problem formulation and in Sec. 12.3 we present the algorithm of Richards and How [14, 13] and discuss its main features. In Sec. 12.4 we recall the controllability based stability analysis for NMPC schemes from [8, 7]. Section 12.5 contains the main result of this chapter, i.e., a distributed version of this controllability condition and the corresponding stability result. In Sec. 12.6 we investigate a simple test example theoretically and numerically. Section 12.7 concludes the chapter and presents some ideas for future extensions of our main result.

12.2 Problem Set-Up and Preliminaries

We consider $P \in \mathcal{N}$ control systems described by the discrete time dynamics

$$x_p(k+1) = f_p(x_p(k), u_p(k)) \tag{12.1}$$

for $p = 1, \dots, P$, with $x_p(k) \in X_p$, $u_p(k) \in U_p$ and $f_p : X_p \times U_p \to X_p$, where X_p are arbitrary metric spaces and U_p are sets of admissible control values for $p = 1, \dots, P$. The solution of Eq. (12.1) for initial value $x_p(0) = x_p^0$ and control sequence $u_p(k) \in U_p$, $k = 0, 1, 2, \dots$, will be denoted by $x_p^u(k, x_p^0)$, i.e., we will omit the subscript p in u_p in order to simplify the notation. The combined state space of all systems will be denoted by

$$X = X_1 \times \cdots \times X_P.$$

Our goal is to stabilize each system at a desired equilibrium point $x_p^* \in X_p$. This means we are looking for feedback controllers $\mu_p(x_p(k), I_p(k)) \in U_p$ which render the respective equilibria asymptotically stable. Here the additional argument $I_p(k)$ of the controller μ_p denotes information from the other systems. We assume that for the purpose of exchanging such information the individual systems can communicate over a network with negligible delay. The precise definition of $I_p(k)$ and the controller μ_p are given in Definition 12.2 and Formula (12.5), below. The closed-loop solutions of Eq. (12.1) with controller μ_p, i.e., the solutions of

$$x_p(k+1) = f_p(x_p(k), \mu_p(x_p(k), I_p(k))) \tag{12.2}$$

will be denoted by $x_p(k)$, i.e., in order to simplify the notation we will not explicitly include the controller μ_p, the initial value $x_p(0)$ and the additional information I_p in the notation.

Beyond ensuring stability, we want to design the controllers such that the combined state $x(k) = (x_1(k), \dots, x_P(k))$ of the closed-loop systems satisfies state constraints of the form

$$x(k) = (x_1(k),\dots,x_P(k)) \in \mathbf{X} \subseteq X, \tag{12.3}$$

i.e., the state constraints are defined via a state constraint set $\mathbf{X}$. Note that these constraints induce a coupling between the — otherwise independent — systems which induces the need for passing information $I_p(k)$ between the subsystems.

Example 12.1. As an example which will be used in order to illustrate our concepts throughout this chapter, we consider a very simple model of $p = 1,\dots,P$ autonomous agents moving in the plane[1] $\mathbb{R}^2$ with state $x_p = (x_{p,1}(k), x_{p,2}(k))^T \in X_p = \mathbb{R}^2$, control $u_p \in U_p = [-\bar{u},\bar{u}]^2 \subset \mathbb{R}^2$ for some $\bar{u} > 0$ and dynamics

$$x_p(k+1) = x_p(k) + u_p(k).$$

Thinking of $x_p(k)$ as the position of the individual agent in the plane, the state constraints can be used in order to avoid collisions of the agents. To this end, for some desired distance $\delta > 0$ we define

$$\mathbf{X} := \{(x_1,\dots,x_P)^T \in \mathbb{R}^{2P} \mid \|x_{p_1} - x_{p_2}\| \geq \delta,\ \forall p_1, p_2 = 1,\dots,P \text{ with } p_1 \neq p_2\},$$

where $\|\cdot\|$ denotes an arbitrary norm in $\mathbb{R}^2$. If we use a specific norm in the subsequent computations then this will always be explicitly stated.

Clearly, in order to be able to maintain the state constraints in closed loop, i.e., to avoid collisions in the example, the individual controllers need to have some information about the other systems and to this purpose we will use the so far undefined information $I_p(k)$. In order to define what kind of information $I_p(k)$ the systems should exchange, we first need to specify the control algorithm we are going to use. In this chapter, we propose to use a model predictive (or receding horizon) control approach. To this end, at each time instant k for its current state $x_p(k)$ each agent solves the optimal control problem

$$\text{minimize } J_p^N(x_p^0, u_p) = \sum_{j=0}^{N-1} \ell_p(x_p^u(j, x_p^0), u_p(j)) \quad \text{with initial value } x_p^0 = x_p(k) \tag{12.4}$$

over all admissible control sequences $u_p(\cdot) \in U_p^{N,ad}(k, x_p^0, I_p(k)) \subseteq U_p^N$ on the optimization horizon $N \geq 2$, where the set of admissible control sequences $U_p^{N,ad}$ will be defined in Definition 12.2. Here ℓ_p is a stage cost function which penalizes the distance of the state from the equilibrium and the control effort. For instance, ℓ could be $\ell_p(x_p, u_p) = \|x_p - x_p^*\| + \lambda\|u_p\|$ or $\ell_p(x_p, u_p) = \|x_p - x_p^*\|^2 + \lambda\|u_p\|^2$, where $\lambda > 0$ is a weight parameter.

We denote the optimal control sequence for Eq. (12.4) by

$$u_p^{*,k}(0),\dots,u_p^{*,k}(N-1)$$

[1] The example could be extended to arbitrary dimensions, but for simplicity of exposition we stick to the planar case in this chapter.

and the corresponding predicted optimal trajectory by

$$x_p^{u^{*,k}}(0),\ldots,x_p^{u^{*,k}}(N-1)$$

According to the usual receding horizon construction, the value of the MPC controller is given by the first element of the optimal control sequence $u_p^{*,k}(0)$.

In order to define this MPC feedback law in a rigorous way, we need to define the set of admissible control sequences in the optimization (12.4) for the pth system. To this end, we make use of the following definition.

Definition 12.1. (i) For an index set $\mathcal{P} = \{p_1,\ldots,p_m\} \subseteq \{1,\ldots,P\}$ with $m \in \mathcal{N}$, $m \leq P$ we define the set of *partial states* as

$$X_{\mathcal{P}} := X_{p_1} \times \ldots \times X_{p_m}.$$

Elements of $X_{\mathcal{P}}$ will be denoted by $x_{\mathcal{P}} = (x_{p_1},\ldots,x_{p_m})$. The *partial state constraint set* $\mathbf{X}_{\mathcal{P}} \subset X_{\mathcal{P}}$ is defined as

$$\mathbf{X}_{\mathcal{P}} := \{x_{\mathcal{P}} \in X_{\mathcal{P}} \,|\, \text{there is } \tilde{x} \in \mathbf{X} \text{ with } \tilde{x}_{p_i} = x_{p_i} \text{ for } i = 1,\ldots,m\}.$$

(ii) Given an index set $\mathcal{P}$, an element $x_{\mathcal{P}} \in X_{\mathcal{P}}$, an element $x_p \in X_p$ with $p \notin \mathcal{P}$ and a subset $\mathcal{Q} = \{q_1,\ldots,q_l\} \subset \mathcal{P}$ we write

$$(x_p,(x_q)_{\mathcal{Q}}) := (x_p,x_{q_1},\ldots,x_{q_l}) \in X_{\{p\}\cup\mathcal{Q}}.$$

The admissible control sequences over which we optimize in Eq. (12.4) are now defined via the information available from the other agents according to the following definition.

Definition 12.2. (i) We assume that at time instant k when optimizing (12.4) for $x_p^0 = x_p(k)$ the pth agent knows prediction sequences $x_q^{k_q}(\cdot) = (x_q^{k_q}(0),\ldots,x_q^{k_q}(N-1))$ for $q \in \{1,\ldots,P\} \setminus \{p\}$ computed at time instant $k_q \leq k$ from the other agents. We define

$$I_p(k) := \{(k_q,x_q^{k_q}(\cdot)) \,|\, q \in \{1,\ldots,P\} \setminus \{p\}\}.$$

Note that $I_p(k)$ lies in the set

$$\mathcal{I}_p := (\mathcal{N}_0 \times X_1^N) \times \cdots \times (\mathcal{N}_0 \times X_{p-1}^N) \times (\mathcal{N}_0 \times X_{p+1}^N) \times \cdots \times (\mathcal{N}_0 \times X_P^N).$$

(ii) Given a time $k \in \mathcal{N}_0$ and $I_p \in \mathcal{I}_p$ with $k_q \leq k$ for all k_q contained in I_p, we define the set of admissible control sequences for system p at time k as

$$\begin{aligned} U_p^{N,ad}(k,x_p^0,I_p) := \{u_p(\cdot) \in U_p^N \mid & (x_p^u(j,x_p^0),(x_q^{k_q}(j+k-k_q))_{\mathcal{Q}_p(k,j)}) \\ & \in \mathbf{X}_{\{p\}\cup\mathcal{Q}_p(k,j)}, \qquad \forall\ j = 0,\ldots,N-1\} \end{aligned}$$

with

$$\mathcal{Q}_p(k,j) = \{q \in \{1,\ldots,P\} \setminus \{p\} \,|\, j+k-k_q \leq N-1\}.$$

The trajectories $x_p^u(\cdot,x_p^0)$ for $u \in U_p^{N,ad}(k,x_p^0,I_p)$ are called *admissible trajectories*.

In words, this definition demands that the minimization of (12.4) is performed over those trajectories which satisfy the state constraints together with the known predictions from the other systems for $j = 0,\ldots,N-1$.

The resulting feedback law μ_p thus depends on the current state $x_p(k)$ of the pth closed-loop system and on the other systems' predictions

$$x_q^{k_q}(\cdot),\; q \neq p$$

available at time k. For $I_p(k) \in \mathcal{I}_p$ the resulting MPC controller is hence given by the map

$$\mu_p : (x_p(k), I_p(k)) \mapsto u_p^{*,k}(0), \tag{12.5}$$

where $u_p^{*,k}(\cdot)$ is the optimal control sequence minimizing (12.4). For later use we define the associated optimal value function as

$$V_p^N(x_p^0, I_p) := \min_{u_p \in U_p^{N,ad}(k,x_p^0,I_p)} J_p^N(x_p^0, u_p).$$

In order not to overload the notation, the expression as written does not reflect the implicit k-dependence of μ_p and V_p^N. Moreover, for simplicity of exposition, throughout the chapter we assume that the minimum of this expression exists whenever

$$U_p^{N,ad}(k,x_p^0,I_p) \neq \emptyset$$

The important questions to be analyzed for this system are the following:

- Do the resulting closed-loop systems (12.2) maintain the state constraints (12.3)?
- Are the optimization problems feasible in each step, i.e., is the set of admissible control sequences $U_p^{N,ad}(k,x_p^0,I_p(k))$ in the minimization of (12.4) nonempty?
- Is the closed-loop system (12.2) asymptotically stable; in particular, do the trajectories $x_p(k)$ converge to the fixed points x_p^* as $k \to \infty$?

These are the questions we want to investigate in this chapter. Clearly, the precise way of how the information $I_p(k)$ is constructed is crucial for answering these questions. To this end, in the following section we investigate an algorithm in which the construction of the sets $I_p(k)$ implies that feasibility is sufficient for maintaining the state constraints, cf. Proposition 12.1.

12.3 The Scheme of Richards and How

In this section we define how the information $I_p(k)$ is constructed and according to which schedule the information is passed from one system to another. To this end, we use the sequential scheme introduced by Richards and How in [14, 13]. It should be noted that the general setting in these references is different from ours: on the one hand, only linear dynamics are considered in these references; on the other hand, perturbations are explicitly included in the models considered in [14, 13] and the MPC scheme is designed to be robust against perturbations.

The main idea of the way the distributed optimization takes place, however, is independent from these details. Using the notation introduced in the last section, this idea is described in the following algorithm. This scheme is sequential in the

---

Let $(x_1(0),\dots,x_P(0)) \in \mathbf{X}$ be given initial values.

(0) **Initialization for** $k = 0$.

Find control sequences $u_p \in U_p^N$ such that the corresponding trajectories satisfy

$$(x_1^u(j,x_1(0)),\dots,x_P^u(j,x_P(0))) \in \mathbf{X} \quad \text{for} \quad j = 0,\dots,N-1. \tag{12.6}$$

for $p = 1,\dots,P$:

  Set $k_p = 0$, $x_p^{k_p}(j) = x_p^u(j)$ for $j = 0,\dots,N-1$ and send $(k_p, x_p^{k_p}(\cdot))$ to all other systems

  Apply the control value $\mu_p(x_p^0) = u_p(0)$ in the first step.

end of p-loop

(1) **Control loop for** $k \geq 1$.

for $k = 1,2,\dots$:

  for $p = 1,\dots,P$:

    set

$$I_p(k) := ((k,x_1^k(\cdot)),\dots,(k,x_{p-1}^k(\cdot)),(k-1,x_{p+1}^{k-1}(\cdot)),\dots,(k-1,x_P^{k-1}(\cdot)))$$

    and minimize (12.4) for $x_p^0 = x_p(k)$ with respect to $u_p \in U_p^{N,ad}(k, x_p^0, I_p(k))$.

    Denote the resulting optimal control by $u_p^{*,k}$, set $k_p = k$, $x_p^{k_p}(j) = x_p^{u^{*,k}}(j)$.

    for $j = 0,\dots,N-1$ and send $(k_p, x_p^{k_p}(\cdot))$ to all other systems

    Apply the control value $\mu_p(x_p^0, I_p(k)) = u_p^{*,k}(0)$ in the kth step

  end of p-loop

end of k-loop

---

sense that in step (1) the individual systems perform their optimization one after the other before the control values are eventually applied in all systems. Note that system p always uses the most recent available predictions of the other systems in order to construct the set of admissible control sequences $U_p^{N,ad}$, i.e., for $q < p$ the predictions x_q^k made at time k are used and for $q > p$ the predictions x_q^{k-1} computed at time instant $k-1$ are used in $I_p(k)$. In case of a large number P of systems this sequential optimization may cause rather long waiting times, which may not be available in case of fast sampling. While one goal of future research will thus be to relax the strict sequential structure (see also Sec. 12.7, below), we remark that the scheme is well applicable for small values of P and, as pointed out in [13, Sec. 7], even for large P the scheme considerably reduces the numerical effort compared to a centralized solution of the optimization problem in each time instant.

The main advantage of the sequential scheme is that once the initialization step (0) has been performed successfully, the validity of the state constraints for the closed-loop solution follows from feasibility. This is made precise in the following proposition.

Proposition 12.1. *Assume that in Algorithm 12.3 the initialization step (0) is successful in finding $u_p \in U_p^N$ satisfying* (12.6) *and that in step (1) the optimal control problems are feasible, i.e., that $U_p^{N,ad}(k,x_p(k),I_p(k)) \neq \emptyset$ holds for all $p = 1,\dots,P$ and all $k \geq 1$. Then, the closed-loop system maintains the state constraints* (12.3) *for all $k \geq 0$.*

Proof. *Condition* (12.6) *and the definition of μ_p in step (0) immediately imply Eq.* (12.3) *for $k = 1$. Now we proceed by induction over k. Assume that Eq.* (12.3) *holds for some $k \geq 1$ and that $U_p^{N,ad}(k,x_p(k),I_p(k)) \neq \emptyset$ holds for all $p = 1,\dots,P$. Then each μ_p defined in step (1) is well defined and the definition of $U_P^{N,ad}(k,x_P(k),I_P(k))$ implies*

$$(x_1^{u^{*,k}}(1,x_1(k)),\dots,x_P^{u^{*,k}}(1,x_P(k))) \in \mathbf{X}.$$

By definition of the μ_p and (12.2) *we obtain*

$$x_p(k+1) = f_p(x_p(k),\mu_p(x_p(k),I_p(k))) = f_p(x_p(k),u_p^{*,k}(0)) = x_p^{u^{*,k}}(1,x_p(k))$$

for all $p = 1,\dots,P$ and thus

$$(x_1(k+1),\dots,x_P(k+1)) = (x_1^{u^{*,k}}(1,x_1(k)),\dots,x_P^{u^{*,k}}(1,x_P(k))) \in \mathbf{X}.$$

This shows Eq. (12.3) *for $k+1$.* □

In order to ensure $U_p^{N,ad}(k,x_p(k),I_p(k)) \neq \emptyset$, in [13] a condition involving terminal constraints sets is used. The following assumption summarizes this condition in our notation and does so without the additional constructions needed for the robust design in [13].

Assumption 12.1. *There exist closed neighborhoods T_p, $p = 1,\dots,P$, of the equilibria x_p^* satisfying the following conditions.*

(i) $T_1 \times \cdots \times T_P \subset \mathbf{X}$.
(ii) On each T_p there exists a stabilizing controller K_p for x_p such that T_p is forward invariant for the closed-loop system using K_p.
(iii) The control functions u_p in the initialization step (0) and in the optimization of (12.4) *in step (1) are such that $x_p^u(N,x_p(k)) \in T_p$ holds. In the optimization, this amounts to adding $x_p^u(N,x_p(k)) \in T_p$ as a further condition to the definition of the admissible control sequences $U_p^{N,ad}(k,x_p^0,I_p(k))$.*

The benefit of this condition is that if the computation of $u_1,\ldots,u_P$ satisfying (12.6) in step (0) is successful at time $k=0$, then $U_p^{N,ad}(k,x_p^0,I_p(k)) \neq \emptyset$ is ensured for all subsequent times $k \geq 1$ and all $p=1,\ldots,P$. In order to see this, consider the control sequence $u_p^{*,k-1}$ from the previous time step $k-1$ in step (1) for $p=1,\ldots,P$. Then the construction of $I_q(k-1)$ for $q>p$ and $I_q(k)$ for $q<p$ ensures

$$u_p^{*,k-1}(\cdot+1) \in U_p^{N-1,ad}(k,x_p^0,I_p(k))$$

Since

$$x_p^{u^{*,k-1}}(N-1,x_p(k)) = x_p^{u^{*,k-1}}(N,x_p(k-1)) \in T_k$$

by setting $u_p(j) = u_p^{*,k-1}(j+1)$ for $j=0,\ldots,N-2$ and $u_p(N-1) = K_p x_p^{u^{*,k-1}}(N-1,x_p(k))$ we obtain $x_p^u(N,x_p(k)) \in T_p$. Since the predictions of all other systems $q \neq p$ also end up in their respective sets T_q and $T_1 \times \cdots \times T_P \subset \mathbf{X}$, we obtain $u_p \in U_p^{N,ad}(k,x_p^0,I_p(k))$.

Beside ensuring feasibility, Assumption 12.1 also ensures stability. Indeed, a standard MPC stability proof (cf. [10] or [12, Sec. 2.4]) shows that under a compatibility condition between the stage cost ℓ_p and a suitably chosen terminal cost which is defined on T_p and added to J^N in (12.4), the optimal value function V_p becomes a Lyapunov function of the system, which proves stability. For this reason, the sets T_p in Assumption 12.1 are usually called *stabilizing terminal constraints*.

In the context of Example 12.1, the stabilizing terminal constraints demand that already in the initialization step (0) we have to plan collision-free trajectories for all systems from the initial value $x_p(0)$ to a neighborhood T_p of x_p^*. On the one hand, this implies that we may need to use rather large optimization horizons N if we consider initial conditions $x_p(0)$ far away from the terminal sets T_p. On the other hand, and more importantly in our distributed setting, Assumption 12.1 implies that all conflicts, i.e., possible collisions, until the "safe" terminal constraint sets T_p are reached, have to be resolved in the initialization step (0). Although in each iteration in step (1) the optimization algorithm is allowed to replan the trajectory, condition (12.6) is crucial in order to ensure feasibility for $k=1$ and thus—via Proposition 12.1—to ensure that the state constraints are maintained for all $k \geq 1$.

The goal of this chapter is now to relax these two drawbacks. While we will keep using Algorithm 12.3, we will not use Assumption 12.1 and in particular we will not require the solutions to end up in terminal constraint sets T_p. The hope is that this will enable us to obtain an MPC scheme which is stable and maintains the state constraints with considerably smaller optimization horizon N and—in the context of Example 12.1—which is able to solve possible conflicts at the times $k \geq 1$ when they become visible and not necessarily in the initialization step 0.

To this end, in the next section we first revisit a stability condition for NMPC schemes without stabilizing terminal constraints.

12.4 Stability of NMPC without Stabilizing Terminal Constraints

In this section we recall the stability analysis of NMPC controllers without stabilizing terminal constraints from [8, 7]. We will present the analysis for a single system of type (12.1). In Sec. 12.5, we extend these results to our setting with P systems.

Since in this section we deal with a single system of type (12.1), we will omit the index p in all expressions as well as the dependence of V^N and μ on information from the other systems. Analogous to Definition 12.2, admissibility for a control sequence $u \in U^N$ and an initial value $x^0 \in \mathbf{X}$ means that $u(j) \in U$ and $x^u(j,x^0) \in \mathbf{X}$ for $j = 0,\dots,N-1$, i.e., that the state constraints are maintained. Since in this section we do not consider couplings between different systems, Definition 12.2(ii) simplifies to

$$U^{N,ad}(x^0) := \{u(\cdot) \in U^N \,|\, x^u(j,x^0) \in \mathbf{X} \text{ for all } j = 0,\dots,N-1\}. \tag{12.7}$$

We assume that for each $x \in \mathbf{X}$ and each $N \in \mathcal{N}$ this set satisfies $U^{N,ad}(x) \neq \emptyset$, which means that the state constraint set $\mathbf{X} \subset X$ is forward invariant or viable. This assumption provides the easiest way to ensure feasibility of the resulting NMPC scheme and is used here in order to simplify the exposition. If desired, it can be relaxed in various ways; see, e.g., [6, Sec. 8.2–8.3] or [11, Theorem 3].

Stability of the NMPC closed loop is established by showing that the optimal value function V^N is a Lyapunov function for the system. More precisely, we aim at giving conditions under which for all $x \in \mathbf{X}$ we can establish the inequalities

$$\alpha_1(\|x - x^*\|) \leq V^N(x) \leq \alpha_2(\|x - x^*\|) \tag{12.8}$$

and

$$V^N(f(x,\mu(x))) \leq V^N(x) - \alpha \ell(x,\mu(x)) \tag{12.9}$$

for $\alpha_1, \alpha_2 \in \mathcal{K}_\infty$ and $\alpha \in (0,1]$. Then, under the additional assumption that

$$\alpha_3(\|x - x^*\|) \leq \ell^*(x) \leq \alpha_4(\|x - x^*\|) \tag{12.10}$$

holds for all $x \in \mathbf{X}$, suitable $\alpha_3, \alpha_4 \in \mathcal{K}_\infty$ and $\ell^*(x) := \min_{u \in U} \ell(x,u)$, we can conclude asymptotic stability as stated by the following theorem.

Theorem 12.1. *Assume that the inequalites* (12.8), (12.9) *and* (12.10) *hold for the optimal value function* V^N *and the stage cost* ℓ *of the optimal control problem* (12.4) *for one system, i.e., for* $p = P = 1$. *Then the closed-loop system* (12.2) *with the NMPC feedback* (12.5) *is asymptotically stable on* $\mathbf{X}$.

Proof. *The proof follows from by standard Lyapunov function arguments using* V^N *as a Lyapunov function; see [8, Theorem 5.2].* □

The inequalities (12.8) and (12.9) can be ensured by an asymptotic controllability condition of the equilibrium x^*. Here we work with the special case of exponential controllability; more general versions can be found in [8, 7].

Assumption 12.2. *Given constants $C > 0$, $\sigma \in (0,1)$, for each $x \in X$ and each $N \in \mathcal{N}$ there exists an admissible control function $u_x \in U^{N,ad}(x)$ satisfying*

$$\ell(x^{u_x}(j,x), u_x(j)) \leq C\sigma^j \ell^*(x)$$

for all $j \in \{0,\ldots,N-1\}$ with ℓ^* from (12.10).

Observe that the controllability condition is defined here in a slightly weaker form than in [8, 7] in the sense that the control function u_x is implicitly allowed to depend on N while in [8, 7] the existence of one u_x for all $N \in \mathcal{N}$ is assumed. However, it is straightforward to see that the weaker condition given here is sufficient for all arguments used in the proofs in these references. Note that the constant $C > 0$ allows for an increase of $\ell(x^{u_x}(j,x), u_x(j))$ for small j before it must eventually decrease. In particular, ℓ does not need to be a control Lyapunov function for the system.

Example 12.2. Consider Example 12.1 with only one system, which in particular implies that the state constraint $\mathbf{X}$ does not include any coupling terms. Instead, we use the state constraint set $\mathbf{X} = [-1,1]^2$. As stage cost we use $\ell(x,u) = \|x - x^*\|^2 + \lambda\|u\|^2$ for some $x^* \in [-1,1]$ and some $\lambda \geq 0$. Moreover, let $c := \max_{x \in \mathbf{X}} \|x - x^*\|$ denote the maximal distance in $\mathbf{X}$ from x^*.

We inductively define a control $u \in U^{N,ad}(x)$ by

$$u(k) = \kappa(x^* - x^u(k,x)) \quad \text{with} \quad \kappa = \min\{\bar{u}/c, \rho\}$$

for some design parameter $\rho \in (0,1)$. Note that the choice of κ implies $u(k) \in [-\bar{u}, \bar{u}]^2$ for $x^u(k,x) \in \mathbf{X}$. Moreover, this definition implies

$$x^u(k+1,x) = x^u(k,x) + \kappa(x^* - x^u(k,x)) \tag{12.11}$$

and, as a consequence,

$$\|x^u(k+1,x) - x^*\| = (1-\kappa)\|x^u(k,x) - x^*\|. \tag{12.12}$$

Due to the convexity of $\mathbf{X}$ and $\kappa \in (0,1)$, the identity (12.12) ensures feasibility of $x^u(\cdot)$. Using the definition of $u(k)$ and (12.12) yields

$$\begin{aligned}
\ell(x^u(k,x), u(k)) &= \|x^u(k,x) - x^*\|^2 + \lambda\|u(k)\|^2 \\
&= (1 + \lambda\kappa^2)\|x^u(k,x) - x^*\|^2 \\
&= (1 + \lambda\kappa^2)(1-\kappa)^{2k}\|x^u(0,x) - x^*\|^2 \\
&= (1 + \lambda\kappa^2)(1-\kappa)^{2k}\ell^*(x^u(0,x))
\end{aligned}$$

which shows Assumption 12.2 for $u_x = u$ with $C = 1 + \lambda\kappa^2$ and $\sigma = (1-\kappa)^2$. Analogously, one obtains Assumption 12.2 for $\ell(x,u) = \|x - x^*\| + \lambda\|u\|$ with $C = 1 + \lambda\kappa$ and $\sigma = 1 - \kappa$.

Under Assumption 12.2, the following properties hold.

Lemma 12.1. *Let Assumption 12.2 hold and define*

$$B^N(r) := \sum_{n=0}^{N-1} C\sigma^n r = C\frac{1-\sigma^N}{1-\sigma} r.$$

Then for each $x \in \mathbf{X}$ the following properties hold.

(i) The inequality

$$V^N(x) \leq J^N(x, u_x) \leq B^N(\ell^*(x)) \tag{12.13}$$

holds.

(ii) Let u^* be an optimal control sequence for (12.4). *Then for each $k = 0, 1, \ldots, N-2$, the inequality*

$$J^{N-k}(x^{u^*}(k,x), u^*(k+\cdot)) \leq B^{N-k}(\ell^*(x^{u^*}(k,x))) \tag{12.14}$$

holds.

(iii) Let u^* be an optimal control sequence for (12.4). *Then for each $j = 0, 1, \ldots, N-2$ the inequality*

$$V^N(x^{u^*}(1,x)) \leq J^j(x^{u^*}(1,x), u^*(1+\cdot)) + B^{N-j}(\ell^*(x^{u^*}(1+j,x))) \tag{12.15}$$

holds.

Proof.

(i) This follows immediately from Assumption 12.2.

(ii) This inequality follows from (i) applied to $x = x^{u^*}(k,x)$ using the fact that by the dynamic programming principle tails of optimal trajectories are again optimal trajectories; see [8, Lemma 3.4] for details;

(iii) Follows from the inequality $V^N(x^{u^*}(1,x)) \leq J^N(x^{u^*}(1,x), \tilde{u})$ using the control function

$$\tilde{u}(n) = \begin{cases} u^*(1+n), & n \leq j-1 \\ u_x(n), & n \geq j \end{cases}$$

with u_x from Assumption 12.2 with $x = x^{u^*}(1+j,x)$ and (i); for details see [8, Lemma 3.5].

□

Remark 12.1. Lemma 12.1 (i) yields that under Assumption 12.2 the inequalities in (12.10) imply (12.8). Indeed, the inequality

$$V^N(x) = J^N(x, u^*) \geq \ell(x, u^*(0)) \geq \ell^*(x) \geq \alpha_3(\|x - x^*\|)$$

implies the lower inequality in (12.8) with $\alpha_1 = \alpha_3$ and

$$V^N(x) \le B^N(\ell^*(x)) = C\frac{1-\sigma^N}{1-\sigma}\ell^*(x) \le C\frac{1-\sigma^N}{1-\sigma}\alpha_4(\|x-x^*\|)$$

implies the upper inequality in (12.8) with $\alpha_2 = C\frac{1-\sigma^N}{1-\sigma}\alpha_4$.

It remains to establish (12.9) for which we use Lemma 12.1(ii) and (iii) in the following way.

Proposition 12.2. *Assume Assumption 12.2 and consider $N \ge 1$, a sequence $\lambda_n > 0$, $n = 0,\dots,N-1$, and a value $\nu > 0$. Let $x \in \mathbf{X}$ and let $u^* \in U^N$ be an optimal control sequence for* (12.4) *such that $\lambda_n = \ell(x^{u^*}(n,x),u^*(n))$ holds for $n = 0,\dots,N-1$. Then*

$$\sum_{n=k}^{N-1} \lambda_n \le B^{N-k}(\lambda_k), \quad k = 0,\dots,N-2 \tag{12.16}$$

holds. Furthermore, if $\nu = V^N(x^{u^*}(1))$ holds, then

$$\nu \le \sum_{n=0}^{j-1} \lambda_{n+1} + B^{N-j}(\lambda_{j+1}), \quad j = 0,\dots,N-2 \tag{12.17}$$

holds.

Proof. *If the stated conditions hold, then λ_n and ν must meet inequalities* (12.14) *and* (12.15)*, which are exactly* (12.16) *and* (12.17)*, respectively.* □

The conditions (12.16) and (12.17) lead to the following sufficient condition for (12.9).

Theorem 12.2. *Let $N \ge 1$, assume that Assumption 12.2 holds and that the optimization problem*

$$\alpha := \inf_{\lambda_0,\dots,\lambda_{N-1},\nu} \frac{\sum_{n=0}^{N-1}\lambda_n - \nu}{\lambda_0}$$
subject to the constraints (12.16), (12.17) *and*
$$\lambda_0 > 0, \lambda_1,\dots,\lambda_{N-1}, \quad \nu \ge 0 \tag{12.18}$$

has an optimal value $\alpha \in (0,1]$. Then (12.9) *holds for this α for each $x \in \mathbf{X}$.*

Proof. *The optimization objective in* (12.18) *implies that for all values $\lambda_1,\dots,\lambda_{N-1}$, ν satisfying* (12.16), (12.17)*, the inequality*

$$\nu \le \sum_{n=0}^{N-1} \lambda_n - \alpha\lambda_0$$

holds. Proposition 12.2 then implies that for each optimal trajectory starting in some arbitrary $x \in \mathbf{X}$ the values $\lambda_n = \ell(x^{u^*}(n,x),u^*(n))$ and $\nu = V^N(x^{u^*}(1,x))$ satisfy (12.16) *and* (12.17)*, which yields*

$$
\begin{aligned}
V^N(x^{u^*}(1,x)) &\leq \sum_{n=0}^{N-1} \ell(x^{u^*}(n,x),u^*(n)) - \alpha\ell(x^{u^*}(0,x),u^*(0)) \\
&= V^N(x) - \alpha\ell(x,u^*(0)).
\end{aligned}
$$

Since by definition of the MPC feedback law we obtain $\mu(x) = u^*(0)$ and thus $f(x,\mu(x)) = x^{u^*}(1,x)$, this proves (12.9). □

The characterization of α via the optimization problem (12.18) is particularly useful because it admits the following explicit analytic solution.

Theorem 12.3. *Under Assumption 12.2 the optimization problem* (12.18) *has the solution*

$$
\alpha = 1 - \frac{(\gamma_N - 1)\prod_{k=2}^{N}(\gamma_k - 1)}{\prod_{k=2}^{N}\gamma_k - \prod_{k=2}^{N}(\gamma_k - 1)} \quad \text{with } \gamma_k = C\frac{1-\sigma^k}{1-\sigma} \tag{12.19}
$$

for $C > 0$ and $\sigma \in (0,1)$ from Assumption 12.2. Furthermore, for each pair of values $C > 0$ and $\sigma \in (0,1)$ the value α in (12.19) *satisfies $\alpha \to 1$ as $N \to \infty$.*

Proof. *Formula* (12.19) *follows from [7, Theorem 5.4] and the convergence $\alpha \to 1$ from [7, Corollary 6.1].* □

Remark 12.2. An inspection of the proof of [7, Theorem 5.4] shows that some inequalities provided by Lemma 12.1 are not needed in order to prove (12.19) since in this proof a relaxed problem [7, Problem 5.3] with fewer constraints was used. It turns out that the inequalities not needed in this relaxed problem are exactly (12.14) for $k = 1,\ldots,N-2$ or, equivalently, (12.16) for $k = 1,\ldots,N-2$, see [6, Remark 6.35]. While this has no consequence for the analysis in this section since we get all inequalities in (12.14) "for free" from Assumption 12.2, this observation will turn out very useful in the next section.

Combining Theorems 12.1, 12.2 and 12.3 yields the following corollary.

Corollary 12.1. *Consider a single system of type* (12.1) *and the NMPC feedback law* (12.5) *for some $N \geq 2$. Let Assumption 12.2 and* (12.10) *hold and assume that $\alpha > 0$ holds for α from* (12.19)*. Then, the closed-loop system* (12.2) *is asymptotically stable on* **X**.

Using the convergence $\alpha \to 1$ for $N \to \infty$, we can use this corollary in order to conclude that when (12.10) and Assumption 12.2 hold, then asymptotic stability can be guaranteed for each sufficiently large optimization horizon N. Beyond this asymptotic result, however, the condition $\alpha > 0$ in (12.19) also gives a useful stability criterion for small optimization horizons N, as the following example shows.

Example 12.3. We reconsider Example 12.2 with $N = 2$. Formula (12.19) simplifies to $\alpha = 1 - (C + \sigma C - 1)^2$. Since $\kappa \in (0,1)$, $C = 1 + \lambda\kappa^2$, $\sigma = (1-\kappa)^2$ we obtain with $\lambda \in (0,1)$

$$C+\sigma C-1=(1+\lambda\kappa^2)(1-\kappa^2)\leq(1+\kappa)(1-\kappa)^2=(1-\kappa^2)(1-\kappa)<1$$

which implies $\alpha > 0$. For instance, for $\lambda = 0.1$, $\rho = 0.5$ we obtain $\alpha \approx 0.8102$ or $\alpha \approx 0.9209$ for the Euclidean and the ∞-norm respectively. This shows that the MPC closed loop is asymptotically stable for $N = 2$ which is the shortest possible optimization horizon, given that the sum in (12.4) only includes the states $x^u(j,x^0)$ for $j = 0,\ldots,N-1$.

More complex examples of this kind, including infinite-dimensional PDE models, can be found, e.g., in [8, Sec. 6 and 7] or [1, 2, 6]. Finally, we remark that α also allows us to estimate the performance of the MPC feedback law μ in terms of an infinite horizon optimization criterion; for details see, e.g., [8, Theorem 4.2].

12.5 Stability of Distributed NMPC without Stabilizing Terminal Constraints

In this section we adapt the results of the previous section to the distributed MPC setting introduced in Sec. 12.2 using Algorithm 12.3. The goal is to adapt Assumption 12.2 to the distributed setting. This way we derive a sufficient condition for distributed NMPC without stabilizing terminal constraints which ensures feasibility of the optimal control problems in Algorithm 12.3 step(1)—and thus via Proposition 12.1 guarantees that the state constraints are maintained—and stability of the NMPC closed loop. Stability will be guaranteed by showing that each optimal value function V_p^N will satisfy the inequalities (12.8) and (12.9), i.e., that each V_p^N is a Lyapunov function for the corresponding system.

Comparing the distributed setting of Sec. 12.2 with the nondistributed setting of Sec. 12.4, the main difference is that the set of admissible control sequences $U_p^{N,ad}$ in Definition 12.2(ii) changes with time k due to the fact that the information $I_p(k)$ in Algorithm 12.3(1) also changes with time. In contrast to this, the set $U^{N,ad}$ in (12.7) is constant over time. In order to include the time-dependence in the controllability assumption we make use of sets of admissible control sequences according to the following definition.

Definition 12.3.

(i) For $m_1 > m_2 \in \mathcal{N}$ and a control sequence $u = (u(0),\ldots,u(m_1-1)) \in U_p^{m_1}$ we define the restriction

$$u|_{m_2} := (u(0),\ldots,u(m_2-1)) \in U_p^{m_2}.$$

(ii) A family of sets $W_p^m \subset U_p^m$, $m \in \{1,\ldots,N\}$, $N \geq 2$, of admissible control sequences is called *nested* if for all $m_1, m_2 \in \{1,\ldots,N\}$ with $m_1 > m_2$ and all $u \in U_p^{m_1}$ the implication

$$u \in W_p^{m_1} \quad \Rightarrow \quad u|_{m_2} \in W_p^{m_2}$$

holds.

(iii) For a nested family of admissible control sequence sets $W_p^m \subset U_p^m$, integers $l, m \in \mathcal{N}$, $m \in \{1,\dots,N\}$, $l+m \leq N$, and a control sequence $u \in W_p^l$ we define

$$W_p[u,l,m] := \{\tilde{u} \in U_p^m \,|\, (u(0),\dots,u(l-1),\tilde{u}(0),\dots,\tilde{u}(m-1)) \in W_p^{l+m}\}.$$

Recalling that in our setting the admissible control sequences are derived from the state constraint sets $\mathbf{X}$ and the predicted trajectories of the other systems contained in I_p via Definition 12.2(ii), a little computation reveals that for each time instant $k \geq 0$ the sets $W_p^m = U_p^{m,ad}(k, x_p^0, I_p)$, $m \in \mathcal{N}$ are nested and that this choice of W_p^m implies

$$W_p[u,l,m] = U_p^{m,ad}(k+l, x_p^u(l,x_p^0), I_p).$$

Another issue we take into account when adapting Assumption 12.2 is that in the distributed setting it is quite demanding to assume that controllability holds for all possible initial values. Instead, we will formulate the respective condition for fixed initial conditions. The following theorem presents this variant in an abstract setting with nested admissible control sequence sets W_p^m and $\widetilde{W}_p^m$. In Theorem 12.5 we will then show how this condition fits into Algorithm 12.3.

Theorem 12.4. *Consider some $p \in \{1,\dots,P\}$, two families of nested admissible control sequence sets $W_p^m, \widetilde{W}_p^m \subseteq U_p^m$, $m \in \{1,\dots,N\}$ for $N \geq 2$, a point $x_p^0 \in X_p$ and the optimal values*

$$V^N := \min_{u_p \in W_p^N} J_p^N(x_p^0, u_p) \quad \textit{and} \quad \widetilde{V}^N := \min_{\tilde{u}_p \in \widetilde{W}_p^N} J_p^N(\tilde{x}_p, \tilde{u}_p)$$

with $\tilde{x}_p = f_p(x_p^0, u_p^*(x_p^0))$ where $u_p^* \in W_p^N$ denotes the optimal control for V^N, i.e., $V^N = J_p^N(x_p^0, u_p^*)$.

For given constants $C > 0$, $\sigma \in (0,1)$ assume that the following holds:

(i) The inequality $V^N \leq B^N(x_p^0)$ holds for B^N from Lemma 12.1;

(ii) The optimal control $u_p^* \in W_p^N$ satisfies $(u^*(1),\dots,u^*(N-1)) \in \widetilde{W}_p^{N-1}$;

(iii) For each $j = 0,\dots,N-2$ there exists $\tilde{u} \in \widetilde{W}_p[u^*(1+\cdot), j, N-j]$ with

$$\ell_p(x_p^{\tilde{u}}(s, x_p^{u^*}(1+j, x_p^0)), \tilde{u}(s)) \leq C\sigma^s \ell_p^*(x_p^{u^*}(1+j, x_p^0)), \quad s = 0,1,\dots,N-j-1.$$

Then, the inequality

$$\widetilde{V}^N \leq V^N - \alpha \ell_p(x_p^0, u^*(0))$$

holds for α from 12.19.

Proof. *It is sufficient to show (12.13)–(12.15) for $x = x_p^0$, $V^N(x) = V^N$ and $V^N(x^{u^*}(1,x)) = \widetilde{V}^N$. This implies that $\lambda_n = \ell_p(x^{u^*}(n, x_p^0), u^*(n))$ and $\nu = \widetilde{V}^N$ satisfy*

(12.16), (12.17) *and by the same argument as in the proof of Theorem 12.2 we obtain the assertion when we use Theorem 12.3 in order to solve* (12.18). *By Remark 12.2 it is sufficient to show* (12.14) *for* $k = 0$.

In order to prove these inequalities, observe that (12.13) *and* (12.14) *for* $k = 0$ *follow directly from (i). In order to prove* (12.15) *we use that (iii) implies*

$$J_p^{N-j}(x_p^{u^*}(1+j,x_p^0),\tilde{u}) \leq B^{N-j}(\ell_p^*(x_p^{u^*}(1+j,x_p^0))), \quad j = 0,1,\ldots,N-2, \tag{12.20}$$

for B^N *from Lemma 12.1. Observe that (ii) is needed in order to ensure that* $\widetilde{W}_p[u^*(1+\cdot),j,N-j]$ *in (iii) is well defined.*

Now (12.15) *follows from* (12.20) *using the inequality*

$$V^N(x_p^{u^*}(1,x_p^0)) \leq J_p^j(x_p^{u^*}(1,x_p^0),u^*(1+\cdot)) + J_p^{N-j}(x_p^{u^*}(1+j,x_p^0),\tilde{u}),$$

which holds, since (ii) and $\tilde{u} \in \widetilde{W}_p[u^*(1+\cdot),j,N-j]$ *imply* $(u^*(1), \ldots, u^*(j), \tilde{u}(0), \ldots, \tilde{u}(N-j-1)) \in \widetilde{W}_p^N$. □

The following theorem incorporates this condition into Algorithm 12.3.

Theorem 12.5. *Consider Algorithm 12.3 with optimization horizon* $N \in \mathcal{N}$ *in 12.4, let* $C > 0$ *and* $\sigma \in (0,1)$ *and assume that the stage costs* ℓ_p *satisfy 12.10 for all* $p \in \{1,\ldots,P\}$ *and suitable* α_3, $\alpha_4 \in \mathcal{K}_\infty$. *Assume that step (0) of the algorithm is successful and denote the resulting control functions by* $u_p^{*,0}$. *Assume, furthermore, that in step (1) of the algorithm for each* $k \geq 1$ *and each* $p \in \{1,\ldots,P\}$ *condition (iii) of Theorem 12.4 holds with* $u^* = u_p^{*,k-1}$, $x_p^0 = x_p(k-1)$ *and*

$$\widetilde{W}_p^m = U_p^{m,ad}(k,x_p(k),I_p(k)), \quad m = 1,\ldots,N.$$

Then, the closed-loop solutions maintain the state constraints 12.3 and there exists $\alpha_1, \alpha_2 \in \mathcal{K}_\infty$ *such that the optimal value functions* V_p^N *satisfy*

$$\alpha_1(\|x_p(k) - x_p^*\|) \leq V_p^N(x_p(k),I_p(k)) \leq \alpha_2(\|x_p(k) - x_p^*\|) \tag{12.21}$$

for all $k \geq 1$ *and the inequality*

$$V_p^N(x_p(k+1),I_p(k+1)) \leq V_p^N(x_p(k),I_p(k)) - \alpha\ell(x_p(k),\mu_p(x_p(k),I_p(k))) \tag{12.22}$$

holds for α *from* (12.19) *and all* $k \geq 1$.

In particular, if $\alpha > 0$ *(which always holds for* $N > 0$ *sufficiently large) then the* V_p^N *are Lyapunov functions for the closed loop systems for* $k \geq 1$ *and thus asymptotic stability of the equilibria* x_p^* *follows.*

Proof. *We show that for each* $k \geq 2$ *the assumptions of Theorem 12.4 hold with*

$$W_p^m = U_p^{m,ad}(k-1,x_p(k-1),I_p(k-1)), \quad \widetilde{W}_p^m = U_p^{m,ad}(k,x_p(k),I_p(k))$$

with

$$x_p^0 = x_p(k-1),\ \tilde{x}_p = x_p(k)\ and\ u^* = u_p^{*,k-1}$$

To this end, first observe that in the discussion after Assumption 12.1 we have shown that in step (1) the relation

$$u_p^{*,k-1}(1+\cdot) \in U_p^{N-1,ad}(k, x_p(k), I_p(k))$$

holds, which implies that condition (ii) of Theorem 12.4 is satisfied.

Condition (iii) of Theorem 12.4 holds by assumption, and condition (i) of Theorem 12.4 at time k follows from condition (iii) for $j = 0$ *at time* $k-1$*, since* $x_p(k) = x_p^{u^{*,k-1}}(1, x_p(k-1))$ *and* W_p^m *at time k equals* $\widetilde{W}_p^m$ *at time* $k-1$*.*

Thus, Theorem 12.4 is applicable which proves (12.22).

Inequality (12.21) *is then obtained with the same arguments as in Remark 12.1. Finally, since the assumed condition (iii) of Theorem 12.4 in particular demands* $U_p^{N,ad}(k, x_p(k), I_p(k)) \neq \emptyset$*, Proposition 12.1 yields feasibility of the problem and implies that the closed-loop solutions satisfy the state constraints* (12.3). □

The central assumption in this theorem is that condition (iii) of Theorem 12.4 holds. In words, this assumption requires two things: first, $U_p^{N,ad}(k, x_p(k), I_p(k))$ needs to be nonempty, which means that given the predictions of the other systems x_q^u, $q \neq p$, contained in I_p there is still enough space to "squeeze in" a solution x_p^u. Second, the condition requires that starting from any point on the optimal open-loop trajectory from the last time instant, there are solutions which approach the equilibrium x_p^* sufficiently fast in the sense of the controllability assumption. The important fact in this condition is that when the pth system selects its control it knows the other systems' predictions. For this reason this rather technical condition can be rigorously verified at least for simple systems, as the example in the following section shows.

Note that even though step (0) remains formally identical to Algorithm 12.3, without the additional terminal condition from Assumption 12.1(iii) and with smaller N it is much easier to satisfy (12.6). This is illustrated in the numerical simulations at the end of the next section, in which for most of the systems the state constraints only become relevant after several steps of the algorithm.

12.6 An Example

In this section we first verify that Example 12.1 satisfies the conditions of Theorem 12.5 for $P = 2$ under suitable conditions. Afterwards, we numerically illustrate the performance of the scheme for this example with $P = 2$ and $P = 4$.

In order to verify the conditions of Theorem 12.5, we consider Example 12.1 with $P = 2$ and show that the conditions hold for $p = 1$ and all initial values x_1^0 which are bounded by $\|x_1^0\| \leq K$ for some $K > 0$. Analogous arguments then show the condition for $p = 2$. Without loss of generality we may assume $x_1^* = 0$. Since a priori it is not clear how the predictions $x_2^{u^{*,k_2}}$ contained in $I_1(k)$ defining the sets

$U_p^{m,ad}$ in Theorem 12.5 look, we show the stronger property, that the conditions hold for $p=1$ for all possible trajectories x_2^u. The only thing we have to exclude here is that x_2^u stays too close to the equilibrium x_1^*, because then it will never be possible for x_1^u to reach x_1^* without collision and thus to reduce $\ell_1(x_1^u(k),u_1(k))$ to 0. Hence, in what follows we consider all possible trajectories $x_2^u(k)$ that stay outside a neighborhood around $x_1^*=0$.

We show the following lemma, which implies the conditions of Theorem 12.5 whenever the trajectory x_2^{*,k_2} contained in $I_1(k)$ remains outside the neighborhood with radius $R+\delta$ around x_1^*. In order to streamline the exposition, in the following lemma the norm $\|\cdot\|$ is either the Euclidean or the ∞-norm. Without loss of generality we furthermore assume $\delta>\bar{u}$; otherwise, we can restrict ourselves to smaller control values than actually allowed.

Lemma 12.2. *We consider the stage cost $\ell_1(x_1,u_1)=\|x_1\|^2+\lambda\|u_1\|^2$ and the state constraint set $\mathbf{X}$ from Example 12.1 for $P=2$. Given $K>0$ and $R>\delta>\bar{u}$ there exists $C>0$, $\sigma\in(0,1)$ such that for each trajectory $x_2^u(k)$ satisfying $\|x_2^u(k)\|_\infty\geq R+\delta$ or $\|x_2^u(k)\|_2\geq\sqrt{2}(R+\delta)$ for all $k\in\mathcal{N}_0$ and each initial value x_1^0 with $\|x_1^0\|\leq K$ and $(x_1^0,x_2^u(0))\in\mathbf{X}$, there exists a control sequence $u_1(k)\in[-\bar{u},\bar{u}]^2$ with $(x_1^u(k,x_1^0),x_2^u(k))\in\mathbf{X}$ and*

$$\ell_1(x_1^u(k,x_1^0),u(k))\leq C\sigma^k\ell_1^*(x_1^0),\quad \forall k\in\mathcal{N}_0 \tag{12.23}$$

Proof. *For $x_1^0\notin T$, the fact that $\|x_2^u(k)\|_\infty\geq R+\delta$, whenever $x_1^0\in T=[-R,R]^2$ implies that the control $u_1=u$ constructed in Example 12.2 satisfies* (12.23) *for suitable $\widetilde{C}$ and σ since the resulting trajectory remains in T and thus $(x_1^u(k),x_2^u(k))\in\mathbf{X}$ holds for all $k\in\mathcal{N}_0$.*

For $x_1^0\notin T$, Lemma 12.3 applied with $x^u(\cdot)=x_1^u(\cdot)$ and $y(\cdot)=x_2^u(\cdot)$ shows the existence of a constant $\bar{k}$ and a control $u_1=u$ such that $x_1^u(k^*)\in T$ for a $k^*\leq\bar{k}$; cf. Remark 12.4 for the Euclidean norm. Since $\|u_1(k)\|_\infty$ is bounded by $\bar{u}$, the trajectory $x_1^u(k,x_1^0)$ from (12.3) *is bounded in the ∞-norm by $K+\bar{k}\bar{u}$ and thus $\ell(x_1^u(k,x_1^0),u_1(k))$, $k=0,\ldots,\bar{k}$, is bounded by some constant L independent of x_1^0. Using u from Example 12.2 from time k^* on, the resulting overall control sequence satisfies $(x_1^u(k,x_1^0),x_2^u(k))\in\mathbf{X}$ for all $k\geq 0$,*

$$\ell_1(x_1^u(k,x_1^0),u(k))\leq L\leq\frac{L\|x_1^0\|^2}{R^2}=(L/R^2)\ell^*(x_1^0)\leq\frac{L\sigma^{-\bar{k}}}{R^2}\sigma^k\ell^*(x_1^0)$$

for $k=0,\ldots,\bar{k}$, and

$$\ell_1(x_1^u(k,x_1^0),u(k))\leq\widetilde{C}\sigma^{k-\bar{k}}\ell_1^*(x_1^u(\bar{k},x_1^0))\leq\widetilde{C}\sigma^{k-\bar{k}}2\ell_1^*(x_1^0),$$

for $k\geq\bar{k}$. Here the last inequality follows from $\ell^*(x_1^u(\bar{k},x_1^0))=\|x_1^u(\bar{k},x_1^0)\|^2\leq 2\|x_1^u(\bar{k},x_1^0)\|_\infty^2\leq 2R^2\leq 2\|x_1^0\|_\infty^2\leq 2\|x_1^0\|^2=2\ell_1^*(x_1^0)$. Together this yields

$$\ell_1(x_1^u(k,x_1^0),u(k)) \le \max\{\frac{L\sigma^{-\bar{k}}}{R^2}, 2\widetilde{C}\sigma^{-\bar{k}}\}\sigma^k \ell_1^*(x_1^0)$$

and thus (12.23) *with* $C = \max\{L\sigma^{-\bar{k}}/R^2, 2\widetilde{C}\sigma^{-\bar{k}}\}$. □

Remark 12.3. (i) The construction used in the proof of Lemma 12.3 and, thus, in this example heavily relies on the knowledge of x_2^u. Indeed, without this knowledge the construction of k^* and u_1 would not be possible. Hence, the communication of I_p in the algorithm is crucial for ensuring the conditions of Theorem 12.5 in this example.

(ii) The additional condition $\|x_2^u(k) - x_1^*\|_\infty \ge R + \delta$ (and vice versa for $x_1^u(k) - x_2^*$) could be ensured by including it into the state constraint set **X**. However, when x_1^* and x_2^* are sufficiently far apart, then there is no incentive for the finite horizon optimal trajectory x_2^u to stay near x_1^* and vice versa. This is the situation in the following examples in which we did not explicitly include this condition in the optimization.

The following numerical simulation of the MPC control of this example confirms that the scheme yields a feasible and stable closed loop. The numerical examples were performed with MATLAB[2] using the stage cost $\ell_p(x_p,u_p) = \|x_p - x_p^*\| + 0.1\|u_p\|$, converting the optimal control problems in static optimization problems, which are solved with MATLAB's `fmincon`-routine. Here the control functions in step (0) of Algorithm 12.3 are computed using optimal control similar to step (1), where the admissible control sequences for the pth system are defined via the state constraints induced by the predictions $x_1^u(k),\ldots,x_{p-1}^u(k)$. In all examples we have added the additional constraints $x_p \in [-1,1]^2$ to **X**, and we have chosen the initial conditions $x_p(0)$ at the boundary of $[-1,1]^2$ and the desired equilibria on the opposite side of $[-1,1]^2$, i.e., $x_p^* = -x_p(0)$. This implies that the agents meet far away from the boundary of $[-1,1]^2$ and from the equilibria x_p^*; thus the theoretical analysis from the first part of this section remains valid. Figure 12.1 shows a corresponding simulation in which $N = 3$ turns out to be sufficient for stability. Further numerical simulations show that the scheme is stable also for a larger number P of agents. However, in this case it becomes more difficult to control the individual systems to their equilibria, which is reflected by the fact that for $P = 4$ systems and the initial values from Fig. 12.2—which extend those of Fig. 12.1—stability is only obtained for $N \ge 8$. However, even with $N = 8$ the horizon is considerably shorter than the optimization horizon needed in order to find predictions that end up in stabilizing terminal constraint sets around the equilibria. Our final numerical experiment shows that the optimization horizon N needed in order to obtain stability of the closed loop heavily depends on the initial values, i.e., on the way the individual agents meet and in which direction they are heading when they meet. Due to the fact that the control constraints $u_p \in [-\bar{u},\bar{u}]$ are box constraints, which allow the systems to move faster and more flexibly in diagonal direction, it is not surprising

[2] All MATLAB-Files are available on www.math.uni-bayreuth.de/~lgruene/publ/disNMPC.html

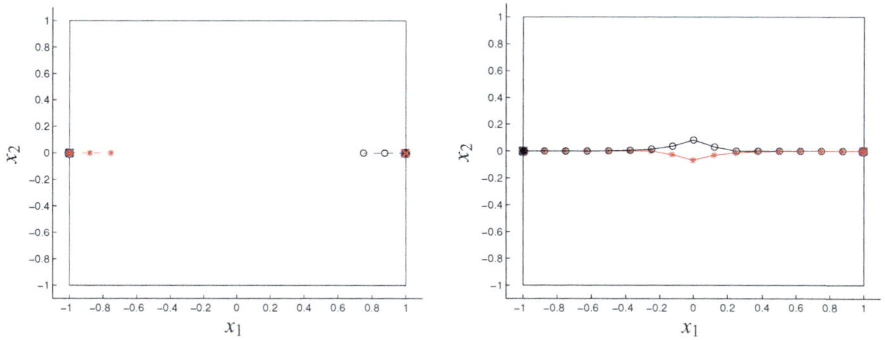

Fig. 12.1 MPC for Example 12.1 with $P = 2$ systems with initial values $x_1(0) = (1,0)^T$, $x_2(0) = (-1,0)^T$ and optimization horizon $N = 3$. Prediction for $k = 0$ (left) and final trajectories at $k = 16$ (right).

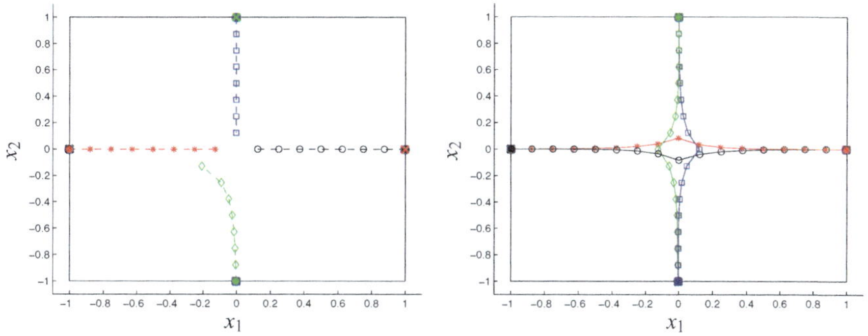

Fig. 12.2 MPC for Example 12.1 with $P = 4$ systems with initial values $x_1(0) = (1,0)^T$, $x_2(0) = (-1,0)^T$, $x_3(0) = (0,1)^T$, $x_4(0) = (0,-1)^T$ and optimization horizon $N = 8$. Prediction for $k = 0$ (left) and final trajectories at $k = 16$ (right)

that avoiding conflicts is easier when the agents are approaching each other in a diagonal way. This explains why resolving the conflict is easier in the situation of Fig. 12.3, in which the optimization horizon $N = 3$ is sufficient in order to stabilize all $P = 4$ agents. A further interesting observation in this example is that the resulting trajectories are not symmetric as in Figs. 12.1 and 12.2. Rather, here one sees the effect of the sequential ordering, since x_1 and x_3 approach their respective equilibria directly (with x_3 performing a shorter step at time $k = 7$ in order to avoid the collision with x_1), while x_2 and x_4 are forced to take small detours. Due to the short horizons N, most of the conflicts, i.e., the possible collisions, are resolved by the optimization at runtime in step (1) and not in the initialization in step (0) of Algorithm 12.3. The only exception is the fourth system starting at $x_4(0) = (0,-1)^T$ in Fig. 12.2, which at the time of its first optimization in step (0) already knows all other systems' predictions. Hence, the simulations nicely illustrate our schemes' ability to resolve conflicts at runtime.

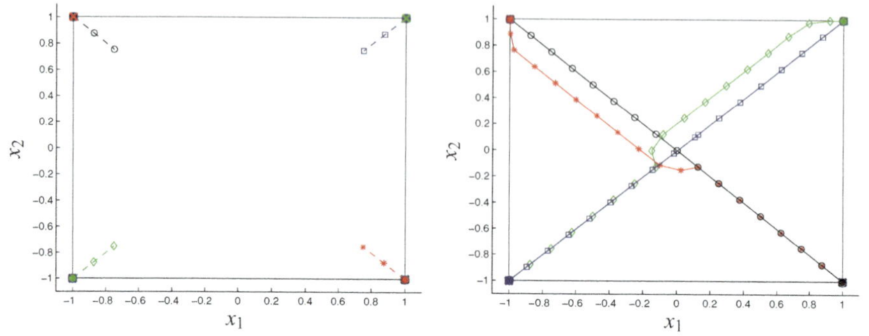

Fig. 12.3 MPC for Example 12.1 with $P = 4$ systems with initial values $x_1(0) = (-1,1)^T$, $x_2(0) = (1,-1)^T$, $x_3(0) = (1,1)^T$, $x_4(0) = (-1,-1)^T$, optimization horizon $N = 3$ and different initial condition as in Fig. 12.2. Prediction for $k = 0$ (left) and final trajectories at $k = 18$ (right)

12.7 Conclusion and Future Work

In this chapter we have shown that the noncooperative distributed model predictive control scheme from [14, 13] can be formulated without stabilizing terminal constraints. An extension of the controllability-based NMPC stability analysis from [8, 7] yields a sufficient distributed controllabilty condition, ensuring stability and feasibility. Numerical examples show that the resulting scheme is able to stabilize a test system with small optimization horizons N and illustrate the schemes' ability to resolve conflicts at runtime.

We regard the analysis in this chapter as a first step, which can and needs to be improved in many ways. The controllability conditions (i) and (ii) required in Theorem 12.5 are certainly difficult to check for systems more complex than Example 12.1, and even for Example 12.1 with large P a rigorous verification currently appears out of reach. In fact, the inequality (12.22) resulting from the controllability condition is a quite strong property by itself in the sense that in each sampling period each optimal value function V_p is decreasing. This property is most likely too demanding in many applications in which one would rather expect that in each step only some of the V_p decrease while others may even increase. ISS small-gain arguments for large systems as in [3] may be suitable for handling such situations; however, so far the controllability properties needed to ensure the appropriate small gain properties of the NMPC closed-loop systems remain unknown.

Another extension will be to relax the requirement of strict sequential optimization imposed in Algorithm 12.3. A first step in this direction could be to exploit the fact that stability can also be expected if the optimization is performed only in each mth sampling period for $m \in \{1,\ldots,N-1\}$; cf. [7]. This enables us to set up a cyclic scheme in which in each sampling period only a subset of systems performs an optimization. While this reduces the waiting time, it does not completely remove the sequential order of the optimization. The development of a scheme in which the optimization is performed completely in parallel remains a major challenge.

12.8 Appendix

In this section we prove an auxiliary result needed for the analysis of the example in Sec. 12.6. In Lemma 12.3 we assume that the norm used in the definition of $\mathbf{X}$ is the maximum norm $\|\cdot\|_\infty$. For the Euclidean norm, see Remark 12.4.

Lemma 12.3. *Let $K > 0$, $R > \delta > \bar{u}$ and a trajectory $y(\cdot)$ with $\|y(k)\|_\infty \geq R+\delta$, $k \in \mathbb{N}_0$ be given and define the set $\mathbf{X}$ according to Example 12.1 for $P = 2$. Then there exists $\bar{k} \in \mathcal{N}$ such that for each $x^0 \notin T = [-R,R]^2$ with $\|x^0\| \leq K$ and $(x^0, y(0)) \in \mathbf{X}$, there is a finite number $k^* \in \mathcal{N}$ with $k^* \leq \bar{k}$ and a control $u(\cdot)$ satisfying $(x^u(k,x^0), y(k)) \in \mathbf{X}$ for all $k = 0,\ldots,k^*$ and $x^u(k^*,x^0) \in T$.*

Proof. *Note that $(x^0,y(0)) \in \mathbf{X}$ implies $\|x^0 - y(0)\|_\infty \geq \delta$. We construct $u(\cdot)$ in three steps and start by finding a control sequence such that*

$$\min\{|x_1^u(k_1)|, |x_2^u(k_1)|\} < \bar{u} \tag{12.24}$$

holds for some $k_1 \in \mathcal{N}$. If this inequality holds for $k = 0$, then there is nothing left to do. Otherwise, at each time $k = 0,1,2,\ldots$ such that (12.24) does not hold we choose one of the following control sequences. If $|x_1^0 - y_1(0)| \geq \delta$ then we set

$$u(k) = \begin{pmatrix} y_1(k+1) - y_1(k) \\ -\bar{u}\, sign(x_2^u(k)) \end{pmatrix}.$$

which implies $|x_1^u(k) - y_1(k)| = |x_1^0 - y_1(0)| \geq \delta$ and $|x_2^u(k)| = |x_2^0| - k\bar{u}$ for $k \leq k_1$. Thus, the trajectory is admissible and (12.24) is satisfied for $k_1 = \lfloor |x_2^0/\bar{u}| \rfloor$. Note that $|u_1(k)| \leq \bar{u}$ since $y_1(k+1) - y_1(k)$ equals the control applied at time k in order to generate $y(\cdot)$. If $|x_1^0 - y_1(0)| < \delta$ then $\|x^0 - y(0)\|_\infty \geq \delta$ implies that $|x_2^0 - y_2(0)| \geq \delta$ holds and we can do the analogous construction exchanging x_1^0 and x_2^0, which implies (12.24) for $k_1 = \lfloor |x_1^0/\bar{u}| \rfloor$. All in all, this shows that (12.24) can always be satisfied with $k_1 \leq \lfloor \|x^0\|_\infty/\bar{u} \rfloor$. Furthermore, note that $\|x^u(k_1)\|_\infty \leq \|x^0\|_\infty + \lfloor \|x^0\|_\infty/\bar{u} \rfloor \bar{u} \leq 2\|x^0\|_\infty$ holds.

In the following second step we construct $u(k_1)$, $u(k_1+1)$, …, $u(k_2)$ such that we approach T as fast as possible until we either reach T or are blocked by $y(\cdot)$. To this end, we assume without loss of generality $|x_1^u(k_1)| < \bar{u}$ and $x_2^u(k_1) > 0$; otherwise we use symmetric arguments in the sequel. We set

$$u(k) = \begin{pmatrix} -x_1^u(k) \\ -\bar{u} \end{pmatrix}$$

for $k = k_1, k_1+1, \ldots, k_2$ where $k_2 \geq k_1$ is the minimal time at which either $x^u(k_2) \in T$ or $\|x^u(k_2+1) - y(k_2+1)\|_\infty < \delta$ holds. Note that since $x_1^u(k_2) \in (-\bar{u},\bar{u}) \subset [-R,R]$ and $x_2^u(k_2) = x_2^u(k_1) - (k_2 - k_1)\bar{u} \in [-R,R]$ for $(k_2 - k_1)\bar{u} \in [x_2^u(k_1) - R, x_2^u(k_1) + R]$, we obtain $k_2 \leq k_1 + \lfloor \|x^u(k_1)\|_\infty/\bar{u} \rfloor \leq k_1 + \lfloor 2\|x^0\|_\infty/\bar{u} \rfloor$.

If $x^u(k_2) \in T$ then we set $k^* = k_2$ and are done; otherwise, we continue with the third part of our construction for $u_1(k_2), u_1(k_2+1), \ldots$. To this end we distinguish *two cases. The first case is that* $x_2^u(k_2) \geq y_2(k_2) + \delta$ *holds. In this case we set*

$$u(k_2) = \begin{pmatrix} -x_1^u(k_2) \\ y_2(k_2+1) + \delta - x_2^u(k_2) \end{pmatrix}$$

which implies $x_2^u(k_2+1) = y_2(k_2+1) + \delta$ *and thus* $\|x^u(k_2+1) - y(k_2+1)\|_\infty = \delta$. *Furthermore we have* $\|x^u(k_2+1)\|_\infty \leq \|x^u(k_2)\|_\infty + \bar{u}$. *Observe that by choice of* k_2 *the relation* $x^u(k_2) \notin T$ *implies* $x_2^u(k_2) > y_2(k_2) - \delta$.

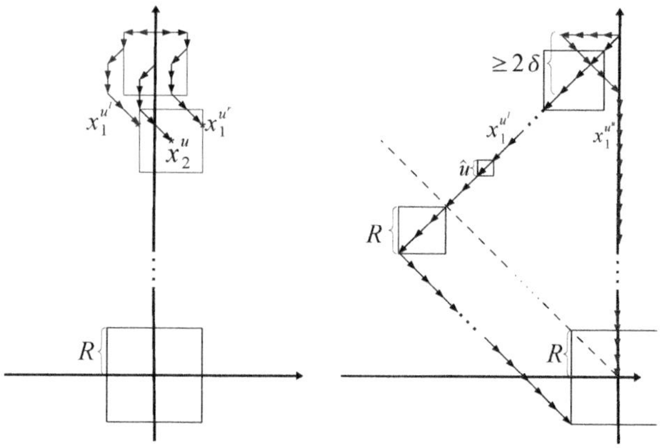

Fig. 12.4 Illustration of the trajectories used in the construction of x_1^u.

Now we continue by constructing two alternative trajectories $x^{u^r}(k)$ *and* $x^{u^l}(k)$ *for* $k \geq k_2 + 1$*; cf. Fig. 12.4 (left). At least one of them is admissible and reaches T in a predetermined number of steps. The first elements of the corresponding control sequences are*

$$u^r(k) = \begin{pmatrix} \bar{u} \\ y_2(k+1) - y_2(k) \end{pmatrix} \quad \text{and} \quad u^l(k) = \begin{pmatrix} -\bar{u} \\ y_2(k+1) - y_2(k) \end{pmatrix}.$$

As soon as $|x_1^{u^j}(k+1) - y_1(k+1)| \geq \delta$ *holds for* $j \in \{l, r\}$ *and the following control values, we change the control inputs to*

$$u_1^r(k) = \begin{cases} -\min\{\bar{u}, \max\{x_1^{u^r}(k) - \delta, -\bar{u}\}\}, & y_1(k) + \delta \leq \delta \\ y_1(k+1) + \delta - x_1^u(k), & \text{otherwise} \end{cases}$$

and $u_2^r(k) = -\bar{u}$ *if* $x_2^{u^r}(k) > R$ *and* $u_2^r(k) = 0$ *otherwise, and*

$$u_{11}^l(k) = \begin{cases} \min\{\bar{u}, \max\{-x_1^{u^r}(k) - \delta, -\bar{u}\}\}, & y_1(k) - \delta \geq -\delta \\ y_1(k+1) - \delta - x_1^u(k), & \textit{otherwise,} \end{cases}$$

and $u_2^l(k) = -\bar{u}$ *if* $x_2^{u^l}(k) > R$ *and* $u_2^l(k) = 0$*, otherwise, respectively. We denote the times at which this is done as* k_3^l *and* k_3^r*. Observe that for at least one* $j \in \{l, r\}$ *we have* $k_3^j \leq k_2 + \lceil \delta/\bar{u} \rceil$ *and for the corresponding trajectory we get* $x_2^{u^j}(k_3^j) \leq x_2^u(k_2) + \delta + \bar{u}$*. Moreover, the choice of* $u_1^j(k)$ *implies that* $u^j(k)$ *is admissible for all* $k \geq k_3^j$*,* $j \in \{r, l\}$ *and that for each* $k \geq k_2 + 1$ *we have* $x_1^{u^j} \in [-\delta, \delta]$ *for at least one* $j \in \{r, l\}$*. Since after the times* k_3^j *the trajectories move down as fast as possible, we obtain* $x_2^{u^j}(k) \in [-R, R]$ *for all* $k \geq k_4^j$ *with*

$$k_4^j \leq k_3^j + \left\lfloor \frac{x_2^u(k_2) + \delta + \bar{u}}{\bar{u}} \right\rfloor \leq k_3^j + \left\lfloor \frac{2\|x^0\|_\infty + \delta + \bar{u}}{\bar{u}} \right\rfloor$$

Without loss of generality let $k_4^l \leq k_4^r$*, which implies*

$$k_4^l \leq k_2 + \lceil \delta/\bar{u} \rceil + \left\lfloor \frac{2\|x^0\|_\infty + \delta + \bar{u}}{\bar{u}} \right\rfloor$$

Then, by construction of u^l *we have that either* $x^{u^l}(k_4^l) \in T$ *or* $y_1(k_4^l) < 0$*. In the latter case our construction implies* $k_3^r \leq k_4^l$*. This, in turn, implies*

$$k_4^r \leq k_4^l + \left\lfloor \frac{2\|x^0\|_\infty + \delta + \bar{u}}{\bar{u}} \right\rfloor + k_4^l - k_2 \leq k_2 + 3 \left\lfloor \frac{2\|x^0\|_\infty + \delta + \bar{u}}{\bar{u}} \right\rfloor + 2\lceil \delta/\bar{u} \rceil$$

At time k_4^r *we have* $x_2^{u^j}(k_4^r) \in [-R, R]$ *for all* $j \in \{r, l\}$*, which together with* $x_1^{u^j} \in [-\delta, \delta]$ *for at least one* $j \in \{r, l\}$ *implies* $x_1^{u^j}(k_4^r) \in T$ *for at least one* $j \in \{r, l\}$*. Hence, at least one of the trajectories reaches* T *in the time*

$$k^* = k_4^r \leq \left\lfloor \frac{\|x^0\|_\infty}{\bar{u}} \right\rfloor + \left\lfloor \frac{2\|x^0\|_\infty}{\bar{u}} \right\rfloor + 3 \left\lfloor \frac{2\|x^0\|_\infty + \delta + 2\bar{u}}{\bar{u}} \right\rfloor + 2\lceil \delta/\bar{u} \rceil$$

It remains to deal with the case $x_2^u(k_2) < y_2(k_2) + \delta$ *and* $|x_1^u(k_2)| \leq \bar{u}$*. Again we construct two alternative trajectories for* $k \geq k_2$*, cf. Fig. 12.4 (right). Recall that from the construction of* k_2 *and* $x_1^u(k_2) \notin T$ *the inequality* $x_2^u(k_2) > y_2(k_2) - \delta$ *follows. Since* $\|x^u(k_2) - y(k_2)\|_\infty \geq \delta$ *this implies* $|x_1^u(k_2) - y_1(k_2)| \geq \delta$*. Without loss of generality we assume* $x_1^u(k_2) \leq y_1(k_2) - \delta$*. We define two different control sequences for* $k \geq k_2$ *whose first elements are given as*

$$u^u(k) = \begin{pmatrix} -\bar{u} \\ 0 \end{pmatrix} \quad \textit{and} \quad u^l(k) = \begin{pmatrix} -\bar{u} \\ -\bar{u} \end{pmatrix}.$$

Since x^{u^j}, $j \in \{u,l\}$, moves left with maximal speed, $x_1^{u^j}(k) \le y_1(k) - \delta$ holds, implying feasibility, i.e., $(x^{u^j}(k), y(k)) \in \mathbf{X}$. Furthermore, this choice implies $x_2^{u^u}(k_3) - x_2^{u^l}(k_3) \ge 2\delta$ at time $k_3 = k_2 + \lceil 2\delta/\bar{u} \rceil$. Then, for $k \ge k_3$, we use the controls

$$u^u(k) = \begin{pmatrix} \min\{-x_1^{u^u}(k), \bar{u}\} \\ -\bar{u} \end{pmatrix} \tag{12.25}$$

and

$$u_1^l(k) = \begin{cases} -\bar{u}, & -x_1^{u^l}(k) \le x_2^{u^l}(k) + 2(R - \bar{u}) \\ \bar{u}, & \text{otherwise} \end{cases}, \qquad u_2^l(k) = -\bar{u}. \tag{12.26}$$

By k^l we denote the minimal time at which $u_1^l(k^l) = \bar{u}$ holds. Provided that the resulting trajectories are admissible, they both reach T before time $k_4 \le k_2 + \lceil (2\|x^0\| + R)/\bar{u} \rceil$. Furthermore, they satisfy $x_2^{u^u}(k) - x_2^{u^l}(k) \ge 2\delta$ for $k \in \{k_3, \ldots, k_4\}$. Since x^{u^l} moves left with maximal speed for $k = k_3, \ldots, k^l$ and satisfies $x_1^{u^l}(k_3) \le y_1(k_3) - \delta$, we obtain that the trajectory is admissible at least until time $k = k^l$.

Now assume that there exists $\tilde{k} \in \{k^l + 1, \ldots, k_4 - 1\}$ at which x^{u^l} is not admissible, i.e., at which $\|x^{u^l}(\tilde{k}) - y(\tilde{k})\|_\infty < \delta$ holds. Since x^{u^l} moves downwards with maximal speed, this is only possible if $y_2(k) \le x_2^{u^l}(k) + \delta$ holds for $k \in \{k_3, \ldots, \tilde{k}\}$. Since this implies that $x_2^{u^u}(k) \ge x_2^{u^l}(k) + 2\delta \ge y_2(k) + \delta$ holds, $x^{u^u}(k)$ is admissible for $k = k_3, \ldots, \tilde{k}$. On the other hand, $\|x^{u^l}(\tilde{k}) - y(\tilde{k})\|_\infty < \delta$ for $\tilde{k} \ge k^l + 1$, and a little computation based on the definition of k^l reveals that $y_1(k) \le -\delta$ holds as long as $x^{u^u}(k) \notin T$. Hence, x^{u^u} is admissible until it reaches T and consequently, at least one of the trajectories is admissible and reaches T in a time

$$k^* \le k_4 \le \left\lfloor \frac{\|x^0\|_\infty}{\bar{u}} \right\rfloor + \left\lfloor \frac{2\|x^0\|_\infty}{\bar{u}} \right\rfloor + \left\lceil \frac{2\|x^0\| + R}{\bar{u}} \right\rceil$$

Summarizing, we have shown that u and a finite number k^* with the required properties required exist. Since $\|x_1^0\|_\infty$ is bounded from above by K, the corresponding time k^* satisfies $k^* \le \bar{k}$ for some suitable $\bar{k}$ depending on K and R which completes the proof. □

Remark 12.4. It is also possible to prove Lemma 12.3 using the Euclidean norm $\|\cdot\|_2$. To this end, one has to assume $\|y(k)\|_2 \ge \sqrt{2}(R + \delta)$, which implies $\|y(k)\|_\infty \ge (R + \delta)$. Moreover, an additional prelimary step is required since $\|x^0 - y(0)\|_2 \ge \delta$ does not imply $\|x^0 - y(0)\|_\infty \ge \delta$. However, since $y(\cdot)$ is known, this can be easily obtained in a finite number of steps. Then, one may proceed as in the proof of Lemma 12.3, since $\|x^u(k) - y(k)\|_\infty \ge \delta$ implies $(x^u(k), y(k)) \in \mathbf{X}$ for the Euclidean norm $\|\cdot\|_2$ as well. Apart from this initialization step, the term $\|x^0\|_\infty$ in the estimates for k^* remains the same.

References

1. Altmüller, N., Grüne, L., Worthmann, K.: Performance of NMPC schemes without stabilizing terminal constraints. In: M. Diehl, F. Glineur, E. Jarlebring, W.M. (Eds.) (eds.) Recent Advances in Optimization and Its Applications in Engineering, pp. 289–298. Springer, London (2010)
2. Altmüller, N., L., Grüne, Worthmann, K.: Receding horizon optimal control for the wave equation. In: Proc. 49th IEEE Conf. Decision and Control (CDC2010), pp. 3427–3432. Atlanta, GA (2010)
3. Dashkovskiy, S., Rüffer, B.S., Wirth, F.R.: An ISS small gain theorem for general networks. Math. Control Signals Systems **19**(2), 93–122 (2007)
4. Giselsson, P., Rantzer, A.: Distributed model predictive control with suboptimality and stability guarantees. In: Proc. 49th IEEE Conference on Decision and Control (CDC2010), pp. 7272–7277. Atlanta, GA (2010)
5. Grimm, G., Messina, M.J., Tuna, S.E., Teel, A.R.: Model predictive control: For want of a local control Lyapunov function, all is not lost. IEEE Trans. Automatic Control pp. 546–558 (2005)
6. Grüne, L., Pannek, J.: Nonlinear Model Predictive Control. Theory and Algorithms. Springer, London (2011)
7. Grüne, L., Pannek, J., Seehafer, M., Worthmann, K.: Analysis of unconstrained nonlinear MPC schemes with varying control horizon. SIAM J. Control Optimization **48**, 4938–4962 (2010)
8. Grüne, L.: Analysis and design of unconstrained nonlinear MPC schemes for finite and infinite dimensional systems. SIAM J. Control Optimization **48**, 1206–1228 (2009)
9. Jadbabaie, A., , J.H.: On the stability of receding horizon control with a general terminal cost. IEEE Trans. Automatic Control **50**(5), 674–678 (2005)
10. Mayne, D.Q., Rawlings, J.B., Rao, C.V., Scokaert, P.O.M.: Constrained model predictive control: stability and optimality. Automatica **36**, 789–814 (2000)
11. Primbs, J., Nevistić, V.: Feasibility and stability of constrained finite receding horizon control. Automatica **36**(7), 965–971 (2000)
12. Rawlings, J.B., Mayne, D.Q.: Model Predictive Control: Theory and Design. Nob Hill Publishing, Madison (2009)
13. Richards, A., How, J.P.: Robust distributed model predictive control. Int. J. Control **80**(9), 1517–1531 (2007)
14. Richards, A., How, J.P.: A decentralized algorithm for robust constrained model predictive control. In: Proc. American Control Conf. (ACC2004), pp. 4261–4266. Boston, MA (2004)
15. Scattolini, R.: Architectures for distributed and hierarchical model predictive control—A review. J. Process Control **19**(5), 723–731 (2009)

Chapter 13
A Set-Theoretic Method for Verifying Feasibility of a Fast Explicit Nonlinear Model Predictive Controller

Davide M. Raimondo, Stefano Riverso, Sean Summers, Colin N. Jones, John Lygeros and Manfred Morari

Abstract In this chapter an algorithm for nonlinear explicit model predictive control is presented. A low complexity receding horizon control law is obtained by approximating the optimal control law using multiscale basis function approximation. Simultaneously, feasibility and stability of the approximate control law is ensured through the computation of a capture basin (region of attraction) for the closed-loop system. In a previous work, interval methods were used to construct the capture basin (feasible region), yet this approach suffered due to slow computation times and high grid complexity.

In this chapter, we suggest an alternative to interval analysis based on zonotopes. The suggested method significantly reduces the complexity of the combined function approximation and verification procedure through the use of DC (difference of

Davide M. Raimondo
Automatic Control Laboratory, ETH, Physikstrasse 3, 8092 Zürich, Switzerland
e-mail: davide.raimondo@control.ee.ethz.ch

Stefano Riverso
Laboratorio di Identificazione e Controllo dei Sistemi Dinamici, Università degli Studi di Pavia, Dipartimento di Informatica e Sistemistica, Via Ferrata, 1, 27100 Pavia, Italy
e-mail: stefano.riverso@unipv.it

Sean Summers
Automatic Control Laboratory, ETH, Physikstrasse 3, 8092 Zürich, Switzerland
e-mail: ssummers@control.ee.ethz.ch

Colin N. Jones
Automatic Control Laboratory, ETH, Physikstrasse 3, 8092, Zürich, Switzerland
e-mail: cjones@control.ee.ethz.ch

John Lygeros
Automatic Control Laboratory, ETH, Physikstrasse 3, 8092 Zürich, Switzerland,
e-mail: lygeros@control.ee.ethz.ch

Manfred Morari
Automatic Control Laboratory, ETH, Physikstrasse 3, 8092 Zürich, Switzerland
e-mail: morari@control.ee.ethz.ch

R. Johansson and A. Rantzer (Eds.): Distributed Decision Making and Control, LNCIS 417, pp. 289–311.
springerlink.com

convex) programming, and recursive splitting. The result is a multiscale function approximation method with improved computational efficiency for fast nonlinear explicit model predictive control with guaranteed stability and constraint satisfaction.

13.1 Introduction

This chapter proposes a method of approximate explicit model predictive control (MPC) for nonlinear systems. While it is possible to compute the optimal control law offline for a limited number of cases (e.g., affine or piecewise affine dynamics [5, 20, 29]), it is in general necessary to approximate, and therefore validation techniques are required for the resulting approximate closed-loop system. In this chapter, we present a new technique for approximation and certification of stability and recursive feasibility for explicit NMPC controllers, in which the control law is precomputed and verified offline in order to speed online computation.

The control law is approximated via an adaptive interpolation using second order interpolets, which results in an extremely fast online computation time and low data storage. The resulting suboptimal closed-loop system is verified by computing an inner approximation of the capture basin and an algorithm is proposed that iteratively improves the approximation where needed in order to maximize the size of the capture basin. The key novelty of this chapter is the use of difference of convex (DC) programming and zonotope approximation in order to significantly improve both the computational performance and efficacy of the calculation of the capture basin.

Methods for the approximation of explicit solutions of nonlinear model predictive control (NMPC) problems have been addressed recently by various authors (e.g., see [8, 19]). In [8], the authors compute an approximate control law $\tilde{u}(x)$ with a bound on the controller approximation error ($\|u^*(x) - \tilde{u}(x)\|$), from which performance and stability properties are derived using set membership (SM) function approximation theory. In [19] the authors use multiparametric nonlinear programming to compute an explicit approximate solution of the NMPC problem defined on an orthogonal structure of the state-space partition. An additional example of the approximation of explicit solutions of NMPC can be found in [26].

In almost all cases, the suboptimality of the resulting control law (and as a consequence the stability of the feedback system) is valid under various strong assumptions. Examples include the approximation of the Lipschitz constant ([8]) and the availability of global optimization tools ([19]). While these approaches often work well in practice, in many problems the stability of the closed-loop system (and the resulting region of attraction) cannot be guaranteed. Thus, in this chapter we exploit advances in reachability analysis and adaptive interpolation to construct an approximate explicit control law that encompasses the strengths of the recent works ([8, 19]) while guaranteeing stability and feasibility and preserving a minimal representation of the control law.

Extending the results of [30, 31], in this chapter we introduce a constructive algorithm for the approximation of an explicit receding horizon NMPC control law. We approximate the optimal control law by adaptive interpolation using second order interpolets, while concurrently verifying feasibility and stability of the resulting feedback system via the computation of an inner approximation of the capture basin (see, e.g., [12]). In contrast to the capture basin computational method considered in [12, 31], we develop a mechanism for computing the capture basin using zonotopes [22, 11, 33] and DC programming [3] that significantly reduces the complexity of the combined function approximation and verification procedure. Using zonotopes and DC programming rather than interval analysis [25, 6] additionally leads to an approximate control law with less storage requirements and a larger verifiable region of attraction. With the approach we propose, we are able to construct a sparse approximation of the optimal control law while taking into consideration performance loss and the feasibility and stability of the feedback system. Further, since the solution is defined on a gridded hierarchy, the online evaluation of the control law is extremely fast, see [30].

The rest of the chapter is arranged as follows: Section 13.2 introduces the NMPC problem. Section 13.3 provides background on multiscale sparse function approximation and Secs. 13.4 and 13.5 discuss reachability analysis and the proposed method of calculating the capture basin of an approximation NMPC explicit control law. Section 13.6 provides a numerical example indicating the effectiveness of the approach.

13.2 Nonlinear Model Predictive Control

Consider the following finite horizon optimal control problem (NMPC):

$$\begin{aligned} J^*(x) = \min_{u_0,\ldots,u_{N-1}} \quad & J(u_0,\ldots,u_{N-1},x_0,\ldots,x_N) \\ \text{subject to} \quad & x_{i+1} = f(x_i,u_i), \quad \forall i = 0,\ldots,N-1 \\ & (x_i,u_i) \in \mathcal{X} \times \mathcal{U}, \quad \forall i = 0,\ldots,N-1 \\ & x_N \in \mathcal{X}_F, \\ & x_0 = x, \end{aligned} \tag{13.1}$$

where $x_i \in \mathbb{R}^n$ is the state of the system, $u_i \in \mathbb{R}^m$ is the control input of the system, and N is the prediction horizon length. The cost function J takes the form

$$J(u_0,\ldots,u_{N-1},x_0,\ldots,x_N) := V_N(x_N) + \sum_{i=0}^{N-1} L(x_i,u_i) \ , \tag{13.2}$$

where L is the running (stage) cost and V_N is the terminal cost.

The system dynamics $f : \mathbb{R}^n \times \mathbb{R}^m \to \mathbb{R}^n$ is a continuous and differentiable function, and the objective is to regulate the state of the system to the origin under state and control input constraints represented by the (compact) sets $\mathcal{X} \subseteq \mathbb{R}^n$, $\mathcal{U} \subseteq \mathbb{R}^m$, respectively. We assume that the terminal set $\mathcal{X}_F$ is compact and positively control invariant under a known stabilizing feedback law κ_F. For the sake of simplicity (as in [8, 19]), it is assumed that the control input constraint set $\mathcal{U}$ is convex, although the following results can be extended to the nonconvex control input constraint setting.

A dual-mode NMPC control approach is taken, in which the optimal NMPC control law $\kappa^\star(x)$ is defined as

$$\kappa^\star(x) := \begin{cases} \kappa_F(x), & \text{if } x \in \mathbb{X}_F \\ u_0^\star(x), & \text{otherwise} \end{cases} \tag{13.3}$$

where $u^*(x) = u_0^\star(x), \ldots, u_{N-1}^\star(x)$ is an optimal sequence of inputs of NMPC problem (13.1) for the state x. Following [8, 19], we assume in this chapter that the optimal receding horizon control law $u_0^*(x)$ asymptotically stabilizes the origin of the closed-loop system.

Remark 13.1. Note that the proposed approximation and analysis techniques can be applied to any optimal control problem that generates a smooth control law. We here use the MPC cost function given in (13.2) because it is a common target for approximation and because sufficiently fine approximation will result in a stabilizing control law by construction.

13.3 Multiscale Function Approximation

The method we propose for approximating $u_0^\star(x)$ relies on coarsely gridding the state space, and then regridding with increasing resolution only the regions which have not been approximated sufficiently. At the same time, we keep only the grid points that play a significant role in the function approximation [30].

Define the one-dimensional scaling function with support $[-1,1]$ by

$$\phi(x) := \begin{cases} 1-|x| \ , & \text{if } x \in [-1,1], \\ 0 \ , & \text{otherwise.} \end{cases} \tag{13.4}$$

In one dimension, we consider a dyadic discretization on the unit interval $\Omega = [0,1]$. The resulting grid Ω_l is characterized by the level of discretization l and the index i. At level l the distance between points is $h_l = 2^{-l}$ and the number of points is $N = 2^l + 1$. The index i determines the location of the grid points according to the equation

$$x_{l,i} := i \cdot h_l, \quad 0 \leq i \leq 2^l.$$

Given function (13.4), via translation and dilation we get

$$\phi_{l,i}(x) = \phi\left(\frac{x - i \cdot h_l}{h_l}\right) \tag{13.5}$$

where $\phi_{l,i}$ represents a family of basis functions with support $[x_{l,i} - h_l, x_{l,i} + h_l]$. The family of univariate multiscale functions $\psi_{l,i}$ that make up the hierarchical basis is given as

$$\psi_{l,i} = \phi_{l,i}, i \in I_l$$

where

$$I_l = \begin{cases} \{i \in \mathbb{N}_0 | 1 \leq i \leq 2^l - 1, i \text{ odd}\}, & l > l_0, \\ \{i \in \mathbb{N}_0 | 0 \leq i \leq 2^l\}, & l = l_0. \end{cases}$$

A multivariate multiscale basis on the unit cube $\Omega^d = [0,1]^d$, where d is the dimension, can be constructed by tensor product expansion of the one-dimensional multivariate functions $\psi_{l,i}$, i.e.

$$\psi_{l,\mathbf{i}} = \prod_{j=1}^{d} \psi_{l,i_j} \tag{13.6}$$

with the d-dimensional multi-index $\mathbf{i} \in I_l^d$ and

$$I_l^d = \begin{cases} \{\mathbf{i} \in \mathbb{N}_0^d | \mathbf{0} \leq \mathbf{i} \leq \mathbf{2}^l\} \setminus \{\mathbf{i} \in \mathbb{N}_0^d | \mathbf{0} \leq \mathbf{i} \leq \mathbf{2}^l, i_j \text{ even } \forall j \in [1,d]\} & l > l_0 \\ \{\mathbf{i} \in \mathbb{N}_0^d | \mathbf{0} \leq \mathbf{i} \leq \mathbf{2}^l\} & l = l_0 \end{cases}$$

I_l^d is the full grid less those points seen at previous levels, as depicted in Fig. 13.1. The d-dimensional hierarchical function spaces of piecewise d-linear functions can be defined as $W_l^d = \text{span}\{\psi_{l,\mathbf{i}} : \mathbf{i} \in I_l^d\}$. Let

$$\phi_{l,\mathbf{i}}(x) = \prod_{j=1}^{d} \phi_{l,i_j}(x_j)$$

the family of d-dimensional nodal basis functions and $V_l^d = \text{span}\{\phi_{l,\mathbf{i}} : \mathbf{0} \leq \mathbf{i} \leq \mathbf{2}^l\}$ the d-dimensional nodal function space. It holds that $V_l^d = \bigoplus_{k \leq l} W_k^d$ where $\bigoplus$ denotes the direct sum. Therefore, any function $u_l \in V_l^d$ can be represented in the hierarchical basis by

$$u_l(x) = \sum_{k=l_0}^{l} \sum_{\mathbf{i} \in I_k^d} w_{k,\mathbf{i}} \cdot \psi_{\mathbf{k},\mathbf{i}}(x)$$

where coefficients $w_{k,\mathbf{i}} \in \mathbb{R}$ (hierarchical details) correspond to the difference between the true function value $u_l(x_{k,\mathbf{i}})$ and the value of the approximate function one level below.

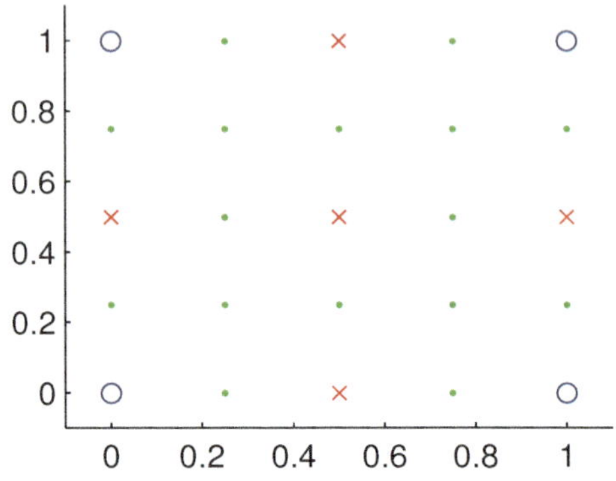

Fig. 13.1 Grid points for subspaces W_0^2 (circles), W_1^2 (x's), and W_2^2 (dots).

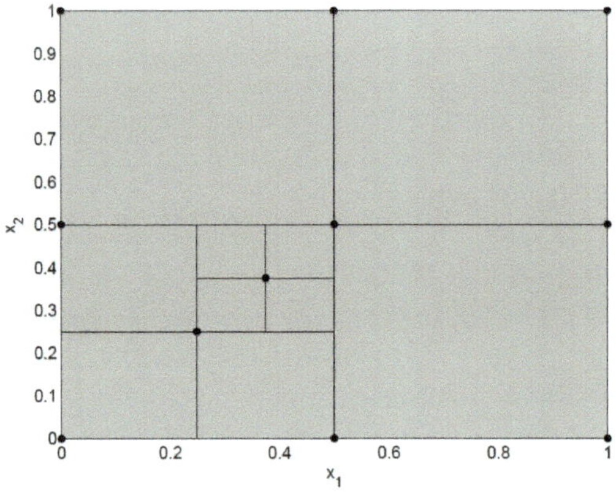

Fig. 13.2 Set of hypercubic regions $\bar{R}$.

In [30], it has been shown that the function approximation by adaptive hierarchical basis function expansion generates a grid (of hypercubes (Fig. 13.2)) spanned by an interpolation by barycentric coordinates. Given a set $\bar{R}$ of hypercubic regions, for each hypercube R it holds that

$$\hat{u}(x) = \sum_{v \in \mathrm{extr}(R)} \hat{u}(v) f_v(x), \qquad \text{if } x \in R. \tag{13.7}$$

where $\mathrm{extr}(R)$ are the extreme points of R and $f_v(x)$ are compactly supported basis functions of the form (13.6) centered at the corners of the hypercube R.

Theorem 13.1. *Given any hypercube R of $\bar{R}$, if $\hat{u}(v) \in U$, $\forall v \in extr(R)$ and U is convex, then $\hat{u}(x) \in U$, $\forall x \in R$. Moreover, if problem (13.1) is convex, then, feasibility* *at vertexes v of R is necessary and sufficient in order to get feasibility for all $x \in R$, i.e., $g(x, \hat{u}(x)) \leq 0$, $\forall x \in R$.*

Proof: The result is obtained by exploiting the barycentricity of the interpolation. See [30] for details.

Note that, in the general nonlinear case, the constraint satisfaction at the vertexes of R is not sufficient in order to prove their satisfaction in the entire box. This is because g and h are in general nonconvex. However, if U is convex, then control constraint satisfaction at the vertexes of R guarantees their satisfaction in R.

Summarizing, we need an alternative method for verifying the stability and state constraints satisfaction of system $x_{i+1} = f(x_i, u_i)$ in closed loop with $\hat{u}(x)$. In the following section, we present reachability analysis of nonlinear systems as a possible solution.

13.4 Reachability

The exact computation of the set

$$\Phi = f(\Omega) \tag{13.8}$$

given an initial set $\Omega \subseteq \mathbb{R}^n$ and a map $f : \mathbb{R}^n \to \mathbb{R}^n$, is not possible in general. Taking this into account, the objective is to construct an outer approximation of Φ in such a way that the set is representable on a computer and the overestimation is kept as small as possible [22]. Several solutions have been proposed in the literature, and in the following sections we provide an overview of existing methods that apply for nonlinear functions f before proposing a new method for guaranteeing tight overbounds.

13.4.1 Interval Arithmetic

Given $S_1, S_2 \subseteq \mathbb{R}^n$ the Minkowski sum is defined as $S_1 \oplus S_2 = \{x+y : x \in S_1, y \in S_2\}$. Given $a, b \in \mathbb{R}$, with $a \leq b$, $[a,b]$ denotes the interval $\{x : a \leq x \leq b\}$. The center of the interval $[a,b]$ is denoted by $\text{mid}([a,b]) = \frac{a+b}{2}$. Let $\mathbb{I}$ be the set of all intervals $[a,b]$, i.e. $\mathbb{I} = \{[a,b] : a, b \in \mathbb{R}, a \leq b\}$. The set of all interval vectors in $\mathbb{R}^n$ is denoted by $\mathbb{I}^n$. The unitary interval $[-1,1]$ is denoted by $\mathbf{B}$. A box is an interval vector and a unitary box, denoted by $\mathbf{B}^m$, is a box composed of m unitary intervals. With a slight abuse of notation, when the superscript is not indicated, $\mathbf{B}$ denotes a generic unitary box. Given a box $\mathbf{X} = [a_1, b_1] \times [a_2, b_2] \ldots \times [a_n, b_n]$, $\text{mid}(\mathbf{X}) = (\text{mid}([a_1, b_1]), \ldots, \text{mid}[a_n, b_n])^T$ denotes the midpoint (or center) of $\mathbf{X}$, $\text{diam}(\mathbf{X}) = (b_1 - a_1, \ldots, b_n - a_n)^T$ and $\text{rad}(\mathbf{X}) = \text{diam}(\mathbf{X})/2$. Interval arithmetic is based on operations applied to intervals. An operation $\bullet$ can be extended from real numbers to intervals, i.e., given $\mathbf{X}_1, \mathbf{X}_2 \in \mathbb{I}$, $\mathbf{X}_1 \bullet \mathbf{X}_2 = \{x_1 \bullet x_2 : x_1 \in \mathbf{X}_1, x_2 \in \mathbf{X}_2\}$. The four basic interval operations as well as the interval extension of standard functions (sin, cos, tan, arctan, exp, ln, $|,|$ sqrt), are defined in [25].

Definition 13.1. (Natural interval extension [21]) If $f : \mathbb{R}^n \to \mathbb{R}^n$ is a function computable as an expression, algorithm or computer program involving the four elementary arithmetic operations and standard functions, then a *natural interval extension* of f, denoted by $\Box f$, is obtained by replacing the occurrence of each variable by the corresponding interval variable and by executing all operations.

Theorem 13.2 ([21]). *Given a function $f : \mathbb{R}^n \to \mathbb{R}^n$ and any box $\mathbf{X} \subseteq \mathbb{R}^n$ within the domain of f, a natural interval extension $\Box f : \mathbb{I}^n \to \mathbb{I}^n$ of f satisfies $f(\mathbf{X}) \subseteq \Box f(\mathbf{X})$.*

Definition 13.2. (Taylor interval extension of degree k) Let $f : \mathbb{R}^n \to \mathbb{R}^n$ be a $k+1$ times differentiable function, $\mathbf{X} \subseteq \mathbb{R}^n$ any box within the domain of f and $y \in \mathbf{X}$. The *Taylor interval extension of f of degree k* is given by

$$\Box^k f(\mathbf{X}) = \sum_{i=0}^{k} \frac{1}{i!} \nabla^i f(y) \cdot (\mathbf{X} - y)^i + \Box r_k(\mathbf{X}, \mathbf{X}, y)$$

where $\nabla^i f(y)$ is the ith order differential of f at the point y and $\Box r_k$ is an interval extension of the Taylor remainder

$$r_k(x, \xi, y) = \frac{1}{(k+1)!} \nabla^{k+1} f(\xi) \cdot (x - y)^{k+1}.$$

By substituting $\mathbf{X}$ for ξ we obtain an overestimation of the remainder. Usually, y is chosen to be the midpoint of the box $\mathbf{X}$, and natural interval extension is used to bound the remainder.

Theorem 13.3. *Let $f : \mathbb{R}^n \to \mathbb{R}^n$ be a $k+1$ times differentiable function and $\mathbf{X} \subseteq \mathbb{R}^n$ any box within the domain of f. A Taylor interval extension of f of degree k satisfies $f(\mathbf{X}) \subseteq \Box^k f(\mathbf{X})$.*

Because of the special form of r_k, in practice the Taylor remainder usually decreases as $|x-y|^{k+1}$. Hence if $|x-y|$ is chosen to be small, then the interval extension of the Taylor remainder gets smaller for increasing k, i.e., higher order Taylor interval extensions yield better enclosures on small boxes [24]. A comparison between natural interval extension and Taylor interval extension of degree 0 (yellow) and 9 (green) is given in Figs. 13.3 and 13.4. While Fig. 13.3 shows the advantage of Taylor interval extension on a small box, Fig. 13.4 shows its limits over big boxes. In this work, interval arithmetic has been implemented using INTLAB [27].

The main drawback of interval analysis is that it always outer bounds the image of a box with a box. The use of more complex domains can reduce the conservatism. For this reason, the use of zonotopes as the class of approximates is considered in the next section.

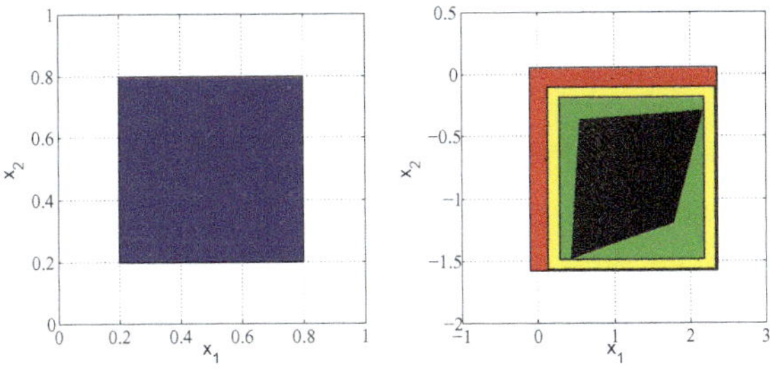

Fig. 13.3 In this example we considered the function f defined in (13.14). The starting set $\mathbf{X}$ is depicted on the left. On the right, samples of $f(\mathbf{X})$ are depicted in black while the red, yellow and green boxes represent the natural interval extension, the Taylor interval extension of degree 0 and 9 ($y = \text{mid}(\mathbf{X})$) respectively.

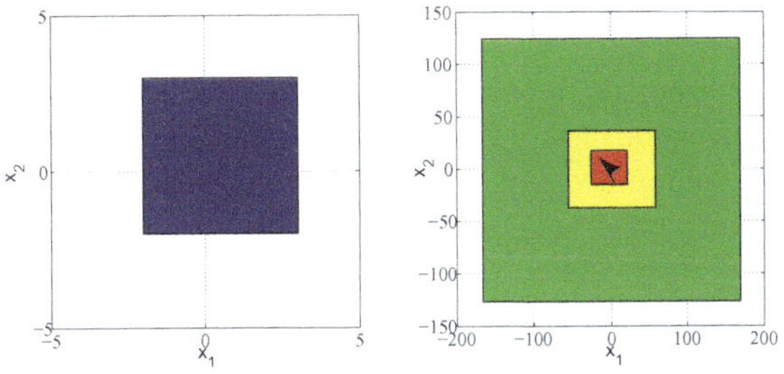

Fig. 13.4 In this example we considered the function f defined in (13.14). The starting set $\mathbf{X}$ is depicted on the left. On the right, samples of $f(\mathbf{X})$ are depicted in black while the red, yellow and green boxes represent the natural interval extension, the Taylor interval extension of degree 0 and 9 ($y = \text{mid}(\mathbf{X})$) respectively.

13.4.2 Zonotopes

Zonotopes are centrally symmetric convex polytopes. Given a vector $p \in \mathbb{R}^n$ and a matrix $H \in \mathbb{R}^{n \times m}$, the zonotope $\mathbf{Z}$ of order $n \times m$ is the set

$$\mathbf{Z} = p \oplus H\mathbf{B}^m = \{p + Hz | z \in \mathbf{B}^m\}.$$

The zonotope $\mathbf{Z}$ is the Minkowski sum of the line segments defined by the columns of the matrix H translated to the central point p. $\mathbf{Z}$ can be described as the set spanned by the column vectors of H

$$\mathbf{Z} = \{p + \sum_{i=1}^{m} \alpha_i h_i | -1 \leq \alpha_i \leq 1\}$$

where h_i, also called the *line segment generator*, is the ith column of H. When the matrix H is diagonal, the zonotope is a box composed of n intervals. The construction of zonotopes based on the Minkowski addition of convex polytopes is described in [16] and here adopted and implemented using the MPT toolbox [23]. An example of a two-dimensional (2D) zonotope is depicted in Fig. 13.5.

The image of a zonotope $\mathbf{Z}$ under a nonlinear function is not, in general, a zonotope. Kühn developed a procedure that guarantees a tight approximation by bounding $f(\mathbf{Z})$ with a zonotope [22]. The following theorem introduces the zonotope inclusion operator that is needed for computing Kühn's zonotope extension.

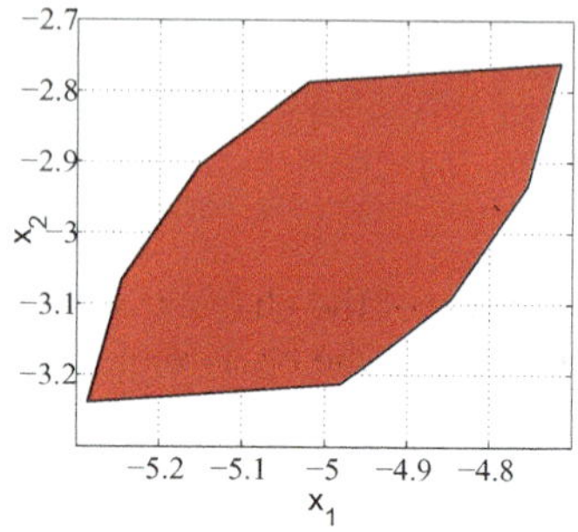

Fig. 13.5 A 2D zonotope.

Theorem 13.4. *(Zonotope inclusion) Consider a family of zonotopes represented by* $\mathbf{Z} = p \oplus \mathbf{M}\mathbf{B}^m$*, where* $p \in \mathbb{R}^n$ *is a real vector and* $\mathbf{M} \in \mathbb{I}^{n\times m}$ *is an interval matrix. A zonotope inclusion, denoted by* $\diamond(\mathbf{Z})$*, is defined by*

$$\diamond(\mathbf{Z}) = p \oplus [mid(\mathbf{M}) \quad G] \begin{bmatrix} \mathbf{B}^m \\ \mathbf{B}^n \end{bmatrix} = p \oplus J\mathbf{B}^{m+n},$$

where $G \in \mathbb{R}^{n\times n}$ *is a diagonal matrix that satisfies*

$$G_{ii} = \sum_{j=1}^{m} \frac{diam(\mathbf{M}_{ij})}{2}, \; i = 1,\ldots,n.$$

Under these definitions it results that $\mathbf{Z} \subseteq \diamond(\mathbf{Z})$.

Theorem 13.5. *(Zonotope extension) Consider a function* $f : \mathbb{R}^n \to \mathbb{R}^n$ *with continuous derivatives and a zonotope* $\mathbf{Z} = p \oplus H\mathbf{B}^m$*. Given an interval matrix* $\mathbf{M} \in \mathbb{I}^{n\times m}$ *such that* $\mathbf{M} \supseteq \nabla f(\mathbf{Z})H$*, it results that*

$$f(\mathbf{Z}) \subseteq f(p) \oplus \diamond(\mathbf{M}\mathbf{B}^m).$$

This theorem is a particular case of Kühn's method (see Proof in [2]). Note that, $\nabla f(\mathbf{Z})H$, multiplication of a matrix of sets by a matrix, is a matrix of sets. A possible outer bound is $\mathbf{M} = \Box\nabla f(\Box\mathbf{Z})H$.

The zonotope extension represents a Taylor extension of order 0, where the remainder has been evaluated on $\mathbf{Z}$. A comparison between zonotope Taylor extension of order 0 and Taylor interval extension of order 0 is given in Fig. 13.6. As expected, zonotopes better approximate the output set.

A first order Taylor zonotope extension was proposed in [11]

$$f(\mathbf{Z}) \subseteq f(p) \oplus \nabla f(p)(\mathbf{Z} - p) \oplus c_R \oplus [Z_Q \; Z_H]\mathbf{B}$$

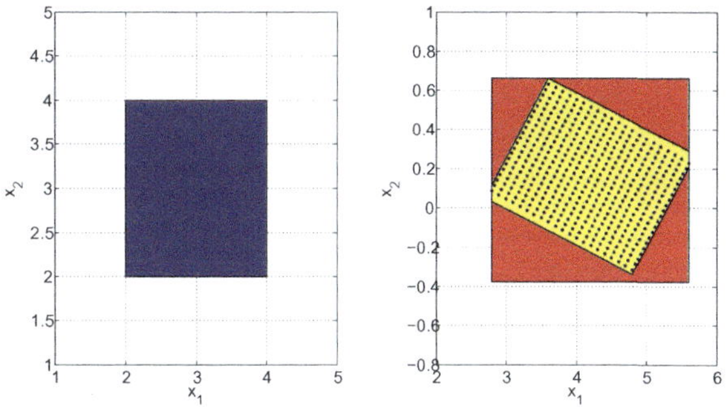

Fig. 13.6 In this example we considered the function f defined in (13.14). The starting set $\mathbf{Z}$ is depicted on the left. On the right, samples of $f(\mathbf{Z})$ are depicted in black while the red and yellow zonotopes represent the Taylor interval extension of degree 0 and the zonotope extension of degree 0 respectively.

where c_R and $[Z_Q\ Z_H]$ provide a less conservative approximation of the remainder by making use of some interval analysis properties (see Eq. (14) in [11] for the definition of c_R and $[Z_Q\ Z_H]$).

A Taylor zonotope extension of order k can been obtained as follows:

$$f(\mathbf{Z}) \subseteq f(p) \oplus \nabla f(p)(\mathbf{Z}-p) \oplus c_R \oplus Z_Q\mathbf{B} \oplus \sum_{i=3}^{k} \frac{1}{i!}\nabla^i f(p) \cdot (\Box\mathbf{Z}-p)^i \oplus \Box r_k(\Box\mathbf{Z}, \Box\mathbf{Z}, p).$$

Note that $c_R \oplus Z_Q\mathbf{B}$ represents the second order Taylor expansion term computed at p, center of the zonotope $\mathbf{Z}$ (see [11] for details). The higher order terms have been obtained by applying the Taylor interval extension. Due to over-approximation of zonotopes with boxes (i.e. $(\mathbf{Z}-p)^i$ is replaced by $(\Box\mathbf{Z}-p)^i$), there is no guarantee that higher order Taylor zonotope extensions will produce tighter enclosures than low order ones.

A comparison between zonotope Taylor extension of order 0 (red) and order 1 (yellow) (computed as in [11]) is given in Figs. 13.7 and 13.8. In Fig. 13.7, as one would expect, the first-order extension is better, but in Fig. 13.8 it is the opposite. This depends on the dynamics f (Eq. (13.13) in the case of Fig. 13.7 and Eq. (13.14) in Fig. 13.8) and the size of the starting set $\mathbf{Z}$, i.e., the larger the set, the worse the approximation. Furthermore, the use of higher order extensions does not guarantee improvement a priori. In Figs. 13.9 and 13.10, starting sets that are boxes (and not generic zonotopes) have been considered. In the first figure, higher-order extensions do better while in the second the opposite is observed. Again, this depends on the dynamics f and the size of the starting set $\mathbf{Z}$.

It is important to note that if a linear system is considered, zonotopes provide an exact description of the reachable set, unless a bound on the number of line generators is imposed. If the system is instead nonlinear, and the starting set is a small box, then high order extensions are generally better than low order ones, although the dynamics of the system plays an important role in determining which approach provides the best approximation. The drawback of high-order extensions is the need to compute the derivatives $\nabla^i f,\ i = 1,\cdots,k+1$. For this reason, we are interested in finding other techniques that are less computationally expensive while still rea-

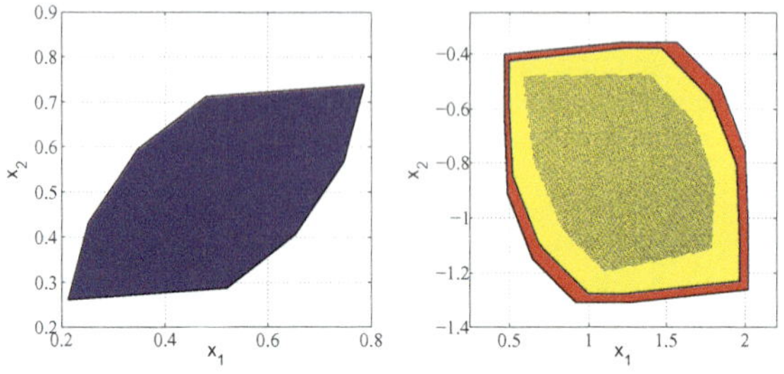

Fig. 13.7 In this example we considered the function f defined in (13.13). The starting zonotope $\mathbf{Z}$ is depicted on the left. On the right, samples of $f(\mathbf{Z})$ are depicted in black while the red and yellow zonotopes represent the Taylor zonotope extension of degree 0 and 1 (computed as in [11]) respectively.

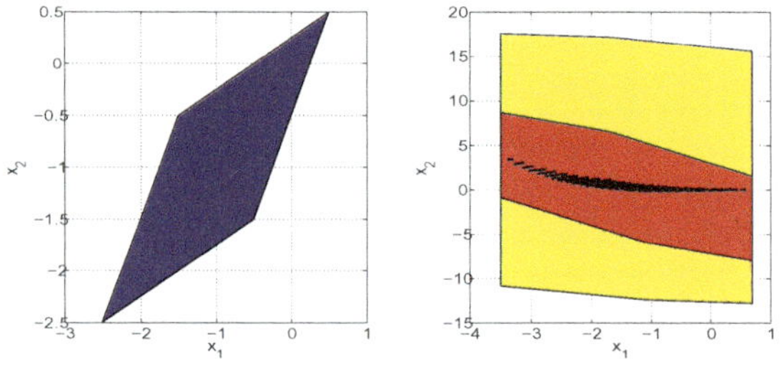

Fig. 13.8 In this example we considered the function f defined in (13.14). The starting zonotope $\mathbf{Z}$ is depicted on the left. On the right, samples of $f(\mathbf{Z})$ are depicted in black while the red and yellow zonotopes represent the Taylor zonotope extension of degree 0 and 1 (computed as in [11]), respectively.

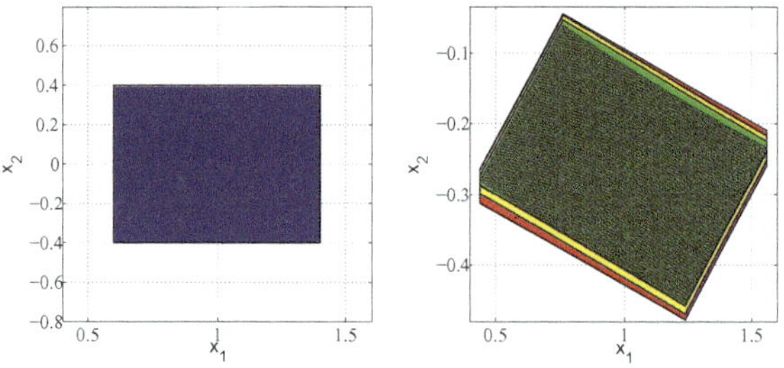

Fig. 13.9 In this example we considered the function f defined in (13.14). The starting box $\mathbf{Z}$ is depicted on the left. On the right, samples of $f(\mathbf{Z})$ are depicted in black while the red, yellow and green zonotopes represent the Taylor zonotope extension of degree 0, 1 (computed as in [11]), and 5, respectively.

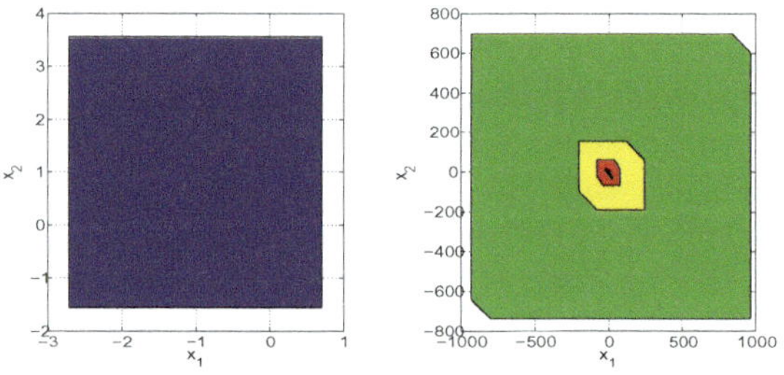

Fig. 13.10 In this example we considered the function f defined in (13.13). The starting box $\mathbf{Z}$ is depicted on the left. On the right, samples of $f(\mathbf{Z})$ are depicted in black while the red, yellow and green zonotopes represent the Taylor zonotope extension of degree 0, 1 (computed as in [11]), and 8, respectively.

sonably good in approximating the real set. In next section, we consider a method based on DC programming that provides a better first order zonotope Taylor extension than the one proposed in [11] when the function f is C^2.

13.4.2.1 DC Programming

Definition 13.3 (DC function). [17] Let Ω be a convex subset of $\mathbb{R}^n$. A real-valued function $f : \Omega \to \mathbb{R}$ is called *DC (difference of convex)* on Ω, if there exist two convex functions $g, h : \Omega \to \mathbb{R}$ such that f can be expressed in the form

$$f(x) = g(x) - h(x).$$

If $\Omega = \mathbb{R}^n$, then f is simply called a *DC function*.

Every continuous function can be approximated by a difference of two convex functions (DC functions) ([17]) and every C^2-function is a DC function ([32]). Even if f is a C^2-function, finding a DC decomposition of f—namely, the functions g and h—is often a challenging problem, although suitable procedures exist. In this work we use the method described in [1]. Given $f : \Omega \to \mathbb{R}$ and recalling that a C^2-function is convex in Ω if and only if $\nabla^2 f(x) \geq 0, \forall x \in \Omega$, we search for a parameter $\alpha \geq 0$ such that $\nabla^2 f(x) > -2\alpha I, \forall x \in \Omega$. Then $f(x) = g(x) - h(x)$ is a DC function with $g(x) = f(x) + \alpha x^T x$ and $h(x) = \alpha x^T x$.

Programming problems dealing with DC functions are called *DC programming problems*. In [3] the authors propose a DC programming-based method for constructing a tight outer approximation of $f(\mathbf{Z})$ (where f is a nonlinear C^2 function and $\mathbf{Z}$ a zonotope). First they linearize f and compute the image of $\mathbf{Z}$ under this linearization. A parallelotope that bounds the approximation error between f and its linearization for the given set $\mathbf{Z}$ is then added to the image. By exploiting the convexity of $g(x)$ and $h(x)$, a tighter approximation than the one achievable by simply bounding the remainder of the first-order Taylor approximation is thus obtained. The results are summarized in the following.

Definition 13.4 ([3]). Let $f(x) : \mathbb{R}^n \to \mathbb{R}^n$ be a nonlinear differentiable function and $\mathbf{Z} = p \oplus H\mathbf{B}^m$ a zonotope. Given $f^L(x) = f(p) + \nabla f(p)(x - p)$, the error set ε is defined as

$$\varepsilon = \{e \in \mathbb{R}^n : e = f(x) - f^L(x), x \in \mathbf{Z}\}.$$

Lemma 13.1 ([3]). *Let $\mathbf{Z} = p \oplus H\mathbf{B}^m$ be a zonotope and $f(x) : \mathbb{R}^n \to \mathbb{R}^n$ a nonlinear DC function, i.e. $f(x) = g(x) - h(x)$, with g and h convex. Let $g_i^L(x) = g(p) + \nabla g(p)(x - p)$, $h_i^L(x) = h(p) + \nabla h(p)(x - p)$ and define the parallelotope $\bar{\varepsilon}$ as*

$$\bar{\varepsilon} = \{x \in \mathbb{R}^n : \gamma_i^- \leq x_i \leq \gamma_i^+, \quad i = 1, \ldots, n\}, \tag{13.9}$$

with

$$\gamma_i^+ = \max_{x \in \mathbf{Z}} (g_i(x) - h_i^L(x) - f_i^L(x))$$
$$\gamma_i^- = \min_{x \in \mathbf{Z}} (g_i^L(x) - h_i(x) - f_i^L(x)).$$

Then, the parallelotope $\bar{\varepsilon}$ is an outer bound of the set ε, i.e., $\varepsilon \subseteq \bar{\varepsilon}$.

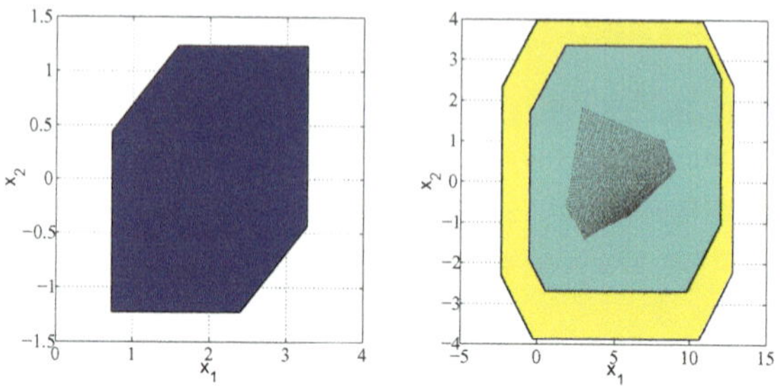

Fig. 13.11 Comparison of a DC zonotope (cyan) with a 1st order one (yellow, computed as in [11]) calculated for a big starting box (blue). The black dots represent sampling of $f(\mathbf{Z})$, with f defined in (13.14).

Theorem 13.6 ([3]). *Let* $\mathbf{Z} = p \oplus H\mathbf{B}^m$ *be a zonotope and* $f(x) : \mathbb{R}^n \to \mathbb{R}^n$ *a DC function. Then*

$$f(\mathbf{Z}) \subseteq f^L(\mathbf{Z}) \oplus \bar{\varepsilon}.$$

Theorem 13.6 is the main result of [3] and shows how to bound the image of a zonotope under a nonlinear DC function. A comparison betweeen DC programming (cyan) and first-order Taylor zonotope extension (yellow, computed as in [11]) is provided in Fig. 13.11. In this case, the DC programming-based approach produces a better approximate. Several experiments confirmed this tendency.

Summarizing, DC programming is generally better than first order Taylor zonotope extension. If the starting set is small, then high order zonotope extensions could be better. The main drawbacks of the latter approach are the computational complexity and the efficacy just for small starting sets. The reason is that Taylor approximation is a local property, i.e., it is accurate around mid($\mathbf{Z}$) but gets worse as soon as the set becomes large. In any case, zonotopes have an intrinsic limitation in that, being symmetric and convex, they are poor at bounding nonconvex sets, while, in general, the output of $f(\mathbf{Z})$ is nonconvex.

In this chapter, we use reachability analysis in order to evaluate the image of the sets $R \in \bar{R}$ introduced in Sec. 13.3 through the nonlinear dynamics of the system in closed-loop with the approximate MPC control law. What is important to note is that R can be arbitrarily big a priori. For this reason, in the next section we propose a method for partitioning R in subsets. A similar idea has been proposed in [13]. We apply DC programming to each one of the subsets and the union of the output sets is used to contain $f(R)$. This approach should provide better performance because the starting sets are smaller and the output, being a union of zonotopes (hence not necessarily convex anymore), could be a tighter approximation to the real set. DC programming is preferred to high order zonotope extensions for computational reasons.

13.4.3 Splitting the Starting Set

First of all, we introduce the operator bisect$_k(\cdot)$ (see [33] for details) that splits a zonotope into two. Given $\mathbf{Z} = p \oplus H\mathbf{B}^m$ the operator bisect$_k(\mathbf{Z})$ generates two

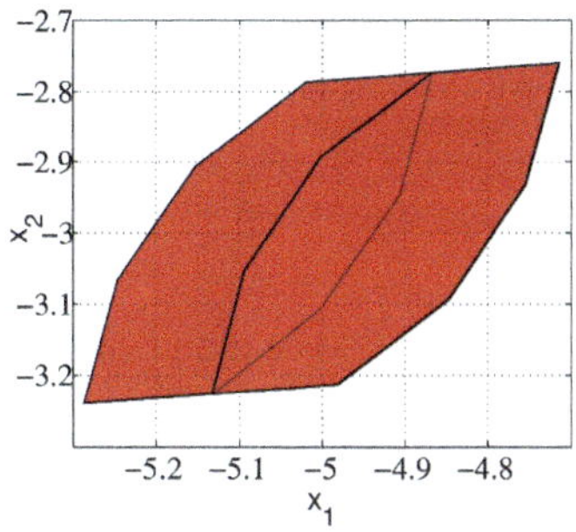

Fig. 13.12 An example of a split. bisect(**Z**).

subzonotopes

$$\mathbf{Z}^L = (p - \frac{h_k}{2}) \oplus [h_1 \dots \frac{h_k}{2} \dots h_m]\mathbf{B}^m$$

$$\mathbf{Z}^R = (p + \frac{h_k}{2}) \oplus [h_1 \dots \frac{h_k}{2} \dots h_m]\mathbf{B}^m$$

where h_k is the kth column of H. Figure 13.12 shows an example of the operator $\text{bisect}_k(\mathbf{Z})$ applied to the zonotope $\mathbf{Z}$ in Fig. 13.5. In this case, $\mathbf{Z}^L$ and $\mathbf{Z}^R$ intersect. This is because the line segment generators $h_1, \dots, h_m$ are not linearly independent. If $H \in \mathbb{R}^{m \times m}$, with $\text{rank}(H) = m$, then, $\text{bisect}_k(\mathbf{Z})$ provides two subzonotopes that do not overlap. As stated above, given a generic nonlinear function f, zonotope extensions as well as DC programming work better on smaller boxes. This is because Taylor approximation is a local property, i.e., it is effective for a neighborhood of $\text{mid}(\mathbf{Z})$. How big this neighborhood is depends on how close to linear the system is. For this reason, the approach we propose consists of splitting more where the system in more nonlinear, i.e., where the Taylor approximation is less effective. We evaluate this by considering $\bar{\varepsilon}$, the remainder of the first order extension with the DC programming-based approach. The procedure is summarized in Algorithm 13.1. A comparison between DC programming with $\mathbf{Z}$ split into 5 and 10 subzonotopes is reported in Table 13.1 and is depicted in Fig. 13.13. The case with 10 zonotopes produces, as expected, a tighter approximation (smaller volume) at the expense of a higher computational time and a higher number of evaluations.

Table 13.1 Algorithm 13.1. Comparison between the case of 5 and 10 splits. The algorithm has been applied in combination with the DC programming-based approach.

	# of $\mathbf{Z}_i$ in output	# of evaluated $\mathbf{Z}_i$	Comp. time (s)	Initial Volume	Final Volume
DC prog.	5	17	2.28	181.1	6.47
DC prog.	10	37	5.01	181.1	3.06

Algorithm 13.1 Splitting the starting set by using the $\bar{\varepsilon}$ criteria

Require: Nonlinear function $f : \mathbb{R}^n \rightarrow \mathbb{R}^m$, a starting set $\mathbf{Z}$ and n_s, number of required splits.
Ensure: Union of n_s zonotopes, which is an outer approximation of $f(\mathbf{Z})$.

Compute $\mathbf{Z_O}$, outer approximation of $f(\mathbf{Z})$ using DC programming.
$Stack = \mathbf{Z}$, $Stack_{out} = \mathbf{Z}_O$.
while $length(Stack_{out}) \leq n_s$ **do**
  Find in $Stack_{out}$ the zonotope $\mathbf{Z}_{O_i}$ with the biggest $\bar{\varepsilon}$ (see (13.9) for the definition) in terms of volume.
  Select from $Stack$ $\mathbf{Z}_i = p_i \oplus H_i \mathbf{B}^p$, the starting zonotope associated with $\mathbf{Z}_{O_i}$.
  for $j = 1$ to p **do**
    Generate $\mathbf{Z}_{ij}^L$ and $\mathbf{Z}_{ij}^R$ by applying the operator bisect$(\cdot)$ to the jth column of matrix H_i.
    Compute $\mathbf{Z}_{O_{ij}}^L$ and $\mathbf{Z}_{O_{ij}}^R$, outer approximations of $f(\mathbf{Z}_{ij}^L)$ and $f(\mathbf{Z}_{ij}^R)$.
    Calculate the volumes of $\bar{\varepsilon}_{ij}^L$ and $\bar{\varepsilon}_{ij}^R$.
  end for
  Find the index j with the smallest sum of volumes of $\bar{\varepsilon}_{ij}^L$ and $\bar{\varepsilon}_{ij}^R$.
  $Stack = (Stack \backslash \{\mathbf{Z}_i\}) \cup \{\mathbf{Z}_{ij}^L, \mathbf{Z}_{ij}^R\}$.
  $Stack_{out} = (Stack_{out} \backslash \{\mathbf{Z}_{O_i}\}) \cup \{\mathbf{Z}_{O_{ij}}^L, \mathbf{Z}_{O_{ij}}^R\}$.
end while

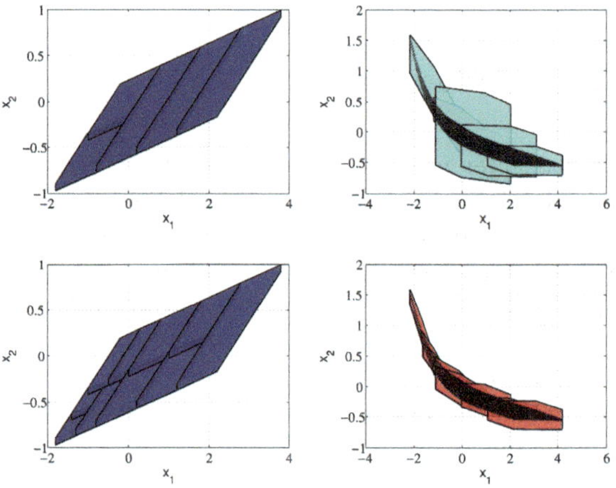

Fig. 13.13 In this example we considered the function f defined in (13.14). The starting sets $\mathbf{Z}$ are depicted on the left. In one case, $\mathbf{Z}$ has been split into 5 while in the other into 10. The splits have been done according to Algorithm 13.1. The results are depicted on the right, in cyan the output obtained with 5 splits while in red the one with 10 splits.

13.5 Capture Basin

Consider the following system

$$x_{i+1} = \begin{cases} f(x_i, \kappa_f(x)), & \forall x \in \mathcal{X}_F \\ \bar{f}_R(x_i), & \forall x \in R, R \nsubseteq \mathcal{X}_F \end{cases} \tag{13.10}$$

the system $x_{i+1} = f(x_i, u_i)$ in closed-loop with the approximated control law (13.7), for all sets $R \in \bar{R}$ under the dual mode approach. Given (13.10), we define with R_s the capture basin [4, 28], i.e., the set of initial states such that the terminal invariant

Algorithm 13.2 Computation of the capture basin

Require: Discrete time system (13.10) and $\overline{R}$, set of all hypercubic regions.
Ensure: Capture basin R_s.
1: Initialize $R_s = \mathcal{X}_F$ and $\Xi = \{R : R \subseteq \bar{R},\ R \not\subseteq R_s\}$.
2: **while** $\Xi \neq \emptyset$ **do**
3: $X_{temp} = R_s$.
4: for all $R \in \Xi$, compute an outer approximation $R_O(1)$ of the 1-step ahead reachable set by using Algorithm 13.1.
5: Add to X_{temp} all the boxes R that satisfy $R_O(1) \subseteq R_s$. X_{temp} represents an inner approximation of the 1-step backward reachable set of R_s.
6: Set $R_s = X_{temp}$.
7: $\Xi_{prec} = \Xi$.
8: Update Ξ, as $\Xi = \{R : R \subseteq \bar{R},\ R \not\subseteq R_s\}$.
9: **if** $\Xi_{prec} == \Xi$ **then**
10: Return R_s.
11: **end if**
12: **end while**

set $\mathcal{X}_F$ is reached in finite time while at the same time satisfying the state and the control constraints $\mathcal{X}$ and $\mathcal{U}$. Note that, since system constraints are satisfied in $\mathcal{X}_F$ and $\mathcal{X}_F$ is by definition invariant for the closed-loop system (13.10), one has $R_s \supseteq \mathcal{X}_F$.

The exact computation of the capture basin R_s is a difficult problem for the case of nonlinear systems. For this reason we suggest Algorithm 13.2 to compute an inner approximation of R_s. Algorithm 13.2 makes use of Algorithm 13.1. With Algorithm 13.2 we check which boxes R belong to R_s. Since the interpolated control law guarantees the satisfaction of the control constraints if $\mathcal{U}$ is convex (see [30]), and state constraint $\mathcal{X}$ are satisfied by requiring $\bar{R} \subseteq \mathcal{X}$, Algorithm 13.2 has just to check from which boxes the set $\mathcal{X}_F$ is attainable in finite time.

13.5.1 Approximate Explicit NMPC

In the following we introduce a recursive algorithm for multiresolution approximation of explicit nonlinear model predictive control laws. The algorithm is initialized with a user-defined coarse uniform grid before a dyadic refinement strategy is used to improve the approximation to a specified accuracy. Exploiting the fact that the state space can be decomposed into a union of hypercubes R (with respect to the approximate receding horizon control law), the algorithm restricts the dyadic refinement to the hypercubes intersecting the current invariant set. In this way the basin of attraction is constructed from the inside out (starting from the terminal set). The procedure is summarized in Algorithm 13.3. The algorithm requires the NMPC problem (13.1) and the NMPC cost function (13.2). The index set Λ is initialized at level l_0 along with all indices and details. The set of stored detail coefficients is given by the set $\mathbf{w}$. When the grid is refined, Λ stores the levels of resolution k and

Algorithm 13.3 Adaptive Hierarchical Approximate NMPC

Require: NMPC problem (13.1), NMPC Cost Function (13.2), l_0, and $l_{\max}$.
Ensure: detail coefficients $\mathbf{w}$ and index set Λ such that the system $x_{i+1} = f(x_i, u_i)$ in closed-loop with the approximate control law $\hat{u}(x)$ (see (13.7)) has guaranteed feasibility and stability over the capture basin R_s.
1: Initialize the index set $\Lambda = \{(k,i) : i \in I_k, k = l_0\}$ and the initial set of hypercubes
2: Initialize the capture basin $R_s = \mathcal{X}_F$ and the set of intersecting hypercubes $R_c = \{R \in R_{\text{active}} : R \cap R_s \neq \emptyset\}$ where R_{active} is the set of hypercubes not contained within R_s
3: Compute the initial details $\mathbf{w} = \{w_{k,i} : (k,i) \in \Lambda\}$ by solving the NMPC problem (13.1) pointwise at the vertices of all $R \in R_{\text{active}}$
4: **while** $R_c \neq \emptyset$ **do**
5: Compute the capture basin R_s with Algorithm 13.2.
6: Recompute the set of candidate refinement hypercubes $R_c = \{R \in R_{\text{active}} : R \cap R_s \neq \emptyset,\ l_R \leq l_{\max}\}$ where l_R is the level of the hypercube
7: Refine all hypercubes $R \in R_c$
8: Update R_{active} and define the set of new vertices as Λ_n
9: Solve the NMPC problem (13.1) at the new vertices and compute the new detail coefficients $\mathbf{w}_n$
10: Update the index set $\Lambda = \Lambda \cup \Lambda_n$
11: Update the detail set $\mathbf{w} = \mathbf{w} \cup \mathbf{w}_n$
12: **end while**

indices corresponding to the set of hierarchical details that are not discarded due to being initial conditions not feasible for problem (13.1). The maximum level of resolution is given as $l_{\max}$. The capture basin is computed using Algorithm 13.2. The set of hypercubes R_c intersecting R_s represents the set of refinement candidate sets. R_{active} is the set of hypercubes not contained within R_s; note that $R_c \subseteq R_{\text{active}}$. See [30] for details about the complexity of the real-time implementation of the approximate control.

The main theorem is now stated. It proves that Algorithm 13.3 always provides a stabilizing receding horizon control law and verifiable region of attraction for the NMPC problem (13.1). Note that we adopt a dual mode strategy, i.e., once the terminal set is attained, the stabilizing feedback law defined in $\mathcal{X}_F$ is applied.

Theorem 13.7. *Let $\hat{u}_0$ be the resulting receding horizon approximate control law computed from Algorithm 13.3 for the NMPC problem with cost (13.2), $l_0 \in \mathbb{N}$, and $l_{\max} \in \mathbb{N}$. The following properties hold for $\hat{u}_0$:*

a) Asymptotic stability to the origin for all $x_0 \in R_s$
b) $\hat{u}_0 \in \mathcal{U}$ for all $x \in R_s$
c) For all $x_0 \in R_s$, $x_i \in \mathcal{X}$ for all $i = 1,2,3,\ldots$
d) $R_s \supseteq \mathcal{X}_F$
e) As $l_{\max} \to \infty$, then $\hat{u}_0 \to u_0^*$ and $R_s \to \mathcal{R}$ where $\mathcal{R}$ is the maximum invariant set for (13.1).

See proof of Theorem 12 in [31].

13.6 Numerical Example

Consider the following two-dimensional continuous-time nonlinear system (e.g., see [10, 19, 9]):

$$\dot{x}_1(t) = x_2(t) + [0.5 + 0.5x_1(t)]u(t) \tag{13.11}$$
$$\dot{x}_2(t) = x_1(t) + [0.5 - 2x_2(t)]u(t) \tag{13.12}$$

It is well known (see [10]) that the origin of the system governed by (13.11) and (13.12) is unstable, and that the linearized system is stabilizable (but not controllable).

In consideration of the NMPC problem (13.1), the system (13.11) and (13.12) is discretized using a forward difference Euler approximation with sampling time $T = 0.1$. The input and state constraint sets are $\mathcal{U} = \{u \in \mathbb{R} : |u| \leq 2\}$ and $\mathcal{X} = \{x \in \mathbb{R}^2 : ||x||_\infty \leq 1\}$. The cost function is defined over a prediction horizon of length $N = 15$ as

$$J(u_0,\ldots,u_{N-1},x_0,\ldots,x_N) := x_N^T P x_N + \sum_{i=0}^{N-1} x_i^T Q x_i + u_i^T R u_i$$

where

$$Q = \begin{bmatrix} 0.01 & 0 \\ 0 & 0.01 \end{bmatrix}, \quad R = 0.01, \quad P = \begin{bmatrix} 19.6415 & 13.1099 \\ 13.1099 & 19.6414 \end{bmatrix}.$$

The terminal penalty matrix P as well as the auxiliary controller $u = -Kx$, are computed using a linear differential inclusion (LDI, see [7]), in place of the original nonlinear system, and thus determine an invariant ellipsoid $\mathcal{X}_F = \{x \in \mathbb{R}^2 : x^T P x \leq 1\}$ for an uncertain linear time-varying system. With $l_0 = 2$ and $l_{\max} = 8$, we compute a stabilizing control law using Algorithm 13.3. In Table 13.2 we compare the results obtained by computing the capture basin with pure interval arithmetic (i.e., the splitting procedure has not been used) and Algorithm 13.1. As we can see, Algorithm 13.1 with 5 and 10 splits provides a capture basin that is sligthly bigger than the one obtained with interval arithmetic but, at the same time, the number of points describing the interpolated control law is 40% less. This motivates the use of Algorithm 13.1 instead of pure interval arithmetic. Note that, by the comparison to Algorithm 13.1 with 5 and 10 splits, we conclude that the use of more than 10 splits will not add further value since it will imply more computational effort for a very small improvement in the volume of the capture basin. It is important to note that the number of splits necessary to better describe the capture basin is problem dependent. In Figure 13.14 and Figure 13.15 the approximate receding horizon control law and an approximation of the capture basin, obtained with Algorithm 13.1 and 10 splits are shown.

ACADO [18] Toolkit has been used in order to solve the pointwise NMPC problem (13.1), while MPT toolbox [23] and the INTLAB interval toolbox [27] have been used to recursively compute the capture basin and the outer approximations.

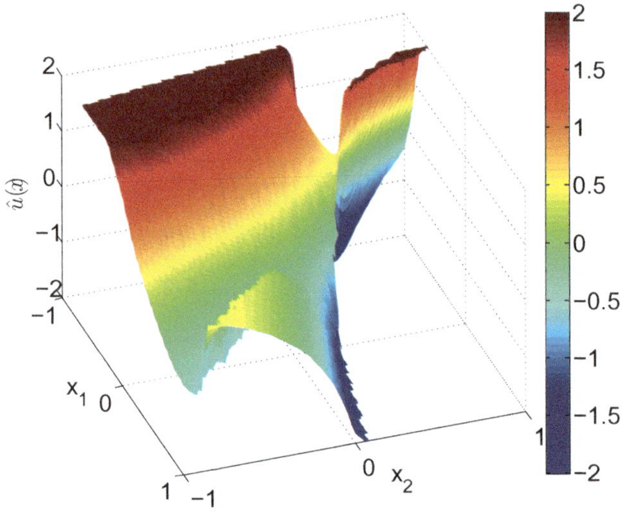

Fig. 13.14 Approximate control law $\hat{u}_0(x)$.

13.7 Conclusion

The approximate explicit NMPC method we have presented combines an adaptive hierarchical gridding scheme with a verification method based on reachability analysis. The approach approximates the optimal control law directly, and because of the basis functions used to build the function approximation, can be tuned in order to control the complexity and accuracy of the solution. This ability to guarantee a level of accuracy at the grid points enables an adaptive approach based on thresholding

Table 13.2 Comparison between pure interval arithmetic and Algorithm 13.1.

	# of points describing $\hat{u}_0(x)$	capture basin volume
Interval analysis	5808	2.5857
5 splits	3571	2.6035
10 splits	3439	2.5996

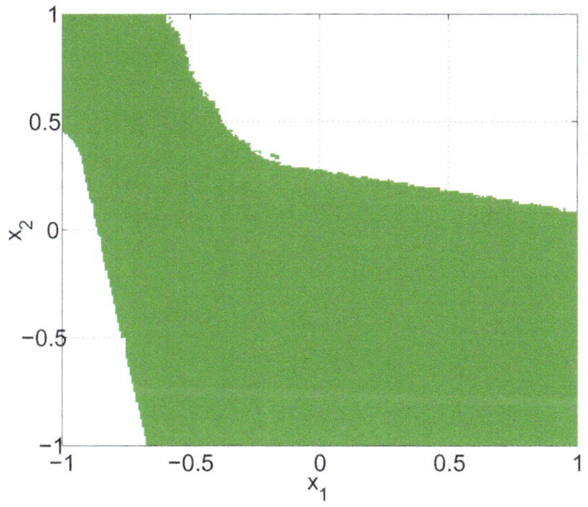

Fig. 13.15 Feasible and stable region.

that can lead to sparse representations of the explicit control law, while preserving guaranteed feasibility and stability of the solution. By employing reachability methods based on zonotopes and DC programming, the complexity of the function approximation and verification procedure is significantly decreased. A direct result of this reduction in complexity is a smaller storage requirement for the receding horizon control law and a larger verifiable region of attraction.

Appendix

13.7.1 Models Used in the Examples

The examples given in Sec. 13.4 are based on the following models.

Model 1:

$$\begin{cases} x_1(k+1) &= 3x_1(k) - \dfrac{x_1(k)^2}{7} - \dfrac{4x_1(k)x_2(k)}{4+x_1(k)} \\ x_2(k+1) &= -2x_2(k) + \dfrac{3x_1(k)x_2(k)}{4+x_1(k)} \end{cases} \tag{13.13}$$

Model 2:

$$\begin{cases} x_1(k+1) &= x_1(k) + 0.4x_2(k) \\ x_2(k+1) &= -0.132e^{(-x_1(k))}x_1(k) - 0.213x_1(k) + 0.274x_2(k) \end{cases} \tag{13.14}$$

Acknowledgements This research was partially supported by the Swiss National Science Foundation under grant 200021-122072 and by the European Commission under the projects Feednetback FP7-ICT-223866 [15] and EMBOCON FP7-ICT-2009-4 248940 [14].

References

1. Adjiman, C., Floudas, C.: Rigorous convex underestimators for general twice-differentiable problems. Journal of Global Optimization **9**(1), 23–40 (1996)
2. Alamo, T., Bravo, J., Camacho, E.F.: Guaranteed state estimation by zonotopes. Automatica **41**(6), 1035–1043 (2005)
3. Alamo, T., Bravo, J., Redondo, M., Camacho, E.: A set-membership state estimation algorithm based on DC programming. Automatica **44**(1), 216–224 (2008)
4. Aubin, J.: Viability Theory. Birkhauser Boston Inc., Cambridge, MA, USA (1991)
5. Bemporad, A., Morari, M., Dua, V., Pistikopoulos, E.N.: The explicit linear quadratic regulator for constrained systems. Automatica **38**(1), 3–20 (2002)
6. Berz, M., Makino, K.: Suppression of the wrapping effect by Taylor model-based verified integrators: Long-term stabilization by shrink wrapping. International Journal of Differential Equations and Applications **10**(4), 385–403 (2005)
7. Boyd, S., El Ghaoui, L., Feron, E., Balakrishnan, V.: Linear matrix inequalities in system and control theory, vol. 15. Society for Industrial and Applied Mathematics (SIAM), Philadelphia, PA (1994)
8. Canale, M., Fagiano, L., Milanese, M.: Set membership approximation theory for fast implementation of model predictive control laws. Automatica **45**(1) (2009)
9. Canale, M., Fagiano, L., Milanese, M., Novara, C.: Set membership approximations of predictive control laws: The tradeoff between accuracy and complexity. In: European Control Conference (ECC2009). Budapest, Hungary (2009)
10. Chen, H., Allgöwer, F.: A quasi-infinite horizon nonlinear model predictive control scheme with guaranteed stability. Automatica **34**(10), 1205–1218 (1998)
11. Combastel, C.: A state bounding observer for uncertain non-linear continuous-time systems based on zonotopes. In: Proc. 44th IEEE Conf. Decision and Control (CDC2005), and 2005 European Control Conf. (CDC-ECC'05), pp. 7228–7234. Seville, Spain (2005)
12. Delanoue, N., Jaulin, L., Hardouin, L., Lhommeau, M.: Guaranteed characterization of capture basins of nonlinear state-space systems. In: Informatics in Control, Automation and Robotics: Selected Papers from the International Conference on Informatics in Control, Automation and Robotics 2007 (ICINCO2007), pp. 265–272. Springer, Angers, France (2008)
13. Donzé, A., Clermont, G., Legay, A., Langmead, C.: Parameter synthesis in nonlinear dynamical systems: Application to systems biology. In: Proc. 13th Annual Int. Conf. Research in Computational Molecular Biology, pp. 155–169 (2009)
14. EMBOCON www.embocon.org
15. Feednetback www.feednetback.eu
16. Fukuda, K.: From the zonotope construction to the Minkowski addition of convex polytopes. Journal of Symbolic Computation **38**(4), 1261–1272 (2004)
17. Horst, R., Thoai, N.: DC programming: Overview. Journal of Optimization Theory and Applications **103**(1), 1–43 (1999)
18. Houska, B., Ferreau, H., Diehl, M.: ACADO toolkit—An open-source framework for automatic control and dynamic optimization. Optimal Control Applications and Methods **32**(3), 298–312 (2011)

19. Johansen, T.A.: Approximate explicit receding horizon control of constrained nonlinear systems. Automatica **40**(2), 293–300 (2004)
20. Johansen, T.A., Petersen, I., Slupphaug, O.: On explicit suboptimal LQR with state and input constraints. In: Proc. 39th IEEE Conf. Decision and Control (CDC2000), pp. 662–667. Sydney, Australia (2000)
21. Kearfott, R.: Rigorous global search: Continuous problems, vol. 13. Kluwer Academic Publishers, Dordrecht-Boston-London (1996)
22. Kuhn, W.: Rigorously computed orbits of dynamical systems without the wrapping effect. Computing **61**(1), 47–67 (1998)
23. Kvasnica, M., Grieder, P., Baotić, M.: Multi-Parametric Toolbox (MPT) (2004). URL http://control.ee.ethz.ch/ mpt/
24. Makino, K.: Rigorous analysis of nonlinear motion in particle accelerators. Ph.D. thesis, Michigan State University (1998)
25. Moore, R.E.: Interval Analysis, vol. 60. Prentice Hall, Englewood Cliffs, NJ (1966)
26. Pin, G., Filippo, M., Pellegrino, A., Parisini, T.: Approximate off-line receding horizon control of constrained nonlinear discrete-time systems. In: Proc. European Control Conference, pp. 2420–2431. Budapest, Hungary (2009)
27. Rump, S.: INTLAB - INTerval LABoratory. In: T. Csendes (ed.) Developments in Reliable Computing, pp. 77–104. Kluwer Academic Publishers, Dordrecht (1999). http://www.ti3.tu-harburg.de/rump/
28. Saint-Pierre, P.: Approximation of the viability kernel. Applied Mathematics and Optimization **29**(2), 187–109 (1994)
29. Seron, M.M., Goodwin, G.C., Dona, J.A.D.: Geometry of model predictive control including bounded and stochastic disturbances under state and input constraints. Tech. rep., University of Newcastle, Newcastle, Australia (2000)
30. Summers, S., Jones, C.N., Lygeros, J., Morari, M.: A multiscale approximation scheme for explicit model predictive control with stability, feasibility, and performance guarantees. In: Proc. IEEE Conf. Decision and Control (CDC2009), pp. 6328–6332. Shanghai, China (2009)
31. Summers, S., Raimondo, D.M., Jones, C.N., Lygeros, J., Morari, M.: Fast explicit nonlinear model predictive control via multiresolution function approximation with guaranteed stability. In: Proc. 8th IFAC Symp. Nonlinear Control Systems (NOLCOS 2010). Bologna, Italy (2010). URL http://control.ee.ethz.ch/index.cgi?page=publications;action=details;id=3557
32. Tuy, H.: DC optimization: theory, methods and algorithms. Handbook of global optimization pp. 149–216 (1995)
33. Wan, J., Vehi, J., Luo, N.: A numerical approach to design control invariant sets for constrained nonlinear discrete-time systems with guaranteed optimality. Journal of Global Optimization **44**(3), 395–407 (2009)

Chapter 14
Towards Parallel Implementation of Hybrid MPC—A Survey and Directions for Future Research

Daniel Axehill and Anders Hansson

Abstract In this chapter parallel implementations of hybrid MPC will be discussed. Different methods for achieving parallelism at different levels of the algorithms will be surveyed. It will be seen that there are many possible ways of obtaining parallelism for hybrid MPC, and it is by no means clear which possibilities should be utilized to achieve the best possible performance. This question is a challenge for future research.

14.1 Introduction

Speed in numerical computations has increased dramatically over a long period of time. This is due partly to increase of processor performance in computers and partly to development of more sophisticated algorithms and methods. However, for the last five years single-core processor performance has not significantly increased. To compensate for this multicore and multiprocessor computers have seen increased use. In addition to this clusters and grids have emerged as another way to speed up computations. Multicore and multiprocessor computers typically have only few cores and processors, whereas clusters and grids can be composed of hundreds of processors distributed over a significant number of computers. It is clear that these new architectures pose new challenges on how algorithms for numerical computations should be designed. If care is not taken, further potential speedup will not be fulfilled.

Daniel Axehill
Automatic Control Laboratory, ETH, Physikstrasse 3, 8092 Zürich, Switzerland,
e-mail: axehill@control.ee.ethz.ch

Anders Hansson
Division of Automatic Control, Linköping University, SE-581 83 Linkping, Sweden
e-mail: hansson@isy.liu.se

R. Johansson and A. Rantzer (Eds.): Distributed Decision Making and Control, LNCIS 417, pp. 313–338.
springerlink.com

Model predictive control (MPC) is a popular control strategy that has been used in many applications for a long time. It is hard to say exactly when MPC was invented, but probably the first patent was granted to Martin-Sanchez in 1976 [62]. An early publication in academia containing the basic ideas was presented in 1963 by Propoi [62]. There are also some methods similar to MPC, but with different names. One of the most wellknown ones is dynamic matrix control (DMC), [31]. Traditionally, MPC has involved computation of the solution to a relatively demanding optimization problem that has to be solved on-line for each sampling interval, [62]. At a fairly high computational expense it is possible for small scale examples to precompute the optimal feedback off-line [16, 20, 36, 37]. The benefit of that approach is that the on-line effort is reduced to the evaluation of a look-up table. The applications of MPC are many, see [46, 71] for surveys. In recent years work has been carried out to generalize MPC to so-called *hybrid systems*, [21]. For these systems the computational demand is even higher. Hybrid systems have applications in, e.g., transportation, logistics, economics, process control, building control, airplane routing and communications.

In recent years there has been a trend in the control community to develop distributed algorithms for control. There are several reasons for this. Often systems to be controlled can be decomposed in a natural way, and then it is reasonable to also distribute the controllers according to this decomposition. This means that each controller is only concerned with a small subsystem. Sometimes the controllers of the different subsystems need to be coordinated in some fashion in order to obtain reasonable overall performance of the controlled system. Despite this cost for coordination it has been seen in many cases that global optimality still can be obtained and at a lower computational cost compared to if there had been one centralized controller [75]. This type of distributed control has much in common with parallel implementations. One of the few references available for the hybrid set-up is [13], where a continuous time hybrid optimal control problem is solved using simulations. In [67] it was shown that an augmented Lagrangian approach could deliver a feasible solution in an example. For other examples, it has been shown in [65, 77] that distributed hybrid MPC is suboptimal in general.

The remaining part of the chapter is organized as follows. First a brief review of hybrid MPC will be given in Sec. 14.2. In Sec. 14.3 different optimization methods for solving hybrid MPC problems will be discussed. Then, in Sec. 14.4, different approaches for parallelization will be reviewed, and their potential applicability to hybrid MPC will be investigated. Finally, in Sec. 14.5, conclusions will be presented together with recommendations for future research. This chapter will only consider the discrete-time setting of hybrid MPC. For references to hybrid control in continuous-time the reader is referred to [2, 30, 60, 86] and the references therein.

14.2 Hybrid MPC

In this section, MPC for hybrid systems will be discussed. Furthermore, some commonly used models for hybrid systems that are useful in the MPC context will be reviewed.

14.2.1 Model Predictive Control

The most commonly used variant of MPC is linear MPC, where the dynamics are linear and often a quadratic performance measure similar to the one used in linear quadratic (LQ) control is used. A difference compared to LQ control is that linear MPC is able to consider linear inequality constraints on states and control signals. A discrete-time linear time-invariant model on state space form is given by

$$\begin{aligned} x(t+1) &= Ax(t) + Bu(t) \\ y(t) &= Cx(t) \end{aligned} \tag{14.1}$$

where $t \in \mathbb{Z}$ is the discrete time, $x(t) \in \mathbb{R}^n$ is the state, $u(t) \in \mathbb{R}^m$ is the control input and $y(t) \in \mathbb{R}^p$ is the controlled output. An example of an objective function, or performance measure, is a quadratic function in the form

$$\begin{aligned} J(t_0) =& \frac{1}{2}\sum_{s=0}^{N-1} \left(\|y(t_0+s) - r(t_0+s)\|_{Q_e}^2 + \|u(t_0+s)\|_{Q_u}^2 \right) \\ &+ \frac{1}{2} \|y(t_0+N) - r(t_0+N)\|_{Q_e}^2 \end{aligned} \tag{14.2}$$

where $Q_e \in \mathbb{S}_+^p$ and $Q_u \in \mathbb{S}_{++}^m$ and $r(t) \in \mathbb{R}^p$ is the reference signal. Other common performance measures for linear MPC are formed by replacing the squared 2-norm in (14.2) with a 1-norm or ∞-norm. Often, the constraints are defined as polyhedral constraints in the form

$$H_u(t)u(t) + H_x(t)x(t) + h(t) \leq 0 \tag{14.3}$$

In MPC, the future behavior of the system is predicted N time steps ahead. In this context, prediction means that a system model like (14.1) is used to calculate how the system will react to control inputs and thereby what will happen in the future if a certain control input is applied to the system. Not surprisingly, N is called the *prediction horizon*, which in practice is chosen long enough to cover a normal transient of the controlled system.

There are several different ways to cast (14.1), (14.2) and (14.3) in the form of a formal optimization problem. The two most common variants are presented in the Appendix. For a more in-depth treatment, see [59]. If the system is linear and the objective function is quadratic, the resulting optimization problem is, for a fixed

Algorithm 14.1 Basic MPC controller

1: Measure or estimate the state of the controlled process $\bar{x}$ at time instant t_0.
2: Obtain $\{u(t_0), u(t_0+1), \ldots, u(t_0+N-1)\}$ by minimizing (14.2) subject to the constraints (14.1), (14.3) and the initial condition $x(t_0) = \bar{x}$.
3: Apply the first element $u(t_0)$ to the controlled process.
4: $t_0 \leftarrow t_0 + 1$
5: Repeat the procedure.

value of the initial state $\bar{x}$, a quadratic programming (QP) problem in the general form

$$\underset{x}{\text{minimize}} \quad \frac{1}{2}x^T H x + f^T x \tag{14.4a}$$

$$\text{subject to} \quad A_{\mathcal{E}} x = b_{\mathcal{E}}, \tag{14.4b}$$

$$A_{\mathcal{I}} x \leq b_{\mathcal{I}}, \tag{14.4c}$$

where x contains the control inputs, states and controlled outputs for the entire prediction horizon stacked. This problem is defined in more detail in Sec. 14.3.1. QP problems in general are well-studied problems for which there exist well-developed optimization routines. Similarly, if the system is linear and a 1-norm or an ∞-norm performance measure is used, the resulting optimization problem becomes a linear programming (LP) problem. Hence, for linear MPC the optimization problem is considered relatively easy to solve. An extension is to not only consider the optimization problem for a single initial state, but for all initial states of interest. The problem then becomes a parametric QP, which is further described in Sec. 14.3.3.

In order to get closed-loop control, the approach above is used in a receding horizon fashion, which means that the prediction interval is moved one step forward after each completed optimization. After the optimization has been performed, only the first control signal in the optimal control signal sequence computed is actually applied to the system. In the next time step, a new optimization is performed and the procedure is repeated. Due to modeling errors and unknown disturbances, the predicted behavior and the actual behavior of the system do not usually completely coincide. Such errors are handled by the feedback in the algorithm. The procedure is visualized in Fig. 14.1 and the conceptual steps are summarized in Algorithm 14.1.

An extension to linear MPC is nonlinear MPC. This extension handles nonlinear systems and a general nonlinear performance measure in the objective function. Unfortunately, the resulting optimization problem is often more difficult to solve as compared to the linear case. A special case of nonlinear MPC is control of systems described partly by logic. These are called hybrid systems and provide a framework for describing processes evolving according to continuous dynamics, discrete dynamics and logic rules [21]. This class of systems is especially important when we analyze and control systems arising in the growing interaction between physical processes and digital controllers. A survey covering both linear and nonlinear MPC is found in [64]. A reference book covering most of MPC is [62].

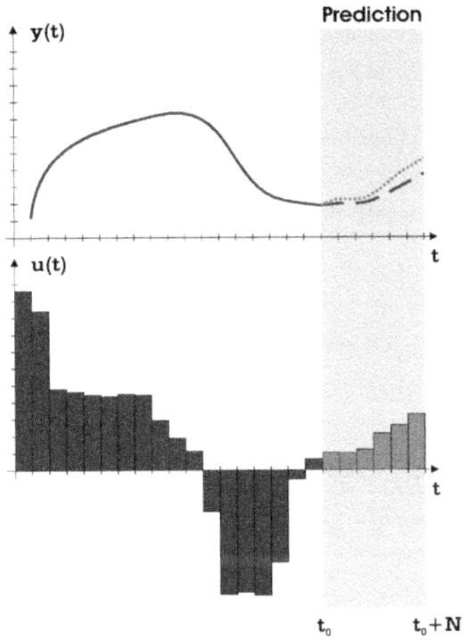

Fig. 14.1 Illustration of receding horizon policy at $t = t_0$ for an MPC controller with control input $u(t)$, controlled output $y(t)$ and prediction horizon N. The dotted line is the predicted output and the dashed line is the actual output.

14.2.2 Modeling Frameworks for Hybrid Systems

In this subsection, some different modeling frameworks for discrete-time hybrid systems are considered. The most important ones for the purpose of this work are reviewed in more detail and are related to each other.

14.2.2.1 Mixed Logical Dynamical Systems

Mixed logical dynamical (MLD) systems is one way of describing an important class of hybrid systems defined by linear dynamic equations subject to linear mixed integer inequalities, that is, inequalities involving both continuous and binary variables. The MLD description is a very general model class capable of describing a broad range of systems.

In [21] an MPC framework used for systems described by physical laws, logic rules and operating constraints is presented. An important part of this framework consists of the definition of MLD systems. This class of systems includes linear hybrid systems, finite state machines, some classes of discrete event systems, constrained linear systems and nonlinear systems that can be exactly or approximately described by piecewise affine functions. There are many applications for MLD systems reported in the literature. Some illustrative examples can be found in [10, 21, 42]. An MLD system can be described by the following linear relations [21]:

$$\begin{aligned} x(t+1) &= A(t)x(t) + B_1(t)u(t) + B_2(t)\delta(t) + B_3(t)z(t) \\ y(t) &= C(t)x(t) + D_1(t)u(t) + D_2(t)\delta(t) + D_3(t)z(t) \\ E_2(t)\delta(t) + E_3(t)z(t) &\leq E_1(t)u(t) + E_4(t)x(t) + E_5(t) \end{aligned} \tag{14.5}$$

where $t \in \mathbb{Z}$. Furthermore, $y(t) \in \mathbb{R}^{p_c} \times \{0,1\}^{p_l}$ denotes the controlled outputs, $x(t) \in \mathbb{R}^{n_c} \times \{0,1\}^{n_l}$ denotes the states of the system, $u(t) \in \mathbb{R}^{m_c} \times \{0,1\}^{m_l}$ denotes the control inputs, $\delta(t) \in \{0,1\}^{r_l}$ denotes the auxiliary binary variables, and $z(t) \in \{0,1\}^{r_c}$ denotes the auxiliary continuous variables. If the desired finite alphabet is not binary as here, it can be coded using binary variables.

In [21], both optimal control and receding horizon estimation for MLD systems is discussed. The control signal at a state $\bar{x}$ is found by minimizing either a linear or a quadratic performance measure similar to the one in (14.2) subject to $x(t_0) = \bar{x}$ and the dynamics in (14.5). This MPC problem can be rewritten as an optimization problem in mixed integer quadratic programming (MIQP) form [21], i.e., in the general form

$$\underset{x}{\text{minimize}} \quad \frac{1}{2}x^T Hx + f^T x \tag{14.6a}$$

$$\text{subject to} \quad \text{Eqs. (14.4b), (14.4c),} \tag{14.6b}$$

$$x_i \in \{0,1\}, \forall i \in \mathcal{B}, \tag{14.6c}$$

where x contains the variables from the system in (14.5) for the entire prediction horizon stacked and $\mathcal{B}$ denotes the set of indices to binary components of x. More details on how to formulate the MPC problem as an optimization problem can be found in the Appendix. The problem data for the MIQP problem in (14.6) are defined in Sec. 14.3.2. An extension is to not only consider the optimization problem for a single initial state, but for all initial states of interest. The problem then becomes a parametric MIQP, which is further described in Sec. 14.3.3.

As with linear MPC, hybrid MPC is implemented in a receding horizon fashion. The difference is that in hybrid MPC it is much more complicated to find the optimal control signal sequence, since the system is neither linear nor smooth, [21]. One way of reducing the computational complexity is to use tailored MIQP solvers. This is further discussed in [4].

14.2.2.2 Piecewise Affine Systems

Piecewise affine (PWA) systems [76] are hybrid systems where the control and state-space is partitioned into different polyhedral regions each implying certain affine dynamics. Mathematically, this can be formulated as [12]:

$$\begin{aligned} x(t+1) &= A^i x(t) + B^i u(t) + f^i \\ \text{if } &\begin{bmatrix} x(t) \\ u(t) \end{bmatrix} \in \mathcal{C}^i,\ i = \{1,\ldots,s\}, \end{aligned} \tag{14.7}$$

where $x \in \mathbb{R}^{n_c} \times \{0,1\}^{n_l}$ denotes the continuous and binary states, $u \in \mathbb{R}^{m_c} \times \{0,1\}^{m_l}$ denotes the continuous and binary control inputs and $\{\mathcal{C}^i\}_{i=1}^{s}$ denotes the polyhedral partition of the set of the state and input space. Often, there are also hard constraints on the inputs and states. These are brought into the framework by adding polyhedral constraints in the form in (14.3). The feasible set for the constrained PWA system is then given by the intersection of the set over which the system is defined in (14.7) and the feasible set of the constraints in (14.3).

The optimal control problem for PWA systems is formulated analogously to the one for MLD problems. An objective function in the form in (14.2) is minimized subject to the constraints in (14.3) and in (14.7). However, in order to bring the optimal control problem into the form of an optimization problem, it is often first reformulated as an MLD system and later solved as an MIQP problem.

14.2.2.3 Discrete Hybrid Automata

Discrete hybrid automata (DHA) are composed by the interconnection of a finite state machine (FSM), a switched affine system (SAS), a mode selector (MS), and an event generator (EG), [79]. The FSM models the discrete dynamics in the DHA. The state update equation for this subsystem is

$$x_l(k+1) = f_l(x_l(k), u_l(k), e(k)) \tag{14.8}$$

where $x_l \in \{0,1\}^{n_l}$ is the logic state, $u_l \in \{0,1\}^{m_l}$ is the logic input, $e \in \{0,1\}^{n_e}$ is a logic event signal from the EG, and $f_l(\cdot)$ is a Boolean function. The SAS can be represented as

$$x_c(k+1) = A_{i(k)}x_c(k) + B_{i(k)}u_c(k) + f_{i(k)} \tag{14.9}$$

where $x_c \in \mathbb{R}^{n_c}$ is the real-valued state, $u_c \in \mathbb{R}^{m_c}$ is the real-valued input and $i(k) \in \{0,1\}^s$ is an input that is used to select the mode in which the SAS is working. The MS computes the mode of the SAS $i(k)$ based on the state in the finite state machine $x_l(k)$, the input $u_l(k)$ and the event signal $e(k)$ according to

$$i(k) = f_{MS}(x_l(k), u_l(k), e(k)) \tag{14.10}$$

where f_{MS} is a Boolean function. The result from the EG is the binary-valued signal $e(k)$, which represents whether specified linear inequalities are satisfied or not. More specifically,

$$[e_j(k) = 1] \longleftrightarrow \left[a_j^T x_c(k) + b_j^T u_c(k) \le c_j\right] \tag{14.11}$$

Optimal control for systems represented as DHA can be performed in at least two ways. First, the DHA model can be transformed into an MLD model and solved as an MPC problem for such systems. Second, the ideas in [18] can be used, where the structure of the DHA is exploited in a more direct way by combining mixed integer optimization with satisfiability solvers and constraint logic programming solvers.

14.2.2.4 Equivalences

As seen in previous sections, there are several different ways of modeling discrete-time hybrid systems. Fortunately, it is possible to show equivalences between these under more or less restrictive assumptions. As a result, derived theoretical properties and computational tools can be transferred from one class to another. For optimal control and state estimation, the MLD description is proposed, while most other hybrid techniques are built on a PWA representation [15]. For an in-depth review of equivalences, see e.g., [15, 17, 21, 33, 52, 53, 79].

14.3 Optimization methods

In this section we will review optimization methods that can be used to solve the optimization problems that were formulated in the previous section. Both on-line methods as branch and bound (BnB) as well as off-line methods as multiparametric programming will be discussed.

14.3.1 Quadratic Programming

In this work, convex QP problems in the form in (14.4) are considered, where $x \in \mathbb{R}^n$, $H \in \mathbb{S}_+^n$, $f \in \mathbb{R}^n$ and the rows in $A_\mathcal{E} \in \mathbb{R}^{p\times n}$ are given by the vectors in $\{a_i \in \mathbb{R}^n \mid i \in \mathcal{E}\}$ and the rows in $A_\mathcal{I} \in \mathbb{R}^{m\times n}$ are given by the vectors in $\{a_i \in \mathbb{R}^n \mid i \in \mathcal{I}\}$. The column vectors $b_\mathcal{E}$ and $b_\mathcal{I}$ are analogously defined. The sets $\mathcal{I}$ and $\mathcal{E}$ are finite sets of indices where $\mathcal{I} \cap \mathcal{E} = \emptyset$. The problem in (14.4) can be solved using, for example, an active set (AS) method or an interior point (IP) method. If the matrix H is zero, the problem is an LP problem. These are usually solved either with a simplex method or an IP method. However, the main focus in this work will be on QP problems. More information about QP and how to solve these problems can be found in, e.g., [68].

14.3.2 Mixed Integer Programming

MIQP is a special case of mixed integer nonlinear programming (MINLP). At first glance, the MIQP problem looks similar to the ordinary QP problem. There is, however, one important difference. Some optimization variables are not allowed to be real-valued, but they are constrained to be integer-valued. This seemingly minor modification turns the easily solved QP problem, into an NP-hard problem, [85]. A common special case of MIQP occurs when the integer variables are constrained to be 0 or 1. To use a precise notation, this problem is called a mixed binary quadratic programming (MBQP) problem. The standard notation for MBQP seems, at least in the control literature, to be MIQP. In what follows, the problem studied will be an MBQP, but to keep the standard notation, it will be denoted MIQP. A survey considering quadratic integer programming (QIP) can be found in [83].

14.3.2.1 Problem Definition

The mathematical formulation of an MIQP problem can be found in (14.6), where $f \in \mathbb{R}^{n_c+n_b}$ and $H \in \mathbb{S}_{+}^{n_c+n_b}$. Furthermore, let $A_{\mathcal{E}}, A_{\mathcal{I}}, b_{\mathcal{E}}$ and $b_{\mathcal{I}}$ be defined as in Sec. 14.3.1 with $n = n_c + n_b$. The difference is that n_b optimization variables indexed by the set $\mathcal{B}$ are not real-valued but binary-valued. As a consequence, the problem is no longer convex. Mixed integer linear programming (MILP) can be seen as a special case of MIQP where H is the zero matrix.

Often, this problem is solved using a branch and bound method, where many QP problems in the form in (14.4) are solved in order to find the optimal solution to the problem in (14.6). The procedure is similar for the MILP case, but the relaxations are of LP type instead of QP type. There also exist other methods for solving these problems. The four most commonly used methods are cutting plane methods, decomposition methods, logic-based methods and branch and bound methods, [21]. Several authors report that branch and bound is the best method for mixed integer programs, [21]. In [44], a branch and bound method is compared to generalized Benders decomposition (GBD), outer approximation (OA) and LP/QP-based branch and bound. The conclusion in this reference is that branch and bound is the best method for solving MIQP problems. With a few exceptions, branch and bound is an order of magnitude faster than any of the other methods. An important explanation to this is that the QP subproblems are very cheap to solve. This is not the case for general MINLP, where the subproblems to be solved in the nodes are more complicated. In the MINLP case there exist important problem classes where branch and bound is not the best method. A review of different methods of solving MIQP problems can be found in [83]. There exist several software for solving MIQP problems. For MATLAB, free software like YALMIP [61] or *miqp.m* [19] can be used. A commonly used commercial software is CPLEX.

14.3.2.2 Branch and Bound Methods

If computational burden is not considered, the most straightforward approach to compute the optimal solution to an optimization problem involving binary variables is to enumerate all possible combinations of the binary variables, and for each such combination kept fixed, compute the optimal solution of any real-valued variables also included in the problem. Thereafter, the objective function values are compared and the solution, or solutions, generating the best objective function value is taken as the optimal solution. However, for problems involving many binary variables the computational burden will become overwhelming, since the number of combinations of the binary variables is 2^{n_b}. Hence, there is a need for a method that can find the optimal solution without enumerating all possible combinations of the binary variables. One such method is branch and bound, where for a majority of problems it is sufficient to explicitly enumerate only *some* of the possible combinations. Unfortunately, the worst case complexity is still exponential, and the number of combinations necessary to enumerate and solve an optimization problem for, is problem dependent. This classical pessimistic complexity bound is improved in [9].

The basics of a branch and bound method will now be discussed. The main part of the presentation follows those in [85] and [45]. The reader is referred to these two references for more detail. The general idea of branch and bound is to split the feasible set $\mathcal{S}$ of the optimization problem into K smaller sets such that

$$\mathcal{S} = \bigcup_{i=1}^{K} \mathcal{S}_i \tag{14.12}$$

This partitioning is performed in several steps and it can be represented using a binary tree structure. The topmost node in the tree is called the *root node* and the nodes at the bottom of the tree are called *leaves*. The rows of nodes in the tree starting with the root node and ending with the leaves are called *levels*. An important property of branch and bound is that the entire tree is not known from the beginning and only the parts of the tree explicitly needed in the solution process are further expanded.

An optimal solution over the set $\mathcal{S}$ can be computed by optimizing over the smaller sets separately according to

$$\begin{aligned} z^{i*} &= \underset{x \in \mathcal{S}_i}{\text{minimize}}\ f_0(x), \quad i \in \{1, \dots, K\} \\ z^* &= \min_{i \in \{1,\dots,K\}} \left\{ z^{i*} \right\} \end{aligned} \tag{14.13}$$

An optimal solution over $\mathcal{S}$ is found as the optimal solution to a subproblem with the lowest optimal objective function value. Note that the leaves of the tree contain the different combinations of the binary variables that have to be investigated if $\mathcal{S}$ is to be explored by explicit enumeration. Hence, if it is necessary to solve all of the problems represented by the leaves, there is no gain from using the branch and bound method.

The key idea in order to reduce the computational effort is to compute upper and lower bounds on the optimal objective function value for the subproblems in the nodes. Often, these bounds can be used to prune entire subtrees, which means that these subtrees do not have to be explicitly considered, since it can be concluded that the optimal solution cannot be found in any of them. Furthermore, these bounds are supposed to be easily computable. Pruning can be interpreted as an implicit enumeration, and is therefore highly desirable. The tree can be pruned if a relaxation in a node

1. is infeasible; the entire subtree below that node is infeasible.

2. is integer feasible; the optimal value for the entire subtree below that node has been found.
3. has an objective function value that is worse than the best known integer solution so far ("dominance"). The objective function value gets worse as the process proceeds further down in the tree. Hence, there is no use in continuation.

To be able to apply the above scheme in practice, one has to decide how to compute the upper and lower bounds. Usually, upper bounds are found from integer feasible solutions and lower bounds are found from relaxations or duality. In MIQP, often QP relaxations are used, where the integer constraints are relaxed to interval constraints. These relaxations are in the form

$$\begin{aligned}
&\underset{x}{\text{minimize}} && \frac{1}{2}x^T Hx + f^T x && \text{(14.14a)}\\
&\text{subject to} && \text{(14.4b)}, \text{(14.4c)} && \text{(14.14b)}\\
& && 0 \leq x_i \leq 1,\ \forall i \in \mathcal{B} && \text{(14.14c)}\\
& && x_i = 0,\ \forall i \in \mathcal{B}_0, \quad x_i = 1,\ \forall i \in \mathcal{B}_1 && \text{(14.14d)}
\end{aligned}$$

where the original integer constraints have been relaxed to interval constraints. Several relaxations with different choices of the sets $\mathcal{B}_0$ and $\mathcal{B}_1$ (where $\mathcal{B}_0 \cap \mathcal{B}_1 = \emptyset$) are ordered and solved in a structured way in the binary search tree. More about relaxations applicable to branch and bound for MIQP in the hybrid MPC application can be found in [11], where also more advanced relaxations of SDP type are considered. Efficient computational methods for computation of these relaxations in the hybrid MPC application are presented in [7] and in [8] for QP relaxations and SDP relaxations, respectively.

In a branch and bound method, there are several parameters and choices that may affect the performance drastically. One important choice to make is to decide which node to solve next. The three most common choices are depth first, breadth first, and best first. In depth first, the next node to solve is chosen as one of the child nodes of the current node. This process is continued until a node is pruned. In breadth first, all nodes at each level are considered before a node in a new level is considered. In best first, the next problem considered is the one with the lowest lower bound so far.

According to [44], solving the subproblems of QP type using a dual AS method offers the most straightforward way to exploit the structure introduced by the branching procedure. After a branch, the solution to the parent problem is in general infeasible in the child problems. But, a dual feasible starting point for the child problems is directly available from the dual solution of the parent problem. Consequently, it is possible to warm start the AS solver using information from the solution to the parent problem. Also, since a dual AS method is an ascent method generating dual feasible points, it can use an upper bound as a cut-off value for terminating the QP solver prematurely [44]. According to [85], AS methods (the reference considers the LP case) are preferable for solving the relaxed problems in branch and bound. For very large problems, IP methods can be used to solve the first subproblem, but in the subsequent subproblems an AS method should be used.

An important step in a commercial branch and bound code is the preprocessing step. The basic operation in preprocessing is to quickly detect and eliminate redundant constraints and variables and to tighten bounds, if it is possible. A smaller and tighter formulation is preferred, since it is necessary to consider the number of nodes, and the dimension of the subproblems might be reduced.

A formal description of a branch and bound algorithm applied to a binary optimization problem P can be found in Algorithm 14.2, where $\bar{z}$ denotes the current upper bound, $\bar{x}$ denotes the solution associated with the current upper bound, $\underline{z}^i$ denotes the optimal objective function value of the relaxation P_i^R to the problem P_i in node i, and $\underline{x}^i$ denotes the optimal solution to P_i^R. The feasible set of P_i and P_i^R is denoted $\mathcal{S}_i$ and $\mathcal{S}_i^R$, respectively. How subproblems are put on the list and retrieved

Algorithm 14.2 Branch and bound for binary variables, [45, 85]

```
z̄ ← +∞
x̄ ← void
Add P to LIST.
while length(LIST) > 0 do
   Pop P_i from LIST.
   Solve relaxation P_i^R ⇒ z_^i and x_^i.
   if S_i^R = ∅ then
      No feasible solution exists for P_i.
   else if z_^i ≥ z̄ then
      There exists no feasible solution of P_i which is better than x̄.
   else if x_^i ∈ S_i then
      x_^i is integer feasible and is therefore optimal also in P_i.
      z̄ ← z_^i
      x̄ ← x_^i
   else
      Split S_i into S_i0 and S_i1.
      Push P_i0 and P_i1 to LIST.
   end if
end while
```

from the list is decided by the choice of the node selection criterion and the branching priority. If it is possible to easily find an upper bound on the optimal objective function value, this bound can be used to initialize the global upper bound $\bar{z}$.

14.3.2.3 Logic-Based Programs

It is possible to modify the branch and bound algorithm for mixed integer programs to logic-based programs as described in [18, 26, 51, 69, 74, 80]. The integer variables which are fixed in a node are used to "infer" new knowledge on other integer variables. In this way the best integer solution can potentially be updated faster and hence reduce the computational time. In [18], the reduction was an order of magnitude for a supply chain management example.

14.3.3 Multi-Parametric Programming

In this subsection, optimization problems that depend on a parameter $\gamma_0 \in \Omega \subset \mathbb{R}^{n_\gamma}$ will be discussed. Throughout the text, the set Ω is assumed to be polyhedral. In general, these problems can be solved using an ordinary optimization algorithm for a *single* value of the parameter γ_0, or they can be solved for *all* parameter values $\gamma_0 \in \Omega$. The latter alternative is called parametric programming. Often a distinction is made between if γ_0 is a scalar or not. If it is not a scalar, it is often called multi-parametric programming which is the case in MPC applications. Many contributions have been published in the area of mp-QP and mp-MIQP. Therefore, the details are not given in this work. For a thorough treatment, see, e.g., [16, 20, 36, 37].

Both multiparametric QP (mp-QP) problems and multiparametric MIQP (mp-MIQP) problems will be considered in this work, mp-QP problems being problems of the form

$$\begin{aligned} \underset{x}{\text{minimize}} \quad & \frac{1}{2}x^T Hx + f^T x \\ \text{subject to} \quad & A_{\mathcal{E}}x = S_{\mathcal{E}}\gamma_0 + b_{\mathcal{E}} \\ & \text{Eq. (14.4c)} \end{aligned} \tag{14.15}$$

There are other equivalent forms, but the basic idea is the same where the linear term in the objective function or the right hand side in the constraints can be changed by a parameter. The mp-MIQP problems considered are in the form

$$\begin{aligned} \underset{x}{\text{minimize}} \quad & \frac{1}{2}x^T Hx + f^T x \\ \text{subject to} \quad & A_{\mathcal{E}}x = S_{\mathcal{E}}\gamma_0 + b_{\mathcal{E}} \\ & \text{Eqs. (14.4c), (14.6c)} \end{aligned} \tag{14.16}$$

where $\gamma_0 \in \Omega$ is the parameter and $\mathcal{B}$ denotes the set of indices to binary-valued components in x.

14.3.4 Other Methods

Other methods that can be used to solve the optimization problems arising in hybrid MPC are methods such as genetic algorithms, simulated annealing and tabu search. Since these methods in general only provide suboptimal solutions, they will not be discussed in more detail. However, it should be stressed that they can be parallelized fairly easily [70].

14.4 Parallel implementation

In this section, possibilities for parallelization at different computational levels are discussed. On the top level, the integer optimization problem is considered directly. The main focus is on branch and bound methods, but logic-based methods and parametric programming methods are alsoconsidered. In order to solve the MIQP problem of interest using a branch and bound method, QP problems should be solved by a convex QP method at an intermediate computational level. At an even lower level, the numerical linear algebra can be found. The reason for the partitioning into levels is that the possibilities and the applicable parallelization methods vary with the level and it is a nontrivial trade-off at what level, or levels, the algorithms should be parallelized. Work should be scheduled such that overhead from, e.g., communication, idle time due to load imbalance, and waiting time for shared resources is minimized [49]. An important feature of parallelism is the way the memory is distributed. Shared address space parallel computers have global memory, which all processors can address directly. This is in contrast to message passing computers, where each processor has its own local memory. For this latter case the processors can communicate only by sending messages over a network. Another important consideration is that the processors should be 'sufficiently loaded' in order to utilize the parallel hardware efficiently according to Amdahl's law. This law states that the speedup from parallelization does not grow linearly with the number of processors for a fixed problem size [49]. This is due to the sequential, nonparallelizable, part of the algorithm that eventually saturates the speedup as the number of processors grows.

14.4.1 Parallel Implementations at High Level

In this section ideas from generic parallel solvers for integer programs and logic-based programs are reviewed. These implementations have immediate potential for hybrid MPC. Also, ideas on how to parallelize multiparametric programming are presented.

14.4.1.1 Branch and bound

It is clear from Sec. 14.3.2.2 that the tree used in branch and bound algorithms can easily be partitioned into smaller parts that can be searched by different processors. Parallel formulations of branch and bound have shown near linear speedup in many cases, [3]. However, because of pruning many of the processors will become idle long before the optimal solution is found. This will result in low efficiency, and therefore it is not dessirable that one perform assignment in a static way. Hence, there is a need for a dynamic load balancing that minimizes communication and processor idling. In this work, parallelization on different computational levels will be

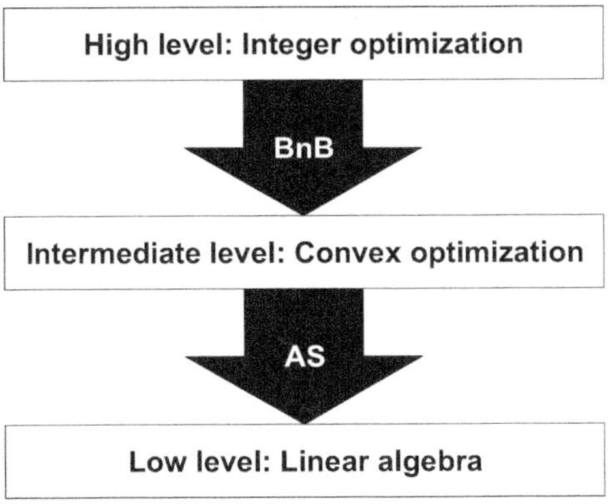

Fig. 14.2 Parallelization for branch and bound can be performed at different computational levels.

discussed. This is illustrated in Figure 14.2. We will only discuss the case of depth-first branch and bound, where the list in Algorithm 14.2 implements a stack. This case is commonly used in hybrid MPC and the case that is most easily parallelized. The other cases are discussed in more detail in [49].

For depth-first branch and bound the load balancing is composed of two phases: task portioning and subtask distribution [49]. In the task portioning typically two different techniques can be employed: so-called stack splitting and node splitting. In stack splitting the stack of a processor is split in two, half of which is kept and half of which is given away to another processor. In node splitting only one node is given away. Node splitting typically results in more work. Subtask distribution involves questions such as when work is split and how it is distributed. In both cases, this effort can be either sender initiated or receiver initiated.

It should be stressed that the best integer solution has to be known to all processors. For shared address space computers this is no problem, but for a message passing computer each processor has to have a local copy of the best integer solution and in case it is updated broadcast it to all other processors.

14.4.1.2 Logic-Based Programming

As explained in the previous section, logic-based programming can be used in a branch and bound algorithm to improve performance. It is not just the branch and bound algorithm that can be parallelized. As explained in [63, 66], logic-based programming can be parallelized also. For certain problems these algorithms can obtain orders of magnitude in speedup over distributed branch and bound search [66].

14.4.1.3 Multiparametric Programming

The explicit hybrid MPC problem can be solved as an mp-MIQP problem in, basically, three conceptually different ways. First, explicit enumeration, where all possible feasible combinations of integer solutions are explicitly enumerated, and one mp-QP problem is solved for each one of these sequences. The second method is a parametric branch and bound approach, where not necessarily all sequences have to be explicitly enumerated, i.e., it is basically the same idea as branch and bound for nonparametric problems, [37, 5]. The third method is a dynamic programming approach [27]. In the first method, parallelization is obvious, since different mp-QP problems can be assigned to different processors. The same applies in principle to the second method; however,problems regarding load balancing similar to those that can potentially occur for the nonparametric case, have to be considered in order for us to to get an efficient algorithm. A difference from the nonparametric case is that the subproblems are mp-QP problems instead of QP problems, which means that the computational effort for each subproblem is significantly higher. This could affect the load balancing strategy, notably because the relative cost for load balancing decreases compared to the nonparametric case, which indicates that more time can be spent on performing a good planning of future computations. The dynamic programming approach also allows us to compute the mp-QP solution for different sequences on different processors. Which approach is the most beneficial seems to depend on the difficulty of the problem. For example, explicit enumeration might be practically impossible, and therefore of no interest at all, for larger problems.

In all three cases, the state-space can also be divided into several smaller parts from the beginning. Given this initial partitioning, which can be, for example, boxes or a general polyhedral one, the solution in different regions of the state-space can be computed by different processors. In this approach, the solution parts have to be merged into one in a final step. This seems also fairly straightforward since the active set and integer sequences can be matched in the different solution parts.

14.4.2 Parallel Implementations at Intermediate Level

In this section, we discuss how the subproblems—the relaxations—in branch and bound can be solved more efficiently using parallelization. The relaxed problems are either of QP type or of SDP type. However, the main focus will be on QP problems.

14.4.2.1 Relaxed Problems in Branch and Bound

As noticed in [4], the QP relaxations used in branch and bound for a hybrid system in MLD form are ordinary linear MPC problems (possibly with an affine system description). In principle, this means that parallelization at this level can be performed as for linear MPC. However, this is not necessarily the most beneficial way, since it might be better to spend the computational power at another computational level in the algorithm. Since the relaxations are linear MPC problems, also work on distribution of such problems are also of interest. Some references considering this topic are [29, 35, 38, 39, 41, 47, 48, 54, 56, 82, 87]. A good overview of distributed MPC can be found in the recent review in [75].

An important key to parallelization of an optimization algorithm is the ability it gives one to, in some way, decompose the problem. This topic is by no means new and dates back to the beginning of the 1960s. Some fundamental and relevant references on this topic are [22, 23, 24, 28, 32, 40, 57, 58] and the references therein.

14.4.2.2 Tailored Solvers for the QP Subproblems

A tailored solver is a solver that in some way exploits properties of the problem to be solved in order to improve the performance. In [4, 6, 7] it was concluded that in order to get good performance for the hybrid MPC problem, tailored, dual active-set-like QP solvers should be used to solve the QP problems at the nodes of branch and bound. AS solvers are preferred over IP solvers, since the former can be efficiently warm-started. Similarly, dual QP solvers can more easily make use of a good initial guess of a (dual) solution to the problem since the geometry of the feasible set in the dual is very simple. The capacity to efficient warm-start is very important in branch and bound since it is often necessary one solve many QP problems to solve one MIQP problem. In order to overcome the well-known limitation of classical AS methods—that one potentially costly iteration is necessary for each addition or removal of a constraint to the working set—a projection method was developed in [7]. In this method, steepest-descent directions and Newton directions are projected onto the dual feasible set and the result is a solver which is able to make large changes to the working set, using only a small number of iterations. This is especially important if large disturbances, or large changes in the reference signal, can occur. The tailoring for MPC is in this algorithm performed at the linear algebra level. The steepest-descent directions and Newton directions are very efficiently computed with $\mathcal{O}(N)$ computational complexity. Furthermore, the line searches are also exploiting the problem structure. This is further explained in Sec. 14.4.3. Some other variants of QP solvers tailored for MPC, or at least designed with this application in mind, can be found in [14, 43, 55, 72, 73, 84].

The algorithms previously developed in [4, 6, 7] are designed mainly for non-parallel architectures. On parallel architectures there might exist other types of algorithms that potentially can be useful. However, we still believe that some form of dual active-set-like QP solver built on projection of first and second order directions is promising. For these solvers, the number of iterations is kept low but the cost for each iteration is higher. This is a result of the fact that large changes of the working set are common, and hence, updates of factorizations are no longer beneficial as they are in a classical AS solver. Instead, the factorization is in each iteration computed from scratch as in an IP method. A smaller number of computationally heavier iterations seems appealing from a parallel point of view, since it seems more promising to parallelize the computations within one iteration, rather than working on several iterations simultaneously.

14.4.2.3 Multiparametric Programming

At the intermediate level, the mp-QP (or mp-LP) solver can be parallelized. In mp-MIQP solvers, often an mp-QP solver is used as a subroutine. One way of parallelizing an mp-QP solver similar to the one presented in [78] is to explore different candidate regions by different processors. Another way is to split the parameter space beforehand, and thus let different processors explore different parts of the state-space. If the latter approach is used, the solution parts have to be merged in a final step. However, it is also true in this case that artificially cut regions can be identified since the active sets coincide. If the first solution is used, this merging is not necessary.

14.4.3 Parallel Implementations at Low Level

At the lowest level of the optimization algorithm, parallelization of the linear algebra is the main focus. It is clear that the linear algebra operations in the solver can in principle be performed more or less concurrently [25]. At this level it is interesting to investigate how these operations can be performed best for the hybrid MPC application, both in cases where there exists some kind of interconnection structure in the system and when such a structure does not exist.

Today, how to tailor most optimization algorithms for the MPC application is well-known. Tailoring usually means in this case to exploit the almost block diagonal structure of the KKT system in order to solve this system of equations efficiently. Examples of algorithms for different types of optimization problems related to MPC that exploit this structure can be found in [4, 6, 7, 8, 14, 34, 50, 55, 72, 81, 1]. The key to solving these equations efficiently is to factor the coefficient matrix for the KKT system efficiently. In this case the factoring can be performed either using a Riccati recursion or some other generic algorithm that is able to exploit the sparsity in the problem. Using either of these two approaches, the system of equations can be solved with a computational complexity that grows as $\mathcal{O}(N)$, which can be compared to applicable generic dense factorizations whose complexity grows as $\mathcal{O}(N^3)$.

A fundamental property of the Riccati recursion is that it factors the coefficient matrix by performing operations on matrices of the size in the order of the state and control signal dimension and it works using recursions backwards and forwards along the prediction horizon. If the number of processors is moderate compared to the state and control signal dimension, it seems reasonable to parallelize the operations that are performed within a single step of the Riccati recursion using standard techniques like those in [25]. However, it is not obvious how to extend the Riccati method to the case where a large number of processors are available, since there is a limit on how much computational power is necessary for each step in the recursion, and the next iteration relies on the result of the previous one. One possible way to limit this bottleneck could be to concatenate the dynamics from several time steps and consider a reformulated problem with higher dimensions of state and control signals, but with a shorter prediction horizon. For the case with a large number of processors, it might be interesting to search for a completely new way to attack the problem designed to fully exploit parallelism, or to investigate how far it is possible to reach with existing generic parallel methods.

In the solver presented in [7], line searches along piecewise affine paths are performed in each iteration. These paths originate from the projection of the search direction onto the feasible set and the result is a piecewise quadratic path defined over one dimension, the step length. Often, these paths are nonconvex and therefore the suboptimal strategy of backtracking is used to find a suitable step length. It is also possible to find the first local optimizer along the path, or even to find the global optimizer; however, this takes more time since all pieces along the piecewise affine path might need to be investigated. Numerical experiments show that the step length calculation takes a fairly large amount of the total time of each iteration and it would be interesting to try to make this step faster using parallelism. For example, one might try different backtracking step lengths simultaneously, or work on several pieces on the piecewise affine path simultaneously in order to quickly get the first local minimizer.

We also expect it will be possible to parallelize the linear algebra in multiparametric solvers, since these solvers perform operations similar to AS solvers. Hence, similar strategies are expected to be useful.

14.5 Conclusion

In this chapter we discussed parallel implementations of hybrid MPC. Different methods for achieving parallelism at different computational levels were surveyed. At the highest level we have discussed how the branch and bound algorithm for solving hybrid MPC problems can be parallelized. Also, brief ideas on ways to parallelize multiparametric programs for hybrid MPC were presented. At an intermediate level we discussed how the relaxed QP problems in branch and bound can be parallelized, and finally at the lowest level we discussed how the linear algebra can be parallelized. It is clear from what has been said that there are many ways of obtaining parallelism for hybrid MPC, and it is by no means clear that one must utilize all possibilities. All possible subsets of parallelism have to be investigated before a conclusive answer can be given.

14.5.1 Future Research

There are many interesting directions for future research in this area. As previously pointed out, the hard question to answer is which of these directions are the most interesting ones to develop further. That question will be left partly unanswered in this work. However, the topics we think are extra promising can be summarized as follows:

- Investigation of parallelization of a branch-and-bound MIQP algorithm like the one in [7]. Branch and bound algorithms have previously been parallelized, which indicates that there already exists a foundation that, hopefully, can be brought one or several steps further. The QP algorithm in [7] has proven to be a very efficient alternative for sequential implementations, and there are several properties that indicate it could benefit from parallelization in a good way.
- Use of logic-based programming, which has previously been used successfully for hybrid MPC. There already exists work where generic variants of these methods have been parallelized. Such work could be a good starting point.
- Use of parametric solvers, which are very promising for parallelization. As outlined in this text, some parts of that work is rather straightforward, other parts are less so. It is clear that everything that can be made to speed up these solvers are of great practical value, since it is highly computationally demanding to solve many of these problems.

Appendix

In this appendix, the MPC problem to minimize the objective function in (14.2) subject to the dynamics in (14.1) and the constraints in (14.3) is cast in the form of a QP problem. This can be done in several ways [59]. For simplicity, the derivation is performed for a linear MPC problem with a zero reference signal. It can be performed analogously for the hybrid case by introducing binary constraints on the desired variables. It is also straightforward to include a reference signal [4, pp. 65–67]. Without any loss of generality, the optimization problems are formulated for time step $t = 0$, which means that $\bar{x}$ is the measured or estimated state of the system at time step $t = 0$.

In this work, two different formulations are used. The main difference between the two is the formulation of the dynamic equations. Before the optimization problems are formulated, some notation is defined:

$$
\begin{aligned}
x &= \begin{bmatrix} x^T(0) & x^T(1) \cdots x^T(N) \end{bmatrix}^T, \; u = \begin{bmatrix} u^T(0) & u^T(1) \cdots u^T(N-1) \end{bmatrix}^T, \\
e &= \begin{bmatrix} e^T(0) & e^T(1) & \cdots & e^T(N) \end{bmatrix}^T, \\
\mathcal{Q}_e &= \operatorname{diag}(Q_e, \ldots, Q_e), \; \mathcal{Q}_u = \operatorname{diag}(Q_u, \ldots, Q_u), \\
\mathcal{H}_x &= \operatorname{diag}(H_x(0), \ldots, H_x(N)), \; \mathcal{H}_u = \begin{bmatrix} \operatorname{diag}(H_u(0), \ldots, H_u(N-1)) \\ 0 \end{bmatrix}, \\
h &= \begin{bmatrix} h^T(0) \cdots h^T(N) \end{bmatrix}^T, \; \mathcal{C} = \operatorname{diag}(C, \ldots, C), \\
\mathcal{A} &= \begin{bmatrix} -I & 0 & \cdots & 0 & 0 \\ A & -I & \cdots & 0 & 0 \\ 0 & A & \cdots & 0 & 0 \\ \vdots & \vdots & \ddots & \ddots & \vdots \\ 0 & 0 & \cdots & A & -I \end{bmatrix}, \; \mathcal{B} = \begin{bmatrix} 0 \\ \operatorname{diag}(B, \ldots, B) \end{bmatrix}, \; b = \begin{bmatrix} -\bar{x} \\ 0 \\ \vdots \\ 0 \end{bmatrix}.
\end{aligned}
\tag{14.17}
$$

Complete Set of Variables

The most straightforward way to cast the MPC problem in the form of a QP problem is to keep the dynamic equations as equality constraints. The MPC problem is then formulated as a QP problem in the form

$$
\begin{aligned}
\underset{x,u,e}{\text{minimize}} \quad & \frac{1}{2} \begin{bmatrix} x^T & u^T & e^T \end{bmatrix} \begin{bmatrix} 0 & 0 & 0 \\ 0 & \mathcal{Q}_u & 0 \\ 0 & 0 & \mathcal{Q}_e \end{bmatrix} \begin{bmatrix} x \\ u \\ e \end{bmatrix} \\
\text{subject to} \quad & \begin{bmatrix} \mathcal{A} & \mathcal{B} & 0 \end{bmatrix} \begin{bmatrix} x^T & u^T & e^T \end{bmatrix}^T = b \\
& \begin{bmatrix} \mathcal{C} & 0 & -I \end{bmatrix} \begin{bmatrix} x^T & u^T & e^T \end{bmatrix}^T = 0 \\
& \begin{bmatrix} \mathcal{H}_x & \mathcal{H}_u & 0 \end{bmatrix} \begin{bmatrix} x^T & u^T & e^T \end{bmatrix}^T + h \leq 0
\end{aligned}
\tag{14.18}
$$

The optimization problem in (14.18) is an optimization problem in the form in (14.4). This formulation requires $(N+1) \cdot (n+p) + N \cdot m$ variables and gives a sparse objective Hessian function and sparse constraint matrices. If this formulation is used, a solver able to take advantage of sparsity should be used.

Reduced Set of Variables

In the second formulation, e and x are eliminated from the problem. This can be done by expressing e as $e = Cx$ and x as a function of the initial state $\bar{x}$ and the control inputs u. From the state update equations it follows that

$$x = S_x\bar{x} + S_u u \tag{14.19}$$

where

$$S_x = \begin{bmatrix} I \\ A \\ A^2 \\ \vdots \\ A^N \end{bmatrix}, \quad S_u = \begin{bmatrix} 0 & 0 & \cdots & 0 \\ B & 0 & \cdots & 0 \\ AB & B & \cdots & 0 \\ \vdots & \vdots & \ddots & \vdots \\ A^{N-1}B & A^{N-2}B & \cdots & B \end{bmatrix}. \tag{14.20}$$

The equality constraints are now eliminated and the objective function can be written as

$$\frac{1}{2}u^T\left(S_u^T Q_x S_u + Q_u\right)u + \left(S_x\bar{x}\right)^T Q_x S_u u + \kappa \tag{14.21}$$

where $\kappa = \frac{1}{2}\left(S_x\bar{x}\right)^T Q_x\left(S_x\bar{x}\right)$ is a constant and $Q_x = C^T Q_e C$. In practice, the constant κ is ignored since the resulting optimization problem is still equivalent. Using (14.19), the inequality constraints can be written as

$$H_x x + H_u u + h = H_x S_u u + H_u u + h + H_x S_x \bar{x} \leq 0. \tag{14.22}$$

The optimization problem to minimize (14.21) subject to (14.22) is an optimization problem of the form in (14.4) without equality constraints. In this formulation, the objective Hessian function and constraint matrix become dense, but only Nm optimization variables are required. This formulation is less relevant for hybrid MPC problems with *binary states* or *binary outputs*, since the binary constrained variables need to be explicitly available (not eliminated) in the formulation.

References

1. Åkerblad, M., Hansson, A.: Efficient solution of second order cone program for model predictive control. International Journal of Control **77**(1), 55–77 (2004)
2. Antsaklis, P.: A brief introduction to the theory and applications of hybrid systems. Proc. IEEE, Special Issue on Hybrid Systems: Theory and Applications **88**(7), 879–886 (2000)
3. Arvindam, S., Kumar, V., Rao, V.N.: Floorplan optimization on multiprocessors. In: Proc. 1989 Int. Conf. Computer Design (1989)
4. Axehill, D.: Integer quadratic programming for control and communication. Ph.D. thesis, Linköping University (2008). URL http://urn.kb.se/resolve?urn=urn:nbn:se:liu:diva-10642
5. Axehill, D., Besselmann, T., Raimondo, D.M., Morari, M.: Suboptimal explicit hybrid MPC via branch and bound. In: Proc. 18th IFAC World Congress (IFAC2011). Milan, Italy (2011)

6. Axehill, D., Hansson, A.: A mixed integer dual quadratic programming algorithm tailored for MPC. In: Proc. 45th IEEE Conf. Decision and Control (CDC2006), pp. 5693–5698. San Diego, CA (2006)
7. Axehill, D., Hansson, A.: A dual gradient projection quadratic programming algorithm tailored for model predictive control. In: Proc. 47th IEEE Conf. Decision and Control (CDC2008), pp. 3057–3064. Cancun, Mexico (2008)
8. Axehill, D., Hansson, A., Vandenberghe, L.: Relaxations applicable to mixed integer predictive control– Comparisons and efficient computations. In: Proc. 46th IEEE Conf. Decision and Control (CDC2007), pp. 4103–4109. New Orleans, LA (2007)
9. Axehill, D., Morari, M.: Improved complexity analysis of branch and bound for hybrid MPC. In: Proc. 49th IEEE Conf. Decision and Control (CDC2010), pp. 4216–4222. Atlanta, GA (2010)
10. Axehill, D., Sjöberg, J.: Adaptive cruise control for heavy vehicles—Hybrid control and MPC. Master's thesis, Linköping University (2003)
11. Axehill, D., Vandenberghe, L., Hansson, A.: Convex relaxations for mixed integer predictive control. Automatica **46**(9), 1540–1545 (2010)
12. Baotic, M.: Optimal control of piecewise affine systems—A multi-parametric approach. Ph.D. thesis, ETH (2005). URL http://control.ee.ethz.ch/index.cgi?page=publications;action=details;id=2235
13. Barth, T., Freisleben, B., Grauer, M., Thilo, F.: Distributed solution of optimal hybrid control problems on networks of workstations. In: Proc. Second IEEE Int. Conf. Cluster Computing (2000)
14. Bartlett, R.A., Biegler, L.T., Backstrom, J., Gopal, V.: Quadratic programming algorithms for large-scale model predictive control. Journal of Process Control **12**, 775–795 (2002)
15. Bemporad, A.: Efficient conversion of mixed logical dynamical systems into an equivalent piecewise affine form. IEEE Trans. Audio and Electroacoustics **49**(5), 832–838 (2004)
16. Bemporad, A., Borrelli, F., Morari, M.: Model predictive control based on linear programming—The explicit solution. IEEE Trans. Automatic Control **47**(12), 1974–1985 (2002)
17. Bemporad, A., Ferrari-Trecate, G., Morari, M.: Observability and controllability of piecewise affine and hybrid systems. IEEE Trans. Automatic Control **45**(10), 1864–1876 (2000)
18. Bemporad, A., Giorgetti, N.: Logic-based solution methods for optimal control of hybrid systems. IEEE Trans. Automatic Control **51**(6), 963–976 (2006)
19. Bemporad, A., Mignone, D.: A Matlab function for solving mixed integer quadratic programs version 1.02 user guide. Tech. rep., Institut für Automatik, ETH (2000)
20. Bemporad, A., Morari, M., Dua, V., Pistikopoulos, E.N.: The explicit linear quadratic regulator for constrained systems. Automatica **38**, 3–20 (2002)
21. Bemporad, A., Morari, M.: Control of systems integrating logic, dynamics, and constraints. Automatica **35**, 407–427 (1999)
22. Benders, J.F.: Partitioning procedures for solving mixed-variables programming problems. Numerische Mathematik **4**(1), 238–252 (1962)
23. Bertsekas, D.P.: Nonlinear Programming. Athena Scientific (1995)
24. Bertsekas, D.P., Tsitsiklis, J.N.: Parallel and Distributed Computation: Numerical Methods. Prentice-Hall (1989)
25. Blackford, L.S., Choi, J., Cleary, A., D'Azevedo, E., Demmel, J., Dhillon, I., Dongarra, J., Hammarling, S., Henry, G., Petitet, A., Stanley, K., Walker, D., Whaley, R.C.: ScaLAPACK Users' Guide. Society for Industrial and Applied Mathematics (1997)
26. Bockmayr, A., Kasper, T.: Branch and infer: a unifying framework for integer and finite domain constraint programming. INFORMS J. Comput. **10**(3), 287–300 (1998)
27. Borrelli, F., Baotić, M., Bemporad, A., Morari, M.: Dynamic programming for constrained optimal control of discrete-time linear hybrid systems. Automatica **41**, 1709–1721 (2005)
28. Boyd, S., Xiao, L., Mutapcic, A., Mattingley, J.: Notes on decomposition methods—Notes for EE364B, Stanford University, winter 2006-2007. Tech. rep., Information Systems Laboratory, Department of Electrical Engineering, Stanford (2008). URL http://see.stanford.edu/materials/lsocoee364b/08-decomposition_notes.pdf

29. Camponogara, E., Jia, D., Krogh, B.H., Talukdar, S.: Distributed model predictive control. IEEE Control Systems Magazine **22**(1), 44–52 (2002)
30. Cassandras, C., Pepyne, D., Wardi, Y.: Optimal control of a class of hybrid systems. IEEE Trans. Automatic Control **46**(3), 398–415 (2001)
31. Cutler, C.R., Ramaker, B.L.: Dynamic matrix control—A computer control algorithm. In: Proc. AIChE National Meeting. Houston, TX (1979)
32. Dantzig, G.B., Wolfe, P.: Decomposition principle for linear programs. Operations Research **8**(1), 101–111 (1960)
33. De Schutter, B., van den Boom, T.J.J.: MPC for continuous piecewise-affine systems. Systems & Control Letters **52**, 179–192 (2004)
34. Diehl, M., Ferreau, H.J., Haverbeke, N.: Nonlinear Model Predictive Control, chap. Efficient Numerical Methods for Nonlinear MPC and Moving Horizon Estimation, pp. 391–417. Springer Berlin / Heidelberg, Berlin, Heidelberg (2009)
35. Du, X., Xi, Y., Li, S.: Distributed model predictive control for large-scale systems. In: Proc. American Control Conference (ACC'01), vol. 4, pp. 3142–3143 (2001)
36. Dua, V., Bozinis, N.A., Pistikopoulos, E.N.: A multiparametric programming approach for mixed-integer quadratic engineering problems. Computers and Chemical Engineering **26**, 715–733 (2002)
37. Dua, V., Pistikopoulos, E.N.: An algorithm for the solution of multiparametric mixed integer linear programming problems. Annals of Operations Research **99**, 123–139 (2000)
38. Dunbar, W.B.: Distributed receding horizon control of dynamically coupled nonlinear systems. IEEE Trans. Automatic Control **52**(7), 1249–1263 (2007)
39. Dunbar, W.B., Murray, R.M.: Distributed receding horizon control for multi-vehicle formation stabilization. Automatica **42**(4), 549–558 (2006)
40. Everett, H.: Generalized Lagrange multiplier method for solving problems of optimum allocation of resources. Operations Research **11**(3), 399–417 (1963)
41. Fawal, H.E., Georges, D., Bornard, G.: Optimal control of complex irrigation systems via decomposition-coordination and the use of augmented Lagrangian. In: Proc. 1998 IEEE Int. Conf. Systems, Man, and Cybernetics, vol. 4, pp. 3874–3879 (1998)
42. Ferrari-Trecate, G., Mignone, D., Castagnoli, D., Morari, M.: Mixed logical dynamical model of a hydroelectric power plant. In: Proc. 4th Int. Conf. Automation of Mixed Processes: Hybrid Dynamic Systems (2000)
43. Ferreau, H.J., Bock, H.G., Diehl, M.: An online active set strategy to overcome the limitations of explicit MPC. International Journal of Robust and Nonlinear Control **18**(8), 816–830 (2008)
44. Fletcher, R., Leyffer, S.: Numerical experience with lower bounds for MIQP branch-and-bound. SIAM Journal on Optimization **8**(2), 604–616 (1998)
45. Floudas, C.A.: Nonlinear and Mixed-Integer Optimization. Oxford University Press (1995)
46. Garcia, C.E., Prett, D.M., Morari, M.: Model predictive control: Theory and practice—A survey. Automatica **3**, 335–348 (1989)
47. Georges, D.: Decentralized adaptive control for a water distribution system. In: Proc. Third IEEE Conf. Control Applications, vol. 2, pp. 1411–1416 (1994)
48. Gómez, M., Rodellar, J., Vea, F., Mantecón, J., Cardona, J.a.: Decentralized predictive control of multi-reach canals. In: Proc. 1998 IEEE Int. Conf. Systems, Man, and Cybernetics, vol. 4, pp. 3885–3890 (1998)
49. Grama, A.Y., Kumar, V.: A survey of parallel search algorithms for discrete optimization problems. ORSA Journal of Computing **7**(4), 365–385 (1995)
50. Hansson, A.: A primal-dual interior-point method for robust optimal control of linear discrete-time systems. IEEE Trans. Automatic Control **45**(9), 1639–1655 (2000)
51. Harjunkoski, I., Jain, V., Grossmann, I.: Hybrid mixedinteger/constraint logic programming strategies for solving scheduling and combinatorical optimization problems. Comp. Chem. Eng. **24**, 337–343 (2000)
52. Heemels, W.P.M.H., De Schutter, B., Bemporad, A.: Equivalence of hybrid dynamical models. Automatica **37**, 1085–1091 (2001)

53. Heemels, W.P.M.H., De Schutter, B., Bemporad, A.: On the equivalence of classes of hybrid dynamical models. In: Proc. 40th IEEE Conf. Decision and Control (CDC2001), pp. 364–369. Orlando, FL (2001)
54. Jia, D., Krogh, B.H.: Distributed model predictive control. In: Proc. American Control Conference (ACC'01), vol. 4, pp. 2767–2772 (2001)
55. Jonson, H.: A Newton method for solving non-linear optimal control problems with general constraints. Ph.D. thesis, Linköpings Tekniska Högskola (1983)
56. Katebi, M.R., Johnson, M.A.: Predictive control design for large-scale systems. Automatica **33**(3), 421–425 (1997)
57. Lasdon, L.S.: Optimization Theory for Large Systems. MacMillan Publishing Co., Inc (1970)
58. Lasdon, L.S.: Optimization Theory for Large Systems. Dover Publications (2002)
59. Lie, B., Díez, M.D., Hauge, T.A.: A comparison of implementation strategies for MPC. Modeling, identification and control **26**(1), 39–50 (2005)
60. Lincoln, B., Rantzer, A.: Optimizing linear system switching. In: Proc. 40th IEEE Conf. Decision and Control, pp. 2063–2068. Barcelona (2001)
61. Löfberg, J.: Yalmip: A toolbox for modeling and optimization in MATLAB. In: Proc. CACSD Conference. Taipei, Taiwan (2004). URL http://control.ee.ethz.ch/ joloef/yalmip.php
62. Maciejowski, J.M.: Predictive Control with Constraints. Pearson Education Ltd. (2002)
63. Mailler, R., Lesser, V.: Solving distributed constraint optimization problems using cooperative mediation. In: Proc. AAMAS, pp. 438–445. New York, USA (2004)
64. Mayne, D.Q., Rawlings, J.B., Rao, C.V., Scokaert, P.O.M.: Constrained model predictive control: Stability and optimality. Automatica **36**, 789–814 (2000)
65. Mestan, E., Turkay, E.M., Arkun, Y.: Optimization of operations in supply chain systems using hybrid systems approach and model predictive control. Ind. Eng. Chem. Res. **45**, 6493–6503 (2006)
66. Modi, P.J., Shen, W.M., Tambe, M.: Adopt: asynchronous distributed constraint optimization with quality guarantees. Artificial Intelligence **161**, 149–180 (2005)
67. Negenborn, R.R.: Multi-agent model predictive control with applications to power networks. Ph.D. thesis, Technische Universiteit Delft, Delft, Netherlands (2007)
68. Nocedal, J., Wright, S.J.: Numerical Optimization—Second Edition. Springer Verlag (2006)
69. Ottosson, G.: Integration of constraint programming and integer programming for combinatorial optimization. Ph.D. thesis, Computer Science Department, Information Technology, Uppsala, Sweden (2000)
70. Pardalos, P.M., Pitsolulis, L., Mavridou, T., Resende, M.G.C.: Parallel search for combinatorial optimization: Genetic algorithms, simulated annealing, tabu search and GRASP. In: P. Sanders (ed.) Parallel Algorithms for Irregularly Structured Problems, *Lecture Notes in Computer Science*, vol. 980, pp. 317–331. Springer, Berlin / Heidelberg (1995)
71. Qin, S.J., Badgwell, T.A.: A survey of industrial model predictive control technology. Control Engineering Practice **11**, 722–764 (2003)
72. Rao, C.V., Wright, S.J., Rawlings, J.B.: Application of interior-point methods to model predictive control. Journal of Optimization Theory and Applications **99**(3), 723–757 (1998)
73. Richter, S., Jones, C.N., Morari, M.: Real-time input-constrained MPC using fast gradient methods. In: Proc. 48h IEEE Conf. Decision and Control (CDC2009) held jointly with 2009 28th Chinese Control Conference, pp. 7387–7393. Shanghai, China (2009)
74. Rodosek, R., Wallace, M., Hajian, M.: A new approach to integrating mixed integer programming and constraint logic programming. Ann. Oper. Res. **86**, 63–87 (1997)
75. Scattolini, R.: Architectures for distributed and hierarchical model predictive control. J. Process Control **19**, 723–731 (2009)
76. Sontag, E.D.: Nonlinear regulation: The piecewise linear approach. IEEE Trans. Automatic Control **26**(2), 346–358 (1981)
77. Tarau, A.N., de Schutter, B., Hellendoorn, J.: Centralized, decentralized, and distributed model predictive control for route choice in automated baggage handling systems. Journal of Control Engineering and Applied Informatics **11**(3), 24–31 (2009)
78. Tøndel, P., Johansen, T.A., Bemporad, A.: An algorithm for multi-parametric quadratic programming and explicit MPC solutions. Automatica **39**, 489–497 (2003)

79. Torrisi, F.D., Bemporad, A.: HYSDEL—A tool for generating computational hybrid models for analysis and synthesis problems. IEEE Trans. Control Systems Technology **12**(2), 235–249 (2004)
80. Tsang, E.P.K.: Foundations of Constraint Satisfaction. Academic Press, London and San Diego (1993)
81. Vandenberghe, L., Boyd, S., Nouralishahi, M.: Robust linear programming and optimal control. Tech. rep., Dept. Electrical Engineering, Univ. California Los Angeles (2002)
82. Venkat, A.N., Hiskens, I.A., Rawlings, J.B., Wright, S.J.: Distributed MPC strategies with application to power system automatic generation control. IEEE Trans. Control Systems Technology **16**(6), 1192–1206 (2008)
83. Volkovich, O.V., Roshchin, V.A., Sergienko, I.V.: Models and methods of solution of quadratic integer programming problems. Cybernetics **23**, 289–305 (1987)
84. Wang, Y., Boyd, S.: Fast model predictive control using online optimization. IEEE Trans. Control Systems Technology **18**(2), 267–278 (2010)
85. Wolsey, L.A.: Integer Programming. John Wiley & Sons, Inc., New York (1998)
86. Xu, X., Antsaklis, P.: An approach to switched systems optimal control based on parameterization of the switching instants. In: Proc. IFAC World Congress. Barcelona, Spain (2002)
87. Zhu, G.Y., Henson, M.A.: Model predictive control of interconnected linear and nonlinear processes. Industrial and Engineering Chemistry Research **41**(4), 801–816 (2002)

Chapter 15
Hierarchical Model Predictive Control for Plug-and-Play Resource Distribution

Jan Bendtsen, Klaus Trangbaek and Jakob Stoustrup

Abstract This chapter deals with hierarchical model predictive control (MPC) of distributed systems. A three-level hierarchical approach is proposed, consisting of a high level MPC controller, a second level of so-called *aggregators*, controlled by an online MPC-like algorithm, and a lower level of autonomous units.

The approach is inspired by smart-grid electric power production and consumption systems, where the flexibility of a large number of power producing and/or power consuming units can be exploited in a smart grid solution. The objective is to accommodate the load variation on the grid, arising on one hand from varying consumption, on the other hand by natural variations in power production, e.g., from wind turbines.

The proposed method can also be applied to supply chain management systems, where the challenge is to balance demand and supply, using a number of storages each with a maximal capacity. The algorithm will then try to balance the risk of individual storages running empty or full with the risk of having overproduction or unsatisfied demand.

The approach presented is based on quadratic optimization and possesses the properties of low algorithmic complexity and of scalability. In particular, the proposed design methodology facilitates plug-and-play addition of subsystems without redesign of any controllers.

Jan Bendtsen
Department of Electronic Systems, Automation and Control, Aalborg University, Fr. Bajers Vej 7C, 9220 Aalborg, Denmark
e-mail: dimon@es.aau.dk

Klaus Trangbaek
Department of Electronic Systems, Automation and Control, Aalborg University, Fr. Bajers Vej 7C, 9220 Aalborg, Denmark
e-mail: ktr@es.aau.dk

Jakob Stoustrup
Department of Electronic Systems, Automation and Control, Aalborg University, Fr. Bajers Vej 7C, 9220 Aalborg, Denmark
e-mail: jakob@es.aau.dk

R. Johansson and A. Rantzer (Eds.): Distributed Decision Making and Control, LNCIS 417, pp. 339–358.
springerlink.com

The method is verified by a number of simulations featuring a three-level smart grid power control system for a small isolated power grid.

15.1 Introduction

We discuss a hierarchical set-up, where an optimization-based high level controller is given the task of following a specific externally generated trajectory of consumption and/or production of a certain resource. The high level controller has a number of units under its jurisdiction, which consume a certain amount of the resource. The flow of resources allocated to each of these units can be controlled, but each unit must at all times be given at least a certain amount of the resource; vice versa, each unit can only consume a certain (larger) amount of the resource.

One can think of various practical examples of systems that fit with this set-up; for instance, a supply chain management system [2], where the challenge is to balance demand and supply, using a number of storages each with a maximal capacity. The algorithm will then try to balance the risk of individual storages running empty or full with the risk of creating overproduction or unsatisfied demand. Other examples include large-scale refrigeration systems (e.g., in supermarkets), where the resource is refrigerant and the consuming units are individual display cases [14]; irrigation systems, where the shared resource is water and the consuming units are adjustable field sprinklers [13]; chemical processes requiring process steam from a common source [5]; or even digital wireless communication systems, where the resource is bandwidth and the consuming units are handheld terminals, e.g., they are connected to a building-wide intranet [1]. See also [6] and [3] for an example of a district heating system that shares some of the physical characteristics outlined here, although the cited papers pursue a decentralized control scheme rather than a centralized one.

Such large scale hierarchical systems are often subject to frequent modifications in terms of subsystems that are added (or removed). This adds an important side constraint to design methodologies for controlling such systems: They should accommodate incremental growth of the hierarchical system in a way that is flexible and scalable. In essence, the design methodology should support a *plug-and-play control* architecture; see e.g., [12].

In many cases, a natural choice for the top level controller is some sort of model-predictive controller (MPC) [11], [7], since systems of the kinds referred to above are multivariable, subject to constraints and often involve considerable delays. Furthermore, some sort of reference estimate is often known in advance, e.g., from 24-hour electric power consumption traces, weather forecasts, purchase orders, etc. Unfortunately, the computational complexity of traditional MPC scales quite poorly with the number of states in the problem ($O(n^3)$); see, e.g., [4]). Refer also to [9] for a recent contribution on MPC control for two-layer hierarchical control systems. In the type of problems considered above, this complexity growth places significant limits on how large systems may be so that a centralized solution can handle them, as also pointed out in, e.g., [10].

In this chapter, we propose a hierarchical control architecture that

- is based on a standard MPC solution at the top level;
- is able to accommodate new units without requiring modifications of the top level controller;
- remains stable for an increasing number of units;
- facilitates plug-and-play addition of units at the bottom level, i.e., new units can be incorporated at the bottom level simply by registering with the unit at the level just above it.

Furthermore, the worst-case complexity is lower than for conventional centralized solutions, which means that the proposed scheme scales more "reasonably" than the centralized solution. As will be illustrated, the involved optimization techniques give rise to quite sparse structures; this sparsity can be exploited to reduce complexity.

By a distributed resource control system we shall understand a system with the following characteristics:

- The system has a number of decentralized storages that can store a certain amount of some resource.
- Each storage can be filled or emptied at some maximal rate(s).
- A central controller has the responsibility of balancing supply and demand by use of the storages.

We illustrate the approach by a specific example, a so-called "smart grid" electric power system, wherein consumers can vary their power consumption within certain bounds by allowing devices to store more or less energy at convenient times [8]. The obvious method to do so physically is by exploiting large thermal time constants in deep freezers, refrigerators, local heat pumps, etc.; extra energy can be stored during off-peak hours, and the accumulated extra cooling can then be used—slowly—by turning compressors and similar devices on less frequently during peak hours. Implementing such schemes is considered a necessity for adoption of large amounts of unpredictable renewable energy sources in the European power grid and requires local measurement and feedback of current energy and power demand. Consumers equipped with such measurement and feedback capabilities are called *intelligent consumers*.

Structural flexibility of large scale systems is important, since subsystems and components may be added, removed or replaced during the system's lifetime. In our example, it is easy to imagine customers wanting to sign up for a contract with a power company, such that the customer is assigned the necessary equipment to become an intelligent consumer. Thus, the top level system should be flexible enough to accommodate new consumers under its jurisdiction without it being necessary to perform significant retuning and/or restructuring every time new consumers appear. Furthermore, it is a basic requirement that the system be stable and provide good performance at all times.

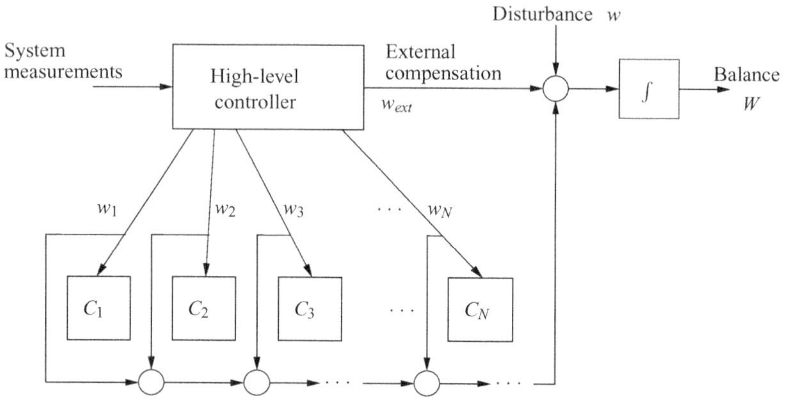

Fig. 15.1 Problem set-up.

The outline of the rest of the chapter is as follows: Section 15.2 explains the problem in a general setting, while Sec. 15.3 presents the proposed algorithm for resource sharing. Section 15.4 shows that the resulting architecture remains stable for increasing numbers of units. Section 15.5 shows a simulation example of the algorithm applied to an electric smart grid with a small number of consumers, and, finally, Sec. 15.6 offers some concluding remarks. Note that, unless otherwise stated, all time-varying quantities (signals) are assumed to be real scalars. Vectors and matrices are indicated with boldface symbols, while sets are written in calligraphic font.

15.2 Problem Formulation

We consider a set-up as depicted in Fig. 15.1. The high level controller is given the task of following a specific, externally generated trajectory of consumption and/or production of a certain resource. The objective is to maintain a certain system-level *balance* (between demand and production); the error in the balance is represented by the scalar signal $W(t)$, which must be driven to zero as the time t tends to infinity. The demand and production must match over time, however, and any disturbance $w(t)$ is hence treated as a short-time change in the balance, whereas $W(t)$ is an integrated error signal. The high level controller can compensate directly for the disturbance $w(t)$ by assigning some of the resource flow $w_{\text{ext}}(t)$ to this task, but at a significant cost. However, the high level controller also has a number of units, which we will in general refer to as *consumers*, C_i, $i = 1, \ldots, N$, under its jurisdiction. Each one of these consumers consumes the resource at a certain, controllable rate $w_i(t)$. The high level controller is able to direct time-varying resources to the consumers, but must ensure that each consumer *on average* receives a specific amount of the resource, and certain upper and lower bounds on the consumption rate, $\underline{w}_i$ and $\overline{w}_i$, may not be exceeded. By doing so, the consumption compensates for some of the disturbance $w(t)$, at a *lower cost* than the direct compensation signal $w_{\text{ext}}(t)$. That is, it is advantageous to utilize the consumers as much as possible, subject to the aforementioned constraints.

This set-up could for instance be interpreted as a supply chain management system, where the challenge is to balance demand and supply by minimizing the excess supply $W(t)$. The demand and supply should in this interpretation be associated with the "disturbance" signal $w(t)$, which can be thought of as short-term market fluctuations, supply irregularities, etc. The system has a number of storages C_i available, each currently being filled at the rate $w_i(t)$. The maximal capacities and maximal filling/emptying rates of each storage should be exploited in such a way that the need for "external compensation" $w_{\text{ext}}(t)$ is minimized. In this interpretation, w_{ext} corresponds, depending on the sign, either to having to rent external storages or to have to buy components from more expensive suppliers. Thus, the goal of the algorithm is to try to balance the risk of individual storages running empty against the risk of having overproduction or unsatisfied demand.

In the following, let $\mathcal{I} = \{1,2,\ldots,N\}$ denote an index set enumerating the consumers. The high level controller must solve the following optimization problem at any given time t:

$$\begin{aligned} \min_{w_i; w_{\text{ext}}} & \int_t^{t+T_h} \phi(W(\tau), w_{\text{ext}}(\tau), \frac{dw_{\text{ext}}}{dt}) d\tau \\ \text{subject to } & \underline{W} \leq W(\tau) \leq \overline{W} \\ & \underline{w}_i \leq w_i(\tau) \leq \overline{w}_i, \quad \forall i \in \mathcal{I} \end{aligned} \tag{15.1}$$

where $\underline{W}$ and $\overline{W}$ are constraints on the balance and $\phi : \mathbb{R} \times \mathbb{R} \times \mathbb{R} \rightarrow \mathbb{R}_+$ is a smooth, convex cost function of the balance error, the external resources that need to be acquired, and the changes in these resources (motivated by the fact that it is often more costly to acquire extra resources on a very short notice); ϕ is typically chosen as a linear or quadratic cost function. T_h is the prediction horizon of the controller. For simplicity, and without loss of generality, the consumption by the consumers is assumed cost-free.

Let $W_i(t)$ denote the amount of resource accumulated in C_i, and $\eta_i \geq 0$ denote a drain rate, respectively; η_i is assumed to be constant for simplicity. Each consumer is characterized by its own linear state equation:

$$\frac{dW_i(t)}{dt} = w_i(t) - \eta_i \tag{15.2}$$

which must satisfy $0 \leq W_i(t) \leq \overline{W}_i$ at all times. Note that this model implies that the consumers are mutually independent. The goal that each consumer receive a specific amount of the resource on average may be expressed as the integral constraint

$$\frac{1}{T_{\text{res}}} \int_0^{T_{\text{res}}} |w_i(\tau) - \eta_i| d\tau = W_{i,\text{ref}} \tag{15.3}$$

where T_{res} is some appropriate time span. Obviously, we must require that $0 \leq W_{i,\text{ref}} \leq \overline{W}_i$.

Note that, since the dynamics contain only pure integrators, (15.1) can easily be approximated by a discrete-time problem of the form

$$\begin{aligned} \min_{w_i, w_{\text{ext}}} \sum_{k=t/T_s+1}^{(t+T_h)/T_s} & \phi(W(kT_s), w_{\text{ext}}(kT_s), w_{\text{ext}}((k-1)T_s)) \qquad (15.4) \\ \text{subject to} & \underline{W} \leq W(kT_s) \leq \overline{W} \\ & \underline{w}_i \leq w_i(kT_s) \leq \overline{w}_i, \quad \forall i \in \mathcal{I} \end{aligned}$$

where T_s is the sampling time. For simplicity, we will often drop T_s from the notation in the sequel, writing, e.g., $w(k)$ as shorthand for $w(kT_s)$.

In order to solve the optimization problem, the high level controller in principle requires access to all states in the system, including the internal states $W_i(t)$. This may lead to a very heavy communication load on distributed systems if the number of consumers is significant. Furthermore, the computational complexity of the optimization problem grows rapidly with the number of consumers as well. This means that adding more consumers into the system may pose significant problems in practice. Thus, a purely centralized solution to the problem may be optimal in terms of maintaining the supply/demand balance but is not desirable from a practical point of view.

15.3 Proposed Architecture

In the following we propose a new architecture for achieving the control objective that requires significantly less system-wide communication, while at the same time being more flexible with respect to changes in the number of consumers. We now consider the modified set-up in Fig. 15.2, where $w(t)$ is an external disturbance and $w_a(t) = \sum_{i=1}^{N} w_i(t)$ is the cumulative rate of resource absorbed by all C_i. As mentioned in the previous section, the main objective of the high level control is to keep the resource balance governed by

$$\frac{dW(t)}{dt} = w(t) - w_{\text{ext}}(t) - w_a(t) \qquad (15.5)$$

at zero. It is assumed that the top level controller can control $w_{\text{ext}}(t)$ directly and is constrained only by a rate limit, but we would like to keep the variations, i.e., the time derivative of $w_{\text{ext}}(t)$, small as well.

Between the controller and $N_A \leq N$ subsets of the intelligent consumers, we introduce a number of so-called *aggregators* $A_j, 1 \leq j \leq N_A$. Together, these aggregators serve as an interface between the top level and the intelligent consumers. To each aggregator A_j we assign a number of consumers identified by an index set $\mathcal{J}^j \subset \mathcal{I}$, where for all $k, j = 1, \ldots, N_A$, we have $\mathcal{J}^j \cap \mathcal{J}^l = \emptyset, l \neq j$, and $\cup_{j=1}^{N_A} \mathcal{J}^j = \mathcal{I}$. Let n^j

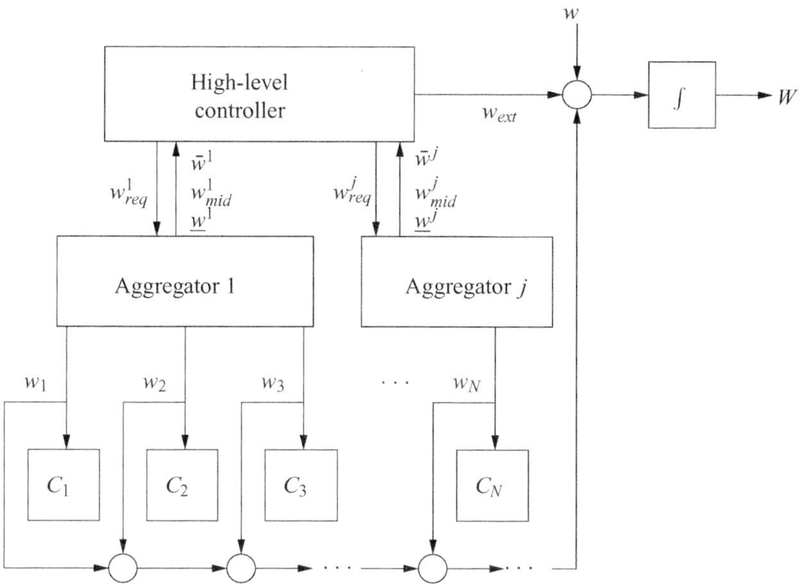

Fig. 15.2 Modified architecture.

denote the cardinality of $\mathcal{J}^j$, i.e., the number of consumers assigned to aggregator A_j. The objective of each aggregator is to make sure that:

- the maximum capacity is available for the upper level at any time instance;
- the resources allocated to consumers are distributed roughly uniformly over the number of consumers;
- the deviation from the nominal consumption is minimized for each consumer;
- the capacity constraint for each consumer is not violated;
- the rate constraint for each consumer is not violated

As in the previous section, we approximate the continuous-time system with a discrete-time version.

The communication between the high level controller is indicated on Fig. 15.2; each aggregator A_j provides the top level with a small number of simple parameters to specify the constraints of the consumers. In particular, the top level is informed of $\overline{w}(k)$ and $\underline{w}(k)$, which are current bounds on the cumulative resource that can be consumed by the consumers assigned to A_j, that is, bounds on

$$w_a^j(k) = \sum_{i \in \mathcal{J}^j} w_i(k)$$

that can be guaranteed over a specified horizon from time t. These limits necessarily depend on both resource storage rate and limitations among the individual consumers and as such depend in a complicated fashion on the horizon length. Several choices can be made with respect to the length of this horizon. A simple choice is to provide the limits for one sample ahead. Various choices could be made here, for instance providing a time-varying profile of limits over the control horizon, or the aggregators could simply provide fixed limits that can be sustained over the entire control horizon, although the latter would tend to be conservative. In addition to these limits, A_j provides $w^j_{\text{mid}}(k)$, a midranging signal that informs the high level controller of the total resource rate would currently be most helpful in bringing the intelligent consumers under its jurisdiction close to their reference resource levels $W_{i,\text{ref}}(k)$.

The aggregator level, as a whole, thus attempts to maintain $w_a(k) = \sum_{j=1}^{N_A} w^j_{\text{req}}(k)$, while the high level controller, in turn, needs to solve the optimization problem

$$\min_{w^j_{\text{req}}, w_{\text{ext}}} \sum_{k=1}^{N_h} \phi(W(k), w_{\text{ext}}(k), w_{\text{ext}}(k-1)) + \beta \sum_{j=1}^{N_A} \sum_{k=1}^{N_h} (w^j_{\text{req}}(k) - w^j_{\text{mid}}(k))^2 \quad (15.6)$$

$$\text{subject to} \underline{W} \le W(k) \le \overline{W}$$

$$\underline{w}^j(k) \le w^j_{\text{req}}(k) \le \overline{w}^j(k), \quad 1 \le j \le N_A$$

which is of significantly lower dimension than (15.1) because the number of decision variables is smaller (since $N_A < N$). The term

$$\beta \sum_{k=1}^{N_A} \sum_{k=1}^{N_h} (w^j_{\text{req}}(k) - w^j_{\text{mid}}(k))^2$$

is introduced to ensure that the high level controller will assign resources to the aggregators such that the intelligent consumers can approach their desired levels of storage, away from their individual limits; β is a constant to be specified later.

That is, in periods where the load is relatively steady, the high level controller will make w^j_{req} approach w^j_{mid}, thereby increasing the short term resource reserves for future load changes (positive or negative).

At each sample, the aggregator A_j solves the simple optimization problem

$$\min_{w_i} \sum_{i \in \mathcal{J}^j} (W_i(k+1) - W_{i,\text{ref}})^2 \quad (15.7)$$

$$\text{subject to} \sum_{i \in \mathcal{J}^j} w_i(k) = w^j_{\text{req}}(k)$$

$$\underline{w}_i \le w_i(k) \le \overline{w}_i$$

$$0 \le W_i(k+1) \le \overline{W}_i$$

with $W_i(k+1) = W_i(k) + T_s w_i(k)$, where T_s is the sampling time.

15.4 Stability, Complexity and Performance Analysis

In this section, we shall discuss stability, complexity and performance of the architecture proposed above.

15.4.1 Stability

First, in order to assess stability, it is of course necessary to establish a sensible definition, especially as the main obstruction to the standard definition is the presence of constraints.

Intuitively, stability for a system of the type described above will mean that

- for constant inputs, all trajectories tend to constant values;
- in steady state, a minimal number of constraints are invoked;
- wind-up behavior of system states is avoided for any bounded input set.

In the following, we will give an outline of a procedure for constructing controllers that stabilize the system in such a way that it satisfies these properties. For ease of the presentation we will only consider one aggregator.

Suppose the system satisfies the following assumptions.

1. The external load is constant;
2. The number of intelligent consumers is non-decreasing;
3. Any new ICs that appear in the system start with an initial amount of resource in storage equal to $W_{i,\text{ref}}$;
4. As long as the sum of all deviations from $W_{i,\text{ref}}$ does not increase, the constraints $\underline{w}$ and $\overline{w}$ do not become narrower (i.e., $\underline{w}$ does not increase, and $\overline{w}$ does not decrease).

The last assumption is technical; we aim to choose the reference levels precisely such that the constraints are as wide as possible, thus making sure that this assumption is satisfied by design. Indeed, in order to accommodate the stability notions introduced above, we will modify the performance objective sligthly, so that we may be able to follow a standard dual mode approach to stability analysis of model predictive control with terminal constraints [7].

First of all, we note that the overall system is a linear, constrained system. Therefore, at the top level we consider the state vector

$$\mathbf{x}(k) = \begin{bmatrix} W(k) \\ w_{\text{ext}}(k) - w(k) \\ W_{\Sigma}(k) \end{bmatrix}$$

where $W_{\Sigma}(k) = \sum_{i \in \mathcal{I}} (W_i(k) - W_{i,\text{ref}})$ denotes the total amount of surplus resources in the ICs. Next, we define the function

$$l(k) = \mathbf{x}(k)^T \mathbf{Q} \mathbf{x}(k) + R \Delta w_{\text{ext}}(k)^2$$

where $\mathbf{Q} \in \mathbb{R}^{3\times 3}$ and $R \in \mathbb{R}_+$ are constant weight factors, and $\Delta w_{\text{ext}}(k) = w_{\text{ext}}(k+1) - w_{\text{ext}}(k)$. If $\mathbf{Q}$ is chosen as a symmetric positive definite matrix, it is easily seen that l is a positive definite, smooth, unbounded function of $\mathbf{x}$ with minimum in $\mathbf{x} = 0$. Based on this function, we define the function

$$V(k_0) = \sum_{k=k_0+1}^{\infty} l(k) \tag{15.8}$$

along with the control optimization

$$\begin{aligned} &\min_{w_{\text{ext}}, w_{req}} V(k_0) \\ &\text{subject to} W_\Sigma(k_0 + N_h) = 0 \\ &\underline{w}(k) \leq w_{\text{req}}(k) \leq \overline{w}(k) \end{aligned} \tag{15.9}$$

where (15.9) is a terminal constraint.

Given the properties of $l(k)$, we see that V can be used as a Lyapunov function, i.e., if we can ensure that it decreases every sample, the closed loop will be stable.

Assuming that the constraints are not active after $k_0 + N_h$, the optimal trajectory will be described by the dynamics $\mathbf{x}(k+1) = \tilde{\mathbf{A}}\mathbf{x}(k)$, where $\tilde{\mathbf{A}}$ can be found as the closed-loop matrix resulting from a standard LQR problem. By construction, all eigenvalues of $\tilde{\mathbf{A}}$ have modulus less than one, so we can find a symmetric positive definite matrix $\bar{\mathbf{Q}}$ that solves the discrete Lyapunov equation

$$\tilde{\mathbf{A}}^T \bar{\mathbf{Q}} \tilde{\mathbf{A}} = \bar{\mathbf{Q}} - \mathbf{Q}$$

We can then write

$$V(k_0) = \sum_{k=k_0+1}^{k_0+N_h} l(k) + \sum_{k=k_0+N_h+1}^{\infty} l(k) = \sum_{k=k_0+1}^{k_0+N_h} l(k) + \mathbf{x}(k_0+N_h)^T \bar{\mathbf{Q}} \mathbf{x}(k_0+N_h) \tag{15.10}$$

which means that we perform the optimization on a finite horizon with an extra weight on the terminal state. In order to realize that V is decreasing, it is enough to note that as the horizon recedes more flexibility is added to the optimization, meaning that $\min V(k_0+1) \leq \min V(k_0) - l(k)$.

The reason that we can assume that constraints are not active at the end of the control horizon follows from Assumption 4 and the terminal constraint (15.9).

Thus, under the assumptions above, we can say the following:

- Each aggregator drives the amount of resource stored in the ICs under its jurisdiction towards their reference values.
- In steady state, the minimal number of constraints are active. This follows from the properties of quadratic optimization; if the number of active constraints is nonminimal, the quadratic cost will always become smaller by shifting load from

one of the subsystems with an active constraint to one or more subsystems with inactive constraints.

- Wind-up behavior of system states is avoided for any bounded input set, since all the individual subsystems are open loop (marginally) stable.

It should finally be noted that the terminal constraint (15.9) was only included in the above to make the argumentation easier; it does not appear to be necessary in practical implementations and has not been included in the examples in Sec. 15.5.

15.4.2 Complexity

In terms of complexity, the proposed design methodology scales in the same way as quadratic optimization, which is $O(N^3)$, where N is the number of consumers.

It should be noted, however, that the optimization problem has a high degree of sparsity. This has not been exploited in the implementation applied in the simulations below, but it should be expected that the complexity could be further reduced by implementing a dedicated quadratic programming solver, which exploits the mentioned sparsity. Further considerations on complexity can be found in Sec. 15.5.2.

15.4.3 Performance

In terms of performance, the following five parameters are decisive:

- the prediction horizon;
- the total installed flexible capacity;
- the instantaneous flexible capacity;
- the total cumulative rate limitation of flexible units;
- the instantaneous cumulative rate limitation of flexible units

The *prediction horizon* is a crucial parameter, since the overall balance W is the result of an integration. This means that for longer horizons the potential of using flexible units becomes much larger, since the slack variables relative to the saturation limits needed for guaranteed stabilization of the integrator becomes much smaller for a longer time horizon.

The *total installed flexible capacity* is the sum of maximal resource storages for all units, i.e., $C_{\text{tot}} = \Sigma_{i=1}^{N} \overline{W}_i$. This capacity clearly scales with the number of units.

The *instantaneous flexible capacity* is the present unexploited part of C_{tot}. Since flexibility is exploited bidirectionally in reaction to either increasing or decreasing load, C_{tot} has to be divided between upwards and downwards movement. The dynamics of this quantity depends on the control algorithm and of the control horizon. Due to the additive nature of the quadratic programming cost function, the instantaneous capacity for the proposed algorithm scales linearly with the number of units, which is clearly optimal.

The *total cumulative rate limitation of flexible units* is the rate limitation experienced by the high level controller and equals $\sum_{i=1}^{N} \overline{w}_i$ for positive load gradients and $\sum_{i=1}^{N} \underline{w}_i$ for negative load gradients. This parameter scales linearly with the number of installed units.

The *instantaneous cumulative rate limitation of flexible units* is current rate limitation experienced by the high level controller and is equal to the sum of individual rate limits for those units, which are not in saturation. Again, due to the additive nature of quadratic programming costs, the instantaneous rate limitation scales linearly with the number of installed units. The average ratio (for a given load pattern) between instantaneous and total cumulative rate limitations is controlled by the weighting factor ρ, which constitutes the trade-off between capacity limitation and rate limitation. For a given load pattern, more average capacity can be obtained at the cost of rate limitation and vice versa, but the quadratic optimization guarantees Pareto optimality.

The sampling times used at aggregator and top levels also influences performance. Since the dynamics consist entirely of pure integrators, there is no approximation in the discretization itself, but of course the flexibility in the optimization will be smaller for a larger sampling time.

15.5 Simulation Example

The example described in this section is inspired by a vision for future Smart Grid technologies called *virtual power plants*, which is depicted in Figure 15.3.

The main objective of the top level control is to keep the energy balance governed by

$$\frac{dE(t)}{dt} = P_{\text{ext}}(t) - P_{\text{load}}(t) - P_{\text{a}}(t) \tag{15.11}$$

at zero. $P_a = \sum_i P_i$ is the power absorbed by the intelligent consumers (ICs). P_{load} is the power absorbed by other consumers, and is considered as a disturbance here. P_{ext} is the power produced by a number of suppliers such as power plants. It is assumed that the top level controller can control P_{ext} directly and is restrained only by a rate limit, but we would also like to keep the time derivative small.

Each intelligent consumer is characterized by its own energy balance

$$\frac{dE_i(t)}{dt} = P_i(t) \tag{15.12}$$

which must satisfy $0 \le E_i(t) \le \overline{E}_i$ at all times. Furthermore, each intelligent consumer can only consume a limited amount of power $\underline{P}_i \le P_i(t) \le \overline{P}_i$. The aggregator serves as an interface between the top level and the ICs. It attempts to maintain $P_a(t) = P_{\text{req}}(t)$ and provides the top level with simple parameters to specify the con-

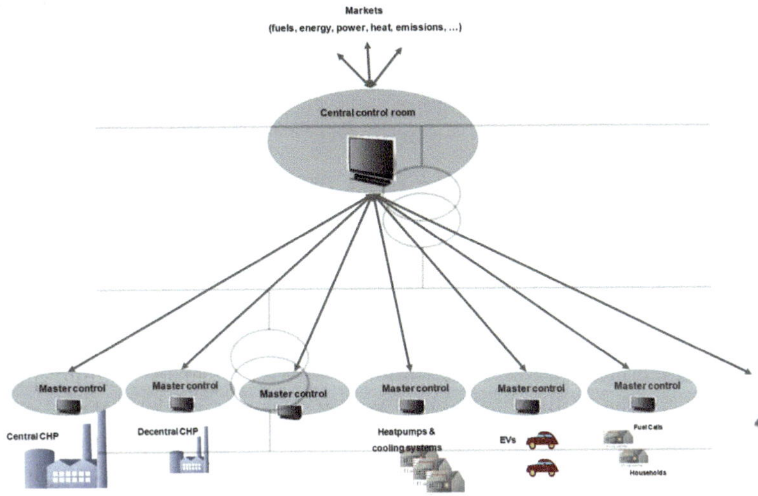

Fig. 15.3 A vision for smart grids: virtual power plants that aggregate producing or consuming units.

straints of the ICs. In particular, the top level is informed of $\overline{P}$ and $\underline{P}$, upper and lower limits on P_a that can be guaranteed over the horizon N_l. These limits depend on both power and energy storage limitations, and as such depend in a complicated fashion on the horizon length. In addition to the limits, the aggregators provide P_{mid}, a midranging signal that tells the top level which P_{req} would be most helpful in bringing the ICs close to their reference energy levels $E_{\text{ref},i}$. In periods where the load is relatively steady, the top level can then prioritize, keeping the energy levels at the reference and thereby increasing the short term reserves for future load changes.

How to choose these reference levels is again a complicated question of the considered horizon. If we consider a long horizon, then we might like to have the same energy reserve in both directions, which would lead to $E_{\text{ref},i} = \overline{E}_i/2$. On the other hand, some ICs have a much higher $\overline{P}$ than $-\underline{P}$, and are therefore much better at providing a positive than negative absorption, while others are better at providing negative absorption. With a short horizon it would make sense to keep the first kind at a low energy level and vice versa. Here, we choose

$$E_{\text{ref},i} = \overline{E}_i \frac{\overline{P}_i}{\overline{P}_i - \underline{P}_i}$$

which corresponds to making the time to fill the energy reserves equal to the time to fully empty it.

At each sample, at time t, the aggregator solves the simple optimization problem

$$\begin{aligned} \min_{P_i} \quad \sum & \quad (E_i(t+T_s) - E_{i,\mathrm{ref}})^2, \\ & \text{subject to} \\ \sum P_i & = P_{\mathrm{req}}, \\ \underline{P}_i & \leq P_i(t) \leq \overline{P}_i, \\ 0 & \leq E_i(t+T_s) \leq \overline{E}_i \end{aligned}$$

with $E_i(t+T_s) = E_i(t) + T_s P_i$, thereby distributing the power in a way that brings the energy levels as close to the reference as possible in a quadratic sense.

The top level control optimizes over a prediction horizon N_p. It minimizes the performance function

$$J_t = \sum_{k=1}^{N_p} E(t+T_s k)^2 + \beta_p \sum_{k=1}^{N_c} (P_{\mathrm{ext}}(t+T_s k) - P_{\mathrm{ext}}(t+T_s(k-1)))^2 + \beta_r \sum_{k=1}^{N_c} (P_{\mathrm{req}}(t+T_s k) - P_{\mathrm{mid}}(t))^2$$

with N_c samples of P_{ext} and P_{req} as decision variables.

The optimization is subject to constraints on the decision variables. There is a rate limit on the power from the power plants:

$$\underline{P}_{\mathrm{ext}} \leq P_{\mathrm{ext}}(t+T_s k) - P_{\mathrm{ext}}(t+T_s(k-1)) \leq \overline{P}_{\mathrm{ext}}$$

As mentioned, the aggregator provides limits on P_a that can be sustained over a horizon N_l. These limits are conservative in the sense that if P_{req} is, for instance, negative for the first part of the horizon, then a positive P_{req} higher than $\overline{P}$ may be feasible for the rest. However, in order to simplify the top level computations, the constraint $\underline{P}(t) \leq P_{\mathrm{req}}(t+kT_s) \leq \overline{P}(t)$ is imposed over the whole horizon. A simulation of this scheme is shown in Fig. 15.4. The controller parameters used are $T_s = 1$, $N_l = N_c = 4$, $N_p = 5$, $\beta_p = 0.1$, $\beta_r = 10^{-4}$. The load is generated by a first order auto-regressive process with a time constant of 100 s, driven by zero-mean Gaussian white noise with unit variance. There are 20 ICs with parameters shown in Table 15.1 becoming available as time passes, making it possible for the aggregator to provide increasingly wider constraints on P_{req}. The result is that the energy balance can be controlled much better while also using a smoother P_{ext}. The requested consumption P_{req} is shown together with $\underline{P}(t)$ and $\overline{P}(t)$, computed by the aggregator. It is noted how the constraints widen as more ICs become available, but will shrink when the reserve is being used. P_{mid} is computed as the P_{req} that would bring the energy levels to the reference in N_l samples, ignoring power limits.

The energy balance of the ICs is shown in Fig. 15.5. The energy constraints and reference are shown by dashed lines. It can be seen that additional consumers are "plugged in," the system automatically incorporates these new consumers and these

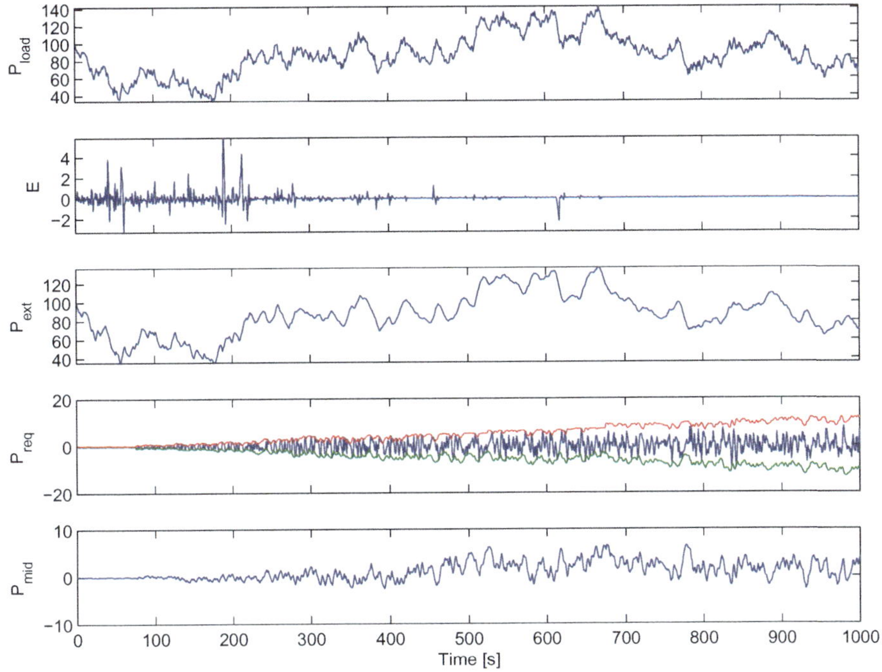

Fig. 15.4 Simulation example.

Table 15.1 Parameters for 20 consumers in simulation

i	1	2	3	4	5	6	7	8	9	10
$\overline{E}_i$	1.0	4.0	4.0	3.0	6.0	10.0	1.0	4.0	9.0	10.0
$\underline{P}_i$	-1.7	-1.4	-0.2	-1.3	-1.6	-1.3	-0.7	-1.9	-1.1	-1.1
$\overline{P}_i$	1.4	0.8	1.8	0.3	0.9	1.1	1.2	0.2	0.2	0.2

i	11	12	13	14	15	16	17	18	19	20
$\overline{E}_i$	9.0	1.0	2.0	10.0	6.0	1.0	9.0	8.0	2.0	9.0
$\underline{P}_i$	-0.2	-1.0	-1.6	-1.3	-0.3	-1.1	-1.9	-0.2	-0.9	-1.6
$\overline{P}_i$	1.1	1.2	1.6	1.9	0.4	0.9	1.8	0.8	0.6	0.5

new resources are exploited throughout the control hierarchy in order to improve the power balance at the top level.

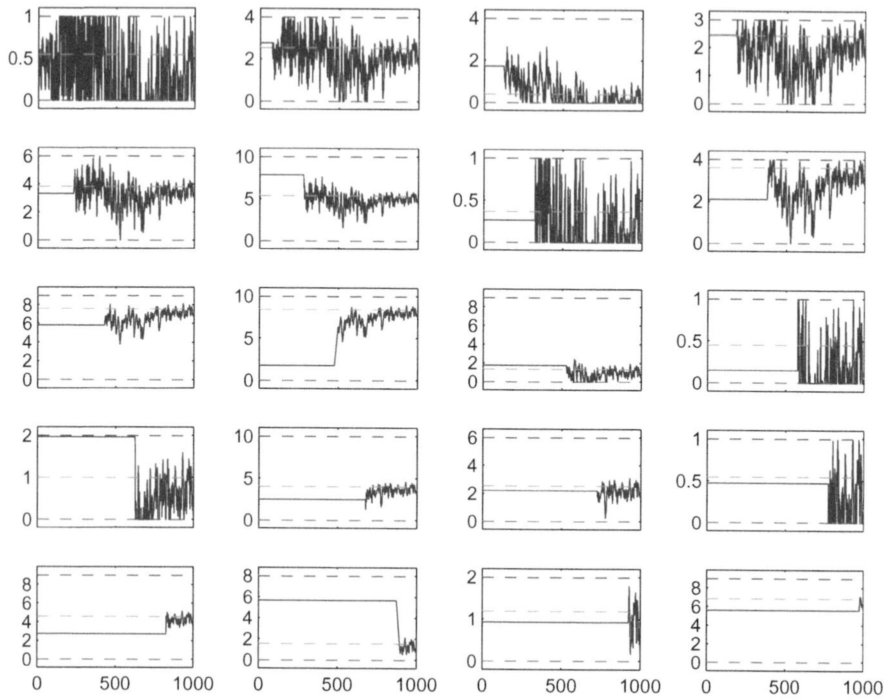

Fig. 15.5 Simulation with aggregators.

15.5.1 Performance Study

The aggregators perform two functions, approximation and aggregation. The main purpose is the aggregation of several consumers into a simple virtual big consumer, thereby simplifying the task at the top level. The approximation, simplifying the power limits into an average over the horizon, is only necessary to facilitate the aggregation. In fact, if each aggregator has only one consumer, then the computation at the top level is not simplified at all. In the next section, the effects of aggregation on the performance will be studied, but before that the question arises of how much the conservative approximation affects performance compared to a centralized scheme as in Fig. 15.1, where the high level controller directly controls the consumers. We compare two control schemes:

Centralized controller: The top level controller optimizes a standard MPC objective

$$\min_{P_i} \sum_{k=t+1}^{t+N_p} (E(k)^2 + \beta_p(P_{\text{ext}}(k) - P_{\text{ext}}(k-1))^2 + \beta_e \sum (E_i(k) - E_{\text{ref},i})^2)$$

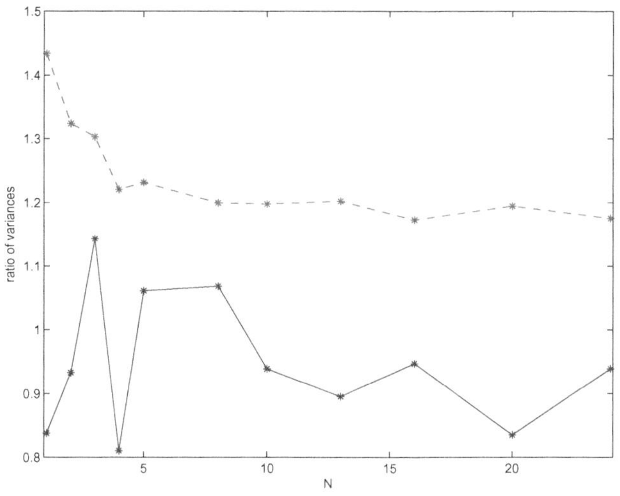

Fig. 15.6 Performance comparison with a centralized control for increasing number of aggregators. Solid: ratio of variances of balance. Dashed: ratio of variances of derivative of external power.

for $i = 1,\dots,N$, over the consumption rates of all consumers over a horizon N_p. The control is not fully optimal, since the horizon is finite. Therefore, the last term is used for keeping the energy levels close to the references if the reserves are not needed immediately.

Approximating control: The scheme described in the previous section, but each aggregator handles only one consumer. In this way, the comparison will reflect the effects of the approximation.

We perform simulations on a system with a small number of consumers, N. The consumer power limits are evenly distributed between $\pm 0.4/N$ and $\pm 2.4/N$, the maximum energy levels between 0.8 and 4.8. $\underline{P}_{\text{ext}} = -0.5$, $\overline{P}_{\text{ext}} = 0.5$. The load follows the same behavior as in the above example.

The approximating control has the same parameters as in the above example. The centralized controller has the same control horizon and performance weights, and $\beta_e = 10^{-4}$.

For each particular N, 100 simulations of 200 samples are performed. For each simulation the ratio of variances is computed. Figure 15.6 shows the mean of these ratios as N grows. The solid line shows the ratio between the variance of E when the approximating control is used and when using the centralized control. The dashed line shows the same but for the variance of $P_{\text{ext}}(k) - P_{\text{ext}}(k-1)$. It is noted that the ratios are quite close to 1, meaning that the performance of the approximating control is almost as good as for the centralized control. Importantly, the ratios seem to decrease towards 1 as N grows. It was not feasible to perform the simulations for higher N, as the computational complexity grew too high. The result leads us to conjecture that the approximation only has a small effect on the performance, and that this effect is unimportant for a large number of consumers.

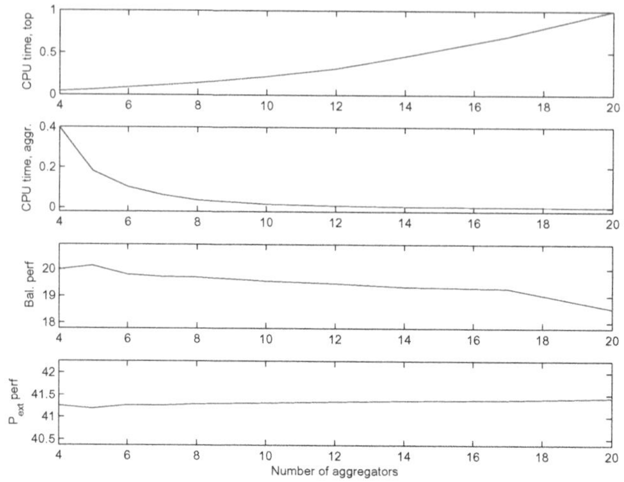

Fig. 15.7 CPU times for simulation with varying numbers of aggregators.

15.5.2 Complexity Study

The aggregators serve as a simplifying interface to the relatively complex top level control, and as such even a configuration with one aggregator is computationally less complex than it would be to let the top level control work on a full model. However, for a large number of ICs, the aggregator can also itself become too complex. It is therefore necessary to allow for more than one aggregator. This also provides the top level with more detailed information and can therefore be expected to yield better performance. On the other hand, more aggregators will make the top level control more complex, so there is a trade-off between complexity at the top and aggregator levels and also with respect to performance.

Here, we examine the effects of the number of aggregators, N_A, through a simulation example. We consider a situation with 800 ICs with the E_is evenly distributed between 0.01 and 0.14, and $\bar{P}_i$s and $-\underline{P}_i$s evenly distributed between 0.01 and 0.06. The other parameters are $T_s = 1$, $N_l = N_c = 4$, $N_p = 5$, $\beta_p = 1$, $\beta_r = 10^{-4}$. The load is generated by a discrete time process $(1-0.99q^{-1})(P_{\text{load}}(k)-100) = e(k)$, where q^{-1} is the delay operator and e is white Gaussian noise with variance 16.

In all the simulations the same 400 sample load sequence was used, only N_A was changed. Figure 15.7 shows the result. The top plot shows the (scaled) time consumption of the top level controller. This grows with N_A^3. The second plot uses the same scaling and shows the average time consumption of each of the aggregators. As the number of ICs handled by each aggregator is inversely proportional to the number of aggregators, this consumption is inversely proportional to N_A^3. It is noted that the computational complexity of the top level and of each of the aggregators is approximately equal with around 6 aggregators, so this may be a sensible choice.

The variance of the balance E and of the derivative of P_{ext} are shown in the next two plots. As expected, more aggregators give better performance, but the difference is rather small.

15.6 Discussion

In this chapter a design methodology for a three level hierarchical control architecture was proposed. The emphasis was on systems that accumulate the production and/or consumption of resources through the levels, exemplified by irrigation systems, sewer systems, or power production and consumption systems.

The presented solution was based on MPC-like algorithms, which themselves are based on on-line quadratic programming solvers. The algorithmic complexity is very low and approximately scales with the number of units in the system to the power of 1.5, even without exploiting a significant sparsity of the optimization problems involved.

The approach has the specific feature that it facilitates online modifications of the topography of the controlled system. In particular, units at the lower level can be added or removed without any retuning of any controllers. This plug-and-play control property is enabled by the modular structure of the involved cost functions of the optimizations.

The proposed methodology was exemplified by a simulation of a control system for a small electrical power production and consumption system, where the power flexibility of a number of consumers was exploited. For this example, a significant improvement of flexibility at the grid level was obtained.

Acknowledgements This work was supported by The Danish Research Council for Technology and Production Sciences.

References

1. Beckmann, C.J.: Modeling integrated services traffic for bandwidth management in next-generation intranets. In: Proc. 1997 5th Int. Symposium on Modeling, Analysis and Simulation of Computer and Telecommunication Systems, pp. 154–160. Haifa, Israel (1997)
2. De Kok, A., Graves, S.: Supply chain management: Design, coordination and operation. In: A. De Kok (ed.) Handbooks in Operations Research and Management Science: Vol. 11. Elsevier (2003)
3. De Persis, C., Kallesoe, C.S.: Quantized controllers distributed over a network: An industrial case study. In: Proc. Mediterranean Control Conference '09, pp. 616–621. Thessaloniki, Greece (2009)
4. Edlund, K., Sokoler, L.E., Jørgensen, J.B.: A primal-dual interior-point linear programming algorithm for MPC. In: Proc. 48th IEEE Conf. Decision and Control (CDC2009), pp. 351–356. Shanghai, China (2009)
5. Karlsson, K., Engstrom, I.: Controlling power and process steam. International Power Generation **26**(2), 29–32 (2003)

6. Knudsen, T.: Incremental data driven modelling for plug-and-play process control. In: Proc. 47th IEEE Conf. Decision and Control (CDC2008), pp. 4683–4688. Cancun, Mexico (2008)
7. Maciejowski, J.M.: Predictive Control with Constraints. Prentice Hall (2002)
8. Moslehi, K., Kumar, R.: A reliability perspective of the smart grid. IEEE Trans. Smart Grid **1**(1), 57–64 (2010)
9. Picasso, B., De Vito, D., Scattolini, R., Colaneri, P.: An MPC approach to the design of two-layer hierarchical control systems. Automatica **46**(5), 823–831 (2010)
10. Rao, C.V., Campbell, J.C., Rawlings, J.B., Wright, S.J.: Control of induction motor drives. In: Proc. Int. Conf. Energy Effective and Smart Motors—Development Perspectives and Research Needs. Aalborg University, Institute of Energy Technology, Aalborg University, Institute of Energy Technology, Pontoppidanstræde 101, 9220 Aalborg Ø (1992)
11. Rossiter, J.A.: Model-based predictive control. CRC Press, Boca Raton, FL (2003)
12. Stoustrup, J.: Plug & play control: Control technology towards new challenges. European J. Control **15**(4), 311–330 (2009)
13. Zhang, Y., Zhang, J., Guan, J.: Study on precision water-saving irrigation automatic control system by plant physiology. In: Proc. IEEE Conf. Industrial Electronics and Applications, pp. 1296–1300. Kuala Lumpur, Malaysia (2009)
14. Zheng, L., Larsen, L.F.S., Izadi-Zamanabadi, R., Wisniewski, R.: Analysis of synchronization of supermarket refrigeration system—Benchmark for hybrid system control. Kybernetes (2011)

Chapter 16
Hierarchical Model-Based Control for Automated Baggage Handling Systems

Alina N. Tarău, Bart De Schutter and Hans Hellendoorn

Abstract This chapter presents a unified and extended account of previous work regarding modern baggage handling systems that transport luggage in an automated way using destination-coded vehicles (DCVs). These vehicles transport the bags at high speeds on a network of tracks. To control the route of each DCV in the system we first propose centralized and distributed predictive control methods. This results in nonlinear, nonconvex, mixed integer optimization problems. Therefore, the proposed approaches will be expensive in terms of computational effort. As an alternative, we also propose a hierarchical control framework where at higher control levels we reduce the complexity of the computations by simplifying and approximating the nonlinear optimization problem by a mixed integer linear programming (MILP) problem. The advantage is that for MILP problems, solvers are available that allow us to efficiently compute the global optimal solution. To compare the performance of the proposed control approaches we assess the trade-off between optimality and CPU time for the obtained results on a benchmark case study.

Alina N. Tarău
Delft University of Technology, Delft Center for Systems and Control, Mekelweg 2, 2628 CD Delft, The Netherlands
e-mail: a.n.tarau@tue.nl

Bart De Schutter
Delft University of Technology, Delft Center for Systems and Control, Mekelweg 2, 2628 CD Delft, The Netherlands
e-mail: b.deschutter@tudelft.nl

Hans Hellendoorn
Delft University of Technology, Delft Center for Systems and Control, Mekelweg 2, 2628 CD Delft, The Netherlands
e-mail: j.hellendoorn@tudelft.nl

R. Johansson and A. Rantzer (Eds.): Distributed Decision Making and Control, LNCIS 417, pp. 359–386.
springerlink.com

Fig. 16.1 DCVs running on a network of tracks (Photo courtesy of Vanderlande Industries)

16.1 Introduction

The state-of-the-art technology used by baggage handling systems at airports to transport the bags in an automated way incorporates scanners that scan the (electronic) baggage tags on each piece of luggage, baggage screening equipment for security scanning, networks of conveyors equipped with junctions that route the bags through the system and destination coded vehicles (DCVs). As illustrated in Fig. 16.1, a DCV is a metal cart with a plastic tub on top. These carts are propelled by linear induction motors mounted on the tracks. The DCVs transport the bags at high speed on a network of tracks. Note that the DCVs are used in large airports only, where the distances between the check-in desks and the endpoints towards which the baggage has to be transported are very large (for these airports the conveyor systems are too slow, and therefore, a faster carrier is required for each bag).

In this chapter we consider a DCV-based baggage handling system. Higher-level control problems for such a system are route assignment for each DCV (and, implicitly, the switch control of each junction), line balancing (i.e., route assignment for each empty DCV such that all the loading stations have enough empty DCVs at any time instant) and prevention of buffer overflows. The velocity control of each DCV is a low level control problem. Low level controllers determine the velocity of each DCV so that a minimum safe distance between DCVs is ensured and so that the DCVs are held at switching points, if required. So, a DCV runs at maximum speed, $v^{\max}$, unless overruled by the local on-board collision avoidance controller. Other low level control problems are coordination and synchronization when a bag is loaded onto a DCV (in order to avoid damaging the bags or blocking the system), and when unloading it to its endpoint. We assume low level controllers are already present in the system, and we focus on the higher-level control problems of a DCV-based baggage handling system, in particular the route assignment of the DCVs.

Currently, the track networks on which the DCVs transport the baggage have a simple structure, with the loaded DCVs being routed through the system using routing schemes based on preferred routes. These routing schemes adapt to respond to the occurrence of predefined events. However, the load patterns of the system are highly variable, depending on, e.g., the season, time of the day, type of aircraft at each gate or number of passengers for each flight [11]. Also, note that the first objective of a baggage handling system is to transport all the checked-in or transfer bags to the corresponding endpoints before the planes have to be loaded. However, due to an airport's logistics, an endpoint is allocated to a plane only within a given time span before the plane's departure. Hence, the baggage handling system performs optimally if each of the bags to be handled arrives at its given endpoint within a specific time window. So, predefined routes are far from optimal. Therefore, we will not consider predefined preferred routes, but instead we will develop and compare efficient control methods to determine the optimal routing of the DCVs.

In the literature, the route assignment problem has been addressed to a large extent for automated guided vehicles (AGVs); see e.g., [9, 12]. Traditionally, the AGVs that execute the transportation tasks are controlled by a central server via wireless communication. Hence, the computational complexity of the centralized routing controller increases with the number of vehicles to be routed. In this context [18] presents a decentralized architecture for routing AGVs through a warehouse. However, even for a small number of AGVs to be used for transportation (12 AGVs), the communication requirements are high. But, in baggage handling systems the number of DCVs used for transportation is large (typically airports with DCV-based baggage handling systems have more than 700 DCVs). Hence, in practice, designing an on-board routing controller for each DCV is not yet a tractable problem. Also, we do not deal with a shortest-path or shortest-time problem, since, due to the airport's logistics, we need the bags at their endpoints within given time windows.

The DCV routing problem has been presented in [3], where an analogy to data transmission via Internet is proposed, and in [8] where a multiagent hierarchy has been developed. However, the analogy between routing DCVs through a track network and transmitting data over Internet has limitations [3], while the latter reference [8] does not focus on control approaches for computing the optimal route of DCVs, but on designing a multiagent hierarchy for baggage handling systems and analyzing the communication requirements. Moreover, the multiagent system of [8] faces major challenges due to the extensive communication required. Therefore, the goal of our work is to develop and compare efficient control approaches (viz., predictive control methods) for routing each DCV transporting bags to its end point. This chapter integrates results of previous work [14, 16, 17], presents all the methods that previously proved to be efficient and compares the obtained results for a benchmark case study over typical and extreme scenarios. Moreover, we address the trade-off between accuracy of the overall performance of the system and the total computational effort required to compute the optimal solution.

This chapter is structured as follows. In Sec. 16.2 we describe the automated baggage handling process. Next, in Sec. 16.3, we present the control objective that will be later on used when solving the DCV routing problem. Furthermore, in Sec. 16.4, we propose several control approaches for determining the optimal route for each bag through the baggage handling network. First we develop and compare centralized and distributed predictive methods that could be used to maximize the per-

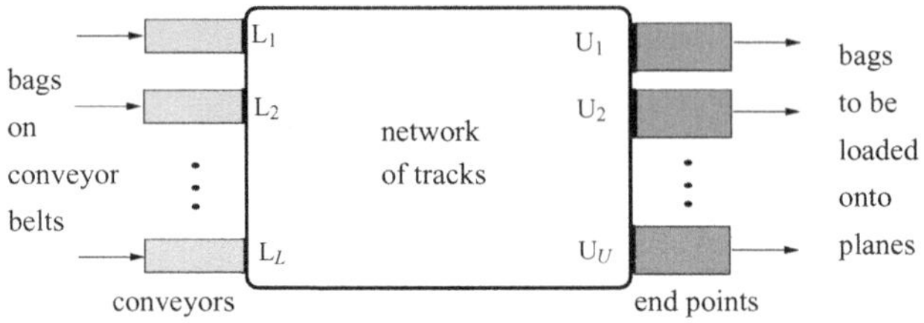

Fig. 16.2 Baggage handling system using DCVs.

formance of the DCV-based baggage handling system. But these methods involve nonlinear, nonconvex, mixed integer optimization problems that are very expensive to solve in terms of computational effort. Therefore, we also propose an alternative approach for reducing the complexity of the computations by simplifying the nonlinear optimization problem and writing it as a mixed integer linear programming (MILP) optimization problem, for which solvers are available that allow us to efficiently compute the global optimal solution. This approach will then be incorporated in a hierarchical control framework for routing the DCVs. The analysis of the simulation results and the comparison of the proposed control methods and control frameworks are elaborated in Sec. 16.5. Finally, in Sec. 16.6 we draw conclusions and present possible directions for future research.

16.2 System Description and Original Model

Now we briefly recapitulate the event-driven route choice model of a baggage handling system that we have developed in [13]. The nodes via which the DCVs enter the track network are called *loading stations*, the nodes via which the DCVs exit the network are called *unloading stations*, while all the other nodes in the network are called *junctions*. The section of track between two nodes is called *link*.

Consider the general DCV-based baggage handling system with L loading stations and U unloading stations sketched in Fig. 16.2. The DCV-based baggage handling system operates as follows: Given a demand of bags and the network of tracks, the route of each DCV (from a given loading station to the corresponding unloading station) has to be computed subject to operational and safety constraints such that the performance of the system is optimized.

The model of the baggage handling system we have developed in [13] consists of a continuous part describing the movement of the individual vehicles transporting the bags through the network and of the following discrete events: loading a new bag onto a DCV, unloading a bag that arrives at its endpoint, updating the position of the switches into and out of a junction and updating the speed of a DCV. The state of the system consists of the positions of the DCVs in the network and the positions of each switch of the network. According to the discrete-event model of [13], as long as there are bags to be handled, the system evolves as follows: We shift the current time to the next event time, take the appropriate action and update the state of the system.

Let DCV_i denote the DCV that transports the ith bag that entered the track network up to the current time instant. According to the model, for each bag that has to be handled, we compute the time instants when each bag enters and exits the track network. Let t_i^{load} denote the time instant when the ith bag that entered the track network is loaded onto a DCV (so, this is DCV_i) and let t_i^{unload} denote the time instant when the same bag is unloaded at its endpoint. Then we consider two models of the baggage handling system which will be used for (1) route control—where we determine a route for each DCV, and consequently, the switch will be positioned so that each DCV travels on the assigned route and (2) switch control—where we determine switch positions over the simulation period, respectively:

$$\mathbf{t} = \mathcal{M}^{\text{route\_ctrl}}\left(\mathcal{T}, \mathbf{x}(t_0), \mathbf{r}\right)$$

or

$$\mathbf{t} = \mathcal{M}^{\text{switch\_ctrl}}\left(\mathcal{T}, \mathbf{x}(t_0), \mathcal{U}\right)$$

where:

- $\mathbf{t} = [t_1^{\text{load}} \ \dots \ t_{N^{\text{bags}}}^{\text{load}} \ t_1^{\text{unload}} \ \dots \ t_{N^{\text{bags}}}^{\text{unload}}]^T$ with N^{bags} the number of bags to be handled in the given simulation period;
- $\mathcal{T}$ is the tuple that consists of the arrival times at loading stations for all the bags to be handled;
- $\mathbf{x}(t_0)$ is the initial state of the system with t_0 the initial simulation time;
- $\mathbf{r}$ is the route control sequence defined as follows: assume that there is a fixed number R of possible routes from a loading station to an unloading station and that the R routes are numbered $1, 2, \dots, R$. Let $r(i) \in \{1, 2, \dots, R\}$ denote the route of DCV_i. Then the route sequence is represented by $\mathbf{r} = [r(1)\, r(2) \cdots r(N^{\text{bags}})]^T$;
- $\mathcal{U}$ is the switch control input for the entire network defined as $\mathcal{U} = (\mathbf{u}_1, \cdots, \mathbf{u}_S)$ with

  $$\mathbf{u}_s = [u_s^{\text{sw\_in}}(1) \cdots u_s^{\text{sw\_in}}(N^{\text{bags}})\, u_s^{\text{sw\_out}}(1) \dots u_s^{\text{sw\_out}}(N^{\text{bags}})]^T \text{ for } s = 1, \dots, S$$

  where S is the number of junctions and where $u_s^{\text{sw\_in}}(j)$ is the position of the switch into junction S_s when the jth bag crosses S_s and $u_s^{\text{sw\_out}}(j)$ is the position of the switch out junction S_s when the jth bag crosses S_s.

  Without loss of generality (i.e., by creating virtual junctions connected by virtual links of zero length) we can assume each junction to have at most 2 incoming links (indexed by the labels 0 and 1) and at most 2 outgoing links (also indexed by 0 and 1). We call the switch that makes the connection between a junction and its incoming links a switch-in, and the switch that makes the connection between a junction and its outgoing links a switch-out.

The operational constraints derived from the mechanical and design limitations of the system are the following: The speed of each DCV is bounded between 0 and $v^{\max}$, while a switch at a junction has to wait at least τ^{switch} time units between two consecutive toggles in order to avoid the quick and repeated back and forth movements of the switch, which may lead to mechanical damage. We assume τ^{switch} to be an integer multiple of τ_{s} where τ_{s} is the sampling time. In this chapter we denote the operational constraints by $\mathcal{C}(\mathbf{t}) \leq 0$.

16.3 Control Objective

Since the baggage handling system performs successfully if all the bags are transported to their endpoint before a given time instant, from a central point of view, the primary objective is the minimization of the overdue time. A secondary objective is the minimization of the additional storage time at the endpoint. This objective is required due to the intense utilization of the endpoints in a busy airport. Let N^{bags} be the number of bags that the baggage handling system has to handle. Hence, one way to construct the objective function J_i^{pen} corresponding to the bag with index i, $i \in \{1,2,\ldots,N^{\text{bags}}\}$, is to penalize the overdue time and the additional storage time. Accordingly, we define the following penalty for bag i:

$$J_i^{\text{pen}}(t_i^{\text{unload}}) = \sigma_i \max(0, t_i^{\text{unload}} - t_i^{\text{close}}) + \lambda_1 \max(0, t_i^{\text{close}} - \tau_i^{\text{open}} - t_i^{\text{unload}}) \quad (16.1)$$

where t_i^{close} is the time instant when the endpoint of bag i closes and the bags are loaded onto the plane, σ_i is the static priority of bag i (the flight priority), and τ_i^{open} is the maximum possible length of the time window for which the endpoint corresponding to bag i is open for that specific flight. The weighting parameter $\lambda_1 > 0$ expresses the penalty for the additionally stored bags. Note that the control actions involved in $\mathbf{r}$ and $\mathcal{U}$ influence J_i^{pen} in the sense that they influence the time instant when bag i is unloaded. Moreover, all approaches that we propose have in common the fact that we are interested in t_i^{unload} with respect to the given unloading time window. Therefore, we have chosen t_i^{unload} as argument of J_i^{pen}.

Moreover, the above performance function has some flat parts, which yield difficulties for many optimization algorithms. Therefore, in order to get some additional gradient and also minimize the energy consumption, we also include the time that a bag spends in the system. This results in (Fig. 16.3):

$$J_i(t_i^{\text{unload}}) = J_i^{\text{pen}}(t_i^{\text{unload}}) + \lambda_2 (t_i^{\text{unload}} - t_i^{\text{load}}) \quad (16.2)$$

where λ_2 is a small weight factor ($0 < \lambda_2 \ll 1$). The final objective function to be used when we compare the proposed control approaches is given by

$$J^{\text{tot}}(\mathbf{t}) = \sum_{i=1}^{N^{\text{bags,sim}}} J_i^{\text{pen}}(t_i^{\text{unload}}) \quad (16.3)$$

where $N^{\text{bags,sim}}$ is the number of bags that reached their endpoint during the simulation period $[t_0, t_0 + \tau^{\text{sim}})$, where t_0 the initial simulation time and τ^{sim} is either

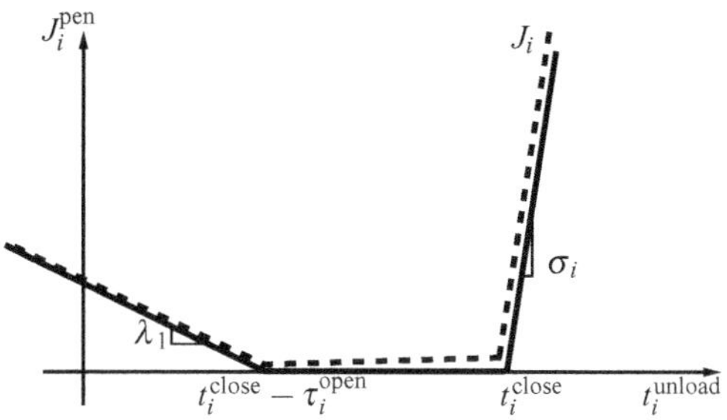

Fig. 16.3 Objective functions J_i^{pen} and J_i.

the time instant when all the bags have been handled (and then $N^{\mathrm{bags,sim}} = N^{\mathrm{bags}}$) or $\tau^{\mathrm{sim}} = \tau^{\mathrm{max\_sim}}$, with $\tau^{\mathrm{max\_sim}}$ the maximum simulation period.

16.4 Control Methods

In this section we develop and compare centralized and distributed predictive methods that could be used to optimize the performance of the system. The centralized control method results in a nonlinear, nonconvex, mixed integer optimization problem that is very expensive to solve in terms of computational effort. Therefore, we also propose an alternative approach for reducing the complexity of the computations by approximating the nonlinear optimization problem by a mixed integer linear programming (MILP) problem. The MILP approach will then be incorporated in a hierarchical control framework.

16.4.1 Centralized MPC

Since later on we will use model predictive control (MPC) for determining the routes of the DCVs in the network, in this section we first briefly introduce the basic concepts of MPC.

MPC is an on-line model-based control design method; see e.g., [10]. It uses a receding horizon principle. In the basic MPC approach, given a horizon N, at step $k \geq 0$, where k is integer-valued, corresponding to the time instant $t_k = k\tau_{\mathrm{s}}$ with τ_{s} the sampling time, the future control sequence $u(k), u(k+1), \ldots, u(k+N-1)$ is computed by solving a discrete-time optimization problem over the period $[t_k, t_k + N\tau_{\mathrm{s}})$ so that a performance criterion defined over the considered period $[t_k, t_k + N\tau_{\mathrm{s}})$ is optimized subject to the operational constraints. After computing the optimal control sequence, only the first control sample is implemented, and subsequently the horizon is shifted. Next, the new state of the system is measured or estimated, and a new optimization problem at time t_{k+1} is solved using this new information.

We define now a variant of MPC, where k is not a time index, but a bag index. If $k > 0$, then bag step k then corresponds to the time instant t_k^{load} when the kth bag has just entered the track network, while bag step $k = 0$ corresponds to the initial simulation time t_0. For this variant of MPC, the horizon N corresponds to the number of bags for which we look ahead, while computing the control inputs $r(k+1)$, $r(k+2)$, ..., $r(k+N)$, where $r(k+j)$ represents the route of DCV_{k+j} (from a given loading station to the corresponding unloading station). Next, we implement all the computed control samples, and accordingly we shift the horizon with N steps. So, once we have assigned a route to a DCV, the route of that DCV cannot be changed later on.

The total objective function of centralized MPC is then defined as

$$J_{k,N}^{\text{Centr\_MPC}}(\mathbf{t}(k)) = \sum_{i=1}^{k+N} J_i(\hat{t}_i^{\text{unload}})$$

where $\hat{t}_i^{\text{unload}}$ is the predicted unloading time of DCV_i depending on the routes of the first $k+N$ bags that entered the network, and $\mathbf{t}(k) = [t_1^{\text{load}} \ldots t_{k+N}^{\text{load}}\ t_1^{\text{unload}} \ldots t_{k+N}^{\text{unload}}]^T$.

Now let $\mathbf{r}(k)$ denote the future route sequence for the next N bags entering the network at bag step k: $\mathbf{r}(k) = [r(k+1)\ r(k+2)\ \ldots\ r(k+N)]^T$. Accordingly, the MPC optimization problem at bag step k is defined as follows:

$$\begin{aligned}
&\min_{\mathbf{r}(k)} J_{k,N}^{\text{Centr_MPC}}(\mathbf{t}(k)) \\
&\text{subject to } \mathbf{t}(k) = \mathcal{M}^{\text{route_ctrl}}\big(\mathcal{T}, \mathbf{x}(t_k^{\text{load}}), \mathbf{r}(k)\big), \\
&\qquad\qquad \mathcal{C}(\mathbf{t}(k)) \leq 0
\end{aligned}$$

When using centralized MPC, at each bag step k the future route sequence $\mathbf{r}(k)$ is computed over an horizon of N bags so that the objective function is minimized subject to the dynamics of the system and the operational constraints.

Centralized MPC can compute on-line the route of each DCV in the network, but it requires a large computational effort, as will be illustrated in Sec. 16.5. Therefore, we also propose distributed control approaches, which offer a trade-off between the optimality of the performance for the controlled system and the time required to compute the solution.

16.4.2 Distributed MPC

One can decrease the computation time required by the *centralized* control approach proposed above by implementing a *distributed* approach that computes local control actions by solving local optimization problems similar to those that we have detailed in [15].

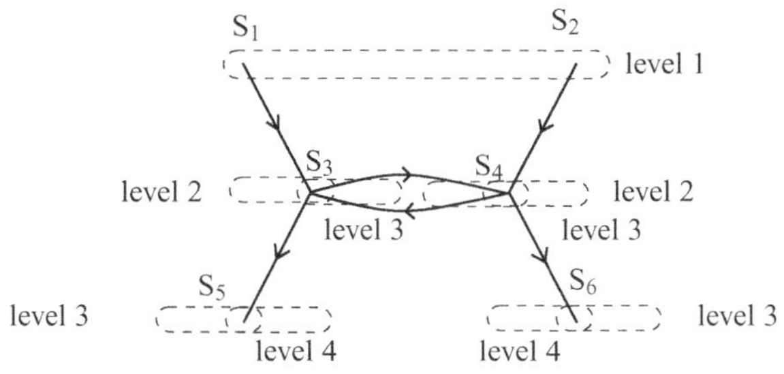

Fig. 16.4 Levels of downstream influence for parallel computation.

16.4.2.1 Levels of Influence

In distributed model predictive route choice control we consider local subsystems, each consisting of a junction S_s with $s \in \{1,2,\ldots,S\}$, its incoming and its outgoing links. But in contrast to decentralized approaches, data is communicated between neighboring junctions, which are characterized by the concept of level of influence. The levels of influence are defined as follows.

Let us first assign one or more levels of *downstream* influence to each junction in the network. We assign downstream influence level 1 to each junction in the network connected via a link to a loading station. Next, we consider all junctions connected to some junction with influence level 1 via an outgoing link, and we assign influence level 2 to them. In this way we recursively assign an influence level to each junction with the constraint that at most $\kappa_{\mathrm{d}}^{\max}$ downstream influence levels are assigned to a given junction.[1] For example, see Fig. 16.4, where we define maximum 2 levels of downstream influence for each junction in the network ($\kappa_{\mathrm{d}}^{\max} = 2$). For this example we have considered the junctions S_1 and S_2 to have been assigned downstream influence level 1. Then S_3 and S_4 are assigned level 2 (since these junctions are connected to S_1 and S_2 via outgoing links). Next, we assign influence level 3 to S_4, S_5, S_3 and S_6 (since they are connected to S_3 and S_4). Note that now S_3 and S_4 have two levels of downstream influence: 2 and 3. Therefore, S_5 and S_6 are also assigned influence level 4 (since they are connected to S_3 and S_4 with influence level 3). Similarly we can also assign levels of *upstream* influence to each junction in the network. We assign upstream influence level 1 to each junction in the network connected via a link to an unloading station. Next, we assign upstream influence level 2 to all the junctions connected to some junction on upstream influence level 1 via its incoming links. Recursively, we then assign levels of upstream influence to each junction with the constraint that at most $\kappa_{\mathrm{u}}^{\max}$ levels of upstream influence are assigned to a given junction.

[1] The constraint that at most $\kappa_{\mathrm{d}}^{\max}$ downstream influence levels are assigned to a junction limits the computational complexity and keeps all levels of influence finite.

16.4.2.2 Distributed MPC with a Single Round of Downstream Communication

Let us now consider distributed MPC with a single round of downstream communication. This means that first the local controller of each junction with influence level 1 solves the local optimal switch control problem.

After computing the optimal switch control sequence, each junction with influence level 1 then communicates to its neighboring junctions at level 2 which bags (out of all the bags over which we make the prediction for the corresponding junction with influence level 1) will enter the incoming link of the junction at level 2 and at which time instant they will do so. Next, we iteratively consider the junctions at levels $2,3,\ldots,K^{\text{downstream}}$, where $K^{\text{downstream}}$ is the largest level of downstream influence assigned in the network. Then, for each junction with influence level larger than 1, we compute a local solution to the local MPC problem as presented next.

Assume S_s with $s \in \{1,\ldots,S\}$ has influence level $\kappa_{\mathrm{d}} > 1$. Let $\mathrm{S}_{s,l}^{\text{prev}}$ denote the neighboring junction of S_s connected via the incoming link[2] $l \in \{0,1\}$ of S_s (so, $\mathrm{S}_{s,l}^{\text{prev}}$ has influence level $\kappa_{\mathrm{d}} - 1$). Then, we compute a local solution for S_s to the local MPC problem defined below over a horizon of

$$N_s = \min\left[N^{\max}, \sum_{l=0}^{1}\left(n_{s,l}^{\text{horizon}} + n_{s,l,0}^{\text{pred\_cross}} + n_{s,l,1}^{\text{pred\_cross}}\right)\right] \tag{16.4}$$

bags, where $N^{\max}$ is the maximum prediction horizon for the local MPC problem, $n_{s,l}^{\text{horizon}}$ is the number of DCVs traveling on link $l \in \{0,1\}$ going into S_s at the time instant when we start optimizing and $n_{s,l,m}^{\text{pred\_cross}}$ is the number of DCVs traveling towards $\mathrm{S}_{s,l}^{\text{prev}}$ on its incoming link m that we predict (while solving the local optimization problem at $\mathrm{S}_{s,l}^{\text{prev}}$) to cross $\mathrm{S}_{s,l}^{\text{prev}}$ and to continue their journey towards S_s ($n_{s,l,m}^{\text{pred\_cross}} < n_{s,l}^{\text{horizon}}$).

Let us now index[3] the bags that successively cross junction S_s during the entire simulation period $[t_0, t_0 + \tau^{\text{max\_sim}})$ as $b_{s,1}, b_{s,2}, \ldots, b_{s,N_s^{\text{bags}}}$, where N_s^{bags} is the number of bags that cross S_s during the simulation period.

Recall that we use a variant of MPC with a bag index. So, in this approach the local control is updated at every time instant when some bag has just entered an incoming link of junction S_s. Let t_s^{crt} be such a time instant.

Then we determine bag index k such that $t_{s,k}^{\text{cross}} \leq t_s^{\text{crt}} < t_{s,k+1}^{\text{cross}}$, where $t_{s,k}^{\text{cross}}$ is defined as the time instant when bag $b_{s,k}$ has just crossed the junction. If no bag has crossed the junction yet, we set $k = 0$.

When solving the local MPC optimization problem for junction S_s, we will use a local objective function $J_{s,k,N_s}^{\text{Distr\_MPC}}$. The local objective function is computed via a simulation of the local system for the next N_s bags that will cross the junction and

---

[2] Recall that we may assume without loss of generality that each junction has at most 2 incoming links.

[3] This order depends on the evolution of the position of the switch-in at junction S_s.

is defined as follows:

$$J^{\text{Distr\_MPC}}_{s,k,N_s}(\mathbf{t}_s(k)) = \sum_{j=1}^{\min(N_s,N_s^{\text{cross}})} J_{k+j}(\hat{t}^{\text{unload},*}_{s,k+j}) + \lambda^{\text{pen}}(N_s - N_s^{\text{cross}})$$

where

- N_s^{cross} is the number of DCVs that actually cross junction S_s during the prediction period;
- $\hat{t}^{\text{unload},*}_{s,k+j}$ is the predicted unloading time instant of bag $b_{s,k+j}$;
- λ^{pen} is a nonnegative weighting parameter;
- $\mathbf{t}_s(k) = [t^{\text{load}}_{s,k+1} \ \dots \ t^{\text{load}}_{s,k+N_s} \ \hat{t}^{\text{unload},*}_{s,k+1} \ \dots \ \hat{t}^{\text{unload},*}_{s,k+N_s}]^T$ with $t^{\text{load}}_{s,k+j}$ the loading time instant of bag $b_{s,k+j}$

The second term $\lambda^{\text{pen}}(N_s - N_s^{\text{cross}})$ of the local objective function is included for the following reasoning. Assume that, at step k, there are no DCVs traveling on the incoming link $l \in \{0,1\}$ of junction S_s, while some DCVs travel on link $1-l$. If this term were not considered, then $J^{\text{Distr\_MPC}}_{s,k,N_s}(\mathbf{t})$ would be minimal when the switch-in is positioned on link l during the prediction period. However, this is obviously not a good solution when the endpoints are open.

The MPC optimization problem at junction S_s for bag k is then defined as follows:

$$\begin{aligned}
&\min_{\mathbf{u}_s(k)} J^{\text{Distr_MPC}}_{s,k,N_s}(\mathbf{t}_s(k)) \\
&\text{subject to } \mathbf{t}_s(k) = \mathcal{M}^{\text{local,switch_ctrl}}\big(\mathcal{T}, \mathbf{x}_s(t^{\text{cross}}_{s,k}), \mathbf{u}_s(k)\big), \\
&\qquad\qquad\ \ \mathcal{C}(\mathbf{t}_s(k)) \le 0
\end{aligned}$$

with N_s given by (16.4).

Note that in this approach $\mathcal{M}^{\text{local,switch\_ctrl}}\big(\mathcal{T}, \mathbf{x}_s(t^{\text{cross}}_{s,k}), \mathbf{u}_s(k)\big)$ describes the local dynamics of junction S_s with its incoming and outgoing links and additional data from neighboring junctions (if any).

After computing the optimal control, only $u_s^{\text{sw\_in}}(k+1)$ and $u_s^{\text{sw\_out}}(k+1)$ are applied. Next, the state of the system is updated. At bag step $k+1$, a new optimization will be then solved over the next N_s bags.

Every time some bag has crossed some junction we update the local control of junctions in the network as follows: Assume that some bag has just crossed junction S_s, which has assigned level κ_{d}. Then, we update the control as follows. We consider a subtree rooted at S_s and consisting of nodes of subsequent levels of influence that are connected via a link to nodes already present in the subtree. So, only the control of the switch-in and switch-out of the junctions in this subtree have to be updated.

16.4.2.3 Distributed MPC with a Single Round of Communication

In order to further improve the performance of the distributed control approach presented above, we now add an extra round of communication and consider distributed MPC with one round of downstream and upstream communication.

So, every time a bag has crossed a junction we compute the local control sequences according to the downstream levels of influence, as explained above. Then for the junctions on level 1 of upstream influence we update the release rate of their incoming links as follows: We take as example junction S_s with $\kappa_u = 1$. For all other junctions we will apply the same procedure. We virtually apply at S_s the optimal control sequence $\mathbf{u}_s^*$ that we have computed when optimizing in the downstream direction. Let $t_s^{\text{last},*}$ be the time instant at which the last bag crossed S_s (out of all the bags over which we make the prediction for S_s). Let τ^{rate} be the length of the time window over which we compute the link release rate. The variable τ^{rate} can be derived using empirical data. Then, if

$$t_s^{\text{last},*} < t_0 + \tau^{\text{rate}}$$

we set $\zeta_{s,l} = \zeta^{\max}$ for $l = 0, 1$ with $\zeta^{\max}$ the maximum number of DCVs per time unit that can cross a junction using maximum speed. Otherwise, if $n_{s,l}^{\text{rate}} > 0$ with $n_{s,l}^{\text{rate}}$ the number of DCVs that left the outgoing link l of S_s within the time window $[t_s^{\text{last},*} - \tau^{\text{rate}}, t_s^{\text{last},*})$, we set

$$\zeta_{s,l} = \frac{n_{s,l}^{\text{rate}}}{\tau^{\text{rate}}}$$

Finally, if $n_{s,l}^{\text{rate}} = 0$ we set $\zeta_{s,l} = \varepsilon$ with $0 < \varepsilon \ll 1$. Now we solve the local MPC problem presented in Sec. 16.4.2.2 using the updated release rates and we compute the local control of all junctions at upstream level $\kappa_u + 1$. Recursively, we compute the local control until reaching level K^{upstream}, where K^{upstream} is the largest level of upstream influence assigned in the network.

By also performing the upstream round of communication, more information about the future congestion is provided via the updated release rate. This information might change the initial intended control actions of each junction. Typically (if one allows sufficient time to compute the solution of each local optimization problem), this new variant of distributed MPC increases the performance of the system, but also the computational effort increases, since we deal with one more round of optimizations.

16.4.3 Hierarchical MPC

In order to efficiently compute the route of each DCV we propose a hierarchical control framework that consists of a multilevel control structure; see Fig. 16.5. The layers of the framework can be characterized as follows:

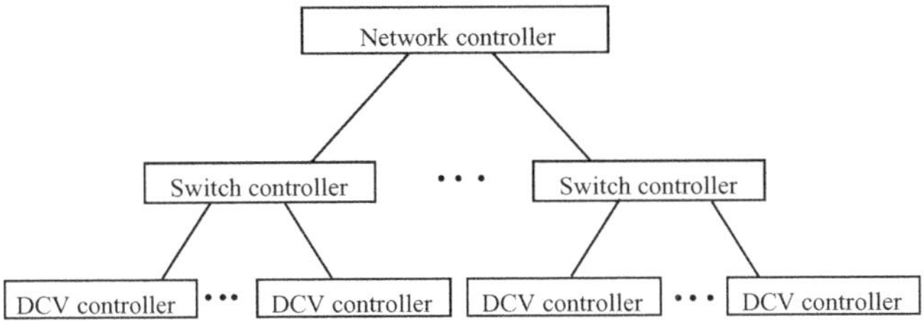

Fig. 16.5 Hierarchical control for DCV-based baggage handling systems.

- The *network controller* considers flows of DCVs instead of individual DCVs. Moreover, the network controller determines reference DCV flow trajectories over time for each link in the network. These flow trajectories are computed so that the performance of the DCV-based baggage handling system is optimized. Then the optimal reference flow trajectories are communicated to switch controllers.
- The *switch controller* present in each junction receives the information sent by the network controller and determines the sequence of optimal positions for its ingoing and outgoing switches at each time step so that the tracking error between the reference flow trajectory and the actual flow trajectory is minimal.
- The *DCV controller* present in each vehicle detects the speed and position of the vehicle in front of it, if any, and the position of the switch into the junction the DCV travels towards to. This information is then used to determine the speed to be used next such that no collision will occur and such that the DCV stops in front of a junction, the switch of which is not positioned on the link on which the DCV travels.

The lower levels in this hierarchy deal with faster time scales (typically in the milliseconds range for the DCV controllers up to the seconds range for the switch controllers), whereas for the higher-level layer (network controller) the frequency of updating is up to the minutes range.

16.4.3.1 Route Control

We now focus on the network controller. In Sec. 16.5 it will be shown that when each DCV is considered individually, the predictive switch control problem in DCV-based baggage handling systems results in a huge nonlinear integer optimization problem with high computational complexity and requirements, making the problem in fact intractable in practice. So, since considering each individual DCV is too computationally intensive we will now consider streams of DCVs instead (characterized by real-valued demands and flows expressed in vehicles per second). In this chapter the routing problem will then be recast as the problem of determining flows on each link. Once these flows are determined, they can be implemented by switch controllers at the junctions. So, the network controller provides flow targets to the

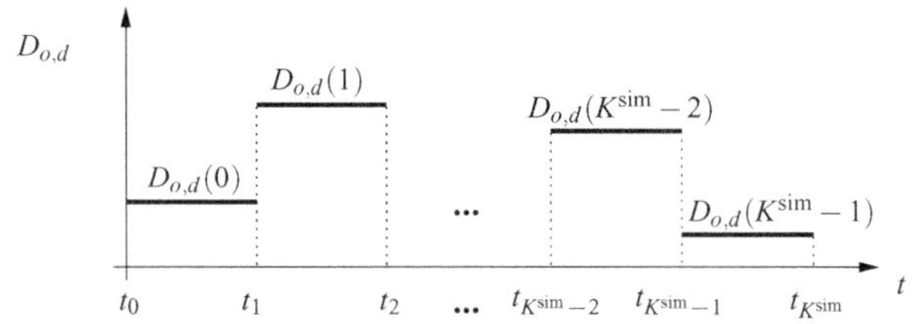

Fig. 16.6 Piecewise constant demand profile $D_{o,d}$.

switch controllers, which then have to control the position of the switch into and out of each junction in such a way that these targets are met as well as possible. This corresponds to blocking flows before a junction whenever necessary and possible, and routing the DCVs towards the outgoing links.

In the literature one can find extensive work addressing the flow-over-time problem; see e.g., [7]. However, in this chapter we propose a nonstandard, but efficient approach to model the flows of DCVs as presented next.

Set-up

We consider the following set-up. We have a transportation network with a set of origin nodes $\mathcal{O}$ consisting of the loading stations, a set of destination nodes $\mathcal{D}$ consisting of the unloading stations and a set of internal nodes $\mathcal{I}$ consisting of all the junctions in the network. We define the set of all nodes as $\mathcal{V} = \mathcal{O} \cup \mathcal{I} \cup \mathcal{D}$. The nodes are connected by unidirectional links. Let $\mathcal{L}$ denote the set of all links.

Furthermore, let the time instant t_k be defined as

$$t_k = t_0 + k\tau^{\mathrm{nc}}$$

with t_0 that time when we start the simulation and τ^{nc} the sampling time for the network controller. Then, for each pair $(o,d) \in \mathcal{O} \times \mathcal{D}$, there is a dynamic, piecewise constant demand pattern $D_{o,d}(\cdot)$ as shown in Fig. 16.6, with $D_{o,d}(k)$ the demand of bags at origin o with destination d in the time interval $[t_k, t_{k+1})$ for $k = 0, 1, \ldots, K^{\mathrm{sim}} - 1$ with K^{sim} the simulation horizon (we assume that beyond $t_{K^{\mathrm{sim}}}$ the demand is zero).

Next, let $\mathcal{L}_d$ be the set of links that belong to some route going to destination d, $\mathcal{L}_d \subseteq \mathcal{L}$. We denote the set of incoming links for node $v \in \mathcal{V}$ by $\mathcal{L}_v^{\mathrm{in}}$, and the set of outgoing links of v by $\mathcal{L}_v^{\mathrm{out}}$. Note that for origins $o \in \mathcal{O}$ we have $\mathcal{L}_o^{\mathrm{in}} = \emptyset$ and for destinations $d \in \mathcal{D}$ we have $\mathcal{L}_d^{\mathrm{out}} = \emptyset$. Also, assume each origin node to have only one outgoing link and each destination node to have only one incoming link.[4] Then $|\mathcal{L}_o^{\mathrm{out}}| = 1$ and $|\mathcal{L}_d^{\mathrm{in}}| = 1$.

[4] If a loading station has more than one outgoing link, then one can virtually expand that loading station into a loading station connected via a link of length zero to a junction with two outgoing links; similarly, one can virtually expand an unloading station with more than one incoming link.

Next, for each destination $d \in \mathcal{D}$ and for each link $\ell \in \mathcal{L}_d$ in the network we will define a real-valued flow $u_{\ell,d}(k)$. The flow $u_{\ell,d}(k)$ denotes the number of DCVs per time unit traveling towards destination d that enter link ℓ during the time interval $[t_k, t_{k+1})$.

The aim is now to compute using MPC, for each time step k, flows $u_{\ell,d}(k)$ for every destination $d \in \mathcal{D}$ and for every link $\ell \in \mathcal{L}_d$ in such a way that the capacity of the links is not exceeded and such that the performance criterion is minimized over a given prediction period $[t_k, t_{k+N})$. Later on we will write a model of the baggage handling system to be used by the network controller, and show that this model can be rewritten as an MILP model. Therefore, in order to obtain an MILP optimization problem one has to define a linear or piecewise affine performance criterion. Possible goals for the network controller that allow linear or piecewise affine performance criteria are reaching a desired outflow at destination d or minimizing the lengths of the queue in the network.

Model

We now determine the model for the DCV flows through the network. Let τ_ℓ denote the free-flow travel time on link ℓ. Recall that the free-flow travel time of link ℓ represents the time period that a DCV requires to travel on link ℓ when using maximum speed. In this subsection we assume the travel time τ_ℓ to be an integer multiple of τ^{nc}, say

$$\tau_\ell = \kappa_\ell \tau^{\text{nc}} \quad \text{with } \kappa_\ell \text{ an integer.} \tag{16.5}$$

In case the capacity of a loading station is less than the demand, queues might appear at the origin of the network. Let $q_{o,d}(k)$ denote the length at time instant t_k of the partial queue of DCVs at origin o going to destination d. In principle, the queue lengths should be integers as their unit is [number of vehicles], but we will approximate them using reals.

For every origin node $o \in \mathcal{O}$ and for every destination $d \in \mathcal{D}$ we now have

$$u_{\ell,d}(k) \leqslant D_{o,d}(k) + \frac{q_{o,d}(k)}{\tau^{\text{nc}}} \quad \text{for } \ell \in \mathcal{L}_o^{\text{out}} \cap \mathcal{L}_d \tag{16.6}$$

with $D_{o,d}(k) = 0$ for $k \geqslant K$. Moreover,

$$q_{o,d}(k+1) = \max\left(0, q_{o,d}(k) + \left(D_{o,d}(k) - \sum \ell \in \mathcal{L}_o^{\text{out}} \cap \mathcal{L}_d u_{\ell,d}(k)\right)\tau^{\text{nc}}\right) \tag{16.7}$$

However, queues can form also inside the network. We assume that the DCVs run with maximum speed along the track segments and, if necessary, they wait in vertical queues before crossing the junction. Let $q_{v,d}(k)$ denote the length at time instant t_k of the vertical queue at junction $v \in \mathcal{I}$, for DCVs going to destination $d \in \mathcal{D}$. Taking into account that a flow on link ℓ has a delay of κ_ℓ time steps before it reaches the end of the link, for every internal node $v \in \mathcal{I}$ and for every $d \in \mathcal{D}$ we have

$$F^{\text{out}}_{v,d}(k) \leqslant F^{\text{in}}_{v,d}(k) + \frac{q_{v,d}(k)}{\tau^{\text{nc}}} \tag{16.8}$$

where $F^{\text{in}}_{v,d}(k)$ is the flow into the queue at junction v, defined as

$$F^{\text{in}}_{v,d}(k) = \sum \ell \in \mathcal{L}^{\text{in}}_{v} \cap \mathcal{L}_d u_{\ell,d}(k-\kappa_\ell) \tag{16.9}$$

and where $F^{\text{out}}_{v,d}(k)$ is the flow out of the queue at junction v, defined as:

$$F^{\text{out}}_{v,d}(k) = \sum \ell \in \mathcal{L}^{\text{out}}_{v} \cap \mathcal{L}_d u_{\ell,d}(k) \ . \tag{16.10}$$

The evolution of the length of the queue for every internal node $v \in \mathcal{I}$ and for every $d \in \mathcal{D}$ is given by

$$q_{v,d}(k+1) = \max\left(0, q_{v,d}(k) + \left(F^{\text{in}}_{v,d}(k) - F^{\text{out}}_{v,d}(k)\right)\tau^{\text{nc}}\right) \tag{16.11}$$

Moreover, for each origin $o \in \mathcal{O}$ and for each junction $v \in \mathcal{I}$ we have the following constraints:

$$\sum d \in \mathcal{D} q_{o,d}(k+1) \leq q_o^{\max} \tag{16.12}$$
$$\sum d \in \mathcal{D} q_{v,d}(k+1) \leq q_v^{\max} \tag{16.13}$$

where $q_o^{\max}$ and $q_v^{\max}$ express respectively the maximum number of DCVs the conveyor belt transporting bags towards loading station o can accommodate and the maximum number of DCVs the track segments of the incoming links of junction v can accommodate.

We also have the following constraint for every link ℓ:

$$\sum d \in \mathcal{D} u_{\ell,d}(k) \leqslant U_\ell^{\max} \tag{16.14}$$

where $U_\ell^{\max}$ is the maximum flow of DCVs that can enter link ℓ.

Then, at time step k, the model of the DCV flows through the network of tracks describing (16.6)–(16.14) can be written as a system of equalities and a system of inequalities as follows:

$$\mathbf{q}_{k+1} = \mathcal{M}^{\text{eq}}(\mathbf{q}_k, \mathbf{u}_k)$$
$$\mathcal{M}^{\text{ineq}}(\mathbf{q}_{k+1}, \mathbf{u}_k) \leq 0$$

where

- $\mathbf{q}_k$ is the vector consisting of all the queue lengths $q_{o,d}(k)$, for all $o \in \mathcal{O}$ and for all $d \in \mathcal{D}$, and of all the queue lengths $q_{v,d}(k)$, for all $v \in \mathcal{I}$ and for all $d \in \mathcal{D}$;
- $\mathbf{u}_k$ is the vector consisting of all the flows $u_{\ell,d}(k)$, for all $d \in \mathcal{D}$ and for all $\ell \in \mathcal{L}_d$.

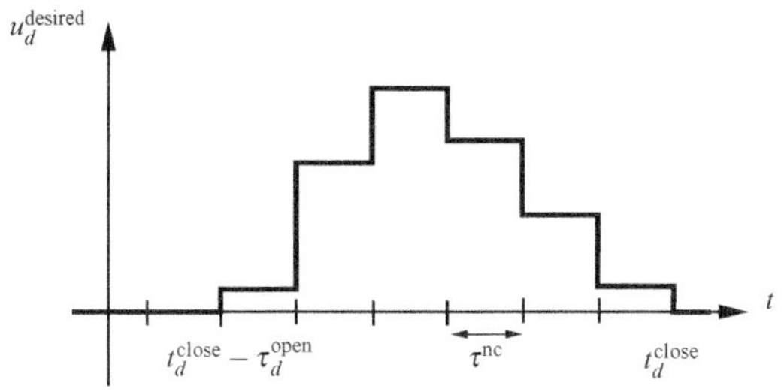

Fig. 16.7 Desired arrival profile at destination d.

Performance criterion

Next we define the performance to be used for computing the optimal routing at step k for a prediction period of N time steps. The objective is to have each bag arriving at its endpoint within a given time interval $[t_d^{\text{close}} - \tau_d^{\text{open}}, t_d^{\text{close}})$ where t_d^{close} is the time instant when the endpoint d closes and τ_d^{open} is the time period for which the end point d stays open for a specific flight. We assume t_d^{close} and τ_d^{open} to be integer multiples of τ_{s}.

Hence, one MPC objective that allows a piecewise affine performance criterion is achieving a desired flow at destination d during the prediction period. Let u_d^{desired} denote the desired piecewise constant flow profile at destination d as sketched in Fig. 16.7, where the area under u_d^{desired} equals the total number of bags out of the total demand that have to be sent to destination d. Note that $u_d^{\text{desired}}(k) = 0$ for all $k < k_d^{\text{open}}$ and all $k \geq k_d^{\text{close}}$ with

$$k_d^{\text{open}} = \frac{t_d^{\text{close}} - \tau^{\text{open}}}{\tau^{\text{nc}}} \text{ and } k_d^{\text{close}} = \frac{t_d^{\text{close}}}{\tau^{\text{nc}}}$$

Let $\kappa_{\ell_d} = \tau_{\ell_d}/\tau^{\text{nc}}$. Hence, one can define the following penalty for flow profiles corresponding to destination $d \in \mathcal{D}$:

$$J_{d,k}^{\text{pen}} = \left| u_d^{\text{desired}}(k) - u_{\ell_d,d}(k + \kappa_{\ell_d}) \right|$$

where ℓ_d is the incoming link of destination d. Later on we will include the penalty term

$$\sum_{i=k}^{k+N-1-\kappa_{\ell_d}} J_{d,i}^{\text{pen}}$$

into the MPC performance criterion for each destination d and for each time step k. Note that we make the summation of these penalization indices only up to $k+N-1-\kappa_{\ell_d}$, since for $i > k+N-1-\kappa_{\ell_d}$ the variable $u_{\ell_d,d}(k+\kappa_{\ell_d})$ is not defined at MPC step k.

Moreover, note that the use of

$$\sum_{i=k}^{k+N-1-\kappa_{\ell_d}} J_{d,i}^{\text{pen}}$$

MPC performance criterion for each destination d and for each time step k could have adverse effects for small prediction horizons. Therefore, to counteract these effects we consider as additional controller goal, that of maximizing the flows of all links that are not directly connected to unloading stations. To this aim, let $\tau_{\ell,d,k}^{\text{link}}$ be the typical[5] time required for a DCV that entered link ℓ in $[t_k, t_{k+1})$ to reach destination d, with $\tau_{\ell,d,k}^{\text{link}}$ an integer multiple of τ_{s}. Also, let

$$\kappa_{l,d} = \frac{\tau_{\ell,d,k}^{\text{link}}}{\tau^{\text{nc}}}$$

Then, one can define the following penalty:

$$J_{\ell,d,k}^{\text{flow}} = \begin{cases} u_{\ell,d}(k), & \text{if } k_d^{\text{open}} - \kappa_{l,d} \leq k < k_d^{\text{close}} - \kappa_{l,d} \\ 0, & \text{otherwise.} \end{cases}$$

Later on this penalty will be used in the MPC performance criterion.

Next, in order to make sure that *all* the bags will be handled in finite time, we also include in the MPC performance criterion the weighted length of queues at each junction in the network, as presented next. Let $\tau_{v,d}^{\text{junc}}$ be the typical[6]time required for a DCV in the queue at junction v to reach destination d, with $\tau_{v,d}^{\text{junc}}(k)$ an integer multiple of τ^{nc}. Also, let $\kappa_{v,d} = \tau_{v,d}^{\text{junc}}(k)/\tau^{\text{nc}}$. Then we define the new penalty:

$$J_{v,d,k}^{\text{overdue}} = \begin{cases} d_{v,d}^{\min} q_{v,d}(k), & \text{if } k \geq k_d^{\text{close}} - \kappa_{v,d} \\ 0, & \text{otherwise} \end{cases}$$

where $d_{v,d}^{\min}$ represents the length of the shortest route from junction v to destination d. Note that $J_{v,d,k}^{\text{overdue}}$ is nonzero only for steps that are larger than or equal to $k_d^{\text{close}} - \kappa_{v,d}$. Moreover, for these steps $J_{v,d,k}^{\text{overdue}}$ is proportional to $d_{v,d}^{\min}$. The reason for this is that we want to penalize more the queues at junctions that are further away from destination d since the DCVs in those queues will need a longer time to travel to d.

Finally, let $\mathcal{L}^{\text{dest}}$ denote the set of links directly connected to unloading stations. Then the MPC performance criterion is defined as follows:

[5] These durations are determined based on historical data.

[6] Ibid.

$$J_{k,N} = \sum d \in \mathcal{D}\Bigg(\sum_{i=k}^{k+N-1-\kappa_{\ell_d}} \lambda_d J^{\text{pen}}_{d,i} + \beta \sum_{i=k}^{k+N-1} \sum v \in \mathcal{I} J^{\text{overdue}}_{v,d,i} - \alpha \sum_{i=k}^{k+N-1} \sum_{\ell \in (\mathcal{L} \setminus \mathcal{L}^{\text{dest}}) \cap \mathcal{L}_d} J^{\text{flow}}_{\ell,d,i} \Bigg)$$

with $\lambda_d > 0$ a weight that expresses the importance of the flight assigned to destination d, $\alpha \ll 1$ and $\beta \ll 1$ nonnegative weighting parameters.

Then the nonlinear MPC optimization problem is defined as follows:

$$\begin{aligned}
\min_{\substack{\mathbf{u}_k,\ldots,\mathbf{u}_{k+N-1},\\ \mathbf{q}_{k+1},\ldots,\mathbf{q}_{k+N}}} \quad & J_{k,N} \\
\text{subject to} \quad & \mathbf{q}_{k+1} = \mathcal{M}^{\text{eq}}(\mathbf{q}_k, \mathbf{u}_k) \\
& \vdots \\
& \mathbf{q}_{k+N} = \mathcal{M}^{\text{eq}}(\mathbf{q}_{k+N-1}, \mathbf{u}_{k+N-1}) \\
& \mathcal{M}^{\text{ineq}}(\mathbf{q}_{k+1}, \mathbf{u}_k) \leq 0 \\
& \vdots \\
& \mathcal{M}^{\text{ineq}}(\mathbf{q}_{k+N}, \mathbf{u}_{k+N-1}) \leq 0
\end{aligned} \tag{16.15}$$

The nonlinear MPC optimization problem defined above is typically complex and it requires large computational effort to solve. Therefore, in the next section we will recast this problem into an MILP for which efficient and fast solvers are available.

MILP optimization problem for the network controller

The general formulation of a mixed integer linear programming problem is:

$$\begin{aligned}
\min_{\mathbf{x}} \quad & \mathbf{c}^T \mathbf{x} \\
\text{subject to} \quad & \mathbf{A}^{\text{eq}} \mathbf{x} = \mathbf{b}^{\text{eq}} \\
& \mathbf{A}\mathbf{x} \leq \mathbf{b} \\
& \mathbf{x}^{\text{low}} \leq \mathbf{x} \leq \mathbf{x}^{\text{up}} \\
& x_i \text{ is an integer, for each } i \in I
\end{aligned} \tag{16.16}$$

where $\mathbf{c}$, $\mathbf{x}$, $\mathbf{x}^{\text{low}}$, $\mathbf{x}^{\text{up}}$, $\mathbf{b}^{\text{eq}}$ and $\mathbf{b}$ are vectors, with $\mathbf{x}^{\text{low}}$ the lower bound for $\mathbf{x}$ and $\mathbf{x}^{\text{up}}$ the upper bound, and where $\mathbf{A}^{\text{eq}}$ and $\mathbf{A}$ are matrices (all these vectors and matrices have appropriate size), and $I \subset \{1,\ldots,n\}$, where n is the number of variables.

Next we transform the dynamic optimal route choice problem presented above into an MILP problem, for which efficient solvers have been developed [4]. To this aim we use the following equivalences, see [2] where f is a function defined on a bounded set X with upper and lower bounds M and m for the function values, δ is a binary variable, y is a real-valued scalar variable, and ε is a small tolerance (typically the machine precision):

P1: $[f(x) \leqslant 0] \iff [\delta = 1]$ is true if and only if

$$\begin{cases} f(x) \leqslant M(1-\delta) \\ f(x) \geqslant \varepsilon + (m-\varepsilon)\delta \ , \end{cases}$$

P2: $y = \delta f(x)$ is equivalent to

$$\begin{cases} y \leqslant M\delta \\ y \geqslant m\delta \\ y \leqslant f(x) - m(1-\delta) \\ y \geqslant f(x) - M(1-\delta) \ . \end{cases}$$

As an example, we will show how Eq. (16.7) of the nonlinear route choice model presented in the previous section can be transformed into a system of linear equations and inequalities by introduction of some auxiliary variables. For the other equations of the route choice model we apply a similar procedure.

We consider now (16.7). This is a nonlinear equation and thus it does not fit the MILP framework. Therefore, we will first introduce the binary variables $\delta_{o,d}(k)$ such that

$$\begin{aligned} &\delta_{o,d}(k) = 1 \ \text{if and only if} \\ &q_{o,d}(k) + \Big(D_{o,d}(k) - \sum l \in \mathcal{L}_o^{\text{out}} \cap \mathcal{L}_d u_{\ell,d}(k)\Big)\tau^{\text{nc}} \leqslant 0 \end{aligned} \tag{16.17}$$

and rewrite (16.7) as follows:

$$q_{o,d}(k+1) = (1-\delta_{o,d}(k)) \cdot (q_{o,d}(k) + (D_{o,d}(k) - \sum l \in \mathcal{L}_o^{\text{out}} \cap \mathcal{L}_d u_{\ell,d}(k))\tau^{\text{nc}}) \tag{16.18}$$

Condition (16.17) is equivalent to (cf. Property **P1**):

$$\begin{cases} f(k) \leqslant (q_o^{\max} + D_{o,d}^{\max}\tau^{\text{nc}})(1-\delta_{o,d}(k)) \\ f(k) \geqslant \varepsilon + (-U^{\max}\tau^{\text{nc}} - \varepsilon)\delta_{o,d}(k) \ , \end{cases}$$

where $f(k) = q_{o,d}(k) + (D_{o,d}(k) - \sum l \in \mathcal{L}_o^{\text{out}} \cap \mathcal{L}_d u_{\ell,d}(k))\tau^{\text{nc}}$, $q_o^{\max}$ is the maximal queue length at origin o, and where $D_{o,d}^{\max} = \max_k D_{o,d}(k)$ is the maximal demand for origin-destination pair (o,d).

However, (16.18) is still nonlinear since it contains a multiplication of a binary variable $\delta_{o,d}(k)$ with a real-valued (linear) function. However, by using Property **P2** this equation can be transformed into a system of linear inequalities.

The rest of the model equations can be transformed, in a similar way, into a system of MILP equations. Next we will transform the MPC performance criterion into its MILP form.

The problem

$$\min \sum_{i=k}^{k+N-1} \sum_{d \in \mathcal{D}} \lambda_d \left| u_d^{\text{desired}}(i) - u_{\ell_d,d}(i+\kappa_{\ell_d}) \right|$$

can be written as:

$$\begin{aligned} \min \quad & \sum_{i=k}^{k+N-1} \sum_{d \in \mathcal{D}} \lambda_d u_d^{\text{diff}}(i) \\ \text{subject to} \quad & u_d^{\text{diff}}(i) \geqslant u_d^{\text{desired}}(i) - u_{\ell_d,d}(i+\kappa_{\ell_d}) \\ & u_d^{\text{diff}}(i) \geqslant -u_d^{\text{desired}}(i) + u_{\ell_d,d}(i+\kappa_{\ell_d}) \\ & \text{for } i = k, \dots, k+N-1 \end{aligned} \tag{16.19}$$

which is a linear programming problem.

If we add the MILP equations of the model, the nonlinear optimization problem of Sec. 16.4.3.1 can be written as an MILP problem.

Several efficient branch-and-bound MILP solvers [4] are available for MILP problems. Moreover, there exist several commercial and free solvers for MILP problems such as, e.g, CPLEX, Xpress-MP, GLPK, or lp_solve; see [1] for an overview. In principle—i.e., when the algorithm is not terminated prematurely due to time or memory limitations—these algorithms guarantee we will find the global optimum. This global optimization feature is not present in the other optimization methods that can be used to solve the original nonlinear, nonconvex, nonsmooth optimization problem. Moreover, if the computation time is limited (as is often the case in on-line real-time control), then it might happen that the MILP solution can be found within the allotted time whereas the global and multistart local optimization algorithm still will not converge to a good solution (as will be illustrated in Sec. 16.5).

16.4.3.2 Switch Control

We now focus on the switch controller for the proposed hierarchy and on how optimal switch positions can be determined.

Recall that at each control step k, the network controller provides optimal flows for each link in the network and for each destination. Let these flows be denoted by $u_{\ell,d}^{\text{opt}}(k), \dots, u_{\ell,d}^{\text{opt}}(k+N-1)$, with $d \in \mathcal{D}$, $\ell \in \mathcal{L} \cap \mathcal{L}_d$ and N the prediction horizon of the network controller. Then the switch controller of each junction has to compute optimal switch-in and switch-out positions such that the tracking error between the reference optimal flow trajectory and the flow trajectory obtained by the switch controller is minimal for each network controller time step $k = 0, \dots, K^{\text{sim}}$.

Recall that the optimal flows

$$u_{\ell,d}^{\text{opt}}(k),\ldots,u_{\ell,d}^{\text{opt}}(k+N-1)$$

are determined for the time window $[t_k, t_{k+N})$ with $t_k = t_0 + k\tau^{\text{nc}}$. In order to determine the switch control action during the time window $[t_k, t_{k+N})$ we will now again use MPC. Next we will refer to one junction $v \in \mathcal{I}$ only. For all other junctions, the switch control actions are determined similarly.

Let τ^{sc} be the switch controller sampling[7] time. Also, let k^{sc} be an integer that expresses the number of switch control actions determined until now. At t_k, k^{sc} is defined as $k^{\text{sc}} = \tau^{\text{nc}}/\tau^{\text{sc}}k$. Then, let $t_{k^{\text{sc}}}^{\text{sw}}$ denote the time instant corresponding to the time step k^{sc} of the switch controller, $t_{k^{\text{sc}}}^{\text{sw}} = t_0 + k^{\text{sc}}\tau^{\text{sc}}$ with t_0 the time instant when we start the simulation.

Furthermore, let $s_v^{\text{in}}(k^{\text{sc}})$ denote the position of the switch-in at junction v during the time interval $[t_{k^{\text{sc}}}^{\text{sw}}, t_{k^{\text{sc}}+1}^{\text{sw}})$ and let $s_v^{\text{out}}(k^{\text{sc}})$ denote the position of the switch-out at junction v during $[t_{k^{\text{sc}}}^{\text{sw}}, t_{k^{\text{sc}}+1}^{\text{sw}})$.

We want to determine the switch control sequence during the time window $[t_k, t_{k+N})$ while using MPC with a prediction period of N^{sc} steps. Hence, at each MPC step k^{sc}, the switch controller solves the following optimization problem:

$$\min_{\mathbf{s}_{v,k^{\text{sc}},N^{\text{sc}}}} J_{v,k^{\text{sc}},N^{\text{sc}}}^{\text{sw}} \tag{16.20}$$

with $\mathbf{s}_{v,k^{\text{sc}},N^{\text{sc}}} = [s_v^{\text{in}}(k^{\text{sc}}) \cdots s_v^{\text{in}}(k^{\text{sc}}+N^{\text{sc}}-1) \cdots s_v^{\text{out}}(k^{\text{sc}}) \cdots s_v^{\text{out}}(k^{\text{sc}}+N^{\text{sc}}-1)]^T$ if junction v has two incoming and two outgoing links and with $\mathbf{s}_{v,k^{\text{sc}},N^{\text{sc}}}$ containing only switch-in or only switch-out positions if junction v has only one outgoing or only 1 incoming link, respectively, and where the local MPC performance criterion $J_{v,k^{\text{sc}},N^{\text{sc}}}^{\text{sw}}$ is defined as:

$$\begin{aligned} J_{v,k^{\text{sc}},N^{\text{sc}}}^{\text{sw}} = & \sum_{\ell \in \mathcal{L}_v^{\text{out}}} \left| X_{\ell,k,k^{\text{sc}},N^{\text{sc}}}^{\text{opt}}(\mathbf{u}_\ell^{\text{opt}}) - X_{\ell,k^{\text{sc}},N^{\text{sc}}}(\mathbf{s}_{v,k^{\text{sc}},N^{\text{sc}}}) \right| \\ & + \gamma\big(n_{k^{\text{sc}},N^{\text{sc}}}^{\text{sw\_in}}(\mathbf{s}_{v,k^{\text{sc}},N^{\text{sc}}}) + n_{k^{\text{sc}},N^{\text{sc}}}^{\text{sw\_out}}(\mathbf{s}_{v,k^{\text{sc}},N^{\text{sc}}})\big) \end{aligned}$$

where

- $X_{\ell,k,k^{\text{sc}},N^{\text{sc}}}^{\text{opt}}(\mathbf{u}_\ell^{\text{opt}})$ denotes the optimal number of DCVs to enter the outgoing link ℓ of junction v during the period $[t_{k^{\text{sc}}}^{\text{sw}}, t_{k^{\text{sc}}+N^{\text{sc}}-1}^{\text{sw}})$, where $\mathbf{u}_\ell^{\text{opt}}$ is the vector consisting of all the flows

  $$u_{\ell,d}^{\text{opt}}(k),\ldots,u_{\ell,d}^{\text{opt}}(k+N) \text{ with } d \in \mathcal{D} \text{ and } \ell \in \mathcal{L} \cap \mathcal{L}_d$$

  The variable $X_{\ell,k,k^{\text{sc}},N^{\text{sc}}}^{\text{opt}}(\mathbf{u}_\ell^{\text{opt}})$ is derived later on (see (16.22)).
- $X_{\ell,k^{\text{sc}},N^{\text{sc}}}(\mathbf{s}_{v,k^{\text{sc}},N^{\text{sc}}})$ is the actual number of DCVs entering link ℓ during the prediction period. The variable $X_{\ell,k^{\text{sc}},N^{\text{sc}}}$ is determined via simulation for a nonlinear (event-based) model similar to the one used in the distributed MPC approach of

[7] We select the sampling time τ^{nc} of the network controller and the sampling time τ^{sc} of the switch controller such that τ^{nc} is an integer multiple of τ^{sc}.

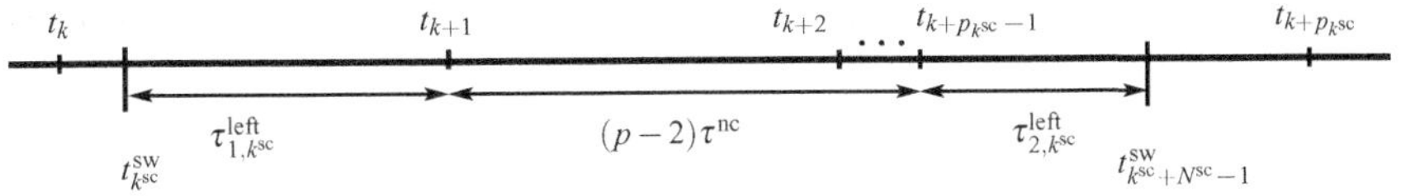

Fig. 16.8 Prediction window $[t^{sw}_{k^{sc}}, t^{sw}_{k^{sc}+N^{sc}-1})$ over which we solve the MPC optimization problem (16.20) illustrated with respect to the window $[t_k, t_{k+p_{k^{sc}}})$ for $p_{k^{sc}} > 2$.

Sec. 16.4.2 (the difference is that now the switch positions $\mathbf{s}_{v,k^{sc},N^{sc}}$ are given for each period $[t^{sw}_{k^{sc}}, t^{sw}_{k^{sc}+1}), \ldots, [t^{sw}_{k^{sc}+N^{sc}-1}, t^{sw}_{k^{sc}+N^{sc}})$ instead of for each of the next N_s DCVs to cross a junction).

- $n^{sw\_in}_{k^{sc},N^{sc}}(\mathbf{s}_{v,k^{sc},N^{sc}})$ and $n^{sw\_out}_{k^{sc},N^{sc}}(\mathbf{s}_{v,k^{sc},N^{sc}})$ represent the number of toggles of the switch-in and of the switch-out respectively during the time period $[t^{sw}_{k^{sc}}, t^{sw}_{k^{sc}+N^{sc}})$ that are obtained from the simulation.
- γ is a nonnegative weighting parameter.

Next, we derive the variable $X^{opt}_{\ell,k,k^{sc},N^{sc}}(\mathbf{u}^{opt}_{\ell})$. To this aim, we first determine how many steps $p_{k^{sc}}$ of the network controller will be involved in solving (16.20), as follows:

$$p_{k^{sc}} = \left\lceil \frac{N^{sc}\tau^{sc}}{\tau^{nc}} \right\rceil$$

where $\lceil x \rceil$ denotes the smallest integer larger than or equal to x (so, $p_{k^{sc}} \geq 1$). Furthermore, note that the index k of the time instant t_k for which $t_k \leq t^{sw}_{k^{sc}} < t_{k+1}$ can be computed as follows:

$$k = \left\lfloor \frac{k^{sc}\tau^{sc}}{\tau^{nc}} \right\rfloor$$

where $\lfloor x \rfloor$ denotes the largest integer less than or equal to x. Figure 16.8 illustrates the prediction window $[t^{sw}_{k^{sc}}, t^{sw}_{k^{sc}+N^{sc}-1})$ with respect to the window $[t_k, t_{k+p_{k^{sc}}})$. The variable $X^{opt}_{\ell,k,k^{sc}}(\mathbf{u}^{opt}_{\ell})$ is given by:

$$X^{opt}_{\ell,k,k^{sc},N^{sc}}(\mathbf{u}^{opt}_{\ell}) = \tau^{left}_{1,k^{sc}} \sum_{d\in\mathcal{D}} u^{opt}_{\ell,d}(k) + \tau^{nc} \sum_{i=k+1}^{k+p_{k^{sc}}-2} \sum_{d\in\mathcal{D}} u^{opt}_{\ell,d}(i) \tag{16.21}$$

$$+\tau^{left}_{2,k^{sc}} \sum_{d\in\mathcal{D}} u^{opt}_{\ell,d}(k+p_{k^{sc}}-1) \tag{16.22}$$

where $\sum_{i=k+1}^{k+j} x(i) = 0$ by definition for $j < 1$ and where

$$\tau^{left}_{1,k^{sc}} = \min(t_{k+1}, t^{sw}_{k^{sc}+N^{sc}-1}) - t^{sw}_{k^{sc}}$$

$$\tau^{left}_{2,k^{sc}} = \begin{cases} t^{sw}_{k^{sc}+N^{sc}-1} - t_{k+p_{k^{sc}}-1} & \text{if } p_{k^{sc}} > 1 \\ 0 & \text{otherwise.} \end{cases}$$

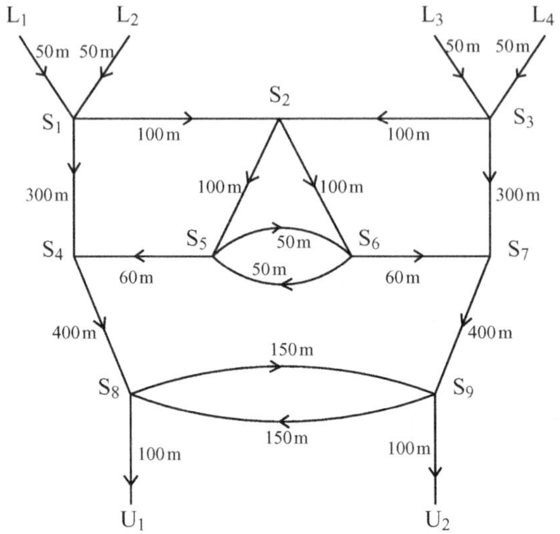

Fig. 16.9 Case study for a DCV-based baggage handling system.

16.5 Simulation Results

In this subsection we compare the performance of the centralized, distributed, and hierarchical MPC based on a simulation example.

Set-up

We consider the network of tracks depicted in Fig. 16.9 with four loading stations, two unloading station, nine junctions, and twenty unidirectional links. Note that this network allows more than four possible routes to each destination from any origin point (e.g., U_1 can be reached from L_1 via junctions S_1, S_4, S_8; S_1, S_4, S_8, S_9, S_8; S_1, S_2, S_5, S_4, S_8; S_1, S_2, S_5, S_6, S_5, S_4, S_8; $S_1, S_2, S_6, S_7, S_9, S_8$, and so on). We consider this network because on the one hand it is simple, allowing an intuitive understanding of and insight to the operation of the system and the results of the control[8], and because on the other hand, it also contains all the relevant elements of a real set-up. We assume that the velocity of each DCV varies between $0\,\mathrm{m/s}$ and $v^{\max} = 20\,\mathrm{m/s}$, and that the minimum time period after we allow a switch toggle is $\tau^{\mathrm{switch}} = 2\,\mathrm{s}$. The lengths of the track segments are indicated in Fig. 16.9.

In order to more rapidly assess the efficiency of our control method we assume that we do not start with an empty network but with a network already populated by DCVs transporting bags.

We consider 6 typical scenarios where 2400 bags will be loaded into the baggage handling system (600 bags arrive at each loading station in the time interval

[8] The proposed control approaches allow the choice of routes containing loops.

Table 16.1 Comparison of average performance of the system and total computation time

Control approach	$J^{\text{approach,avg}}(s)$	Total CPU time (s)
Centralized MPC	$1.13 \cdot 10^6$	$1.95 \cdot 10^5$
Distributed MPC downstream communication	$2.27 \cdot 10^7$	$3.90 \cdot 10^4$
Distributed MPC communication back & forth	$1.90 \cdot 10^7$	$1.46 \cdot 10^5$
Hierarchical MPC	$1.98 \cdot 10^5$	$1.06 \cdot 10^2$

$[t_0, t_0 + 100\,\text{s})$). These scenarios include different classes of demand profiles for each loading station, different initial states of the system, queues on different links and different time criticality measures (e.g., cases where the transportation of the bags is very tight, i.e., the last bag that enters the system can only arrive in time at the corresponding end point if the shortest path is used and its DCV is continuously running with maximum speed, or cases where the timing is more relaxed).

16.5.1 Discussion

In order to solve the nonlinear, nonsmooth MPC optimization problem, one may use specialized search algorithms [5, 6] such as sequential quadratic programming algorithms, pattern search, genetic algorithms, and so on. We have chosen the genetic algorithm `ga` of the MATLAB optimization toolbox *Genetic Algorithm and Direct Search* with multiple runs and "bitstring" population, since simulations show that this optimization technique gives good performance with the shortest computation time.

Based on simulations we now compare, for the given scenarios, the proposed control methods. For all the proposed predictive control methods we set the horizon to $N = 5$ bags. We make this choice since for a larger horizon, the computation time required to obtain a good solution of the local optimization problem increases substantially. Hence, using larger horizons for the considered MPC optimization problems yields a considerably larger total computation time.

Let $J_j^{\text{tot,approach}}$ denote the performance of the baggage handling system corresponding to scenario index j and the considered control approach. Moreover, let $J^{\text{approach,avg}}$ denote the average performance

$$J^{\text{approach,avg}} = \frac{1}{|\Delta|} \sum_{j \in \Delta} J_j^{\text{tot,approach}}$$

with Δ the set of considered scenarios. Then in Table 16.1 we list the average results.

Theoretically, the performance of the baggage handling system obtained when the *centralized* predictive switch control is used is better than when the hierarchical approach is used. However, to obtain the true performance of centralized MPC would require extremely high computational time; see e.g., [13] where the CPU time is over 2 hours for routing 25 bags (on a network with only 2 junctions) when using a prediction horizon $N = 2$. Hence, to obtain the true optimum with centralized MPC requires a too high computational burden—centralized control becomes intractable in practice when the number of junctions is large due to the high computation time required. Therefore, we need to limit the computation time. In these simulations, in order to reduce the computational effort of the route choice control using centralized MPC, we ran the genetic algorithm four times for each optimization problem, while limiting the time allowed to compute a solution to 400 s.

The simulation results indicate that *distributed MPC* gives worse performance than centralized MPC. But this happens due to the time limitations that we have imposed when solving the nonlinear optimization problems. Note that when using distributed MPC we ran the genetic algorithm once for each local optimization problem, while allowing a maximum of three generations of population. We have chosen these options in order to have a balance between the overall performance and the total computation time.

Regarding the hierarchical control approach we note that we have set the control time step for the network controller to 60 s, and the control time step for the switch controller was set to 2 s. The simulation results indicate that using the hierarchical control framework yields a better system performance than using the other predictive methods. But, recall that the solutions of centralized or distributed MPC were returned by the prematurely terminated global and multistart local optimization method. However, even with the computational restrictions mentioned above (we allow a limited amount of time for solving an optimization problem), the total computation time of centralized MPC and of distributed MPC with a single round of downstream and upstream communication is much larger than (over 40 hours) the one of the hierarchical control (an average of 100 s per junction, plus 6 s for solving the MILP optimization problems).

Hence, the advantage of using hierarchical MPC is clear: much better performance and much lower computational effort. To compute centralized or distributed MPC solutions in a more precise way one should route only a few bags on a very simple network. But then all approaches show the same performance; the only advantage of using the hierarchical framework is the low computation time.

16.6 Summary

We considered the baggage handling process in large airports using destination coded vehicles (DCVs) running at high speeds on a network of tracks. Then, for a DCV-based baggage handling system, we developed and compared efficient control methods to determine the optimal DCV routing. In particular, we developed and compared centralized, distributed and hierarchical predictive methods to control the DCV routing.

In practice, centralized model predictive control (MPC) is not suitable for determining the optimal DCV routing due to the high computation time required to solve the route choice optimization problem. The simulation results indicate that distributed MPC yields a worse performance than centralized MPC when hard computational restrictions are imposed in solving the nonlinear optimizations. Simulation results also indicate that the hierarchical control with MILP flow solutions outperforms the other predictive control approaches, where the multistart local optimization method has been terminated prematurely.

In future work we will perform extensive simulations in order to assess the efficiency of these control approaches. Moreover, we will further improve the performance of the distributed MPC by considering multiple up and down rounds of optimizations and by extending the range of communication exchange to more than one level. Also, in order to account for the increased computation and communication time of such an approach, we will extend the local control area to more than one node and assess the efficiency and the balance between performance and computation and communication requirements of such an alternative approach.

Acknowledgements This research was supported by the VIDI project, "Multi-agent Control of Large-Scale Hybrid Systems" (DWV.6188) of the Dutch Technology Foundation STW, by the BSIK project, "Next Generation Infrastructures (NGI)", by the Transport Research Centre Delft, by the Delft Research Center Next Generation Infrastructures and by the European 7th framework STREP project, "Hierarchical and Distributed Model Predictive Control of Large-Scale Systems" (Contract number INFSO-ICT-223854).

References

1. Atamtürk, A., Savelsbergh, M.: Integer-programming software systems. Annals of Operations Research **140**(1), 67–124 (2005)
2. Bemporad, A., Morari, M.: Control of systems integrating logic, dynamics, and constraints. Automatica **35**(3), 407–427 (1999)
3. Fay, A.: Decentralized control strategies for transportation systems. In: Proc. IEEE Int. Conf. Control and Automation, pp. 898–903. Budapest, Hungary (2005)
4. Fletcher, R., Leyffer, S.: Numerical experience with lower bounds for MIQP branch-and-bound. SIAM Journal on Optimization **8**(2), 604–616 (1998)
5. Floudas, C.: Nonlinear and Mixed-Integer Optimization: Fundamentals and Applications. Oxford University Press, New York, USA (1995)
6. Gill, P.E., Murray, W., Wright, M.H.: Practical Optimization. Academic Press, London, UK (1981)
7. Hall, A., Hippler, S., Skutella, M.: Multicommodity flows over time: Efficient algorithms and complexity. Theoretical Computer Science **379**(3), 387–404 (2007)
8. Hallenborg, K., Demazeau, Y.: Dynamical control in large-scale material handling systems through agent technology. In: Proc. 2006 IEEE /WIC/ACM Int. Conf. Intelligent Agent Technology, pp. 637–645. Hong Kong, China (2006)
9. Langevin, A., Lauzon, D., Riopel, D.: Dispatching, routing, and scheduling of two automated guided vehicles in a flexible manufacturing system. International Journal of Flexible Manufacturing Systems **8**(3), 247–262 (1996)
10. Maciejowski, J.: Predictive Control with Constraints. Prentice Hall, Harlow, UK (2002)

11. de Neufville, R.: The baggage system at Denver: Prospects and lessons. Journal of Air Transport Management **1**(4), 229–236 (1994)
12. Taghaboni, F., Tanchoco, J.M.A.: Comparison of dynamic routeing techniques for automated guided vehicle systems. Int. Journal of Production Research **33**(10), 2653–2669 (1995)
13. Tarău, A., De Schutter, B., Hellendoorn, J.: Travel time control of destination coded vehicles in baggage handling systems. In: Proc. 17th IEEE Int. Conf. Control Applications, pp. 293–298. San Antonio, Texas, USA (2008)
14. Tarău, A., De Schutter, B., Hellendoorn, H.: Receding horizon approaches for route choice control of automated baggage handling systems. In: Proc. European Control Conf. (ECC2009), pp. 2978–2983. Budapest, Hungary (2009)
15. Tarău, A., De Schutter, B., Hellendoorn, J.: Decentralized route choice control of automated baggage handling systems. In: Proc. 12th IFAC Symposium on Control in Transportation Systems, pp. 70–75. Redondo Beach, California, USA (2009)
16. Tarău, A., De Schutter, B., Hellendoorn, J.: Predictive route choice control of destination coded vehicles with mixed integer linear programming optimization. In: Proc. 12th IFAC Symposium on Control in Transportation Systems, pp. 64–69. Redondo Beach, California, USA (2009)
17. Tarău, A., De Schutter, B., Hellendoorn, J.: DCV route control in baggage handling systems using a hierarchical control architecture and mixed integer linear programming. In: Proc. 3rd Int. Conf. Information Systems, Logistics and Supply Chain (ILS 2010). Casablanca, Morocco (2010)
18. Weyns, D., Holvoet, T.: Architectural design of a situated multiagent system for controlling automatic guided vehicles. Int. Journal Agent Oriented Software Engineering **2**(1), 90–128 (2008)

Chapter 17
Stability with Uniform Bounds for On-line Dial-a-Ride Problems under Reasonable Load

Sven Oliver Krumke and Jörg Rambau

Abstract In continuously running logistic systems (like in-house pallet transportation systems), finite buffer capacities usually require controls that can achieve uniformly bounded waiting queues (strong stability). Standard stochastic traffic assumptions (arrival rates below service rates) cannot, in general, guarantee these strong stability requirements, no matter which control policy is used. So, the worst-case traffic notion of reasonable load was introduced, originally for the analysis of the on-line dial-a-ride Problem. A set of requests is reasonable if the requests that are presented in a sufficiently large time period can be served in a time period of at most the same length. The rationale behind this concept is that the occurrence of nonreasonable request sets renders the system overloaded, requiring capacity be extended. For reasonable load, there are control policies that can guarantee uniformly bounded flow times, leading to strong stability in many cases. Control policies based on naïve reoptimization, however, can in general achieve neither bounded flow times nor strong stability. In this chapter, we review the concept and examples for reasonable load. Moreover, we present new control policies achieving strong stability as well as new elementary examples of request sets where naïve reoptimization fails.

17.1 Introduction

Consider a distribution center with various floors, connected to a warehouse for standard pallets. Pallets are moving horizontally along conveyor belts and vertically

---

Sven Oliver Krumke
Department of Mathematics University of Kaiserslautern Paul-Ehrlich-Str. 14-434 67663 Kaiserslautern, Germany
e-mail: krumke@mathematik.uni-kl.de

Jörg Rambau
Lehrstuhl für Wirtschaftsmathematik, Universität Bayreuth, D-95440 Bayreuth, Germany
e-mail: joerg.rambau@uni-bayreuth.de

R. Johansson and A. Rantzer (Eds.): Distributed Decision Making and Control, LNCIS 417, pp. 387–412.
springerlink.com

Fig. 17.1 Pallet elevators at the distribution center of Herlitz with a single waiting slot in front of each

in elevators. (One such system can be found in a distribution center of Herlitz, for office supplies, in Falkensee near Berlin.) By design of the microcontrol, a pallet can only move forward on a conveyor belt, if there is enough space in front of it. A section of a conveyor belt completely filled with pallets causes a complete standstill of that belt. In such an event, pallets have to be removed manually in order to resume transportation.

For example, if pallets requesting an elevator stay too long in the subsystem available for waiting pallets, that subsystem will cease to work. Thus, the goal is to control an elevator in such a way that the waiting slot capacities are never exceeded. The number of pallets in such waiting slots can be minimized if the average flow times (also called sojourn times) of the pallets waiting for an elevator are minimized. Another requirement is that the flow time of an individual pallet not be arbitrarily large (infinite deferment), since such a forgotten pallet can hold back the delivery of a large order. Figure 17.1 shows the single waiting slots in front of some pallet elevators in the Herlitz distribution center.

There are several mathematical theories that suggest a framework for the analysis of this system. *Queuing theory* [10] captures particularly well average entities in the steady state of such a system. It utilizes stochastic information on the input stream of requests. However, it is extremely difficult to derive nontrivial control policies from it, since policies are usually an input and not an output of the computations. *Stochastic dynamic programming* in Markov decision processes [19] is, in principle, suitable for finding a control policy for stochastic inputs that is optimal in expectation, but in this case the curse of dimensionality renders a direct application impossible. For example, an elevator system with one elevator of capacity one and with n floors containing w waiting slots each requires a state space of at least n^{nw+1} states. For one waiting slot on each of eight floors—as can be found at Herlitz—this means more than 134 million states, and this does not yet include state information on already accumulated waiting times in the system. *Approximate dynamic programming* [5, 6] yields policies that are reported to perform well in experiments, but performance guarantees cannot be given.

An interest in worst-case results rather than expected performance measures triggered the concept of *competitive analysis* [4]: The worst case results computed by min-max dynamic programming are often meaningless because in the worst-case all policies are equally and maximally bad; see, for example, the famous paging problem [4, 7], where, obviously, each paging policy may have page fault at every page request. To present a principal solution to this dilemma, the more game-theoretic competitive analysis was developed [4, 7] and gave rise to the area of *on-line optimization* in algorithmic theory. In on-line optimization, an *on-line algorithm* is presented a sequence of requests. The on-line algorithm has to answer the requests in such a way that an answer at any point in time is a function of all requests and answers given up to that time; i.e., the on-line algorithm has no clairvoyance. In competitive analysis, the cost of an on-line algorithm on a *particular request sequence* is compared to the cost of an *optimal off-line algorithm* on *the same request sequence*. In contrast to the on-line algorithm, the optimal off-line algorithm is clairvoyant, i.e., it knows the whole sequence in advance and computes an optimal sequence of answers. Both the online- and offline-algorithm are assumed to have unlimited computing power. This way, the performance difference between online- and offline-algorithm is caused only by the different amounts of information they can use: the offline-algorithm has complete information, the online-algorithm has no information about future requests. The supremum of the cost ratios over all request sequences is the *competitiveness* of the on-line algorithm. The infimum of the competitivenesses over all on-line algorithms is the competitiveness of the on-line problem.

However, for many practical problems, the so-called *triviality barrier* is met: All on-line algorithms are equally bad if compared to an optimal off-line algorithm. This is, unfortunately, also the case in our application: the competitiveness of our problem is infinite. The reason for this is simple: There are arbitrarily long request sequences for which the optimal off-line algorithm can always be at the floor exactly when a new request arrives (i.e., no waiting time), whereas this is impossible for any on-line algorithm (i.e., positive waiting time). Thus, the cost ratio depends on the request sequence, i.e., there is no uniform bound.

Recent progress has been made using *stochastic dominance* as a means for performance comparison of on-line algorithms [14]. The statement that one algorithm stochastically dominates another has far-reaching consequences. For many interesting pairs of algorithms, a stochastic-dominance relation does not hold or is very difficult to prove. Thus, the problems that can be tackled by this approach so far are quite elementary.

Despite these difficulties in the theoretical analysis, experience shows of so-called *replan algorithms* in on-line optimization perform satisfactorily. These are on-line algorithms that resemble the paradigm of a receding-horizon in model predictive control (MPC); in a closed loop, a control that is computed as the optimum open-loop control over some finite horizon is used. The development of the system is estimated on the basis of a model of the system behavior and assumptions on the possible future inputs and/or disturbances. Since future requests in on-line optimization are completely unknown and no deterministic prediction is likely enough to become true, replan algorithms in on-line optimization usually perform their computations with no future requests at all. The model prediction then restricts itself to the forecast of future events, when the schedule is carried out with no new requests arriving. We assume in this chapter that this part of the prediction is exact.

In our application the structure of a generic replan algorithm is informally described as follows. At a particular point in time, we consider only pallets known to the system (a *system snapshot*). We then compute an open-loop scheduling that is optimal with respect to carefully engineered constraints and a carefully engineered objective function (the *snapshot problem*). This scheduling is "used" until a new request enters the system. The new request then triggers a new optimization computation. What does it mean to "use" a schedule? In order to interpret a policy utilizing precomputed schedules as a control policy for the elementary elevator movements, the schedules need to be translated in sequences of microscopic controls like "go up/down one floor", "halt at current floor", "let pallet enter/exit". Although in the on-line optimization community this transformation of schedules and routings into low level controls is rarely mentioned explicitly, there are more detailed descriptions of this process in the MPC literature [22]. We call the resulting control policy a *replan policy*. Since we are assuming that carrying out a schedule happens with no disturbances, this transition is straightforward and is not made explicit in the following. So, the terms "policy" and "on-line algorithm" can be understood interchangeably in the following.

The arguably most straightforward replan policy is what we call *naïve reoptimization*. It assumes that there is a so-called *associated off-line version* of the problem. This is an off-line optimization problem that describes the best possible operation of the system given all inputs throughout an evaluation period. Such an evaluation period may be one day for the elevator system in our application. The off-line version should have the property that an optimal solution to it corresponds to a best possible control of the system when the inputs are exactly as predicted. The (non-implementable) policy given by a solution to the off-line version with precisely the actual future requests can be interpreted as a policy used by a clairvoyant controller that optimizes controls under exact predictions of the future on the whole evaluation period. Naïve reoptimization is the (implementable) policy that uses the same constraints and objective as the off-line version does; the inputs, however, are restricted to all currently known inputs. Such a policy is optimal when no further inputs occur. In our case, when no additional requests arrive. Note, that on the controller level this policy still uses nontrivial predictions because carrying out a schedule through all known requests usually takes many stages of low level controls.

In order to utilize sensible stochastic forecasts about incoming requests, all replan policies can in principle be enhanced by incorporating a probability distribution over a limited number of future requests. This, however, leads to extremely hard to solve stochastic optimization models in the snapshot problems, and we know of no real-world example where this stochastic replan technique has been applied.

More common is the addition of further constraints or penalty terms in the objective function to the snapshot problem of naïve reoptimization. These constraints and penalties represent a mostly heuristic safeguard against unwanted system states that are not ruled-out by the off-line optimization model on known requests. One example is the common technique of using an average squared waiting time in the objective of the snapshot problem instead of an average waiting time [18, 16]. The goal of this is to balance the desire for individually not too large waiting times (fairness) with small average waiting times (performance). The main problem with this engineering approach is that in many cases nothing can be proved about the performance of the resulting replan policy.

Our application is an instance of the so-called *on-Line dial-a-ride problem*. For the sake of an easier exposition, we restrict ourselves to the special case of a single server of capacity one. In our application, this means that want to control an idealized elevator of capacity one with one waiting queue of infinite capacity.

We show the following for such a system: The combination of ideas from on-line optimization, model predictive control and queuing theory yields experimentally and theoretically reliable control policies for these elevator systems. These stable policies are *not* defined as REPLAN-policies: our policies ignore some open-loop solutions and keep the solutions computed earlier. The theoretical performance guarantees are dependent on a bound on the combinatorial load in the system—the *reasonability* of the input stream. As we go along, we put known results into context and present some so far unpublished results.

The chapter is structured as follows: Section 17.2 formally defines the problem under consideration. In Sec. 17.3 we define some on-line algorithms for the OLDARP, followed by some known performance results in Sec. 17.4. Our original contribution beyond these results is outlined in Sec. 17.5. The core concept dealt with in this chapter is introduced formally in Sec. 17.6, before we suggest the notion of strong stability in Sec. 17.7. Sections 17.8 and 17.9 present performance guarantees under reasonable load for two competitive algorithms, IGNORE and SMARTSTART, respectively. Sections 17.10 and 17.11 show that such performance guarantees do not exist for the seemingly more natural competitive algorithms REPLAN and AVGFLOWREPLAN. The analysis for the new algorithm DELTAREPLAN is carried out in Sec. 17.12. Section 17.13 concludes the chapter and offers possible further directions.

17.2 Formal Problem Statement

After the informal introduction let us define the problem under consideration more precisely. We are given a metric space (X,d) with a special vertex $o \in X$ (the origin) in which a server of capacity $C \in \mathbb{R}_{\geq 0} \cup \{\infty\}$ moves at unit speed in order to serve transportation requests. Requests are triples $r = (t,a,b)$, where a is the start point of a transportation task, b its endpoint, and t its release time, which is—in this context—the time where r becomes known. If r is a request, we also use $t(r)$ for its release time and $a(r)$ and $b(r)$ for its start and endpoint, respectively.

A *transportation move* is a quadruple $m = (\tau,x,y,R)$, where a is the starting point and b the endpoint and τ the starting time, while R is the set (possibly empty) of requests carried by the move. The *arrival time* of a move is the sum of its starting time τ and $d(x,y)$. A *(closed) transportation schedule* for a sequence σ of requests is a sequence

$$S = (\tau_1,x_1,y_1,R_1),(\tau_2,x_2,y_2,R_2),\ldots,(\tau_\ell,x_\ell,y_\ell,R_\ell)$$

of transportation moves with the following properties:

(i) The $(i+1)$st move starts at the endpoint of the ith move and not earlier than the time that the ith move is completed; that is, $x_{i+1} = y_i$ and $\tau_{i+1} \geq \tau_i + d(x_i,y_i)$ for all i.

(ii) Each move carries at most C requests, that is, $|R_i| \leq C$ for all i.

(iii) For any request $r \in \sigma$, the subsequence of S consisting of those moves (τ_i,x_i,y_i,R_i) with $r \in R_i$ is a contiguous nonempty subsequence

$$S(r) = (\tau_l,x_l,y_l,R_l),\ldots,(\tau_{l+p},x_{l+p},y_{l+p},R_{l+p})$$

of S which forms a transportation from $a(r)$ to $b(r)$, that is, $x_l = a(r)$ and $y_{l+p} = b(r)$. The subtransportation $S(r)$ does not start before r is released, that is, $\tau_l \geq t(r)$.

(iv) The first move starts in the origin o and the last move ends in the origin o.

The time τ_1 and the point $x_1 \in X$ are called the *starting time* and the *starting point* of S. Similarly, the time $\tau_\ell + d(x_\ell,y_\ell)$ and the point y_ℓ are referred to as the *end time* and the *endpoint* of S.

An *on-line algorithm* for OLDARP has to move a server in X so as to fulfill all released transportation tasks without preemption (i.e., once an object has been picked up it is not allowed to be dropped at any other place than its destination; see Condition (ii) above), while it does not know about requests that are presented in the future. In order to plan the work of the server, the on-line algorithm may maintain a preliminary (closed) transportation schedule for all known requests, according to which it moves the server.

A posteriori, the moves of the server induce a complete transportation schedule that may be compared to an off-line transportation schedule that is optimal with respect to some objective function. This is the core of competitive analysis of on-line algorithms.

An on-line algorithm A is called *c-competitive* if there exists a constant c such that for any finite request sequence σ the inequality $\mathsf{A}(\sigma) \leq c \cdot \mathsf{OPT}(\sigma)$ holds. Here, $\mathsf{X}(\sigma)$ denotes the objective function value of the solution produced by algorithm X on input σ and OPT denotes an optimal off-line algorithm. Sometimes we are dealing with various objectives at the same time. We then indicate the objective *obj* in the superscript, as in $\mathsf{X}^{obj}(\sigma)$.

For a detailed set-up that focusses on competitive analysis see [1, 17].

17.3 Known On-line Algorithms

Several on-line algorithms have been suggested in the literature so far. We discuss REPLAN, IGNORE, and SMARTSTART because our stability results refer to them. All these algorithms stem from [1]. The algorithm REPLAN is based on ideas in [3]; the algorithm IGNORE appears in [21] in a more general context. The algorithms can be considered as typical representatives of construction principles for on-line algorithms: REPLAN reoptimizes whenever a new request arrives, IGNORE reoptimizes only when it becomes idle, whereafter it immediately continues to work, SMARTSTART reoptimizes only when idle and stays idle deliberately for a certain amount of time to gather more information about unserved requests.

The on-line algorithm REPLAN for the OLDARP is based on the general idea of a replan algorithm in the introduction in Sec. 17.1.

Definition 17.1 (Algorithm REPLAN). Whenever a new request becomes available, REPLAN computes a preliminary transportation schedule for the set R of all available requests by solving the problem of minimizing the total completion time of R. Then it moves the server according to that schedule until a new request arrives or the schedule is done.

The on-line algorithm IGNORE makes full use of every schedule it computes before it recomputes a new schedule.

Definition 17.2 (Algorithm IGNORE). Algorithm IGNORE works with an internal buffer. It may assume the following states (initially it is IDLE):

IDLE Wait for the next point in time when requests become available. Goto PLAN.

BUSY While the current schedule is in work store the upcoming requests in a buffer ("ignore them"). Goto IDLE if the buffer is empty else goto PLAN.

PLAN Produce a preliminary transportation schedule for all currently available requests R (taken from the buffer) minimizing comp for R. (Note: This yields a feasible transportation schedule for R because all requests in R are immediately available.) Goto BUSY.

The algorithm SMARTSTART was developed to improve the competitive ratios of REPLAN and IGNORE. The idea of this algorithm is basically to emulate the IGNORE algorithm but to make sure that each subtransportation schedule is completed "not too late": if a subschedule would take "too long" to complete then the algorithm waits for a specified amount of time. Intuitively this construction tries to avoid the worst case situation for IGNORE, in which right after the algorithm starts a schedule a new request becomes known.

In this section we use $l(S)$ to denote the length of a schedule (tour) S computed for a (sub)set of requests. SMARTSTART has a fixed "waiting scaling" parameter $\theta > 1$. From time to time the algorithm consults its "work-or-sleep" routine: This subroutine computes an (approximately) shortest schedule S for all unserved requests, starting and ending at the origin. If this schedule can be completed no later than time θt, i.e., if $t + l(S) \leq \theta t$, where t is the current time and $l(S)$ denotes the length of the schedule S, the subroutine returns (S, work); otherwise it returns (S, sleep).

In the sequel it will be convenient again to assume that the "work-or-sleep" subroutine uses a ρ-approximation algorithm for computing a schedule: The approximation algorithm always finds a schedule of length at most ρ times the optimal one.

Definition 17.3 (Algorithm SMARTSTART). The server of algorithm SMARTSTART can assume three states (initially it is IDLE):

IDLE If the algorithm is idle at time T and new requests arrive, it calls "work-or-sleep". If the result is (S, work), the algorithm enters the busy state where it follows schedule S. Otherwise the algorithm enters the sleeping state with wakeup time t', where $t' \geq T$ is the earliest time such that $t' + l(S) \leq \theta t'$ and $l(S)$ denotes the length of the just computed schedule S, i.e., $t' = \min\{t \geq T : t + l(S) \leq \theta t\}$.

SLEEPING In the sleeping state the algorithm simply does nothing until its wakeup time t'. At this time the algorithm reconsults the "work-or-sleep" subroutine. If the result is (S, work), then the algorithm enters the busy state and follows S. Otherwise the algorithm continues to sleep with new wake-up time $\min\{t \geq t' : t + l(S) \leq \theta t\}$.

BUSY In the busy state, i.e, while the server is following a schedule, all new requests are (temporarily) ignored. As soon as the current schedule is completed the server either enters the idle state (if there are no unserved requests) or it reconsults the "work-or-sleep" subroutine, which determines the next state (SLEEPING or BUSY).

17.4 Known Performance Guarantees

Competitive analysis of OLDARP provided the following (see [1]):

- IGNORE and REPLAN are 2.5-competitive for the goal of minimizing the *total completion time* of the schedule; SMARTSTART is 2-competitive for this problem, which is best-possible.
- For the task of minimizing the *maximal (average) waiting time* or the *maximal (average) flow time* there can be no algorithm with constant competitive ratio. In

particular, the algorithms REPLAN, IGNORE, and SMARTSTART have unbounded competitive ratios for this problem.

It should be noted that the corresponding off-line versions with release times (where all requests are known at the start of the algorithm) are $\mathcal{NP}$-hard to solve for the objective functions of minimizing the average or maximal flow time—it is even $\mathcal{NP}$-hard to find a solution within a constant factor from the optimum [15]. The off-line version without release times of minimizing the total completion time is polynomially solvable on special graph classes but $\mathcal{NP}$-hard in general [9, 2, 8, 13].

If we are considering a continuously operating system with continuously arriving requests (i.e., the request set may be infinite), then the total completion time is unbounded anyway, thus meaningless. Thus, in this case, the existing positive results cannot be applied, and the negative results tell us that we cannot hope for performance guarantees that may be relevant in practice. In particular, the performances of the two algorithms REPLAN and IGNORE cannot be distinguished by classical competitive analysis at all (both are 2.5 competitive with respect to the total completion time and not competitive at all with respect to the average or maximal flow time), and the performance of SMARTSTART can not be distinguished from any other algorithm if the average or maximal flow time is the goal.

In order to find theoretical guidance regarding which algorithm should be chosen, the notion of *Δ-reasonable load* was developed [12]. A set of requests is Δ-reasonable if requests released during a period of time $\delta \geq \Delta$ can always be served in time at most δ. A set of requests R is *reasonable* if there exists a $\Delta < \infty$ such that R is Δ-reasonable. That means for nonreasonable request sets we find arbitrarily large periods of time where requests are released faster than they can be served—even if the server has an optimal off-line schedule and all requests can be served immediately. When a system has only to cope with reasonable request sets, we call this situation *reasonable load*. Section 17.6 is devoted to the exact mathematical setting of this idea, because we need it for the new results.

The main historical result based on this idea in [12] is this: For the OLDARP under Δ-reasonable load, IGNORE yields a maximal and an average flow time of at most 2Δ, whereas the maximal and the average flow times of REPLAN are unbounded. The algorithms IGNORE and REPLAN have to solve a number of off-line instances of OLDARP, which is in general $\mathcal{NP}$-hard, as we already remarked. We will show how we can derive results for IGNORE when using an approximate algorithm for solving off-line instances of OLDARP (for approximation algorithms for off-line instances of OLDARP, refer to [9, 2, 8, 13]).

To this end, the notion of reasonable request sets was refined [12], introducing a second parameter that tells us how "fault tolerant" the request set is. In other words, the second parameter tells us, how "good" the algorithm has to be to show stable behavior. Again, roughly speaking, a set of requests is (Δ,ρ)-reasonable if requests released during a period of time $\delta \geq \Delta$ can be served in time at most δ/ρ. If $\rho = 1$, we get the notion of Δ-reasonable as described above. For $\rho > 1$, the algorithm is allowed to work "sloppily" (e.g., employ approximation algorithms) or have breakdowns to an extent measured by ρ and still show a stable behavior.

17.5 Outline of New Contributions

Simulation results [11] show that IGNORE indeed outperforms REPLAN in terms of the maximal flow time, but in terms of the average flow time the behavior of REPLAN is usually much better. This left open the question about whether IGNORE can be improved empirically without losing the performance guarantee. Alternatively, the question is this: is there a version of REPLAN that wins the performance guarantee of IGNORE but stays empirically efficient? As an answer to this question we present the algorithm DELTAREPLAN in Sec. 17.12.

The following results in this chapter have not been published elsewhere before:

- We present a proof that the replan policy SMARTSTART that is optimally competitive for the total completion time (the *makespan*) has bounded flow times under reasonable load as well.
- We show an example for which a replan policy with snapshot objective "minimize the average flow time" produces unbounded maximal and average flow times in the long run.
- We present one particular replan policy DELTAREPLAN that inherits the performance guarantee of IGNORE but is able to yield a better average flow time in simulations.
- We show that using a policy with bounded flow times yields uniformly bounded waiting queues, i.e., strong stability.

17.6 Reasonable Load in Detail

Crucial for the concept of reasonable load is the off-line version of a request set.

Definition 17.4. The *off-line version* of $r = (t, a, b)$ is the request

$$r^{\text{off-line}} := (0, a, b).$$

The *off-line version* of R is the request set

$$R^{\text{off-line}} := \left\{ r^{\text{off-line}} : r \in R \right\}.$$

An important characteristic of a request set with respect to system load considerations is the time period in which it is released.

Definition 17.5. Let R be a finite request set for OLDARP. The *release span* $\delta(R)$ of R is defined as

$$\delta(R) := \max_{r \in R} t(r) - \min_{r \in R} t(r).$$

Provably good off-line algorithms exist for the total completion time and the weighted sum of completion times. How can we make use of these algorithms in order to get performance guarantees for minimizing the maximum (average) waiting (flow) times? We suggest a way of characterizing request sets that we want to consider "reasonable."

In a continuously operating system we wish to guarantee that work can be accomplished at least as fast as it is presented. The idea is stolen from queuing theory where the input rate should not exceed the output rate. In the following we propose a mathematical set-up that models this idea in a worst case fashion. Since we are always working on finite subsets of the whole request set, the request set itself may be infinite, modeling a continuously operating system.

We start by relating the release spans of finite subsets of a request set to the time we need to fulfill the requests.

Definition 17.6. Let R be a request set for the OLDARP. A weakly monotone function

$$f\colon \begin{cases} \mathbb{R} & \rightarrow & \mathbb{R}, \\ \delta & \mapsto & f(\delta) \end{cases}$$

is a *load bound* on R if for any $\delta \in \mathbb{R}$ and any finite subset S of R with $\delta(S) \leq \delta$ the completion time $\mathrm{OPT}^{\mathrm{comp}}(S^{\text{off-line}})$ of the optimum schedule for the off-line version $S^{\text{off-line}}$ of S is at most $f(\delta)$. In formula:

$$\mathrm{OPT}^{\mathrm{comp}}(S^{\text{off-line}}) \leq f(\delta).$$

Remark 17.1. If the whole request set R is finite then there is always the trivial load bound given by the total completion time of R. For every load bound f we may set $f(0)$ to be the maximum completion time we need for a single request, and nothing better can be achieved.

A "stable" situation would easily obtained by a load bound equal to the identity $x \mapsto x$ on $\mathbb{R}$. (By "stable" we mean that the number of unserved requests in the system does not become arbitrarily large.) In that case we would never get more work to do than we can accomplish. If it has a load bound equal to a function id/ρ, where id is the identity and where $\rho \geq 1$, then ρ measures the tolerance of the request set: Assume we have an off-line algorithm at our disposal that produces, in the worst case, a cost of ρ times the cost of an optimal off-line algorithm, then we can still accomplish all the incoming work by using the IGNORE-algorithm: to compute a ρ-approximate schedule for the set R of all released but unserved requests. The load bound and the performance guarantee ensure that the schedule takes no longer than $\rho \cdot \Delta(R)/\rho = \Delta(R)$. Thus, the set of requests that are released in the meantime has a release span no larger than $\Delta(R)$, and we can proceed by computing a ρ-approximate schedule for that set.

However, we cannot expect that the identity (or any linear function) is a load bound for OLDARP because of the following observation: a request set consisting of one single request has a release span of 0, whereas in general it takes nonzero time to serve this request. In the following definition we introduce a parameter describing how far a request set is from being load-bounded by the identity.

Definition 17.7. A load bound f is *(Δ,ρ)-reasonable* for some $\Delta, \rho \in \mathbb{R}$ with $\rho \geq 1$ if

$$\rho f(\delta) \leq \delta \quad \text{for all } \delta \geq \Delta$$

A request set R is *(Δ,ρ)-reasonable* if it has a (Δ,ρ)-reasonable load bound. For $\rho = 1$, we say that the request set is *Δ-reasonable.*

In other words, a load bound is *(Δ,ρ)-reasonable*, if it is bounded from above by $id(x)/\rho$ for all $x \geq \Delta$ and by the constant function with value Δ/ρ otherwise.

Remark 17.2. If Δ is sufficiently small so that all request sets consisting of two or more requests have a release span larger than Δ, then the first-come-first-serve policy is good enough to ensure that there are never more than two unserved requests in the system. Hence, the request set does not require scheduling the requests in order to provide for a stable system.

In a sense, Δ is a measure for the combinatorial difficulty of the request set R. If R is not Δ-reasonable for any $\Delta > 0$, then this indicates that the capacity of the system does not suffice to keep the system stable. Then, the task is not to find the best control but to add capacity first.

Thus, it is natural to ask for performance guarantees for the flow times of algorithms in terms of the reasonability Δ of the input. This is discussed for various algorithms in Sec. 17.8 through 17.12. Before that, we want to argue that flow time bounds guarantee a certain form of stability.

17.7 Strong Stability

We want to find an on-line algorithm for which there is a uniform bound on the number of requests in the system. More formally:

Definition 17.8. An on-line algorithm ALG for OLDARP on (X,d) with origin o is *strongly Δ-stable* if there exists $M \geq 0$ such that for each Δ-reasonable request set R the number of unserved requests in the system controlled by ALG is never larger than M.

In the stochastic setting, Little's formula [10] provides a relation between the traffic, the expected number of requests in the system and the expected flow time of the requests: The expected number of requests in the system equals the average number of requests entering the system times the expected flow time of requests. We would like to replace the traffic intensity by our Δ, but since we are in a worst case setting, the corresponding relation does not always hold.

In contrast to traffic conditions in queuing theory, Δ is only indirectly related to the number of requests in the system. The problem occurs when the service times for requests are not bounded away from zero. The extreme case is an on-line traveling salesman problem, where requests must be visited, nothing else; in that case, the server can serve an unlimited number of requests in arbitrarily little time if it is already very close to the position of the requests. For a short time then, there may be an unlimited number of requests in the system, although serving them requires only an arbitrarily small amount of time, thereby not violating any Δ-reasonability requirement. It is clear that in such a situation no algorithm can achieve strong stability.

The situation is different when serving a request takes at least time $\tau > 0$. In the elevator example this is true, because each pallet has to be transported for at least one floor. We call this variant of OLDARP the *on-line dial-a-ride problem with minimal transport time* τ or τ-OLDARP, for short.

We then obtain the following:

Theorem 17.1. *If the maximal flow time of* ALG *for* τ-OLDARP *is at most* $f(\Delta)$ *for all* Δ*-reasonable request sets, then* ALG *is strongly* Δ*-stable. More specifically, the number of requests in the system is never larger than* $f(\Delta)/\tau$.

Proof. *The time we need to serve a request is at least* τ. *If a request subset with release span at most* Δ *contains more than* Δ/τ *requests, then we need more time than* $\Delta = \tau\Delta/\tau$ *to serve it off-line. The maximal number of requests that can enter the system in time* Δ *is therefore* Δ/τ. *If each request leaves the system after at most* $f(\Delta)$ *time units, then there may be at most*

$$f(\Delta) \cdot \frac{\Delta}{\tau} \cdot \frac{1}{\Delta} \tag{17.1}$$

requests at the same time in the system. □

The result of this (elementary) discussion is that in order to obtain strongly stable on-line algorithms it is sufficient to find on-line algorithms with bounded maximal flow times.

17.8 Bounds for the Flow Times of IGNORE

We are now in a position to prove bounds for the maximal and average flow times, respectively, in the OLDARP for algorithm IGNORE. We assume that IGNORE solves off-line instances of OLDARP employing a ρ-approximation algorithm.

Let us consider the intervals in which IGNORE organizes its work in more detail. The algorithm IGNORE induces a dissection of the time axis $\mathbb{R}$ in the following way: We can assume, without loss of generality, that the first set of requests arrives at time zero. Let $\delta_0 = 0$, i.e., the point in time where the first set of requests is released (these are processed by IGNORE in its first schedule). For $i > 0$ let δ_i be the duration of the time period the server is working on the requests that have been ignored during the last δ_{i-1} time units. Then the time axis is split into the intervals

$$[\delta_0 = 0, \delta_0], (\delta_0, \delta_1], (\delta_1, \delta_1 + \delta_2], (\delta_1 + \delta_2, \delta_1 + \delta_2 + \delta_3], \ldots$$

Let us denote these intervals by $I_0, I_1, I_2, I_3, \ldots$. Moreover, let R_i be the set of those requests that are presented in I_i. Clearly, the complete set of requests R is the disjoint union of all the R_i.

At the end of each interval I_i we solve an off-line problem: All requests to be scheduled are already available. The work on the computed schedule starts immediately (at the end of interval I_i) and is done δ_{i+1} time units later (at the end of interval I_{i+1}). On the other hand, the time we need to serve the schedule is not more than ρ times the optimal completion time $\mathsf{OPT}^{\text{comp}}(R_i^{\text{off-line}})$ of $R_i^{\text{off-line}}$. In other words:

Lemma 17.1. *For all $i \geq 0$ we have*

$$\delta_{i+1} \leq \rho \cdot \mathsf{OPT}^{\text{comp}}(R_i^{\textit{off-line}}).$$

Let us now state and prove the main result of this section, first proved in [12], about the maximal flow time $\mathsf{IGNORE}^{\text{maxflow}}(R)$ incurred by IGNORE on any reasonable request set R.

Theorem 17.2 ([12]). *Let $\Delta > 0$ and $\rho \geq 1$. For all instances of* OLDARP *with (Δ, ρ)-reasonable request sets,* IGNORE *employing a ρ-approximate algorithm for solving off-line instances of* OLDARP *yields a maximal flow time of no more than 2Δ.*

Proof. *Let r be an arbitrary request in R_i for some $i \geq 0$, i.e., r is released in I_i. By construction, the schedule containing r is finished at the end of interval I_{i+1}, i.e., at most $\delta_i + \delta_{i+1}$ time units later than r was released. Thus, for all $i > 0$ we get that*

$$\mathsf{IGNORE}^{\text{maxflow}}(R_i) \leq \delta_i + \delta_{i+1}.$$

If we can show that $\delta_i \leq \Delta$ for all $i > 0$ then we are done. To this end, let $f: \mathbb{R} \to \mathbb{R}$ be a (Δ, ρ)-reasonable load bound for R. Then $\mathsf{OPT}^{\text{comp}}(R_i^{\textit{off-line}}) \leq f(\delta_i)$ because $\delta(R_i) \leq \delta_i$.

By Lemma 17.1, we get for all $i > 0$,

$$\delta_{i+1} \leq \rho \mathsf{OPT}^{\text{comp}}(R_i^{\textit{off-line}}) \leq \rho f(\delta_i) \leq \max\{\delta_i, \Delta\}.$$

Using $\delta_0 = 0$ the claim now follows by induction on i. □

The average flow time of IGNORE is also bounded, because the average is never larger than the maximum.

Corollary 17.1. *Let $\Delta > 0$. For all Δ-reasonable request sets algorithm* IGNORE *yields a average flow time no more than 2Δ.*

17.9 Bounds for the Flow Times of SMARTSTART

The analysis of SMARTSTART under reasonable load was not published before; it essentially parallels that of IGNORE, so we only highlight the differences. The crucial observation needed is formulated in the following lemma.

Lemma 17.2. *For (Δ, ρ)-reasonable request sequences, the server of* SMARTSTART *never sleeps after time $\bar{t} := \Delta/\theta - 1$.*

Proof. *Consider a call to the work-or-sleep routine at an arbitrary time $t \geq \bar{t}$. Let R be the set of requests not served by* SMARTSTART *at time t and let S be a ρ-approximate shortest schedule for R. By the (Δ, ρ)-reasonability of the input sequence, the length of schedule S for R can be bounded from above by*

$$l(S) \leq \rho \cdot \max\left\{\frac{\Delta}{\rho}, \frac{\delta(R)}{\rho}\right\} = \max\{\Delta, \delta(R)\}.$$

Trivially, we have $\delta(R) \leq t$, since all requests in R have been released at time t. Hence, it follows that

$$\begin{aligned} t + l(S) &\leq t + \max\{\Delta, \delta(R)\} \\ &\leq t + \max\{\Delta, t\} && \text{(since } \delta(R) \leq t\text{)} \\ &= t + \max\{(\theta - 1)\bar{t}, t\} && \text{(since } \bar{t} = \Delta/(\theta - 1)\text{)} \\ &\leq \theta t && \text{(since } t \geq \bar{t}\text{)}. \end{aligned}$$

Consequently, the work-or-sleep routine does not return the invitation to sleep.

The same arguments as given above show that, if SMARTSTART *goes to sleep before some time $t < \bar{t}$, the wak-eup time is no later than time $\bar{t}$. Hence, the lemma follows.* □

Let S be the last schedule started by SMARTSTART no later than time $\bar{t}$ and denote by $t_S \leq \bar{t}$ its start time. From Lemma 17.2 we conclude that from time $\bar{t}$ on, SMARTSTART behaves like IGNORE, provided the input sequence is (Δ, ρ)-reasonable. Using the arguments given in the proof of Theorem 17.2 we can conclude that the flow time of any request released after time t_S is bounded from above by 2Δ.

It remains to treat the requests released before time $\bar{t}$. Using again the arguments of Theorem 17.2 we derive that all requests released after time t_S have flow time at most 2Δ and we finally need to consider those requests released until time t_S. Each of these requests is either served by S or by an even earlier schedule. Since, by definition of SMARTSTART, the transportation schedule S is completed no later than time

$$\theta t_S < \theta \bar{t} = \frac{\theta}{\theta - 1}\Delta$$

we obtain the following result:

Theorem 17.3. *Let $\Delta > 0$ and $\rho \geq 1$. For all instances with (Δ,ρ)-reasonable request sets, algorithm* SMARTSTART *employing a ρ-approximation algorithm in its work-or-sleep routine yields a maximal flow time of no more than*

$$\max\left\{\frac{\theta}{\theta-1}\Delta, 2\Delta\right\}$$

In particular, if $\theta \geq 2$, then the maximal flow time provided by SMARTSTART *is bounded from above by 2Δ.* □

As in the case of IGNORE a we can derive a trivial upper bound of 2Δ for the average flow time of SMARTSTART under reasonable load.

17.10 An Example with Unbounded Flow Times for REPLAN

In the sequel, we provide an instance of OLDARP and a Δ-reasonable request set R such that the maximal flow time $\text{REPLAN}^{\text{maxflow}}(R)$ (and thus also the average flow time) of REPLAN is unbounded. This was first proved in [12]. Recall that REPLAN uses a snapshot optimization problem in which the total completion time is minimized. Hence, REPLAN is not a naïve replan policy, since our evaluation objective is the maximal flow time.

Theorem 17.4 ([12]). *There is an instance of* OLDARP *under reasonable load such that the maximal and the average flow time of* REPLAN *is unbounded.*

Proof. *In Fig. 17.2 there is a sketch of an instance for the* OLDARP. *The metric space is a path on four nodes a,b,c,d with origin a; the length of the path is ℓ, the distances are $d(a,b) = d(c,d) = \varepsilon$, and hence $d(b,c) = \ell - 2\varepsilon$. At time 0 a request from a to d is issued; at time $3/2\ell - \varepsilon$, the remaining requests periodically come in pairs from b to a and from c to d, respectively. The time distance between them is $\ell - 2\varepsilon$. We show that for $\ell = 18\varepsilon$ the request set R indicated in the picture is $2\frac{2}{3}\ell$-reasonable. Indeed: it is easy to see that the first request from a to d does not influence reasonability. Consider an arbitrary set R_k of k adjacent pairs of requests from b to a and from c to d, respectively. Then the release span $\delta(R_k)$ of R_k is*

$$\delta(R_k) = (k-1)(\ell - 2\varepsilon).$$

The off-line version ${R_k}^{\textit{off-line}}$ of R_k can be served as follows: First, move the server to c, the starting point of the upper requests. This contributes cost $\ell - \varepsilon$. Next, serve all the upper requests and go back to c: this contributes cost $k \times 2\varepsilon$. Then, go down to b, the starting point of the lower requests. This contributes another $\ell - 2\varepsilon$ to the cost. Now, serve the first lower requests. The additional cost for this is ε. Finally, serve the remaining lower requests at an additional cost of $(k-1)\cdot 2\varepsilon$. In total, we have the following:

$$\text{OPT}^{\text{comp}}({R_k}^{\textit{off-line}}) = 2\ell + (k-1)\cdot 4\varepsilon.$$

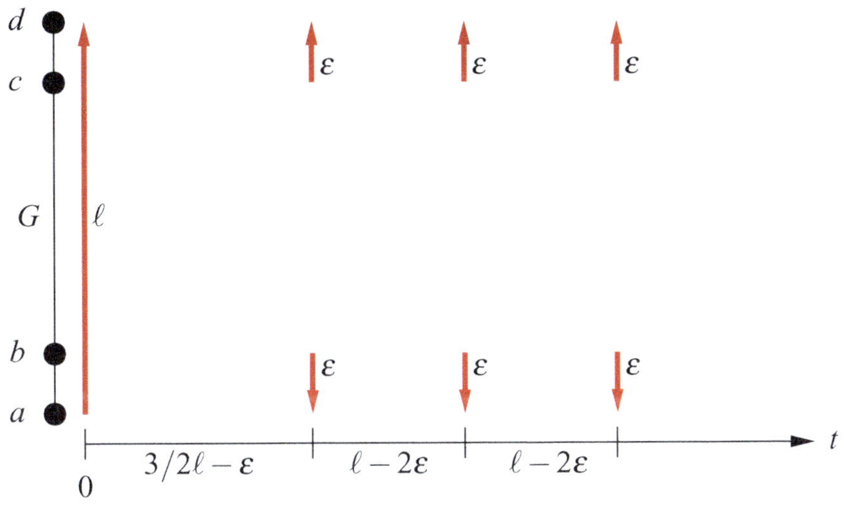

Fig. 17.2 A sketch of a $(2\frac{2}{7}3 \cdot \ell)$-reasonable instance of OLDARP ($\ell = 18\varepsilon$). The horizontal axis holds the time, the vertical axis depicts the metric space in which the server moves. A request is denoted by an arrow from its starting point to its endpoint horizontally positioned at its release time.

In order to find the smallest parameter Δ for which the request set R_k is Δ-reasonable we solve for the integer $k-1$ and get

$$k - 1 = \left\lceil \frac{2\ell}{\ell - 6\varepsilon} \right\rceil = 3.$$

Hence, we can set Δ to

$$\Delta := \mathsf{OPT}^{\text{comp}}({R_4}^{\text{off-line}}) = 2\tfrac{2}{3}\ell.$$

Now, we define

$$f : \begin{cases} \mathbb{R} & \to \quad \mathbb{R}, \\ \delta & \mapsto \quad \begin{cases} \Delta, & \text{for } \delta < \Delta \\ \delta, & \text{otherwise.} \end{cases} \end{cases}$$

By construction, f is a load bound for R_4. Because the time gap after which a new pair of requests occurs is certainly larger than the additional time we need to serve it (off-line), f is also a load bound for R. Thus, R is Δ-reasonable, as desired.

Now: how does REPLAN *perform in this instance? In Fig. 17.3 we see the track of the server following the preliminary schedules produced by* REPLAN *on the request set R.*

The maximal flow time of REPLAN *on this instance is realized by the flow time of the request $(3/2\ell - \varepsilon, b, a)$, which is unbounded.*

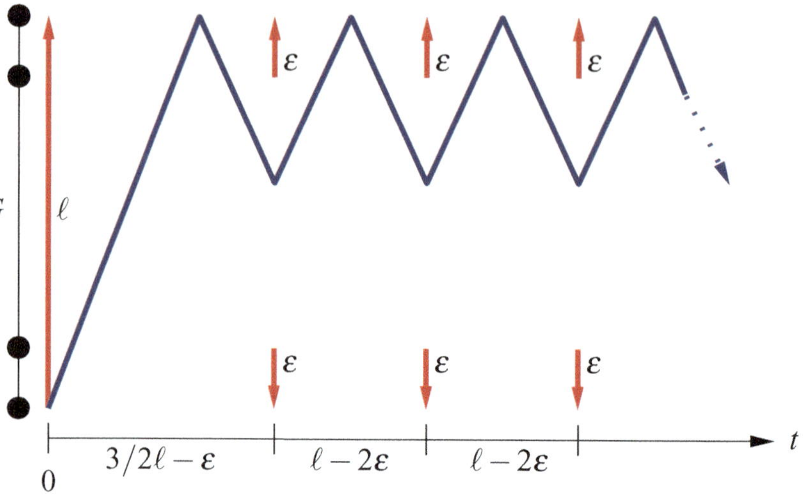

Fig. 17.3 The track of the REPLAN-server is drawn as a line in the diagram. At each point in time t we can read off the position of the server by looking at the height of the line at the horizontal position t. Because a new pair of requests is issued exactly when the server is still closer to the requests at the top, all the requests at the bottom will be postponed in an optimal preliminary schedule. Thus, the server always returns to the top when a new pair of requests arrives.

Moreover, since all requests from b to a are postponed after serving all the requests from c to d we get that REPLAN *produces an unbounded average flow time as well.* □

In Figure 17.4 we show the track of the server under the control of the IGNORE-algorithm. After an initial inefficient phase the server ends up in a stable operating mode. This example also shows that the analysis of IGNORE in Sec. 17.8 is sharp.

17.11 An Example with Unbounded Flow Times for AVGFLOWREPLAN

It is quite a natural question to ask whether modified replan strategies AVGFLOWREPLAN or MAXFLOWREPLAN which use snapshot problems that minimize the average and maximal flow times, respectively, would give a reasonable bound on the maximal and average flow times in the on-line situation. In our taxonomy, MAXFLOWREPLAN implements the naïve replan policy when the evaluation objective is the minimization of the maximal flow time. And AVGFLOWREPLAN corresponds to the naïve replan policy when the evaluation objective is the minimization of the average flow time.

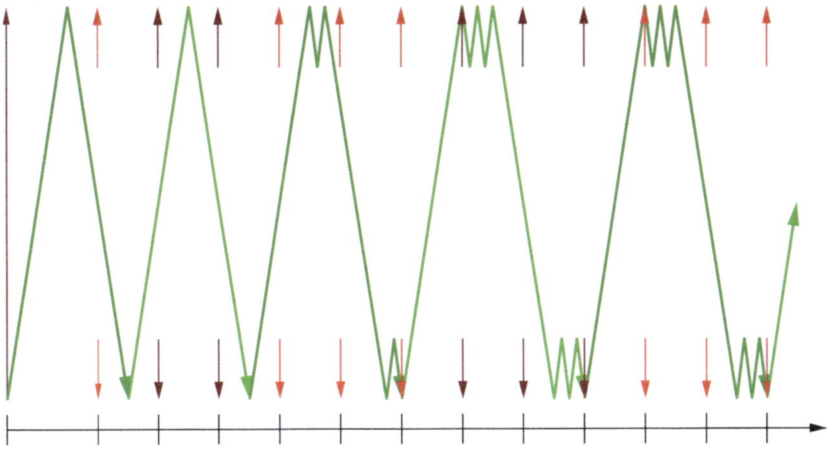

Fig. 17.4 The track of the IGNORE-server.

We mentioned already that the off-line problem of minimizing the average flow time is very hard. In the off-line problem that AVGFLOWREPLAN has to solve, however, all requests have release times in the past. It is then easy to see that the problem is equivalent to the minimization of the average completion time counted from the point in time where the planning takes place. Moreover, since the average flow time is larger by the "average age" of the requests, the performance guarantees of approximation algorithms minimizing the average completion time carry over. Still, in our computational experience minimization of the average completion time takes more time than minimizing the total completion time.

Anyway: the following result shows that even under reasonable load we cannot expect a worst case stable behaviour of AVGFLOWREPLAN, a so far unpublished result.

Theorem 17.5. *There is an instance of* OLDARP *under reasonable load such that the maximal and average flow times of* AVGFLOWREPLAN *are unbounded.*

Proof. *We construct a set of requests in the same metric space as in previous Sec. 17.10 as follows:*

- *At time* 0 *we issue again one request from* a *to* d.
- *At time* $T_0 := 3/2\ell - \varepsilon$ *we issue a pair of requests* R_1^u *from* c *to* d *and* R_1^l *from* b *to* a.
- *At time* $T_{i+1} := T_i + \ell + 2(i-2)\varepsilon$ *we issue:*
  - *a set of* i *"upper" requests* R_{i+1}^u *from* c *to* d, *and*
  - *one "lower" request* R_{i+1}^l *from* b *to* a.

Figure 17.5 sketches the construction.

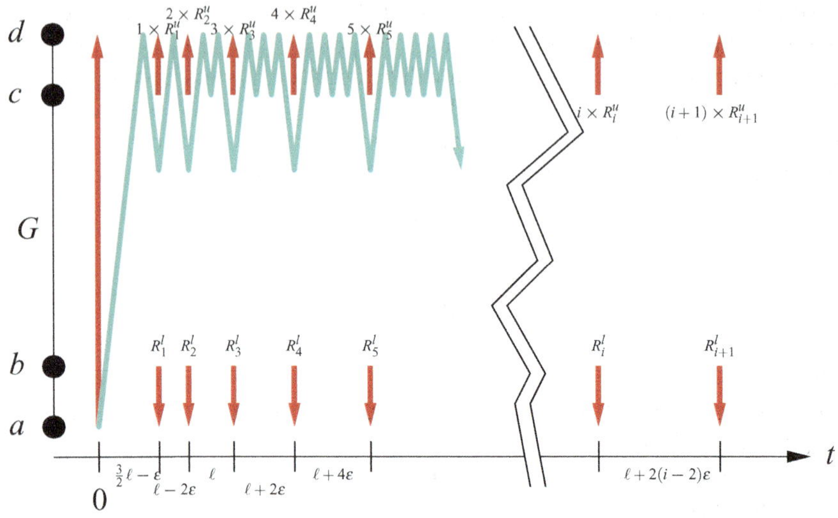

Fig. 17.5 The track of the AVGFLOWREPLAN-server on a the example from Theorem 17.5.

For $\ell = 18\varepsilon$ this request set is again $2\frac{2}{3}\ell$-reasonable, since we have increased the time intervals between the release times of the requests by the additional amount needed to serve the additional copies of upper requests.

At time T_i, for all $i > 0$, AVGFLOWREPLAN *has still to serve as many upper requests as there are lower requests. Thus, at T_i the schedule with minimum average flow time for the currently available requests serves the upper requests first. Hence, the requests at the bottom have to wait for an arbitrarily long period of time.*

In order to prove the assertion concerning the average flow time we consider the result $f(R_N)$ that AVGFLOWREPLAN *produces on the input set R_N, which contains all requests up to time T_N.*

The sum of all flow times $f_\Sigma(R_N)$ is dominated by the waiting times of the lower requests. That is, it is at least

$$f_\Sigma(R_N) \geq \sum_{k=1}^{N} \sum_{i=k}^{N} (\ell + 2(i-2)\varepsilon)$$
$$\geq \sum_{k=1}^{N} \sum_{i=k}^{N} (i-2)\varepsilon.$$

The number of requests $\#R_N$ in R_N is

$$\#R_N = 1 + \sum_{k=1}^{N} (k+1),$$

so that

$$f(R_N) = \frac{f_\Sigma(R_N)}{\#R_N} \overset{N\to\infty}{\longrightarrow} \infty,$$

which completes the proof. □

A policy that minimizes just the maximal flow time does not make a lot of sense since sometimes this only determines which request is to be served first; the order in which all the other requests are scheduled is unspecified. Thus, the most sensible policy in this respect seems to be the following: Consider an off-line instance of the dial-a-ride problem. The vector consisting of all flow times of requests in a feasible solution ordered decreasingly is the *flow vector*. All flow vectors are ordered lexicographically. The on-line policy MAXFLOWREPLAN for the on-line dial-a-ride problem yields the following: Whenever a new request becomes available MAXFLOWREPLAN computes a new schedule of all yet unserved requests minimizing the flow vector.

It is an open problem what the performance of this policy is under Δ-reasonable load. In practice, however, it is probably too difficult to solve the snapshot problem with this objective function.

17.12 Combining the Best of two Ideas: DELTAREPLAN

A closer inspection of the behavior of IGNORE and SMARTSTART, resp., versus the behaviour of REPLAN and AVGFLOWREPLAN, respectively, shows: REPLAN is unstable under Δ-reasonable load because of infinite deferment of requests, which cannot happen in IGNORE, since IGNORE does not replan often enough to defer requests. On the other hand: reoptimizing less frequently means leaving out opportunities to improve, and thus, on average, IGNORE is empirically worse than REPLAN. The key to combine the advantages of both policies is to constrain the reoptimization that REPLAN performs. The result is the following on-line algorithm DELTAREPLAN, so far unpublished, which works as follows:

Whenever a new request becomes available, DELTAREPLAN computes a preliminary transportation schedule for the set R of all available requests by solving the problem of minimizing the total completion time of $R^{\text{off-line}}$ under the restriction that no request in the transportation schedule have predicted flow time more than 2Δ. If the makespan of the optimal transportation schedule is at most Δ, the new schedule is accepted and becomes the active schedule. The new schedule is rejected otherwise, whence the previous schedule is kept active. It then moves the server according to the active schedule until a new request arrives or the schedule is done. Note that the new requests that trigger the reoptimization are not rejected. It is the new schedule that is rejected. Thus, since we do not allow rejection of requests, DELTAREPLAN is only feasible if each request is in an accepted schedule, sooner or later.

Summarized, we define:

Definition 17.9 (Algorithm DELTAREPLAN). Algorithm DELTAREPLAN (Δ,ρ) has parameters $\Delta > 0, \rho > 1$ (indicating that it aims at (Δ,ρ)-reasonable request sets) and works with an internal buffer holding an active schedule and possibly some requests. It may assume the following states (initially it is IDLE):

IDLE Wait for the next point in time when requests become available. Goto PLAN.

PLAN Produce a preliminary transportation schedule for all currently available requests R (taken from the buffer) minimizing comp for $R^{\text{off-line}}$ under the constraint that no request have a predicted flow time exceeding 2Δ, possibly by a ρ-approximation algorithm. If the problem is infeasible or the computed completion time exceeds Δ, reject the new schedule and keep the old one active, thereby buffering the new requests. Otherwise, replace the active schedule by the new one. Goto BUSY.

BUSY Serve requests according to the active schedule. If a new requests is released or the active schedule is done, goto PLAN.

The result is:

Theorem 17.6. *Let $\Delta > 0$ and $\rho \geq 1$. For all instances with (Δ,ρ)-reasonable request sets, algorithm* DELTAREPLAN *(Δ,ρ) employing a ρ-approximation algorithm for reoptimization yields a maximal flow time of no more than 2Δ.*

Proof. *As long as all new schedules are rejected,* DELTAREPLAN *(Δ,ρ) works in the same way as* IGNORE. *Whenever a new schedule is accepted, the constraints on the flow times of the scheduled requests guarantee the bound by construction. Since no schedule of length larger than Δ is accepted, rejection of all optimal schedules thereafter yields a maximal release span for buffered requests of at most Δ. The buffered requests can therefore theoretically be served in time at most Δ/ρ. Because* DELTAREPLAN(Δ,ρ) *employs a ρ-approximation algorithm, it computes a schedule of length at most Δ. Since all requests during the work on a schedule have been ignored, the flow times of them are exactly the flow times* IGNORE *would have produced. Thus, the flow time constraints are satisfied for all of them. Therefore, the first computed schedule after the work on the active schedule has finished will be accepted. Consequently, every request will be in an accepted schedule at some point. Thus, the claim holds.* □

What happens if we do not know how reasonable the request sets are going to be, i.e., if we do not know (Δ,ρ) in advance? Let us restrict to the case with approximation factor $\rho = 1$ in order to concentrate on the core aspect. If DELTAREPLAN is run with a $\Delta' < \Delta$ on a Δ-reasonable request set, then all schedules that would be rejected with DELTAREPLAN (Δ) would also be rejected by DELTAREPLAN (Δ'). A problem may occur that when the active schedule is done: the new schedule has makespan larger than Δ' so that we have to reject it; but then we are stuck. We can then modify DELTAREPLAN in three ways to bypass this problem:

IGNORE-DELTAREPLAN
Accept all schedules that are computed because the old schedule is done.

DOUBLE-DELTAREPLAN

$\Delta'' := 2\Delta'$ as a new estimate of Δ and run DELTAREPLAN (Δ''). This is often called *doubling technique* for parametrized algorithms [4].

DELTAREPLAN

Take the makespan Δ'' of the new schedule (which is at most Δ) as a new estimate of Δ and run DELTAREPLAN (Δ'').

The first option uses IGNORE as a back-up whenever DELTAREPLAN (Δ') fails to produce a schedule. This way, we obtain the same bound 2Δ on the flow times but we may lose some efficiency due to too many rejected schedules.

Theorem 17.7. *Let $\Delta > 0$ and $\rho \geq 1$. For all instances with (Δ, ρ)-reasonable request sets, algorithm* IGNORE-DELTAREPLAN *employing a ρ-approximation algorithm for reoptimization yields a maximal flow time of no more than 2Δ.*

The estimate for Δ in the doubling technique will at some point surpass the true Δ. Then, we still get a bound on the flow times, but only with respect to the over-estimated Δ, i.e., a bound of 4Δ in the worst case.

Theorem 17.8. *Let $\Delta > 0$ and $\rho \geq 1$. For all instances with (Δ, ρ)-reasonable request sets, algorithm* DOUBLE-DELTAREPLAN *employing a ρ-approximation algorithm for reoptimization yields a maximal flow time of no more than 4Δ.* □

Since for DELTAREPLAN the estimates for Δ never exceed Δ and the reoptimization problems as well as the acceptance of new schedules are at least as constrained as for DELTAREPLAN(Δ), we conclude that DELTAREPLAN has flow times bounded by 2Δ, and the loss of efficiency is decreasing as the estimate of Δ gets closer and closer to Δ. We obtain the following result:

Theorem 17.9. *Let $\Delta > 0$ and $\rho \geq 1$. For all instances with (Δ, ρ)-reasonable request sets, Algorithm* DELTAREPLAN *employing a ρ-approximation algorithm for reoptimization yields a maximal flow time of no more than 2Δ.* □

This basic DELTAREPLAN-technique can be applied in much more general situations (see [20] for a sketch). We arrived at an algorithm very much in the spirit of MPC with ingredients from on-line optimization and queuing theory: For a classical problem in on-line optimization, estimate the characteristic difficulty of the input stream in terms of Δ, the definition of which was inspired by queuing theory, and use a cleverly constrained reoptimization model with a suitable objective to obtain a strongly stable system.

17.13 Conclusion

We have shown how naïve reoptimization policies in the control of elevators may lead to unstable systems. Moreover, via the notion of (Δ, ρ)-reasonable load we found a modification of the usual reoptimization policies that achieves strong stability, a new notion aiming at stability in worst case analysis in a queuing system. The new notions and the policies emerge as a combination of paradigms from basic on-line optimization, queuing theory and model predictive control. We conjecture that closing the gap between these fields will lead to interesting, sometimes surprisingly simple, but yet useful innovations.

The analysis under reasonable load is valid in much larger generality. Essentially, every system in which servers have to serve requests can be captured. This encompasses also general dial-a-ride problems. A generic formulation of the principle based on a generic integer linear programming formulation of the off-line version of some on-line problem is presented in [20]. We did not present this here for the sake of a less abstract exposition.

There are several open questions in this area:

- Does MAXFLOWREPLAN produce bounded flow times in terms of Δ under Δ-reasonable load?
- The policies in this chapter are all based on the computation of higher level information, namely a precomputed schedule. On this higher level, there is no immediate notion of a "terminal state." Is there any version of "terminal state constraints" or "terminal costs" for the snapshot problem that can guarantee stability of the corresponding replan policy?
- Of course, since the reasonability Δ is a worst case measure, performance may benefit if Δ is considered as a dynamically changing property of the request set which should be estimated in a time-dependent fashion in order not to use a too large Δ most of the time, especially, when there are traffic peaks. Can one rigorously quantify the benefits of such a dynamic approach?
- We have no nontrivial theoretical guarantees for the expected average flow-times over a distribution of request sets. Does DELTAREPLAN have provably better average flow times than IGNORE, as it seems so empirically?
- Experience shows that minimizing the average *quadratic* flow times in the snapshot problem leads to empirically stable systems. Can one guarantee strong stability for them?

The LCCC theme semester revealed that quite a few types of logistic control problems are attacked by more than one mathematical community; up to now rather in isolation than in cooperation. We would be very happy if this volume—and, in particular, this chapter—motivated a thorough performance comparison. More specifically: What can be achieved, in theory and practice, by the various techniques in queuing theory, model predictive control, stochastic dynamic optimization and on-line optimization on a *common set of problems*?

Acknowledgements We thank two anonymous referees for helpful comments on the presentation of this chapter. The second author is grateful for the opportunity to participate in a very inspiring theme semester at LCCC and the financial support by LCCC.

References

1. Ascheuer, N., Krumke, S.O., Rambau, J.: Online dial-a-ride problems: Minimizing the completion time. In: Proc. 17th Int. Symposium on Theoretical Aspects of Computer Science, vol. 1770, pp. 639–650. Springer (2000)
2. Atallah, M.J., Kosaraju, S.R.: Efficient solutions to some transportation problems with applications to minimizing robot arm travel. SIAM Journal on Computing **17**, 849–869 (1988)
3. Ausiello, G., Feuerstein, E., Leonardi, S., Stougie, L., Talamo, M.: Competitive algorithms for the traveling salesman. In: Proc. 4th Workshop on Algorithms and Data Structures (WADS'95), *Lecture Notes in Computer Science*, vol. 955, pp. 206–217 (1995)
4. Borodin, A., El-Yaniv, R.: Online Computation and Competitive Analysis. Cambridge University Press (1998)
5. Crites, R.H., Barto, A.G.: Improving elevator performance using reinforcement learning. In: S. Touretsky D. C. Mozer M. E. Hasselmo M. (eds.) Advances in Neural Information Processing Systems 8. MIT Press, Cambridge MA (1996)
6. Crites, R.H., Barto, A.G.: Elevator group control using multiple reinforcement learning agents. Machine Learning **33**(2–3), 235–262 (1998)
7. Fiat, A., Woeginger, G.J. (eds.): Online Algorithms: The State of the Art, *Lecture Notes in Computer Science*, vol. 1442. Springer (1998)
8. Frederickson, G.N., Guan, D.J.: Nonpreemptive ensemble motion planning on a tree. Journal of Algorithms **15**, 29–60 (1993)
9. Frederickson, G.N., Hecht, M.S., Kim, C.: Approximation algorithms for some routing problems. SIAM Journal on Computing **7**, 178–193 (1978)
10. Gross, D., Harris, C.: Fundamentals of queueing theory. Wiley Series in Probability and Statistics. Wiley (1998)
11. Grötschel, M., Hauptmeier, D., Krumke, S.O., Rambau, J.: Simulation studies for the online dial-a-ride-problem. Preprint SC 99-09, Konrad-Zuse-Zentrum für Informationstechnik Berlin (1999). URL http://opus4web.zib.de/documents-zib/401/SC-99-09.pdf. Extended abstract accepted for presentation at Odysseus 2000, First Workshop on Freight Transportation and Logistics, Crete, 2000
12. Hauptmeier, D., Krumke, S.O., Rambau, J.: The online dial-a-ride problem under reasonable load. In: Proc. 4th Italian Conference on Algorithms and Complexity, *Lecture Notes in Computer Science*, vol. 1767, pp. 137–149. Springer (2000)
13. Hauptmeier, D., Krumke, S.O., Rambau, J., Wirth, H.C.: Euler is standing in line—Dial-a-ride problems with FIFO-precedence-constraints. Discrete Applied Mathematics **113**, 87–107 (2001)
14. Hiller, B., Vredeveld, T.: Probabilistic analysis of online bin coloring algorithms via stochastic comparison. In: D. Halperin, K. Mehlhorn (eds.) ESA, *Lecture Notes in Computer Science*, vol. 5193, pp. 528–539. Springer (2008)
15. Kellerer, H., Tautenhahn, T., Woeginger, G.: Approximability and nonapproximabiblity results for minimizing total flow time on a single machine. In: Proc. Symposium on the Theory of Computing (1996)
16. Klug, T., Hiller, B., Tuchscherer, A.: Improving the performance of elevator systems using exact reoptimization algorithms. Tech. Rep. 09-05, Konrad-Zuse-Zentrum für Informationstechnik Berlin (2009)
17. Krumke, S.O.: Competitive analysis and beyond. Habilitationsschrift, Technische Universität Berlin (2002)
18. Krumke, S.O., Rambau, J., Torres, L.M.: Realtime-dispatching of guided and unguided automobile service units with soft time windows. In: R.H. Möhring, R. Raman (eds.) Algorithms – ESA 2002, 10th Annual European Symposium, Rome, Italy, September 17–21, 2002, Proceedings, *Lecture Notes in Computer Science*, vol. 2461. Springer (2002)
19. Putermann, M.L.: Markov Decision Processes. Wiley Interscience (2005)
20. Rambau, J.: Deferment control for reoptimization—How to find fair reoptimized dispatches. In: S. Albers, R.H. Möhring, G.C. Pflug, R. Schultz (eds.) Algorithms for Optimization

with Incomplete Information, no. 05031 in Dagstuhl Seminar Proceedings. Internationales Begegnungs- und Forschungszentrum (IBFI), Schloss Dagstuhl, Germany (2005). URL http://drops.dagstuhl.de/opus/volltexte/2005/66 [date of citation: 2010-12-28]
21. Shmoys, D.B., Wein, J., Williamson, D.P.a.: Scheduling parallel machines on-line. SIAM Journal on Computing **24**(6), 1313–1331 (1995)
22. Tarău, A.: Model-based control for postal automation and baggage handling. Ph.D. thesis, Technische Universiteit Delft (2010)

Author Index

Subject Index

Lecture Notes in Control and Information Sciences

Edited by M. Thoma, F. Allgöwer, M. Morari

Further volumes of this series can be found on our homepage:
springer.com

Vol. 417: Johansson, R.; Rantzer, A. (Eds.):
Distributed Decision Making and Control
421 p. 2012 [978-1-4471-2264-7]

Vol. 416: Varga, A.; Hansson, A.;
Puyou, G. (Eds.):
Optimization Based Clearance of Flight
Control Laws
451 p. 2012 [978-3-642-22626-7]

Vol. 412: Fridman, L.; Moreno, J.;
Iriarte R. (Eds.):
Sliding Modes after the First Decade of the
21st Century
595 p. 2011 [978-3-642-22163-7]

Vol. 411: Kaczorek, T.;
Selected Problems of Fractional Systems Theory
344 p. 2011 [978-3-642-20501-9]

Vol. 410: Bourlès, H.; Marinescu, B.;
Linear Time-Varying Systems
637 p. 2011 [978-3-642-19726-0]

Vol. 409: Xia, Y.; Fu, M.; Liu, G.-P.;
Analysis and Synthesis of
Networked Control Systems
198 p. 2011 [978-3-642-17924-2]

Vol. 408: Richter, J.H.;
Reconfigurable Control of
Nonlinear Dynamical Systems
291 p. 2011 [978-3-642-17627-2]

Vol. 407: Lévine, J.; Müllhaupt, P.:
Advances in the Theory of Control,
Signals and Systems with
Physical Modeling
380 p. 2010 [978-3-642-16134-6]

Vol. 406: Bemporad, A.; Heemels, M.;
Johansson, M.:
Networked Control Systems
appro. 371 p. 2010 [978-0-85729-032-8]

Vol. 405: Stefanovic, M.; Safonov, M.G.:
Safe Adaptive Control
appro. 153 p. 2010 [978-1-84996-452-4]

Vol. 404: Giri, F.; Bai, E.-W. (Eds.):
Block-oriented Nonlinear System Identification
425 p. 2010 [978-1-84996-512-5]

Vol. 403: Tóth, R.;
Modeling and Identification of
Linear Parameter-Varying Systems
319 p. 2010 [978-3-642-13811-9]

Vol. 402: del Re, L.; Allgöwer, F.;
Glielmo, L.; Guardiola, C.;
Kolmanovsky, I. (Eds.):
Automotive Model Predictive Control
284 p. 2010 [978-1-84996-070-0]

Vol. 401: Chesi, G.; Hashimoto, K. (Eds.):
Visual Servoing via Advanced
Numerical Methods
393 p. 2010 [978-1-84996-088-5]

Vol. 400: Tomás-Rodríguez, M.;
Banks, S.P.:
Linear, Time-varying Approximations
to Nonlinear Dynamical Systems
298 p. 2010 [978-1-84996-100-4]

Vol. 399: Edwards, C.; Lombaerts, T.;
Smaili, H. (Eds.):
Fault Tolerant Flight Control
appro. 350 p. 2010 [978-3-642-11689-6]

Vol. 398: Hara, S.; Ohta, Y.;
Willems, J.C.; Hisaya, F. (Eds.):
Perspectives in Mathematical System
Theory, Control, and Signal Processing
appro. 370 p. 2010 [978-3-540-93917-7]

Vol. 397: Yang, H.; Jiang, B.;
Cocquempot, V.:
Fault Tolerant Control Design for
Hybrid Systems
191 p. 2010 [978-3-642-10680-4]

Vol. 396: Kozlowski, K. (Ed.):
Robot Motion and Control 2009
475 p. 2009 [978-1-84882-984-8]

Vol. 395: Talebi, H.A.; Abdollahi, F.;
Patel, R.V.; Khorasani, K.:
Neural Network-Based State
Estimation of Nonlinear Systems
appro. 175 p. 2010 [978-1-4419-1437-8]

Vol. 394: Pipeleers, G.; Demeulenaere, B.; Swevers, J.:
Optimal Linear Controller Design for Periodic Inputs
177 p. 2009 [978-1-84882-974-9]

Vol. 393: Ghosh, B.K.; Martin, C.F.; Zhou, Y.:
Emergent Problems in Nonlinear Systems and Control
285 p. 2009 [978-3-642-03626-2]

Vol. 392: Bandyopadhyay, B.; Deepak, F.; Kim, K.-S.:
Sliding Mode Control Using Novel Sliding Surfaces
137 p. 2009 [978-3-642-03447-3]

Vol. 391: Khaki-Sedigh, A.; Moaveni, B.:
Control Configuration Selection for Multivariable Plants
232 p. 2009 [978-3-642-03192-2]

Vol. 390: Chesi, G.; Garulli, A.; Tesi, A.; Vicino, A.:
Homogeneous Polynomial Forms for Robustness Analysis of Uncertain Systems
197 p. 2009 [978-1-84882-780-6]

Vol. 389: Bru, R.; Romero-Vivó, S. (Eds.):
Positive Systems
398 p. 2009 [978-3-642-02893-9]

Vol. 388: Jacques Loiseau, J.; Michiels, W.; Niculescu, S-I.; Sipahi, R. (Eds.):
Topics in Time Delay Systems
418 p. 2009 [978-3-642-02896-0]

Vol. 387: Xia, Y.; Fu, M.; Shi, P.:
Analysis and Synthesis of Dynamical Systems with Time-Delays
283 p. 2009 [978-3-642-02695-9]

Vol. 386: Huang, D.; Nguang, S.K.:
Robust Control for Uncertain Networked Control Systems with Random Delays
159 p. 2009 [978-1-84882-677-9]

Vol. 385: Jungers, R.:
The Joint Spectral Radius
144 p. 2009 [978-3-540-95979-3]

Vol. 384: Magni, L.; Raimondo, D.M.; Allgöwer, F. (Eds.):
Nonlinear Model Predictive Control
572 p. 2009 [978-3-642-01093-4]

Vol. 383: Sobhani-Tehrani E.; Khorasani K.;
Fault Diagnosis of Nonlinear Systems Using a Hybrid Approach
360 p. 2009 [978-0-387-92906-4]

Vol. 382: Bartoszewicz A.; Nowacka-Leverton A.;
Time-Varying Sliding Modes for Second and Third Order Systems
192 p. 2009 [978-3-540-92216-2]

Vol. 381: Hirsch M.J.; Commander C.W.; Pardalos P.M.; Murphey R. (Eds.)
Optimization and Cooperative Control Strategies: Proceedings of the 8th International Conference on Cooperative Control and Optimization
459 p. 2009 [978-3-540-88062-2]

Vol. 380: Basin M.
New Trends in Optimal Filtering and Control for Polynomial and Time-Delay Systems
206 p. 2008 [978-3-540-70802-5]

Vol. 379: Mellodge P.; Kachroo P.;
Model Abstraction in Dynamical Systems: Application to Mobile Robot Control
116 p. 2008 [978-3-540-70792-9]

Vol. 378: Femat R.; Solis-Perales G.;
Robust Synchronization of Chaotic Systems Via Feedback
199 p. 2008 [978-3-540-69306-2]

Vol. 377: Patan K.
Artificial Neural Networks for the Modelling and Fault Diagnosis of Technical Processes
206 p. 2008 [978-3-540-79871-2]

Vol. 376: Hasegawa Y.
Approximate and Noisy Realization of Discrete-Time Dynamical Systems
245 p. 2008 [978-3-540-79433-2]

Vol. 375: Bartolini G.; Fridman L.; Pisano A.; Usai E. (Eds.)
Modern Sliding Mode Control Theory
465 p. 2008 [978-3-540-79015-0]

Vol. 374: Huang B.; Kadali R.
Dynamic Modeling, Predictive Control and Performance Monitoring
240 p. 2008 [978-1-84800-232-6]

Vol. 373: Wang Q.-G.; Ye Z.; Cai W.-J.; Hang C.-C.
PID Control for Multivariable Processes
264 p. 2008 [978-3-540-78481-4]

Vol. 372: Zhou J.; Wen C.
Adaptive Backstepping Control of Uncertain Systems
241 p. 2008 [978-3-540-77806-6]